Residential Wiring

Dedicated to my wife and best friend Diane,
who put up with so much.

Based on the 2002
NATIONAL ELECTRICAL CODE®

Residential Wiring

HARRY W. SORGE

THOMSON

DELMAR LEARNING

Australia Canada Mexico Singapore Spain United Kingdom United States

THOMSON

DELMAR LEARNING

Residential Wiring
Harry W. Sorge

Executive Director:
Alar Elken

Executive Editor:
Sandy Clark

Acquisitions Editor:
Gregory L. Clayton

Development Editor:
Jennifer A. Thompson

Executive Marketing Manager:
Maura Theriault

Channel Manager:
Fair Huntoon

Marketing Coordinator:
Brian McGrath

Executive Production Manager:
Mary Ellen Black

Production Coordinator:
Toni Hansen

Project Editor:
Ruth Fisher

Art/Design Coordinator:
Rachel Baker

Senior Editorial Assistant:
Dawn Daugherty

Cover Design:
Charles Cummings, Advertising

Library of Congress Cataloging-in-Publication Data:
Sorge, Harry W.
  Residential wiring / Harry W. Sorge
    p. cm.
"Based on the 2002 National electrical code."
Includes index.
  ISBN 0-7668-4696-2
  1. Electric wiring, Interior. 2. Dwellings—Electric equipment. I. Title.
  TK3285.S67 2002
  621.319'24—dc21
                                    2002041316

NOTICE TO THE READER

Publisher does not warrant or guarantee any of the products described herein or perform any independent analysis in connection with any of the product information contained herein. Publisher does not assume, and expressly disclaims, any obligation to obtain and include information other than that provided to it by the manufacturer.

The reader is expressly warned to consider and adopt all safety precautions that might be indicated by the activities herein and to avoid all potential hazards. By following the instructions contained herein, the reader willingly assumes all risks in connection with such instructions.

The publisher makes no representation or warranties of any kind, including but not limited to, the warranties of fitness for particular purpose or merchantability, nor are any such representations implied with respect to the material set forth herein, and the publisher takes no responsibility with respect to such material. The publisher shall not be liable for any special, consequential, or exemplary damages resulting, in whole or part, from the readers' use of, or reliance upon, this material.

CONTENTS

FOREWORD

Welcome to the odyssey of electrical construction and the special set of rewards and responsibilities that come with the career choice of electrician. I have been electrically employed since 1975 and still delight in flipping the switch and witnessing the results of having channeled the energy of nature to provide comfort and convenience to those I am paid to serve.

In looking back to the origins of the use of electricity, it must be understood that electricity is, literally, fire in a bottle. It is the job of the electrician to keep the fire in the bottle. When electrical power became publicly available, the fire quite often got out of the bottle, and lives and property were lost. In order to develop this wonderful tool, and also to protect people and property, the National Fire Protection Association (NFPA) eventually became involved. The result is known as the *National Electrical Code* (*NEC*®). The *NEC*® has been an evolving set of rules for keeping the fire in the bottle while providing the electrical resource for human innovation and growing technology. Following the established rules is a must.

I learned my skills in Australia and earned my first electrical license there. My father was an electrician, and when I entered the field, he suggested I consider whether I hoped to have a job, be a tradesman, be a craftsman, or be an artist. That is a question I have asked myself daily for all these years, and that you have to ask yourself each day. I hope you choose wisely.

When I somehow landed in the United States in 1983, it was with extreme good fortune that I crossed paths with H. W. (Wayne) Sorge. Like myself, Wayne not only sought to do electrical installations correctly, but wanted to understand how a circuit functioned and why the rules we follow were put in place. For us, understanding the how and why is as important as doing the work itself. We have shared countless hours exchanging knowledge and ideas in gaining this understanding.

Wayne has been both an electrician and an electrical instructor for over 20 years and has played a pivotal role in the education of hundreds of electricians. This book is your road map to responsibility and reward. Wayne is showing you how to recognize your responsibilities through the proper use and application of the *NEC*® and correct residential wiring practices. The reward comes through absorbing the knowledge, which will not only provide a good living but also give you confidence and esteem by knowing that you are, indeed, keeping the fire in the bottle.

Ian Miller
Master Electrician

PREFACE

This book was born out of the observation that persons newly employed as residential apprentices in the electrical field need the best possible instruction in the basic requirements of the trade. The time available for training of an apprentice by a lead electrician is severely limited by other duties and responsibilities. Although hands-on field experience is the best type of training, it is also the most costly. It is costly in training time expended by otherwise productive employees. It is costly because of mistakes leading to re-work, callbacks, and failed inspections. It is also costly for the career development of the apprentice; advancement is limited by what the apprentice does or does not completely understand. This book strives to remedy this problem by offering the background necessary for the further training that will take place in the field.

INTENDED AUDIENCE

Designed for aspiring electricians in vocational electrical programs, as well as apprentice programs, this book provides the knowledge base needed for installation of residential electrical systems in common use today. Included are the most common challenges likely to be encountered by apprentice electricians during the normal course of their work under the supervision of an experienced residential wireman or journeyman. The essentials of training—required outlets, circuiting, box selection, conductor identification, box make-up, load calculations, and many other topics—as covered in the text must be supplemented by actual hands-on work in the field. In fact, field training should be the larger part, by far, of the apprentice's education.

ABOUT THIS BOOK

In many other residential electrical wiring textbooks, presentation of material on the connection of equipment, or on the requirements set forth by the National Fire Protection Association (NFPA) in the *National Electrical Code*® (*NEC*®), is too detailed, in an attempt to cover every possible eventuality. Excessive detail can deter learning, however, as a "green" apprentice can be overwhelmed by the volume, diversity, and complexity of the electrical field.

This textbook provides a more practical approach: it walks the reader step by step through the wiring of a typical house. With each chaper, text discussion begins with the basics, progressing to more specific applications and expanded as appropriate with illustrated guides presenting the details of important electrical concepts and tricks of the trade. The *NEC*® is referred to often in the text, but the references—which appear in *italic* typeface—are limited to providing broad rules and general installation requirements. On encountering such a reference, the reader should locate and carefully read that particular section within the *NEC*®. Consulting the Code in this manner will allow the reader to begin to understand the basic layout of the *NEC*®, and put the rules in proper context.

Organization of Material

This book follows chronologically the installation of the electrical system of a typical house. The installation process, from the wiring for the temporary service until the final trim and check-out, is examined in a systematic approach that provides the apprentice electrician with the information necessary to be a productive member of an electrical contracting firm.

Chapter 1 provides the background necessary for maintaining a safe and productive workplace. An examination of the sources of available information and a detailed set of plans for a sample house are included. In many cases, this book will be used in a classroom environment where there is not an overabundance of free space. A full set of construction drawings for the sample house is therefore included within the text rather than as separate, full-size blueprints, which would require a large area to use properly. This format makes the book more user friendly in that it allows convenient reference to the construction drawings.

Chapter 2 is concerned with temporary power for the construction of the sample house. The study of temporary power and lighting includes information on services, ground faults, overcurrent protection, and wire sizing as necessary for an understanding of the requirements for temporary power.

Chapter 3 examines the locations in dwellings where receptacle outlets are required and the rules concerning their installation.

Chapter 4 considers the locations for lighting outlets and switches that are required by the Code or that are included in the dwelling to make it more comfortable and convenient.

Chapter 5 details conductors and allowable ampacities of the various conductor insulation types and the reasons behind adjustment and correction factors.

Chapter 6 looks at the various types of boxes commonly used in modern dwelling-unit wiring as well as proper box selection and installation techniques.

Chapter 7 continues with a study in overcurrent. Some knowledge of electrical theory is necessary to understand the four possible causes of overcurrent and the proper procedures to follow when installing overcurrent protection. This is one of the more important sections in the book because most of the rules employed by the *NEC*® are intended to protect the dwelling from, and deal with the problems created by overcurrent.

Chapter 8 examines the various wiring methods with particular attention to those that are used extensively in dwelling units.

Chapter 9 concerns cable installation. The proper spacing of supports; drilling or notching requirements for walls, floors, and ceiling framing members; and conductor make-up are some of the items considered.

Chapter 10 looks at the installation of cables to supply the required receptacle and lighting outlets and switch points. The circuiting of the sample house is examined in detail.

Chapter 11 studies the circuiting for nonrequired outlets such as the dishwasher and range circuits.

Chapter 12 is concerned with feeder and service-conductor load calculations for dwellings. A strong understanding of feeders, branch-circuit conductors, and service conductors as covered in Chapter 7 will be of considerable assistance to the reader.

Chapter 13 details the installation of a dwelling service. Both overhead and underground service connections are studied.

Chapter 14 examines the requirements for grounding the electrical system. Proper grounding is one of the most important skills in the electrician's tool kit, and one of the most misunderstood.

Chapter 15 looks at special systems usually installed by the electrician but that are not power or lighting circuits. These systems include the telephone and television antenna (or cable) wiring systems.

Chapter 16 shows methods and procedures for installing receptacles and switches. The wire make-up system studied in Chapter 9 is used to identify the type of receptacle or the type of switch to install at each outlet and switch point.

Chapter 17 details the installation of the luminaires (fixtures) in the sample house.

Chapter 18 is concerned with the wiring and connection of the various appliances installed in the sample house. These include the kitchen appliances and also nonkitchen appliances such as the furnace.

Chapter 19 involves the testing and troubleshooting of the electrical system. This final phase involves connecting the dwelling's electrical system to the temporary power supply and checking every outlet and switch to ensure they are functioning properly.

Features of the Book

The study of the electrical field, and indeed of the *NEC*® itself, is problematic because it involves circular knowledge. *Circular knowledge* is a term that describes the study of a subject that does not have a clear starting or stopping point—each lesson requires extensive knowledge of many other related subjects for a complete understanding of the material presented. In contrast, the study of mathematics is linear. It begins with numbers and progresses to addition, subtraction, multiplication, division, and so on, one subject building on the foundations of the previous subject, getting progressively more difficult with each new subject. The study of the *NEC*®, on the other hand, does not have a clear starting point. Any passage in the *NEC*® assumes that the reader is already fully versed and competent in all aspects of the electrical trade. Thus, the Code must be considered as a whole, and no part can stand alone. In this regard, circular knowledge can present many challenges to the student and the instructor.

To address this problem, this book contains special illustrations termed "Wireman's Guide." These boxed illustrations contain the following types of information:

- Electrical theory, explaining the "how" and "why" behind difficult concepts in order to promote further understanding

- Hints and tips, consisting of practical advice on certain aspects of field installation

- Wiring details, providing additional insights and challenges for the more advanced student.

These figures allow the instructor to tailor the level of detail to the needs of the individual student.

Study of the material in these boxes is supported by the "Wireman's Guide" Review, provided in addition to the questions for Review that appear at the end of each chapter. The questions in the "Wireman's Guide" Review, which usually require written answers or diagrams, are intended to provoke thought and suggest connections between aspects of the various electrical systems that are not otherwise obvious. Such activity promotes understanding of the subject as a whole and of how that topic relates to other facets of installation requirements. These questions may also be useful in a classroom setting to guide students to consideration of more complex issues.

INSTRUCTOR'S GUIDE

An extensive *Instructor's Guide* (ISBN 0-7668-4697-0) is available to accompany the text and contains the following features for each chapter:

- Objectives: Reviews intent of chapter and provides an outline of topics

- Prerequisite Knowledge: Provides a list of areas students should be competent in so as to get the most benefit from each chapter

- Discussion Topics: Lists other topic suggestions for classroom instruction

- Answers to Review Questions

- Answers to "Wireman's Guide" Review questions

ABOUT THE AUTHOR

Harry W. Sorge is a licensed Master Electrician. He graduated from college in 1970 with degrees in economics and secondary education. He became an apprentice electrician in 1977 after holding several corporate jobs because, as he said in a job interview, "I am 30 years old and I don't know how to do anything." Following completion of an apprenticeship training program, he began teaching electrical classes at night for the Rocky Mountain Chapter of the Independent Electrical Contractors Association (IEC), a nationwide trade organization of independant electrical contractors for all sizes and types of electrical contractors, while working in the field full time. Over the next 14 years he taught first-year, second-year, and fourth-year apprenticeship classes; classes for licensing examination preparation; and classes on *NEC*® changes, motor controls, services, grounding, and many other subjects for the IEC. In 1996, he was asked to take over the functional operation of the apprenticeship training program for the Rocky Mountain Chapter and over the next two years, the school enrollment more than doubled to more than 400 students. The Rocky Mountain Chapter program was chosen as the Outstanding Training Program in 1997 at the IEC's national convention.

Mr. Sorge is a member of the International Association of Electrical Inspectors and the NFPA. He has served on the national committee of the IEC for first-year curriculum.

ACKNOWLEDGMENTS

I want to thank my wife, Diane, and our children, Brett, Meggan, and Luke, for their encouragement and support during the creation of this book. Their unwavering determination was at times stronger than mine. I also want to thank my brother, John, for invaluable assistance during the final draft process and for help with many of the illustrations. Also, Craig Barnes, owner of Barnes Electric and an old friend and former business partner, provided technical assistance for which I am grateful. Special thanks must be given to my good friend and fellow electrician Ian Miller for providing the foreword to this book. He was a sounding board and a source of much help and support for this project.

Jennifer Thompson, my editor at Delmar Learning during the writing process, was very supportive, and kept me on the correct track during a couple of initial blunders. Without her help and patience, this book would not have been possible. I also want to thank Mark Huth at Delmar Learning for his confidence in me.

Most of all, I want to thank all the contractors, journeymen, and apprentices, at whose side I worked during my years as a construction electrician. A wealth of information about the *National Electrical Code®* and installation procedures, about how to work and how to work safely, and about how to get along with other electricians and other trades persons, was over the years freely transferred from one generation to the next—from them to me. This is what I have attempted to do in writing this book: pass along what I have learned to the next generation— you!

A special thanks from the author and publisher goes to those who contributed to the development of this textbook as content reviewers, editors, artists, copy editors, and many other professionals in the publishing industry.

Finally, to the content reviewers, who thoroughly reviewed the manuscript, and provided invaluable insights along the way, we extend our deepest appreciation:

Keith Bunting
Randolph Community College
Ashboro, North Carolina

Robert DiPeiero
Alaska Vo-Tech Center
Seward, Alaska

Keith Elliot
Rockingham Community College
Wentworth, North Carolina

Greg Fletcher
Kennebec Valley Tech College
Fairfield, Maine

Orville Lake
Augusta Tech
Evans, Georgia

Debra Matthews
MS Gulf Coast Community College
Gautier, Mississippi

Wes Mozley
Albuquerque Tech
Albuquerque, New Mexico

Ralph Potter
Bowling Green Technical College
Bowling Green, Kentucky

Cliff Redigar
Rocky Mountain Chapter
Independent Electrical Contractors
Denver, Colorado

Rodney Stanley
Morehead State University
Morehead, Kentucky

Thank you all very much!

Harry W. Sorge
Broomfield, Colorado

CHAPTER 1

The Construction Process

OBJECTIVES

On completion of this chapter, the student will be able to:

☑ Explain the safety rules for working on a construction site, emphasizing the importance of following proper safety procedures at all times.

☑ Explain why all equipment must be installed to the manufacturer's instructions.

☑ Describe the construction process and the relationships among the various trade workers and others working on the job site, including officials from the authority having jurisdiction (AHJ) and those involved in the permit and inspection processes.

☑ Describe the general organization of the *NEC®* and locate information in the *NEC®* when given a reference number (chapter and article number).

☑ Use both English and SI (metric) units of measurement.

☑ Describe the system of listing and labeling by testing agencies.

☑ Use construction drawings, specifications, scope of work documents, and other construction documents as guides in wiring a dwelling.

INTRODUCTION

This chapter provides background information necessary for the electrician to understand the construction process and how a construction project is organized and controlled. Subjects such as the *National Electrical Code®* (*NEC®*), the relationship with the authority having jurisdiction (AHJ), on-the-job safety, the business and authority relationship on the job site, and technical information availability are covered within the context of a residential construction project (*National Electrical Code®* and *NEC®* are registered trademarks of the National Fire Protection Association, Inc., Quincy, MA 02269). A complete set of construction plans and other documents for the sample house studied in this book are included in this chapter.

1.1: THE CONSTRUCTION PROCESS

Before any study of residential wiring can be undertaken, some background on the construction industry as a whole, and on the role of the electrician within that process, is necessary. The successful completion of a building involves many skilled individuals from many different trades. The workers in each trade, such as carpenters, masons, and electricians, have certain procedures, or processes, that they must follow in completing their work. They also have certain requirements of other workers that must be met before they can complete their work.

KEY TERMS

Construction process Collectively, the procedures followed by the various trades for the successful completion of the construction project. The construction process is usually coordinated by a general contractor or a construction management firm.

Authority having jurisdiction (See *NEC® Article 100*): The local governmental authority that is charged with the regulation of construction projects. The AHJ may be a city, county, or sometimes a state organization, but it is always answerable to a legislative body that assigns the authority.

***National Electrical Code*® (*NEC*®)** A set of minimum rules for the design and installation of electrical systems and devices published by the National Fire Protection Association as document NFPA 70. It is the most widely applied standard for electrical installations in the United States, however, it is not recognized in all areas of the country.

Uniform Building Code (UBC) A set of minimum rules for the installation of building systems and the construction of new buildings. It is the most widely applied standard for construction in the United States, however, it is not recognized in all areas of the country.

The **construction process** involves not only workers in the various construction trades but also professionals such as architects, engineers, financial managers, designers, and designated officials from government regulatory agencies. Local city or county building officials are involved in almost every aspect of the construction planning and the construction process. The agency or organization with the legal authority to approve construction projects is referred to as the **authority having jurisdiction** (AHJ). Almost all AHJs utilize building codes that set minimum standards for quality of construction. These codes are usually national standards such as the *National Electrical Code*® (*NEC*®) and the **Uniform Building Code (UBC).**

These codes are published and maintained for the use of technicians and contractors as minimum standards for proper installations. The codes are usually adopted as law in the jurisdictions.

Building permit A document issued by the AHJ, usually for a fee, that allows the construction of, or addition to, a building or other structure. In most areas of the country, it is not legal to begin construction before a permit has been obtained.

Inspection The process whereby the AHJ enforces established installation and construction standards. The AHJ periodically reviews the progress of construction and either approves or rejects the quality of the construction and ensures compliance with minimum standards. These reviews are accomplished by physically inspecting the building, and the person who performs the inspections is usually referred to as an *inspector.*

The AHJ issues **building permits** for the construction of new homes. The general contractor or builder, and many of the subcontractors also (including electrical subcontractors), must purchase these permits from the AHJ prior to beginning work. The AHJ, or its representatives, performs an **inspection** of the dwelling at critical times during the construction process to ensure that all of the code requirements are being met.

Other people also have an interest in the quality of the construction. The owner of the dwelling may inspect the job to ensure that all of the features that were planned and contracted for are included. In many cases, the owner has obtained a construction loan from a bank or other financial institution to provide funds for the construction of the dwelling. The bank may send inspectors to the project to ensure that the construction loan money is being used properly. Engineering firms that created the plans for the house may sometimes inspect the job to ensure that the building meets their specified engineering standards. Figure 1-1 is a diagram that shows the usual relationships among the many different people and companies involved in the construction process for a dwelling.

1.2: INFORMATION AND INSTRUCTION

Each of the companies involved in the construction of the dwelling must have certain information to complete the project successfully. Literally thousands of details need to be communicated from the designers and the engineers to the trade workers. Furthermore, rules related to the installation of various systems need to be closely followed. Information for the electricians on the job site usually comes from one of the following sources:

- The *National Electrical Code*®
- Construction drawings or plans
- Manufacturers' installation instructions
- The specifications
- Shop drawings, cut sheets, and submittals
- The scope of work document
- Work orders or change orders

1.2.1: *The National Electrical Code*®

The *NEC*® is the document that is used in most of the United States, and in many other places in the

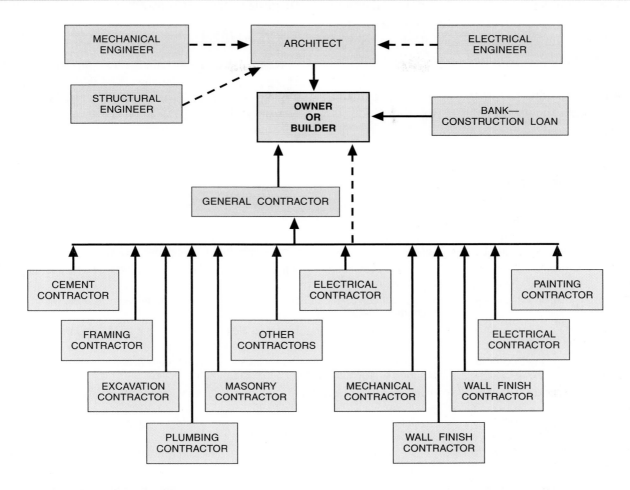

THE OWNER OF THE DWELLING UNDER CONSTRUCTION CAN USE A GENERAL CONTRACTOR TO ORGANIZE AND PACE THE CONSTRUCTION PROCESS, OR THE OWNER MAY CHOOSE TO BE THE BUILDER AND ELIMINATE THE GENERAL CONTRACTOR. SOMETIMES THE BUILDER MAY ALSO EMPLOY ENGINEERING FIRMS FOR SOME ASPECTS OF THE DESIGN. ALL OF THE SUBCONTRACTORS WORK FOR THE GENERAL CONTRACTOR OR BUILDER.

BELOW IS A TIME LINE SHOWING THE APPROXIMATE CONSTRUCTION SCHEDULING FOR A DWELLING UNIT. THE ELECTRICAL CONTRACTOR IS USUALLY THE FIRST ON THE JOB INSTALLING THE TEMPORARY CONSTRUCTION POWER AND IS USUALLY ONE OF THE LAST CONTRACTORS OFF THE JOB FOLLOWING TRIM-OUT AND CHECKOUT.

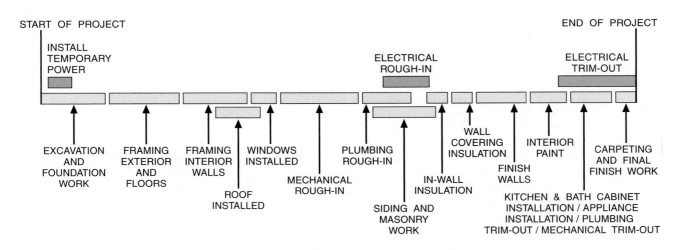

Figure 1-1   Job site organization and time line.

world, as a guide for safe electrical installations. It has been in use by the electrical construction industry, governmental bodies, and electrical equipment and appliance manufacturers as a guide for over 100 years. Before any study of residential wiring can be undertaken, a firm understanding of the methods and organization used by the Code is essential.

Organization of the *NEC*®

> **KEY TERMS**
>
> **National Fire Protection Association (NFPA)** The organization devoted to fire safety and prevention that publishes, along with many other documents, the *National Electrical Code*®.

CHAPTER	ARTICLE/TABLE NUMBERS	CHAPTER TITLE
	Article 80 Article 90	Introduction
1	Articles 100–110	General: Definitions and requirements for electrical installations
2	Articles 200–285	Wiring and Protection: Branch circuits, feeders, services, load calculations, grounding, and overcurrent protection
3	Articles 300–398	Wiring Methods and Materials: Conductors, insulation, ampacity, box fill, wiring methods, panels, services
4	Articles 400–490	Equipment for General Use: Receptacles, lighting fixtures, appliances, motors, heating/air conditioning, switches, generators, transformers
5	Articles 500–555	Special Occupancies: Hazardous locations, health care facilities, places of assembly, theaters, recreational vehicle parks, and floating buildings
6	Articles 600–695	Special Equipment: Signs, office furnishings, elevators, welders, information technology equipment, swimming pools
7	Articles 700–780	Special Conditions: Emergency systems, low-voltage systems, Class 1, Class 2, and Class 3 circuits, instrumentation, fiber optics
8	Articles 800–830	Communication Systems: Radio and TV equipment, CATV, network-powered broadband
9	Tables 1–12(b)	Tables: Conduit fill, sizes of conductors, conductor properties, resistance

Each chapter is divided into articles—for example, *406*. The first number in the article number refers to the *NEC*® chapter number. The other two numbers identify the specific article within that chapter.

Articles are further divided into sections by a dot (.) that separates the article number from the section number—for example, *406.8*.

A section is further divided into subsections. Three levels of subsections may be included in the Code and they are headed with:

- A capital letter inside parentheses, with (A) for the first subsection, (B) for the second subsection, and so on—for example, *406.8(B)*.

- An Arabic number inside parentheses, with (1) for the first sub-subsection, (2) for the second, and so on—for example, *406.8(B)(2)*.

- A small letter inside parentheses, with (a) for the first sub-sub-subsection, (b) for the second, and so on—for example, *406.8(B)(2)(a)*.

Lists are another organizational method used by the *NEC*®. A list is simply a series of related information. Lists are identified by Arabic numbers in parentheses, starting with (1). It is possible to distinguish an item of a list from a second-level subsection (both use a number in parentheses) because first- and second-level subsections begin with a boldface title, and lists do not have titles.

In addition to the articles, sections, subsections, and lists, there are also exceptions. Exceptions appear in the Code in *italic* type. They refer very specifically to the article, or section or subsection, that they immediately follow, and they detail situations in which the previously stated rule does not apply.

If an article is sufficiently large, it can also be divided into parts. These parts group together logically associated material within an article. The different parts of an article are labeled with Roman numerals.

Also included in the text of the *NEC*® are fine print notes (FPNs). FPNs are not enforceable as requirements of the Code. The FPNs present explanatory material that refers the reader to another place in the Code, or to another NFPA publication. FPNs can also provide the reader with guidelines for implementing the requirements of that particular section.

Figure 1-2 Organization of the *National Electrical Code*®.

The *NEC*® is published by the **National Fire Protection Association (NFPA)**. The *NEC*® is publication number NFPA 70 and is only one of many volumes published by the Association concerning fire safety. The *NEC*® is considered the electrician's law book. It provides detailed instructions for what is and what is not allowed concerning electrical systems, materials, and installation requirements. The Code is under continuous review. An upgraded edition of the *NEC*® is published every three years. Each Code edition is known by the year of its adoption. The current *NEC*® is the 2002 edition.

The *NEC*® is divided into nine chapters, plus an introduction. Each chapter covers a broad range of rules that are bound together by a common subject. The most general rules and applications are in *Chapter 1*, and the focus of the Code gets increasingly more specific with each subsequent chapter. The later chapters examine very specific systems and installation environments. The last chapter, *Chapter 9*, includes examples of calculations and tables with useful information. Each chapter is divided into articles, and each article is further divided into sections and subsections. All references to the *NEC*® in this book specify the article and section numbers. See Figure 1-2 for more detailed information on the organization of the *NEC*®.

KEY TERMS

Mandatory Rules Describing the rules and procedures listed in the *NEC*® with the terms "shall" and "shall not." These rules must be followed to comply with the requirements of the *NEC*®.

Permissive Rules Referring to the rules and procedures listed in the *NEC*® that are allowed but that are not strictly required. These rules are identified in the *NEC*® with the terms "shall be permitted" and "shall not be required."

The *NEC*® has two levels of control over electrical installation—mandatory and permissive. **Mandatory rules** are those rules that identify actions that are specifically prohibited or required and are characterized by the use of the terms "shall" and "shall not." *NEC*® *210.50(B)* provides a good example of mandatory rules. **Permissive rules** are defined as those that are allowed but are not required and are characterized by the use of the terms "shall be permitted" and "shall not be required." *NEC*® *210.50(B)* is a good example of permissive rules. Both levels of installation controls can be found in the dwellings under construction shown in Figure 1-3.

Figure 1-3 New dwelling units under construction.

Measurements and the *NEC*®

> **KEY TERMS**
>
> **International System of Units (SI)** The measurement system, sometimes called the *metric system* in the United States, defined by the use of meters and kilograms, that uses base 10 unit divisions (SI units). The SI system is widely employed in almost all areas of the world except in the United States. The *NEC*® uses the SI system as the primary measurement system, with the English system, defined by the use of feet and pounds, used as a secondary system.

The United States is the only industrialized country in the world that does not use the **International System of Units (SI)**, also known as the metric system, as its primary measurement system. The units of measurement in this system are called SI units (for "Système International"—the system originated in France). The NFPA has maintained a dual measurement system for the *NEC*®. Earlier editions of the Code listed SI units (meters and grams) in parentheses following the English units (feet and pounds). Beginning with the 2002 edition of the Code, the order has been reversed: The Code now lists the SI units first, with the English units following in parentheses.

> **KEY TERMS**
>
> **Trade size** A system employed in the *NEC*® to define certain standard sizes of electrical equipment and raceways. For example, ½-in. (16-mm) internal diameter trade size conduit can actually measure between .526 in. (13.36 mm) and .660 in. (16.76 mm).
>
> **Metric designator** A dimension corresponding to trade size, as employed for the equipment and raceways, measured using the SI system.

For determining the size of a conduit or fitting, the *NEC*® uses a **trade size** system. Conduit of ½ in. trade size diameter is really closer to ⅞ in. in diameter. Trade size ¾ really measures slightly over 1 in. in diameter. The dimensions referred to as ½ in. (0.013 m) and ¾ in. (0.019 m) are trade size measurements. The Code has taken each of the English system electrical trade sizes and assigned it a **metric designator**, thus creating a kind of trade size system for SI units. The metric designator for ½ in. (.016 m)

trade size is 16 mm. The metric designator for trade size ¾ in. (.021 m) is 21 mm. A complete listing of the metric designator conversions can be found in *Table 300.1(C)*. In this book, the English system is used for all calculations and measurements, with the metric conversions in parentheses immediately following.

Listing and Labeling

> **KEY TERMS**
>
> **Utilization equipment** (See *NEC*® *Article 100*): Equipment that requires electricity to function. Utilization equipment includes appliances and luminaires (fixtures) but does not include raceways, boxes, or devices.

The *NEC*® is also concerned with the systems and conductors that supply power to **utilization equipment**. Utilization equipment is anything that operates electrically. Luminaires (fixtures), industrial machinery, automatic dishwashers, gasoline-dispensing pumps, and electrical signs are all examples of utilization equipment. The Code also has some rules for manufacturers of various types of utilization equipment, such as appliances and motors, but for the most part it does not dictate how the equipment is to function, or how it will be constructed. That is left up to the manufacturers of the equipment.

> **KEY TERMS**
>
> **Approved** (See *NEC*® *Article 100*): Referring to equipment, devices, raceways, and other electrical materials that are acceptable to the AHJ for installation and use.

The quality of electrical equipment is controlled by a system of testing agencies. The *NEC*® states, in *110.2*, that the various components, conduit, cable clamps, and other fittings, and equipment connected to the electrical system are permitted to be installed only if they are **approved** by the local AHJ. This approval process for each dwelling, by each jurisdiction, would be an overwhelming task. It would involve disassembling equipment in order to inspect the wiring inside of machinery, luminaires (fixtures), and appliances. It would also involve the close inspection of all wire installed and various tests conducted to ensure that the conductor insulation was safe and effective. Everything would have to be inspected and tested before approval could be given.

> **KEY TERMS**
>
> **Fine print note (FPN)**   A type of entry in the *NEC*® that provides explanatory information but is not formally enforceable as part of the *NEC*®.
>
> **Labeled**   (See *NEC*® *Article 100*): Referring to materials and equipment that have been found to meet certain requirements for safety and function by a testing agency recognized by the AHJ. An identification label, marking, or decal from the testing agency is attached to the materials to identify them as having been tested.
>
> **Listed**   (See *NEC*® *Article 100*): Referring to materials and equipment that have been found to meet certain requirements for safety and function by a testing agency recognized by the AHJ. The materials are placed on a list to identify them as having been tested.

A **fine print note (FPN)** following *110.2* directs the reader to *90.7*. The clear intent of *90.7* is to allow outside testing organizations to approve the internal wiring and construction of electrical equipment. If a particular testing organization is acceptable to the local AHJ, then the inspectors can approve the installation without inspecting the internal wiring. These testing agencies are independent of the manufacturer of the equipment. If the equipment passes the tests, it is given a seal, or label, that certifies it as safe if used according to the manufacturer's instructions. The equipment is said to be **labeled**. The manufacturer can display the label of the testing agency on the equipment or appliance, on the packaging, and in its advertising. The testing agency will also put the product on a formal list to confirm that the product has indeed been tested. The product is said to be **listed**. These lists are available from the testing agency, and they are what the AHJ refers to when approving the products or equipment. There are several testing laboratories, with Underwriters Laboratories (UL) being the largest and the best known. A logo consisting of the letters "UL" in a circle is used on the equipment label to notify interested parties that the equipment has been tested and declared safe under the conditions of the test.

1.2.2: Construction Drawings or Plans

> **KEY TERMS**
>
> **Construction drawings**   A series of drawings, sometimes called *blueprints*, that show the intended design of a building or other structure. There are usually different construction drawings for each of the different trades involved in the construction process.

Construction drawings are a series of drawings that detail the design of a building. Different drawings are produced for each trade involved in the project. Included in each set of plans are drawings that apply specifically to an individual trade, such as plumbing plans or electrical plans. The construction drawings are usually letter coded according to the trade. Thus, civil engineering drawings showing the street, curb, gutters, and sidewalks are designated with the letter C. In referencing information from a particular plan, a particular plan number is used—for example, drawing C-3 would be the third drawing of the civil engineering plans.

However, construction drawings for dwellings are not usually so detailed. It is not uncommon for electricians to have only a drawing of the basic floor plan from which to design the electrical system. The floor plan shows the room arrangement and may also show the locations and types of luminaires (fixtures), but true electrical plans showing receptacle and switching locations are often not included.

Drawings

The following drawings are for the sample house referred to throughout this book. The plans consist of construction drawings C-1 through E-7. There are a total of 16 main construction drawings, designated C for civil, S for structural, A for architectural, M for mechanical, and E for electrical, as follows:

- Construction Drawing C-1 (page 9): Site Development Plan and Proximity Map
- Construction Drawing C-2 (page 10): Building Site Plans Lot 16, Block 7, third Filing
- Construction Drawing S-1 (page 11): Foundation Plan
- Construction Drawing A-1 (page 12): Main Level Floor Plan

- Construction Drawing A-2 (page 13): Basement and Garage Dimensioned Floor Plan

- Construction Drawing A-3 (page 14): Main Floor Dimensioned Floor Plan

- Construction Drawing A-4 (page 15): East and North Elevations and Kitchen Cabinet Plans

- Construction Drawing A-5 (page 16): West and South Elevations, West Kitchen Cabinet Plan, and Bathroom Cabinet Plans

- Construction Drawing M-1 (page 17): Mechanical and Appliance Schedule

- Construction Drawing E-1 (page 18): Electrical Specifications

- Construction Drawing E-2 (page 19): Basement Electrical Plan

- Construction Drawing E-3 (page 20): Main Floor Electrical Plan

- Construction Drawing E-4 (page 21): Main Floor Telephone and Television Outlet Plan

- Construction Drawing E-5 (page 22): Luminaire (Fixture) Schedule

- Construction Drawing E-6 (page 23): Electrical Panel Schedules: Main Panel and Subpanel A

- Construction Drawing E-7 (page 24): Electrical Plan Symbols

These drawings provide all of the necessary information for wiring the dwelling. Notice that Construction Drawings A-2 and A-3 use English units of measurement for the floor plans. Also included, immediately following Construction Drawing E-7, are two additional drawings, Construction Drawings A-6SI and A-7SI (page 25 and page 26) that provide dimensions in SI (metric) units for the basement and main floor plans.

All of the drawings listed constitute the complete set of plans for the house. In addition to this set of construction drawings, four other electrical plans not usually found in a standard set of plans are provided, as Figures 1-4 through 1-7 (pages 27 through 30). Each of these plans is superimposed on a line grid system. The first two drawings are placed on a ⅛-inch grid (⅛ in. equals 1 ft), and the other two are placed on a 10.4-millimeter grid (10.4 mm equals 1 m). These line grid drawings are intended

to assist the reader with comprehending the scale of the floor plan dimensions.

Also included in this chapter are electrical plans for the sample house showing circuiting (Figures 1-8 and 1-9 [page 31 and page 32]). These plans are included to show the wiring installed in the house. It should be noted that the wiring scheme presented in these plans is only one of many possible plans available to the installing electricians.

Symbols

> **KEY TERMS**
>
> **Symbols** Icons used on construction drawings to represent various design features such as switches, receptacles, and luminaires (fixtures).

Construction plans use **symbols** to represent actual switches and receptacles. The symbols used in the plans for the sample dwelling are standard drawing symbols used throughout the country, but construction documents may include an example of each symbol with an explanation of what the symbol represents. This listing of the symbols used is called a *key*. The key to the symbols used for these drawings is presented in Construction Drawing E-7.

Notes

> **KEY TERMS**
>
> **Flag note** An icon used on construction drawings that alerts the construction team to the existence of additional information or requirements listed elsewhere on the drawings.

In many cases, additional information, beyond what the standard symbol can represent, must be provided to the installing electricians. To notify the electrician that additional information is required, a **flag note** is placed on the drawing adjacent to the subject equipment or device. The flag notes are numbered, and the numbers direct the electrician to a listing of explanations for each flag note located somewhere on that particular drawing. The flag notes usually apply only to the drawing that they appear on, rather than to the entire set of plans. These notes are called "flag notes" because the symbol looks like a flag or pennant. There are several flag notes on Construction Drawings E-2 and A-3.

SITE DEVELOPMENT PLAN AND PROXIMITY MAP

RUNNING BROOKE RANCH
THIRD FILING

CONSTRUCTION PLAN C-1

C-1

BUILDING SITE PLAN

LOT 16, BLOCK 7, THIRD FILING
RUNNING BROOKE RANCH SUBDIVISION

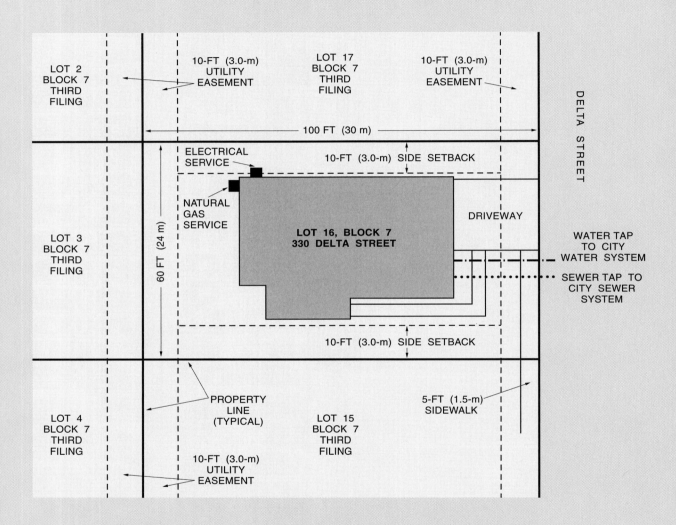

C-2

CONSTRUCTION DRAWING C-2

FOUNDATION PLAN

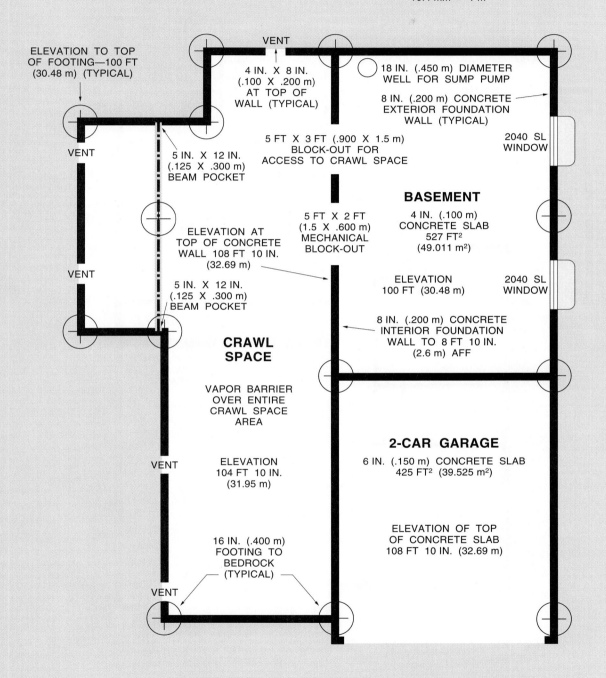

SCALE: 1/8 IN. = 1 FT
10.4 mm = 1 m

ELEVATION TO TOP
OF FOOTING—100 FT
(30.48 m) (TYPICAL)

VENT

4 IN. X 8 IN.
(.100 X .200 m)
AT TOP OF
WALL (TYPICAL)

18 IN. (.450 m) DIAMETER
WELL FOR SUMP PUMP

8 IN. (.200 m) CONCRETE
EXTERIOR FOUNDATION
WALL (TYPICAL)

5 FT X 3 FT (.900 X 1.5 m)
BLOCK-OUT FOR
ACCESS TO CRAWL SPACE

2040 SL
WINDOW

5 IN. X 12 IN.
(.125 X .300 m)
BEAM POCKET

VENT

BASEMENT

5 FT X 2 FT
(1.5 X .600 m)
MECHANICAL
BLOCK-OUT

4 IN. (.100 m)
CONCRETE SLAB
527 FT^2
(49.011 m^2)

ELEVATION AT
TOP OF CONCRETE
WALL 108 FT 10 IN.
(32.69 m)

VENT

ELEVATION
100 FT (30.48 m)

2040 SL
WINDOW

5 IN. X 12 IN.
(.125 X .300 m)
BEAM POCKET

8 IN. (.200 m) CONCRETE
INTERIOR FOUNDATION
WALL TO 8 FT 10 IN.
(2.6 m) AFF

CRAWL
SPACE

VAPOR BARRIER
OVER ENTIRE
CRAWL SPACE
AREA

2-CAR GARAGE

6 IN. (.150 m) CONCRETE SLAB
425 FT^2 (39.525 m^2)

VENT

ELEVATION
104 FT 10 IN.
(31.95 m)

ELEVATION OF TOP
OF CONCRETE SLAB
108 FT 10 IN. (32.69 m)

16 IN. (.400 m)
FOOTING TO
BEDROCK
(TYPICAL)

VENT

CONSTRUCTION DRAWING S-1

S-1

MAIN LEVEL FLOOR PLAN

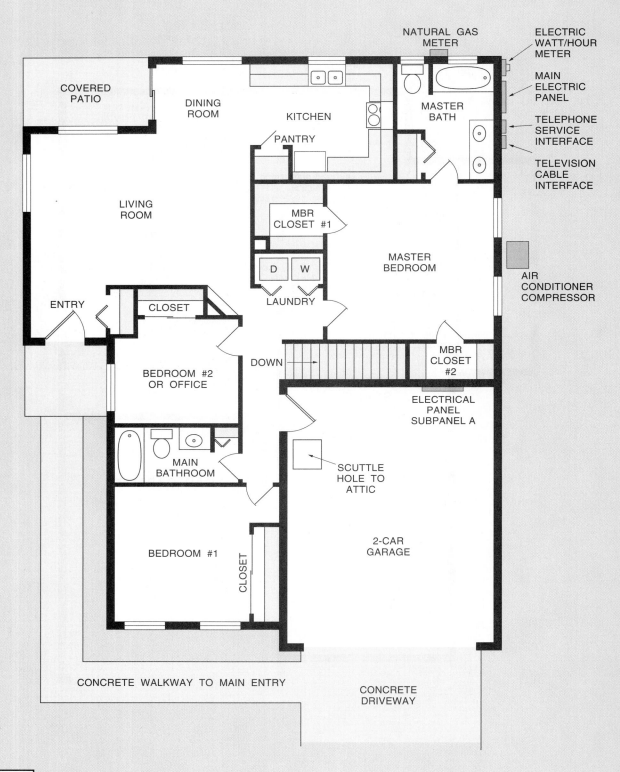

COVERED PATIO

DINING ROOM

KITCHEN

NATURAL GAS METER

ELECTRIC WATT/HOUR METER

MAIN ELECTRIC PANEL

MASTER BATH

TELEPHONE SERVICE INTERFACE

TELEVISION CABLE INTERFACE

PANTRY

LIVING ROOM

MBR CLOSET #1

MASTER BEDROOM

AIR CONDITIONER COMPRESSOR

ENTRY

CLOSET

D   W

LAUNDRY

BEDROOM #2 OR OFFICE

DOWN

MBR CLOSET #2

ELECTRICAL PANEL SUBPANEL A

MAIN BATHROOM

SCUTTLE HOLE TO ATTIC

BEDROOM #1

CLOSET

2-CAR GARAGE

CONCRETE WALKWAY TO MAIN ENTRY

CONCRETE DRIVEWAY

A-1

CONSTRUCTION DRAWING A-1

BASEMENT AND GARAGE
DIMENSIONED FLOOR PLAN

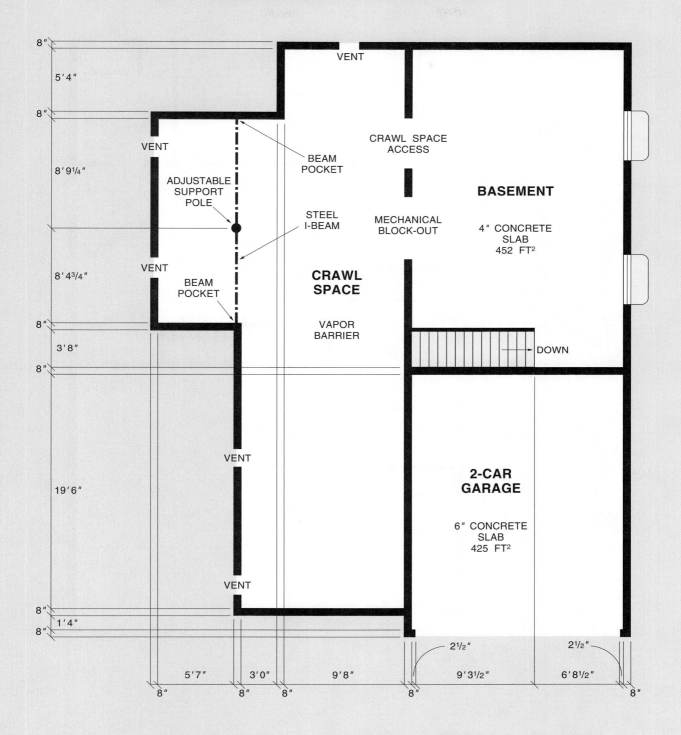

CONSTRUCTION DRAWING A-2

A-2

MAIN FLOOR
DIMENSIONED FLOOR PLAN

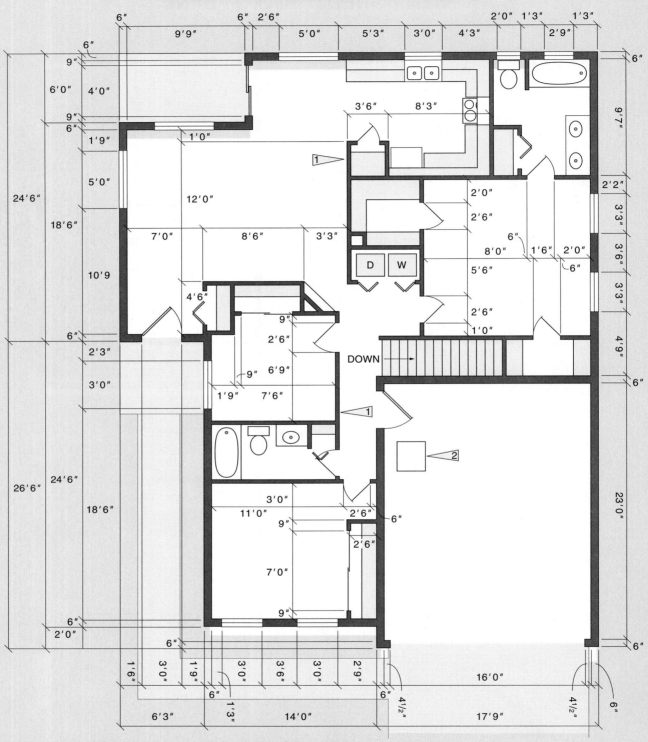

1 ALL INTERIOR WALLS ARE 5 IN. DEEP UNLESS SHOWN OTHERWISE.
OUTSIDE WALLS ARE 6 IN. DEEP. ALL WALLS TO HAVE ¾-IN. SHEETROCK.

2 SCUTTLE HOLE TO ATTIC

A-3

CONSTRUCTION DRAWING A-3

EAST AND NORTH ELEVATIONS AND KITCHEN CABINET PLANS

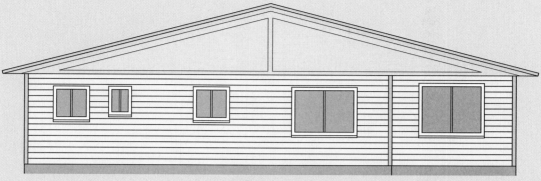

EAST ELEVATION

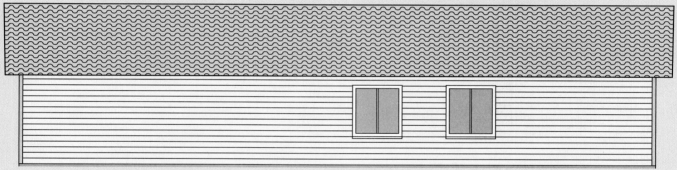

NORTH ELEVATION

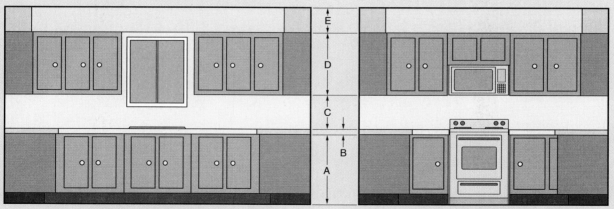

EAST KITCHEN CABINET PLAN

NORTH KITCHEN CABINET PLAN

A = 36 IN. (.900 m)
B = 2 IN. (.050 m)
C = 12 IN. (.300 m)
D = 36 IN. (.900 m)
E = 10 IN. (.250 m)

CONSTRUCTION DRAWING A-4

A-4

WEST AND SOUTH ELEVATIONS, WEST KITCHEN CABINET PLAN, AND BATHROOM CABINET PLANS

WEST ELEVATION

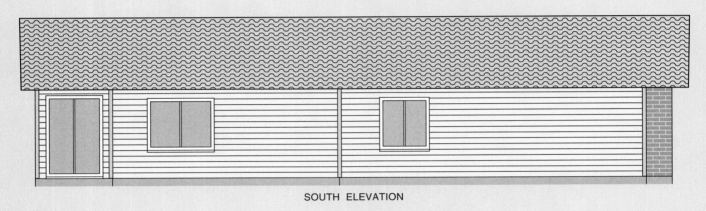

SOUTH ELEVATION

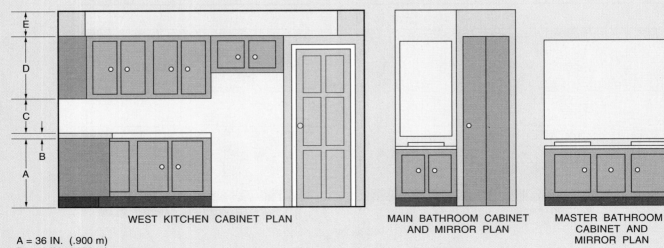

WEST KITCHEN CABINET PLAN

MAIN BATHROOM CABINET
AND MIRROR PLAN

MASTER BATHROOM
CABINET AND
MIRROR PLAN

A = 36 IN. (.900 m)
B =   2 IN. (.050 m)
C = 12 IN. (.300 m)
D = 36 IN. (.900 m)
E = 10 IN. (.250 m)

A-5

CONSTRUCTION DRAWING A-5

MECHANICAL AND APPLIANCE SCHEDULE

APPLIANCE	MANUFACTURER	MODEL NUMBER	VOLTS	FLA	VA/WATTS	HP	MIN CIRCUIT AMPS
Dryer	Franklin Ind	55MFT240-2-1-F	120/240		5500		
Range	Sargent Electric	TRP11.5-7687	120/240		11500		
Microwave	Sargent Electric	NMO-5634-A	120	14.7	1764		
Disposal	Kl Klienne	998-00034-76Y	120	12.0		¾	
Dishwasher	Kl Klienne	102-00198-22P	120	13.5		¹⁄₁₀	
Garage Door Opener	Door Best	2354-2	120	12.3	1475	¾	
Sump Pump	Water Whip	99-12075	120			¾	
Baseboard Heater	Heatease of Ohio	240200T-0-0	240		2000		
Air Conditioner	Wall and Miller	9348UYTEHDG47	240				28.8 35a Fuse
Furnace	Heatease of Ohio	120G110G-2-3	120	9.2	1100	¾	

CONSTRUCTION DRAWING M-1

M-1

ELECTRICAL SPECIFICATIONS

1. ALL SERVICE EQUIPMENT, DISTRIBUTION PANELS, AND DISCONNECT SWITCHES SHALL BE MANUFACTURED BY THE SAME COMPANY. ALL EQUIPMENT SHALL BE NEW AND MANUFACTURED BY ONE OF THE FOLLOWING COMPANIES:

   GENERAL ELECTRIC
   SQUARE D
   CUTTLER-HAMMER
   SIEMENS

2. ALL WIRING DEVICES ARE TO BE SUPPLIED AND INSTALLED BY THE ELECTRICAL CONTRACTOR.

3. ALL WIRING DEVICES ARE TO BE MANUFACTURED BY LITECHTRONIX, OR EQUAL.

4. ALL SWITCHES, EXCEPT FOR SWITCHES ABOVE COUNTERTOPS, SHALL BE INSTALLED AT 44 IN. (1.118 m) ABOVE THE FINISHED FLOOR (AFF) TO THE BOTTOM OF THE BOX.

5. ALL RECEPTACLE OUTLETS, EXCEPT FOR ABOVE COUNTER RECEPTACLES IN THE KITCHEN AND THE BATHROOMS SHALL BE 13 IN. (.330 m) AFF TO THE BOTTOM OF THE BOX. RECEPTACLE OUTLETS IN THE GARAGE SHALL BE 44 IN. (1.118 m) AFF.

6. ALL RECEPTACLES INSTALLED ABOVE KITCHEN OR BATHROOM COUNTERS SHALL BE 6 IN. TO THE BOTTOM OF THE BOX ABOVE THE COUNTER TOP (AC).

7. ALL LUMINAIRES (FIXTURES) SHALL BE PROVIDED BY THE GENERAL CONTRACTOR AND INSTALLED BY THE ELECTRICAL CONTRACTOR, EXCEPT FOR RECESSED LUMINAIRE (FIXTURE) HOUSINGS, RECESSED LUMINAIRE (FIXTURE) TRIMS, AND LAMPS FOR RECESSED LUMINAIRES (FIXTURES), WHICH SHALL BE PROVIDED AND INSTALLED BY THE ELECTRICAL CONTRACTOR.

8. ALL KITCHEN APPLIANCES SHALL BE SUPPLIED AND INSTALLED BY THE GENERAL CONTRACTOR. WHIPS, PIGTAILS, OR CORD NEEDED TO CONNECT THE APPLIANCES TO THE ELECTRICAL SYSTEM WILL BE PROVIDED AND INSTALLED BY THE ELECTRICAL CONTRACTOR.

9. ALL SWITCHES, RECEPTACLES, AND OTHER DEVICES SHALL BE WHITE. ALL COVER PLATES WILL BE WHITE PLASTIC (NONMETALLIC).

10. ALL RECEPTACLES SHALL BE INSTALLED WITH THE GROUND PRONG (CONTACT) TO THE TOP OF THE DEVICE. IF THE DEVICE IS INSTALLED HORIZONTALLY, THE TOP BLADE OF THE RECEPTACLE SHALL BE THE GROUNDED CONDUCTOR BLADE.

11. ALL FIXED-IN-PLACE KITCHEN APPLIANCES SHALL BE ON THEIR OWN INDIVIDUAL CIRCUITS SIZED TO THE REQUIREMENTS OF THE PARTICULAR APPLIANCE.

12. A SUMP PUMP WILL BE INSTALLED IN THE BASEMENT. GENERAL CONTRACTOR WILL PROVIDE THE PUMP. THE LOCATION OF THE PUMP IS PER CONSTRUCTION DRAWING S-1. THE PUMP WILL BE INSTALLED BY THE MECHANICAL CONTRACTOR AND WIRED BY THE ELECTRICAL CONTRACTOR PER THE REQUIREMENTS AND THE MOTOR NAMEPLATE.

13. THE INSTALLATION OF ALL WIRE, BOXES, DEVICES, LUMINAIRES (FIXTURES), AND OTHER ELECTRICAL EQUIPMENT SHALL BE IN STRICT ACCORDANCE WITH THE *NATIONAL ELECTRICAL CODE*®, LOCAL JURISDICTIONAL REQUIREMENTS, MANUFACTURERS' INSTRUCTIONS, AND THESE SPECIFICATIONS.

14. THE ELECTRICAL CONTRACTOR SHALL PROVIDE TEMPORARY POWER, AND IF NECESSARY TEMPORARY LIGHTING, FOR THE USE OF ALL TRADES ON THE JOB SITE.

15. THE ELECTRICAL CONTRACTOR SHALL PROVIDE AND PAY FOR ANY PERMITS THAT MAY BE REQUIRED BY THE CITY BUILDING INSPECTOR'S OFFICE. THE ELECTRICAL CONTRACTOR WILL SUCCESSFULLY OBTAIN ALL REQUIRED INSPECTIONS.

E-1   CONSTRUCTION DRAWING E-1

BASEMENT ELECTRICAL PLAN

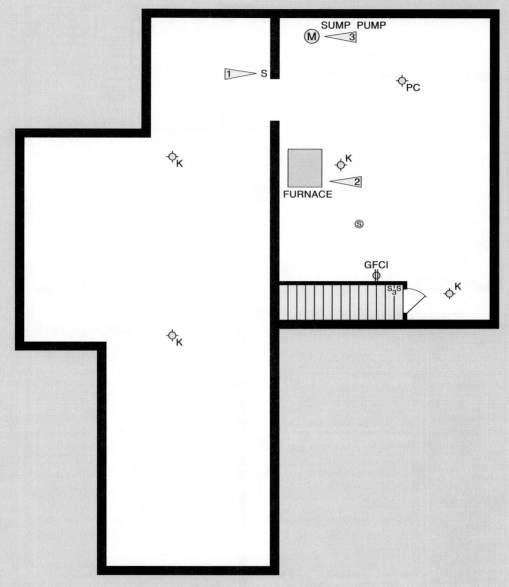

SUMP PUMP

FURNACE

GFCI

1 > PLACE SWITCH WITHIN EASY REACH FROM
CRAWL SPACE ENTRANCE.

2 > DOOR BELL TRANSFORMER TO BE POWERED BY
THE FURNACE CIRCUIT. INSTALL ON THE LINE
SIDE OF THE FURNACE DISCONNECT SWITCH.

3 > SUMP PUMP, 1/2 HP, 120 VOLTS, INDIVIDUAL
20-AMPERE CIRCUIT

CONSTRUCTION DRAWING E-2

E-2

MAIN FLOOR ELECTRICAL PLAN

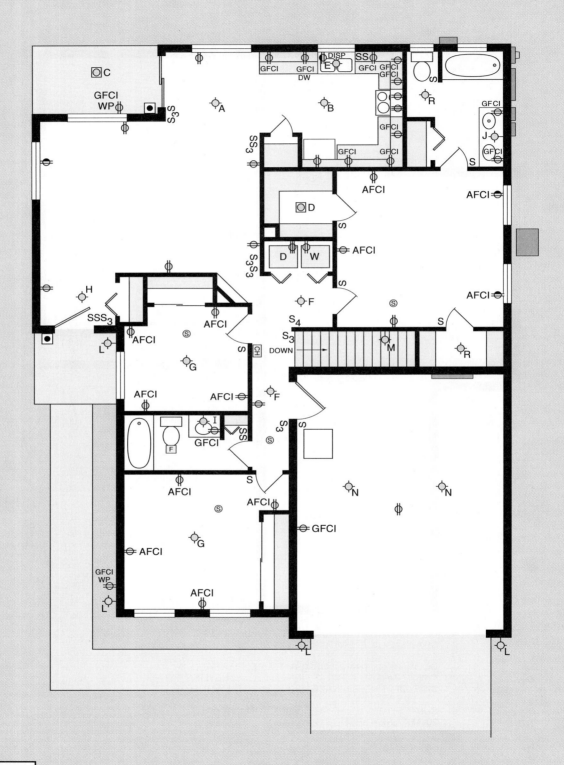

CONSTRUCTION DRAWING E-3

MAIN FLOOR TELEPHONE AND
TELEVISION OUTLET PLAN

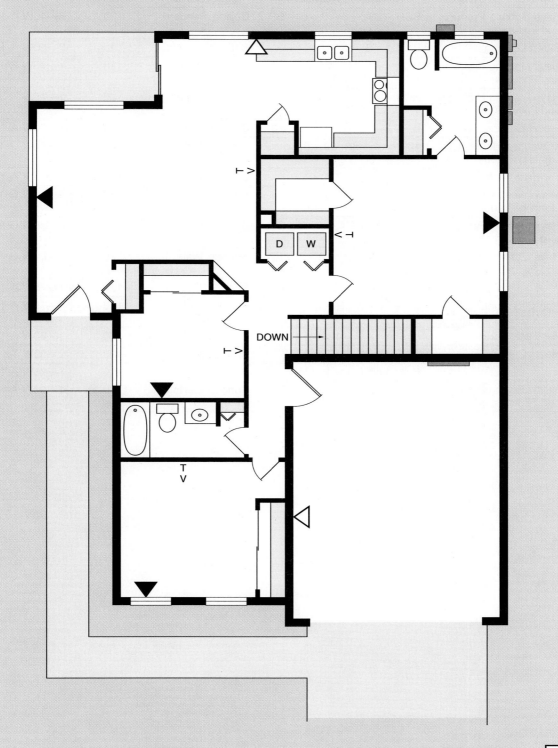

CONSTRUCTION DRAWING E-4

E-4

LUMINAIRE (FIXTURE) SCHEDULE

FIXTURE TYPE	MANUFACTURER	CATALOG NUMBER	MOUNTING	VOLTAGE	LAMPS
A	Hays Lighting Products and Fixtures	SCH – 231 – 0 – 9 – G – BR	Ceiling. Pendant.	120	6—40W A-17
B	Elbert Lighting	B – 84 – 3	Surface. Ceiling.	120	3—60W A-19
C	Hays Lighting Products and Fixtures	RCA – 7.5 – IC Housing T – 7.5 WP – 3 Trim	Recess. Water-proof trim.	120	1—75W PAR 38
D	Hays Lighting Products and Fixtures	RCA – 7.5 – IC Housing T – 7.5 OP – 2 Trim	Recess. Black baffle trim.	120	1—75W R-30
E	Elbert Lighting	B – 71 – 1	Surface. Ceiling.	120	1—60W A-17
F	Milwaukee Home Lighting Products	120 – 40 – BA – 0 – 8	Surface. Ceiling.	120	4—40W G-10
G	Far East Fixtures	75 – AD – R – 120 – CR	Surface. Ceiling.	120	2—60W A-19
H	Milwaukee Home Lighting Products	120 – 40 – BA – T – 8	Surface. Ceiling.	120	4—40W G-10
I	Hays Lighting Products and Fixtures	SWS – 414 – 4 – 2 – K – BR	Surface. Wall.	120	4—60W G-28
J	Hays Lighting Products and Fixtures	SWS – 414 – 6 – 2 – K – BR	Surface. Wall.	120	6—60W G-28
K		Keyless Lamp Holder	Surface. Ceiling.	120	1—75W A-19
L	Howell-Simon	DF – 02387 – P – 7	Surface. Outdoor wall mount.	120	1—75W A-19
M	Far East Fixtures	60 – AR – R – 120 – CR	Surface. Wall.	120	1—60W A-19
N	HeavyLite	120 – 2 – 40 – ST – 0 – 1	Surface. Ceiling.	120	2—32W T-8 CW
PC		Pull Chain Lamp Holder	Surface. Ceiling.	120	1—75W A-19
R	Hays Lighting Products and Fixtures	SCS – 213 – 1 – 2 – L – BR	Surface. Ceiling.	120	1—60W A-19

E-5

CONSTRUCTION DRAWING E-5

ELECTRIC PANEL SCHEDULES

MAIN CIRCUIT BREAKER

150 AMPERES
120/240 VOLTS

PANEL BUSSING 200 AMPERES

MAIN PANEL

CIRCUIT NUMBER	DESCRIPTION	CIR BKR		CIR BKR	DESCRIPTION	CIRCUIT NUMBER
1	AIR CONDITIONER	30				2
3	AIR CONDITIONER	30				4
5	SUBPANEL A	125				6
7	SUBPANEL A	125				8

MAIN LUGS

PANEL BUSSING 200 AMPERES

SUBPANEL A

CIRCUIT NUMBER	DESCRIPTION	CIR BKR		CIR BKR	DESCRIPTION	CIRCUIT NUMBER
1	REFRIGERATOR	15		20	SMALL APPLIANCE	2
3	KIT, DR, LR, LIGHTING / LR, OUTSIDE RECEPT.	15		20	SMALL APPLIANCE	4
5	BR LIGHTING & RECEPT.	15		20	MICROWAVE / HOOD	6
7	MBR, MBR BATH LIGHTING & RECEPT.	15		20	DISHWASHER	8
9	RANGE	40		20	DISPOSAL	10
11	RANGE	40		20	BATHROOM RECEPTACLES	12
13	DRYER	30		15	GARAGE / OUTSIDE FRONT RECEPT.	14
15	DRYER	30		20	CLOTHES WASHER	16
17	FURNACE	20		15	BSMT LIGHTING & RECEPT.	18
19	SMOKE DETECTOR AND ALARMS	20		15	BASEBOARD HEATER	20
21	GARAGE DOOR OPENER	15		15	BASEBOARD HEATER	22
23				20	SUMP PUMP	24

CONSTRUCTION DRAWING E-6

E-6

ELECTRICAL PLAN SYMBOLS

Symbol	Description	Symbol	Description
φ	INDIVIDUAL RECEPTACLE, 120 VOLTS, 15 OR 20 AMPERES	S	SINGLE POLE SWITCH
φ	DUPLEX RECEPTACLE, 120 VOLTS, 15 OR 20 AMPERES	S_3	3-WAY SWITCH
φ	SPLIT WIRED DUPLEX RECEPTACLE, 120 VOLTS, 15 OR 20 AMPERES	S_4	4-WAY SWITCH
φ GFCI	GFCI-PROTECTED DUPLEX RECEPTACLE, 120 VOLTS, 15 OR 20 AMPERES	⌀	CEILING-MOUNTED LUMINAIRE (FIXTURE)
φ WP	WATERPROOF DUPLEX RECEPTACLE, 120 VOLTS, 15 OR 20 AMPERES	⌀	WALL-MOUNTED LUMINAIRE (FIXTURE)
φ GFCI WP	WATERPROOF AND GFCI-PROTECTED DUPLEX RECEPTACLE, 120 VOLTS, 15 OR 20 AMPERES	▢	RECESSED LUMINAIRE (FIXTURE)
φ AFCI	AFCI-PROTECTED DUPLEX RECEPTACLE, 120 VOLTS, 15 OR 20 AMPERES	Ⓢ	SMOKE DETECTOR, 120 VOLTS W/BATTERY BACKUP
φ AFCI	SPLIT WIRED AFCI-PROTECTED DUPLEX RECEPTACLE, 120 VOLTS, 15 OR 20 AMPERES	F	BATH FAN
φ	INDIVIDUAL RECEPTACLE, 240 VOLTS, 30 OR 50 AMPERES	CH	DOORBELL CHIME, 12 VOLTS
		▣	DOORBELL PUSHBUTTON, 12 VOLTS
		$\frac{T}{V}$	CATV (TELEVISION) OUTLET
		▼	TELEPHONE OUTLET—DESK
		▽	TELEPHONE OUTLET—WALL

E-7

CONSTRUCTION DRAWING E-7

BASEMENT AND GARAGE
DIMENSIONED FLOOR PLAN: SI UNITS

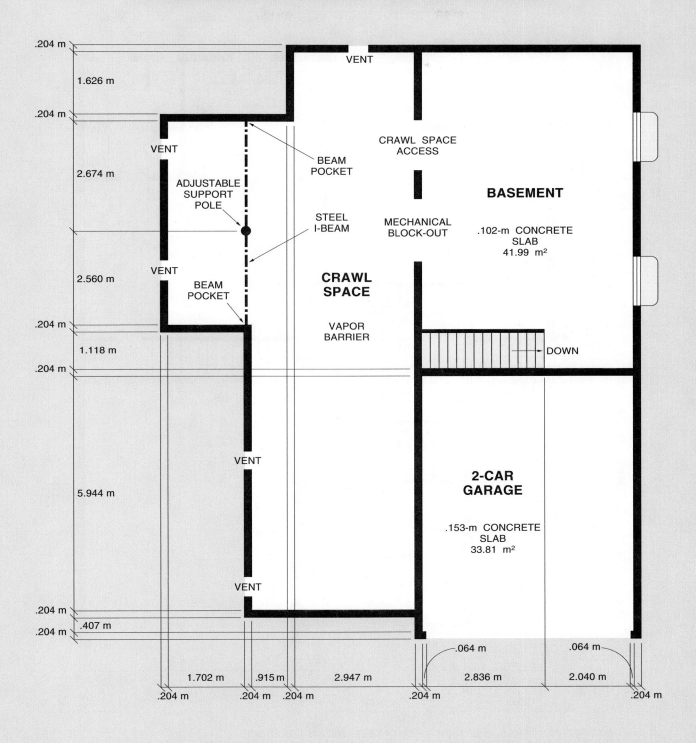

VENT

.204 m

1.626 m

.204 m

VENT

2.674 m

CRAWL SPACE
ACCESS

BEAM
POCKET

ADJUSTABLE
SUPPORT
POLE

STEEL
I-BEAM

MECHANICAL
BLOCK-OUT

BASEMENT

.102-m CONCRETE
SLAB
41.99 m^2

2.560 m

VENT

BEAM
POCKET

**CRAWL
SPACE**

VAPOR
BARRIER

.204 m

1.118 m

.204 m

DOWN

5.944 m

VENT

**2-CAR
GARAGE**

.153-m CONCRETE
SLAB
33.81 m^2

VENT

.204 m

.204 m   .407 m

.064 m   .064 m

1.702 m   .915 m   2.947 m   2.836 m   2.040 m

.204 m   .204 m   .204 m   .204 m   .204 m

CONSTRUCTION DRAWING A-6SI

A-6SI

MAIN FLOOR
DIMENSIONED FLOOR PLAN: SI UNITS

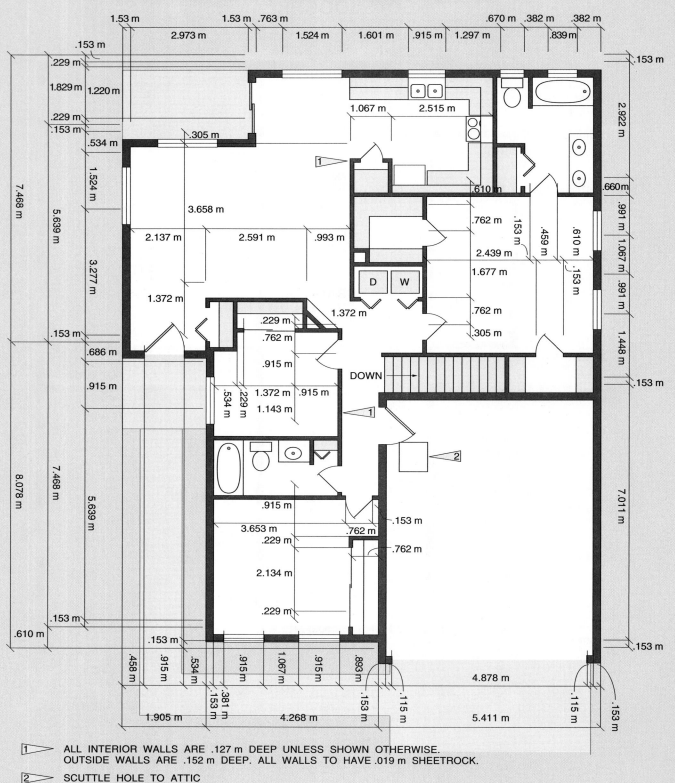

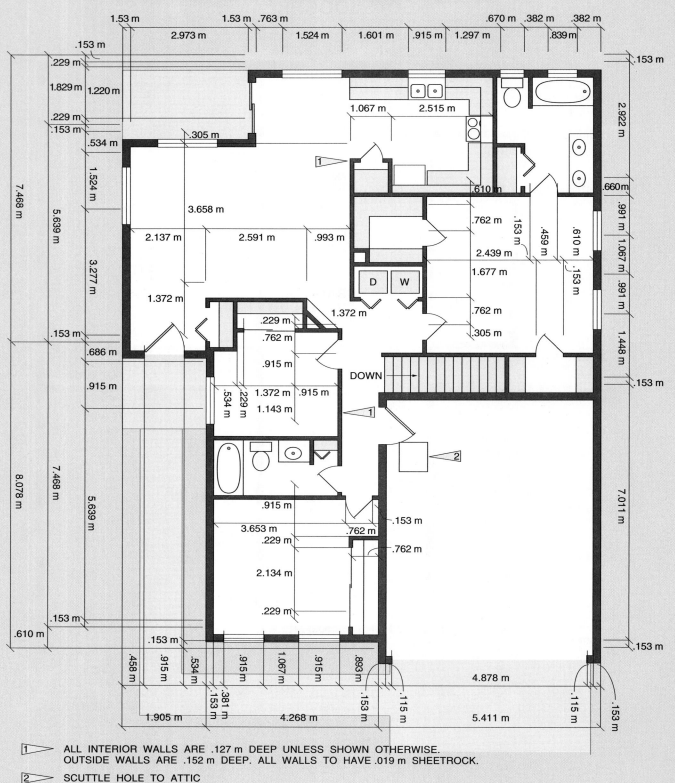► ALL INTERIOR WALLS ARE .127 m DEEP UNLESS SHOWN OTHERWISE.
OUTSIDE WALLS ARE .152 m DEEP. ALL WALLS TO HAVE .019 m SHEETROCK.

► SCUTTLE HOLE TO ATTIC

A-7SI   CONSTRUCTION DRAWING A-7SI

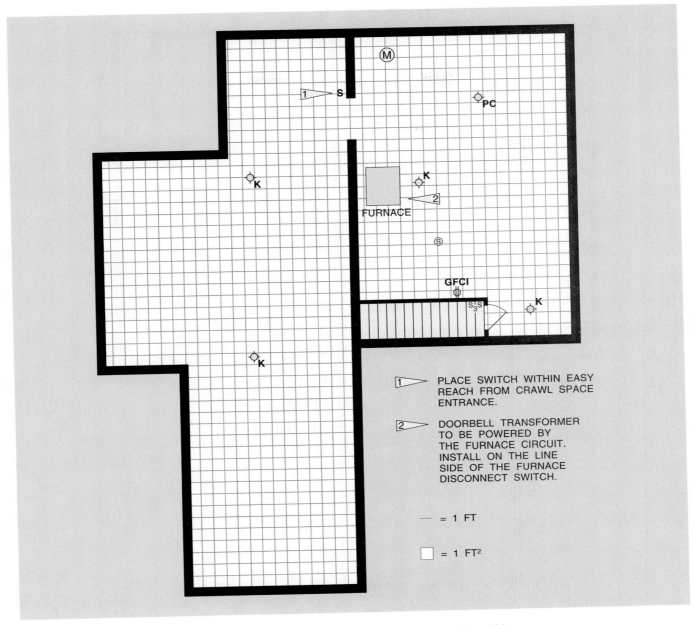

Figure 1-4   Basement electrical plan with 1-ft grid.

Schedules

Schedule   A layout, usually in the form of a table, that is part of the construction drawings and provides detailed information concerning materials, devices, or equipment to be installed as part of the construction process. For example, detailed information about the various luminaires (fixtures) to be installed is often conveyed using a schedule.

Construction drawings may also include other information. Sometimes a chart is drawn onto one or more of the drawings that lists all of the equipment being installed on the various circuits. These charts are called **schedules**. Schedules are used to organize the information and present it in one concise place, thus making it easier to find. Schedules can be used for mechanical equipment, appliances, or luminaires (fixtures), but can also be used for doors, wall finish, floor finish, and others. A luminaire (fixture) schedule and a mechanical and appliance schedule for the sample house are presented in Construction Drawings E-5 and M-1, respectively.

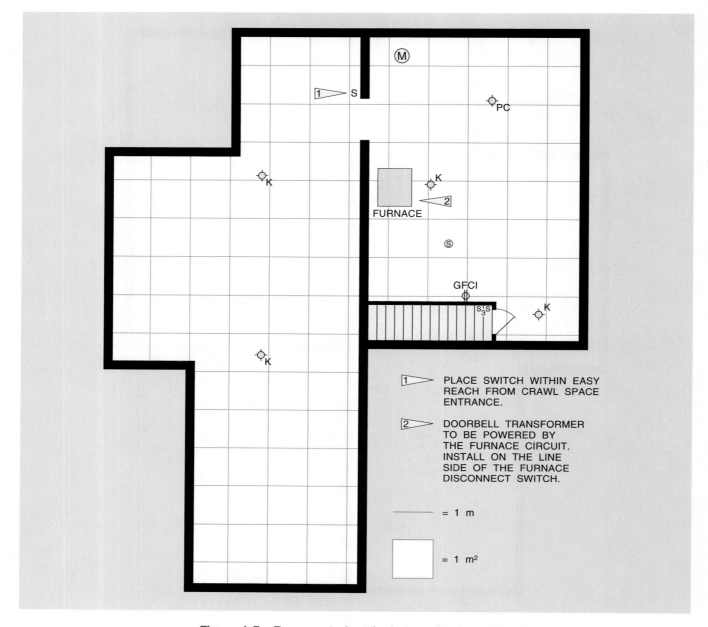

Figure 1-5 Basement electrical plan with 1-m grid.

1.2.3: Manufacturers' Installation Instructions

Installation instructions The directions provided by the manufacturer concerning the procedures to be followed in preparing electrical equipment, devices, or other materials for use. These instructions must be closely followed in installing electrical materials; otherwise, unsatisfactory operation may result in fire or other safety hazards.

NEC® 110.3(B) says that the equipment shall be installed according to all instructions on the listing. These instructions are usually the manufacturer's **installation instructions** that accompany the product at time of sale. These instructions must be closely followed in installing that system component. The intent of the Code is to assure the local AHJ that an installation is safe, if the equipment is listed or labeled, and if it is installed according to the manufacturer's instructions. The installing electrician needs to understand the importance of installing

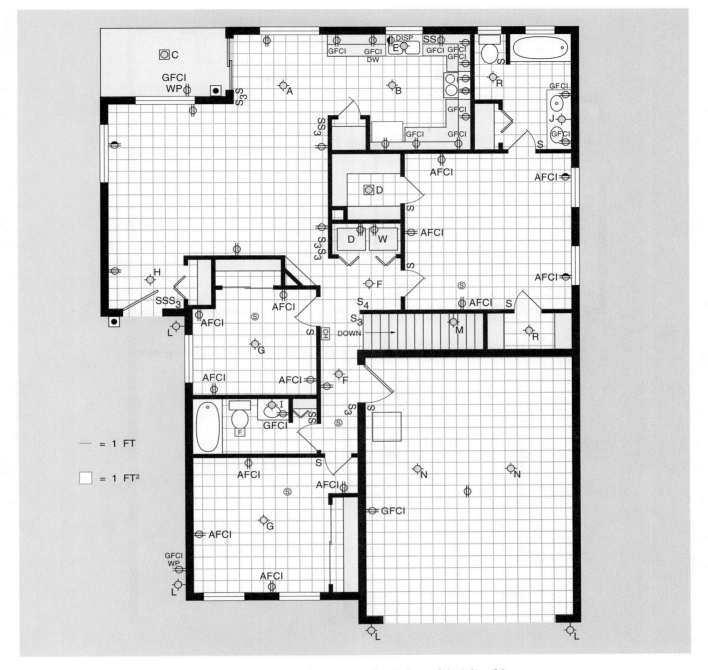

Figure 1-6   Main floor electrical plan with 1-ft grid.

the equipment according to the manufacturer's instructions. Any changes to the required installation method or any misuse of the equipment may void the testing laboratory's certification and make the equipment unlawful (not approved) to install and use. Some of the actions that may void UL certification, for example, are removing parts of the equipment, drilling holes in the case or housing, and installing in unapproved locations. An unapproved

location may be a wet location, for example, when the equipment is not listed for installation in a wet location.

Electricians must be careful to install the equipment only as directed in the manufacturer's instructions. These installation instructions should not be destroyed, but should be filed with the remainder of the project's paperwork as proof of a proper installation should any question arise. Underwriters

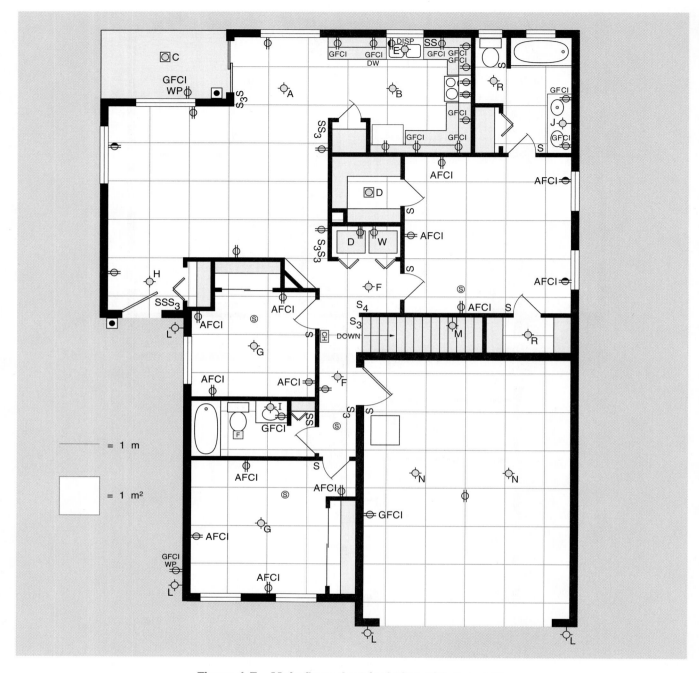

Figure 1-7   Main floor electrical plan with 1-m grid.

Laboratories or any of the other testing agencies approved by the AHJ can supply additional information on listing and labeling. Underwriters Laboratories can be reached at:

Underwriters Laboratories Corporate
  Headquarters
333 Pfingsten Rd.
Northbrook, IL 60062-2096
Telephone: 847-272-8800
Fax: 847-272-8129
Internet address: www.ul.com

1.2.4: The Specifications

Specifications   A construction document that usually controls the quality of the construction installations. It may list the brand names of the materials to be used during construction, certain procedures to be employed, and directions about communications between the various construction trades and the management team.

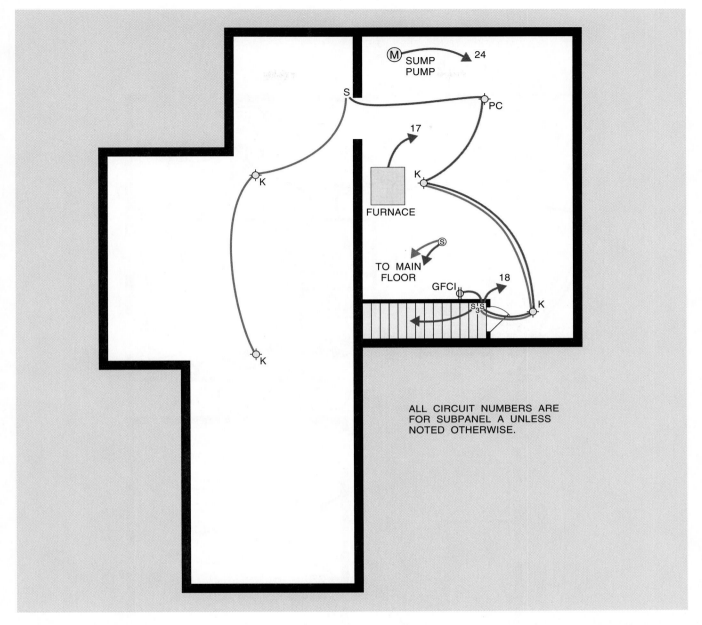

Figure 1-8　Basement electrical plan with circuiting. This plan is the same as in Construction Drawing E-2 except that it shows the details of the basement wiring.

Specifications are used extensively to control the quality of the installation. The architect or the owner may want certain products or wiring methods to be used on the project. Requirements for these preferences are published in the specifications. Such requirements are also sometimes employed to exclude certain products or techniques from use. The specifications may require, for example, that conductors of 12 AWG minimum size must be used in wiring the house. This requirement excludes the use of 14 AWG conductors.

Specifications are sometimes listed on construc-

tion drawings. This approach is the most common in residential construction, but a separate document (sometimes a book-length document) is commonly used also. Specifications for the house are presented in Construction Drawing E-1.

1.2.5: Shop Drawings, Cut Sheets, and Submittals

KEY TERMS

Shop drawings/cut sheets Drawings, usually provided by the manufacturer of equipment or materials

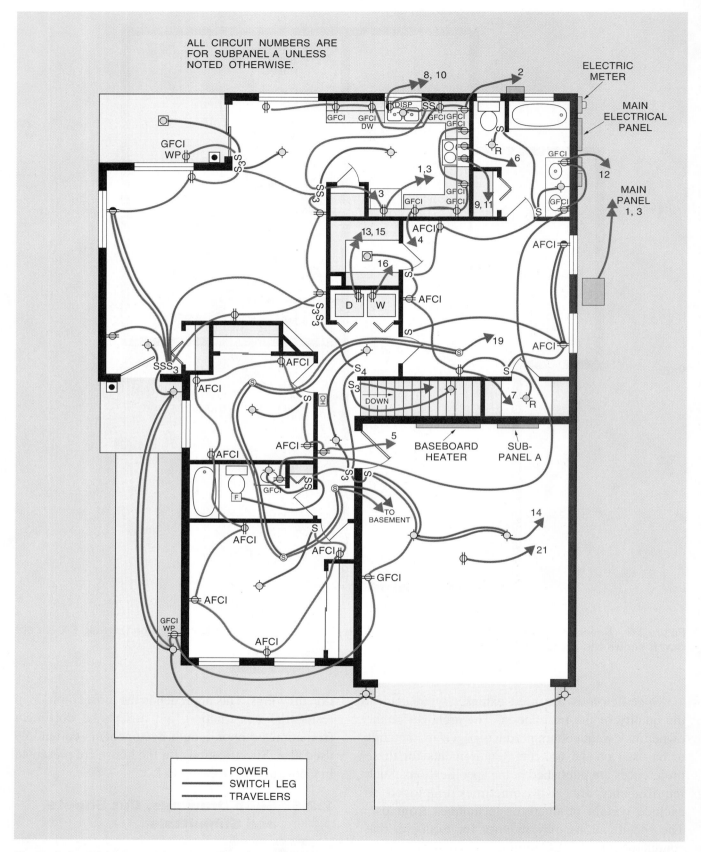

Figure 1-9 Main floor electrical plan with circuiting. This plan is the same as in Construction Drawing E-3 except that it shows the details of the main floor wiring.

Much of the power consumed by a structure is used by utilization equipment other than luminaires (fixtures). Appliances of all types, sizes, and electrical requirements—from exhaust fans to microwave ovens—are installed by trade workers who are not electricians. In order for the electrician to know the exact method and location of the connection and other necessary information about the size and shape of the equipment, the trades that furnish the appliances or equipment should provide the electrician with **shop drawings** or **cut sheets**. Many times the building specifications will require that each trade provide cut sheets or shop drawings to the architect. These cut sheets are also provided to the general contractor and the architect for approval. When they are provided to the general contractor or the architect, these documents are called **submittals**. They are submitted to the architect for approval prior to installation.

1.2.6: The Scope of Work Document

The **scope of work** is a document that lists the work to be accomplished, in a broad sense. The scope lists those areas that the contractor is responsible for completing and answers questions about responsibility for providing and/or installing luminaires (fixtures), appliances, and other equipment. The electrical contractor uses the scope of work to help create an estimate of the installation costs. The lead person on the job uses the scope of work to determine what is included in the contract and what is not included. The general contractor uses the scope of work to ensure that the subcontractors are delivering what they contracted to install. The owner uses the scope of work to determine if there will be extra charges for some change or addition they are considering. The bank or other institution that provides construction financing uses the scope of work to determine if the dwelling will, in fact, be worth the amount loaned to the owner for construction costs.

1.2.7: Work Orders or Change Orders

Many times during the course of the building's construction, changes will be made to the original design. The reasons for these changes are many and varied, but regardless of the cause, changes to the project can cause havoc on the job site. Many procedures are employed by various contractors to attempt to minimize the impact of changes. The general contractor puts in place, at the start of a project, procedures to allow each contractor working on the job to know of the proposed changes, to submit an estimate of the cost impact of those changes, and to convey accurate information to the installation crew to allow for a timely and cost-effective completion of the changes. These procedures are different from one general contractor to another, but all general contractors have some sort of change order system. The changes are usually presented to the subcontractor in the form of a **work order** or a **change order**. A work order authorizes the subcontractor to perform a certain package of work. To a large degree, a change order or work order can be viewed as a change to the scope of work document.

1.3: PERMITS AND INSPECTIONS

The contractor usually must obtain a building permit before beginning construction of a dwelling. Because the permit process varies widely throughout the country, a generalized explanation of the procedure is not possible. Usually, however, a separate permit is required for the electrical installation. A series of three inspections is usually required during the life of the permit. The first is usually the rough inspection. A rough inspection includes the wiring within the walls, ceiling, and any other location that will be covered or otherwise not accessible later during the construction process. A second inspection is usually required for the dwelling when it is completed and ready for occupancy. This is sometimes called the *final inspection* or *trim inspection*. All of the electrical luminaire (fixture) outlets, devices, and other systems must be installed and operating properly for the dwelling to pass this inspection. The third inspection involves the building's main electrical service. The building's electrical service is also usually required to be inspected at the time of, or before, the final inspection. The systems employed by the local AHJ to record and control the inspection process also vary widely. In many cases, the AHJ issues an inspection card to be kept on the job site and available to the inspectors at all times. As a phase of inspection is completed and approved, the inspector notes on the card what type of inspection was completed and what area of the structure was included in the inspection.

1.4: REGIONAL DIFFERENCES

The United States is a large country, and it is very difficult to make general statements that apply to the entire country without exceptions. This is particularly true when dealing with the construction industry. Physical conditions such as weather, availability of materials, local customs, and a host of other factors allow for many variations in installation procedures and wiring methods.

The information presented in this book applies to the majority of the country. Figure 1-10 shows some of the more likely regional differences and some of the reasons for the differences. Electricians must always be aware of the particular rules and procedures followed in their local areas, and adjustments to the information presented in this book should be made accordingly.

1.5: INTRODUCTION TO JOB SITE SAFETY

KEY TERMS

Occupational Safety and Health Administration (OSHA)   A branch of the federal government that is charged with improving workplace safety and encouraging the establishment of safe workplace practices.

Nothing is as important as job site safety—nothing! Not money, not completion date, and not lunch, or break, or quitting time! Safety on the job site must be the first thing in each worker's and each foreperson's mind. The federal government believes in on-the-job safety so much that it established the **Occupational Safety and Health Administration (OSHA)** to oversee safety policy in the workplace. OSHA publishes safety standards and inspects places of employment, including construction sites, periodically (and when requested) to ensure compliance with these standards. Employers found to be out of compliance with the established standards can be heavily fined, as can employees of such firms in some cases.

KEY TERMS

Personal Protective Equipment (PPE)   Safety equipment, such as hardhats, gloves, safety glasses, and work boots, that is provided by individual construction workers for their own use.

Employers also provide written safety policies and a company safety manual. Each employee must read, understand, and adhere to these safety policies. It is also the responsibility of each person on the site to be familiar with the safety rules established for that site by the general contractor. The wearing of **personal protective equipment (PPE)**, such as hard hats and safety glasses, is very commonly required of everyone on the site. Wearing work boots, sometimes with steel toe protectors, and properly sized rugged clothing is also a common requirement. However, all of these safety rules and regulations are of no value unless they are scrupulously followed. Unless every worker on the site obeys the safety rules, it is just a matter of time before a major accident happens. The construction site is inherently a noisy, crowded, and very dangerous place, and injury or death awaits the unprepared. Construction

"WIREMAN'S GUIDE"
REGIONAL DIFFERENCES IN WIRING METHODS

WIRING METHODS AND CONDUCTOR TYPES ACCEPTABLE TO THE LOCAL AHJ CAN VARY WIDELY FROM ONE AREA OF THE COUNTRY TO ANOTHER. IT IS VERY IMPORTANT THAT ELECTRICIANS UNDERSTAND THAT SOME OF THE PRACTICES USED IN THEIR AREA OF THE COUNTRY MAY NOT BE ACCEPTABLE IN OTHER AREAS. THE *NATIONAL ELECTRICAL CODE®* ATTEMPTS TO ESTABLISH RULES THAT WILL BE ACCEPTABLE AND EFFECTIVE IN ALL AREAS OF THE COUNTRY, BUT LOCAL CUSTOM OR CLIMATE MAY TAKE PRECEDENCE.

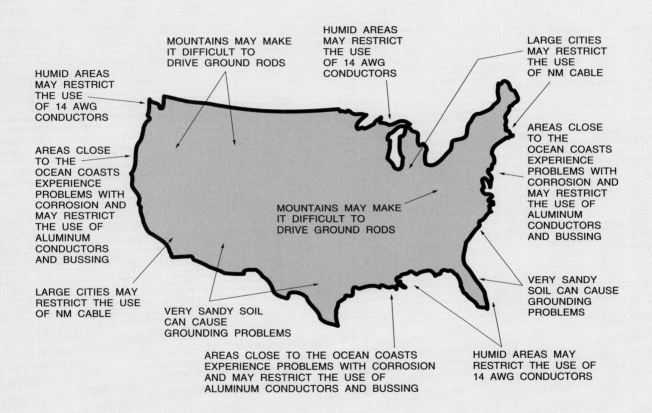

LOCAL CLIMATE AND WEATHER CONDITIONS CAUSE THE MOST VARIATIONS IN INSTALLATION RULES. AREAS OF THE COUNTRY WITH VERY SANDY SOIL MAY REQUIRE GROUND RINGS AND CONCRETE-ENCASED ELECTRODES AT ALL SERVICES TO SUPPLEMENT THE INCOMING METALLIC WATER PIPING SYSTEM, WHEREAS AREAS WITH FIRM CLAY SOILS MAY NEED ONLY A GROUND ROD AS A SUPPLEMENT.

AREAS OF THE COUNTRY THAT ARE EXTREMELY HUMID HAVE PROBLEMS WITH CORROSION. THE MOISTURE IN THE AIR CAN CAUSE BUSSING AND CONDUCTOR TERMINATIONS TO CORRODE AND FAIL. THEREFORE, THE AHJ MAY RESTRICT THE USE OF ALUMINUM CONDUCTORS, LIMIT THE MINIMUM SIZE OF CONDUCTORS TO 12 AWG, AND REQUIRE EXTRA CORROSION PROTECTION FOR CONDUITS AND RACEWAYS.

Figure 1-10

workers must be mindful of the dangers on the site at all times, and they must take actions necessary to properly protect themselves from those dangers. Using the proper personal protective equipment is the first, the most basic, and the most effective step in job site safety.

A major hazard that exists on the job site is electricity. Almost all trade workers use electricity to run their power tools, compressors, and chargers. Furthermore, artificial lighting may be needed in basements or interior rooms with no windows to illuminate the work space. The electrical contractor usually installs temporary power at the construction site and sometimes temporary lighting. There are very strict rules concerning the temporary power that is supplied and how it is supplied. Temporary power to a job site is covered in some detail in Chapter 2 of this book. It is important to remember that workers from all trades will be using this temporary power. Many of these workers have little or no knowledge of the nature of electricity. The temporary wiring at a job site for these workers to use must be not only safe but also relatively foolproof.

There will be times when an electrician must work on a system while it is energized. Working a circuit "hot" is something to be avoided at almost any cost. The term *hot* in the electrical trade is used to describe a circuit that is energized—that is, connected to the electrical supply. Work on an energized circuit poses needless risk of injury or death. As a rule, electricians should not work on hot circuits. Some type of disconnecting switch is required by the Code on all circuits for all hot conductors, except service-entrance conductors. The disconnecting switch may be a circuit breaker in a panel or may be a separate switch. These disconnecting switches must be easily accessible by the electrician and must be either able to be locked in the off position or within easy sight of the individuals working on the circuit. Just de-energizing a system or circuit is not good enough for proper safety procedures. If the system or equipment could be re-energized without

the technician's being aware of it, a *lockout* and *tag-out* procedure must be employed. All electrical contractors have a written tag-out and lockout procedure that should be familiar to each employee.

However, sometimes under the right set of conditions, it is unavoidable that an electrician must work on a system that is energized. In these cases, only qualified persons can work on the circuit or equipment. *NEC®* 100 defines a qualified person as *"one familiar with the construction and operation of the equipment and the hazards involved."* A qualified person is one who has the training, experience, and knowledge to determine the hazards involved. Asking an inexperienced or untrained person to work on energized circuits is a recipe for disaster. A simple admonition "Be careful" or "Don't touch that" certainly does not constitute enough training to make an individual qualified. Electrical contractors also have a written policy with the proper procedures to follow when working on a hot system. Electricians must know and follow these safety procedures explicitly.

SUMMARY

There are many different documents that provide information for the job. Specification, schedules, construction drawings, scope of work documents, cut sheets, and other construction documents make up the available information for the project and contain that information essential to a satisfactory installation. The local government is involved with the construction project through the permit and inspection process, through the listing and labeling process, and it can make changes to the rules established by the *NEC®* or the UBC.

Safety is the first, the middle, and the last thing on a construction site. There are many dangers present on a construction site that are not present elsewhere, and the workers need to be constantly aware of what they are doing and what others around them are doing. Safety rules must be followed at all times.

REVIEW

1. The authority having jurisdiction (AHJ) is _____ .
   a. usually the local city or county government
   b. a federal government representative through the National Standards Building Council (NSBC)
   c. the owner's representative on the job site
   d. the owner of the building

2. The *National Electrical Code*® is published by the _____ .
   a. National Standards Building Council (NSBC)
   b. National Electrocution Prevention Association (NEPA)
   c. National Association of Electrical Contractors (NASC)
   d. National Fire Protection Association (NFPA)

3. The construction drawings that contain the dimensions for walls and rooms are the _____ .
   a. civil plans
   b. architectural plans
   c. electrical plans
   d. plumbing plans

4. The most important thing on a job site is _____ .
   a. safety
   b. safety
   c. safety
   d. all of the above

5. Specifications are intended to control the _____ of the work on the structure.
   a. quality
   b. pace
   c. material storage location
   d. size of the building

6. The panel schedule includes information about the _____ .
   a. type of wall finish
   b. board of governors that oversees the job site completion schedule
   c. electrical distribution equipment
   d. instructions from the general contractor to the various subcontractors

7. The government agency charged with improving job site safety is the _____ .

   a. Board of Safety Commissioners (BSC)

   b. National Construction Safety Administration (NCSA)

   c. Occupational Safety and Health Administration (OSHA)

   d. Construction Safety Council

CHAPTER 2

Temporary Power and Lighting

OBJECTIVES

On completion of this chapter, the student will be able to:

☑ State the reasons for temporary construction power.

☑ Explain the importance of maintenance of the construction power panel.

☑ State the dangers associated with construction power.

☑ Explain the importance of ground-fault circuit-interrupter (GFCI) protection for temporary construction power.

☑ Describe the particular dangers with construction power as a service that do not exist with feeders and branch circuits.

☑ Identify the events that may cause a GFCI device to open the circuit.

☑ Identify GFCI protective devices and test them for proper operation.

INTRODUCTION

KEY TERMS

Construction power   Temporary electrical power that is provided for use by the trade workers and others working on the construction site. The power is usually provided and maintained by the electrical contractor.

This chapter is concerned with temporary construction power and lighting. Before construction can begin on a new dwelling, power must be provided for the use of the many trade workers and others who will eventually work on the site. For example, carpenters need electrical power for saws and compressors and plumbers need power for drills. This power system is known as **construction power**. Construction power is usually provided by the electrical contractor from the permanent distribution system that the utility power provider installed as part of the land development process. The power source may be available at the building site itself, or it may be some distance from the site. Construction wiring is considered temporary because when the structure is complete and the AHJ has completed the final inspections and issued the document (sometimes called a *certificate of occupancy*) allowing habitation, the permanent power is connected and the temporary construction power is removed. Temporary power is considered in *527* of the *NEC®*.

Because building sites vary to a large degree, it is difficult to cover all eventualities in which temporary construction power is concerned. Aspects of temporary construction power are common to all installations and are briefly considered in this chapter. Additional information on faults protection and overcurrent can be found in Chapter 7 of this book.

2.1: TEMPORARY SERVICE

The wiring system for temporary construction power is as dangerous as that for the permanent power (or more so). Although services are covered more completely in Chapter 13, a brief discussion of the nature of services is presented here as background for the study of temporary construction power.

KEY TERMS

Service (See *NEC® Article 100*): The raceways, conductors, and equipment that receive the electrical power from the local power-providing utility. It is important to note that electrical services are, for all intents and purposes of the *NEC®*, unprotected against overcurrent due to faults.

Overcurrent protective device: A device, consisting of fuses or circuit breakers, that will eliminate the flow of power if the current becomes too large for too long a period of time.

Services are not protected against overcurrent caused by short circuits or ground faults. If there is a problem with the wiring of a service, there is no **overcurrent protective device** or switch to open that will stop the current flow. The power-providing utility may have overcurrent protection for its own system conductors, transformers, and other equipment, but these cannot be relied on to provide protection for service conductors. When the service exists on a construction site, it is in constant danger of being physically damaged by construction equipment and delivery trucks; therefore, the service conductors should be kept as short as possible to minimize the dangers. They also must be protected against physical damage by a wiring method approved for use in services. Figure 2-1 offers more information about temporary services.

KEY TERMS

Electrode The conductor or other material that physically connects the electrical system to the earth, or ground. Several types of electrodes are recognized by the *NEC®*; included are underground metallic water piping systems, rod and pipe electrodes, ground rings, and concrete-encased electrodes.

The service must have an **electrode** to provide a ground reference for fault current for the feeders and branch circuits. The usual electrodes are not available before the structure is substantially completed; therefore, a ground rod is most often used as the electrode. Grounding is considered in detail in Chapter 14. In many localities, the AHJ may require a 4 AWG grounding electrode conductor to connect the electrode and the grounded conductor bus because with temporary services this conductor is subject to damage.

The temporary service must provide a means for metering the electricity used at the construction site. The power-providing utility usually installs the meter into a contractor-provided meter enclosure on final inspection and approval of the installation by the inspector.

The temporary service must also provide a means for disconnecting the electrical supply to the site. This switch can be one handle, or up to six individual handles that are grouped. The switch(s) must be mounted where it is readily accessible and must be permanently identified as the service-disconnecting means.

Many contractors use factory-assembled temporary construction panels similar to that shown in Figure 2-2. The panel has provisions for metering as well as receptacle outlets. A waterproof cover is included to protect the receptacle outlets from the weather while the cords are connected. The drawback of these panels is that they are not very flexible and cannot be easily supplemented by the addition of receptacle outlets. Contractors may also construct temporary services from surplus panels and waterproof boxes and covers, as shown in Figure 2-1. These services are very flexible and can easily be added to if the demand arises.

Temporary services can be supplied from the power provider by an overhead service drop or an underground service lateral, as shown in Figure 2-3. The local power provider should be consulted to identify any detailed requirements for installation of these services. Most power providers have support, height, and clearance requirements that may or may not be more restrictive than those imposed by the *NEC®*. Figure 2-4 (page 44) shows a typical underground temporary construction panel on a dwelling unit construction site.

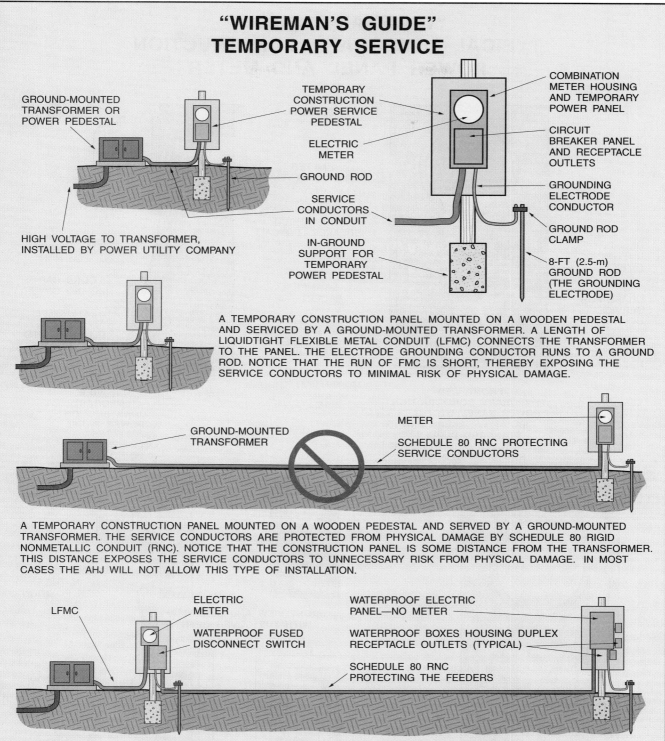

"WIREMAN'S GUIDE" TEMPORARY SERVICE

GROUND-MOUNTED TRANSFORMER OR POWER PEDESTAL

TEMPORARY CONSTRUCTION POWER SERVICE PEDESTAL

ELECTRIC METER

GROUND ROD

SERVICE CONDUCTORS IN CONDUIT

HIGH VOLTAGE TO TRANSFORMER, INSTALLED BY POWER UTILITY COMPANY

IN-GROUND SUPPORT FOR TEMPORARY POWER PEDESTAL

COMBINATION METER HOUSING AND TEMPORARY POWER PANEL

CIRCUIT BREAKER PANEL AND RECEPTACLE OUTLETS

GROUNDING ELECTRODE CONDUCTOR

GROUND ROD CLAMP

8-FT (2.5-m) GROUND ROD (THE GROUNDING ELECTRODE)

A TEMPORARY CONSTRUCTION PANEL MOUNTED ON A WOODEN PEDESTAL AND SERVICED BY A GROUND-MOUNTED TRANSFORMER. A LENGTH OF LIQUIDTIGHT FLEXIBLE METAL CONDUIT (LFMC) CONNECTS THE TRANSFORMER TO THE PANEL. THE ELECTRODE GROUNDING CONDUCTOR RUNS TO A GROUND ROD. NOTICE THAT THE RUN OF FMC IS SHORT, THEREBY EXPOSING THE SERVICE CONDUCTORS TO MINIMAL RISK OF PHYSICAL DAMAGE.

GROUND-MOUNTED TRANSFORMER

METER

SCHEDULE 80 RNC PROTECTING SERVICE CONDUCTORS

A TEMPORARY CONSTRUCTION PANEL MOUNTED ON A WOODEN PEDESTAL AND SERVED BY A GROUND-MOUNTED TRANSFORMER. THE SERVICE CONDUCTORS ARE PROTECTED FROM PHYSICAL DAMAGE BY SCHEDULE 80 RIGID NONMETALLIC CONDUIT (RNC). NOTICE THAT THE CONSTRUCTION PANEL IS SOME DISTANCE FROM THE TRANSFORMER. THIS DISTANCE EXPOSES THE SERVICE CONDUCTORS TO UNNECESSARY RISK FROM PHYSICAL DAMAGE. IN MOST CASES THE AHJ WILL NOT ALLOW THIS TYPE OF INSTALLATION.

LFMC

ELECTRIC METER

WATERPROOF FUSED DISCONNECT SWITCH

WATERPROOF ELECTRIC PANEL—NO METER

WATERPROOF BOXES HOUSING DUPLEX RECEPTACLE OUTLETS (TYPICAL)

SCHEDULE 80 RNC PROTECTING THE FEEDERS

A TEMPORARY CONSTRUCTION PANEL, CONSISTING OF A WATERPROOF PANEL AND WATERPROOF BOXES HOUSING THE DUPLEX RECEPTACLES MOUNTED TO A WOODEN PEDESTAL, IS SERVED BY A FUSED DISCONNECT SWITCH ADJACENT TO THE TRANSFORMER. THE SERVICE CONDUCTORS FROM THE TRANSFORMER TO THE DISCONNECT SWITCH ARE PROTECTED BY LFMC AND THE FEEDER CONDUCTORS FROM THE DISCONNECT SWITCH TO THE CONSTRUCTION PANEL ARE PROTECTED FROM PHYSICAL DAMAGE BY SCHEDULE 80 RNC. THE FUSES IN THE DISCONNECT SWITCH PROVIDE THE SERVICE OVERCURRENT PROTECTION FOR THE INSTALLATION AND THEREFORE MAKES THE CIRCUIT FROM THE DISCONNECT TO THE CONSTRUCTION PANEL FEEDERS INSTEAD OF SERVICE CONDUCTORS. THEY PROTECT THE FEEDERS FROM OVERCURRENT AND WILL OPEN THE CIRCUIT IF A FAULT OCCURS SOMEWHERE IN THE CIRCUIT.

Figure 2-1

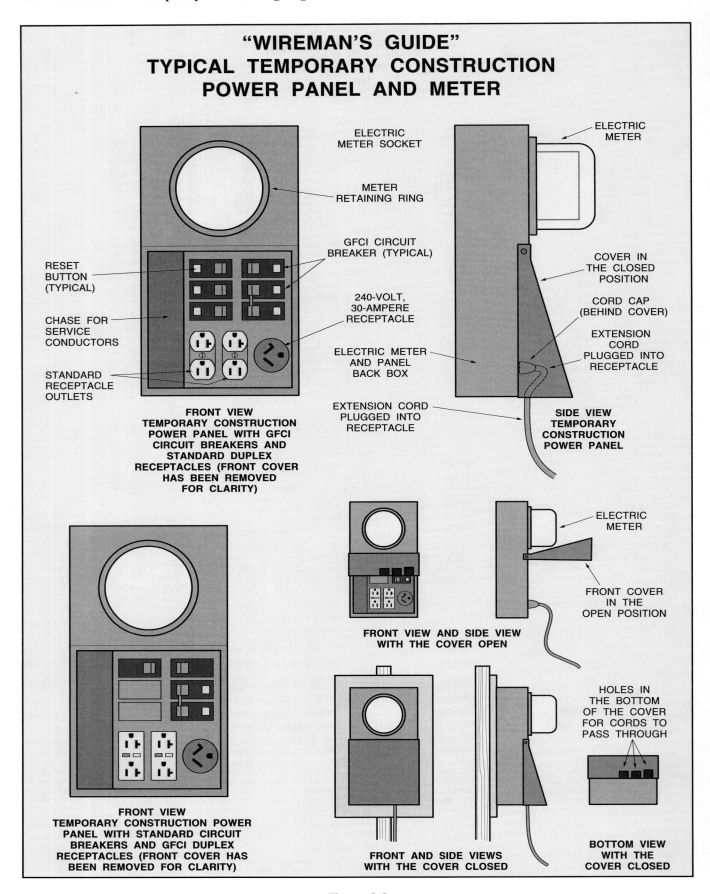

"WIREMAN'S GUIDE"
TYPICAL TEMPORARY CONSTRUCTION
POWER PANEL AND METER

ELECTRIC
METER SOCKET

ELECTRIC
METER

METER
RETAINING RING

GFCI CIRCUIT
BREAKER (TYPICAL)

RESET
BUTTON
(TYPICAL)

240-VOLT,
30-AMPERE
RECEPTACLE

CHASE FOR
SERVICE
CONDUCTORS

STANDARD
RECEPTACLE
OUTLETS

ELECTRIC METER
AND PANEL
BACK BOX

EXTENSION CORD
PLUGGED INTO
RECEPTACLE

COVER IN
THE CLOSED
POSITION

CORD CAP
(BEHIND COVER)

EXTENSION
CORD
PLUGGED INTO
RECEPTACLE

**FRONT VIEW
TEMPORARY CONSTRUCTION
POWER PANEL WITH GFCI
CIRCUIT BREAKERS AND
STANDARD DUPLEX
RECEPTACLES (FRONT COVER
HAS BEEN REMOVED
FOR CLARITY)**

**SIDE VIEW
TEMPORARY
CONSTRUCTION
POWER PANEL**

ELECTRIC
METER

FRONT COVER
IN THE
OPEN POSITION

**FRONT VIEW AND SIDE VIEW
WITH THE COVER OPEN**

HOLES IN
THE BOTTOM
OF THE COVER
FOR CORDS TO
PASS THROUGH

**FRONT VIEW
TEMPORARY CONSTRUCTION POWER
PANEL WITH STANDARD CIRCUIT
BREAKERS AND GFCI DUPLEX
RECEPTACLES (FRONT COVER HAS
BEEN REMOVED FOR CLARITY)**

**FRONT AND SIDE VIEWS
WITH THE COVER CLOSED**

**BOTTOM VIEW
WITH THE
COVER CLOSED**

Figure 2-2

"WIREMAN'S GUIDE"
UNDERGROUND AND OVERHEAD
TEMPORARY POWER SUPPLIES

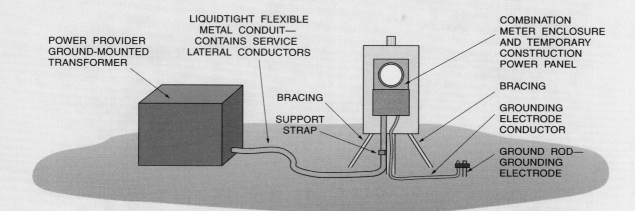

POWER PROVIDER
GROUND-MOUNTED
TRANSFORMER

LIQUIDTIGHT FLEXIBLE
METAL CONDUIT—
CONTAINS SERVICE
LATERAL CONDUCTORS

BRACING

SUPPORT
STRAP

COMBINATION
METER ENCLOSURE
AND TEMPORARY
CONSTRUCTION
POWER PANEL

BRACING

GROUNDING
ELECTRODE
CONDUCTOR

GROUND ROD—
GROUNDING
ELECTRODE

UNDERGROUND SUPPLIED TEMPORARY CONSTRUCTION PANEL
INSTALLATION. THE SERVICE LATERAL CONDUCTORS SHOULD BE
KEPT AS SHORT AS POSSIBLE AND PROTECTED WHERE
POSSIBLE, BECAUSE THEY ARE UNPROTECTED AGAINST FAULTS.

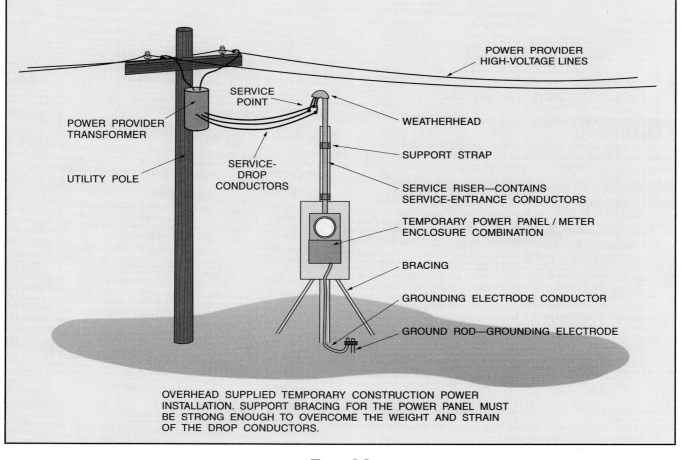

POWER PROVIDER
HIGH-VOLTAGE LINES

SERVICE
POINT

POWER PROVIDER
TRANSFORMER

UTILITY POLE

SERVICE-
DROP
CONDUCTORS

WEATHERHEAD

SUPPORT STRAP

SERVICE RISER—CONTAINS
SERVICE-ENTRANCE CONDUCTORS

TEMPORARY POWER PANEL / METER
ENCLOSURE COMBINATION

BRACING

GROUNDING ELECTRODE CONDUCTOR

GROUND ROD—GROUNDING ELECTRODE

OVERHEAD SUPPLIED TEMPORARY CONSTRUCTION POWER
INSTALLATION. SUPPORT BRACING FOR THE POWER PANEL MUST
BE STRONG ENOUGH TO OVERCOME THE WEIGHT AND STRAIN
OF THE DROP CONDUCTORS.

Figure 2-3

Figure 2-4   A typical underground temporary power panel and meter.

2.2: GROUND-FAULT CIRCUIT-INTERRUPTERS

KEY TERMS

Ground-fault circuit-interrupter (GFCI)   A device that will detect a ground fault and then open the circuit in response to that fault. GFCIs are available as receptacle devices or circuit breaker devices.

Phase conductor   One of the terms used to describe the nongrounded current-carrying conductor of an electrical circuit. The phase conductor is sometimes also called the *hot conductor*.

A **ground-fault circuit-interrupter (GFCI)** is a device that measures the current flow leaving the power source on the **phase conductor** and compares it with the current flow being returned to the source on the grounded circuit, or neutral conductor, as shown in Figure 2-5. In a circuit that is operating normally, these two measurements should be the same. If the current flow measurements differ by as little as 4 to 6 milliamperes (.004 to .006 ampere),

the GFCI device will open the circuit. Figure 2-6 shows several events that can cause such current imbalance.

It is very important for the electricians on the job site to maintain the temporary service and distribution panel in a proper manner. Because many workers from several different trades must use the temporary power at the same time and because the GFCI can be affected by poorly maintained extension cords and other electrical equipment, there is often pressure from other contractors to bypass the GFCI protection. It is not uncommon for other trade workers to open the construction site electrical panel and attempt to rewire the circuits when the electrician is not on site or is busy with other tasks. Such rewiring attempts present obvious danger. Any changes to the circuits must be repaired immediately and actions must be taken to prevent access to the electrical panel by unauthorized persons.

There are several ways in which GFCI protection can be provided, as shown in Figure 2-7. The circuits can be protected by GFCI circuit breakers. The

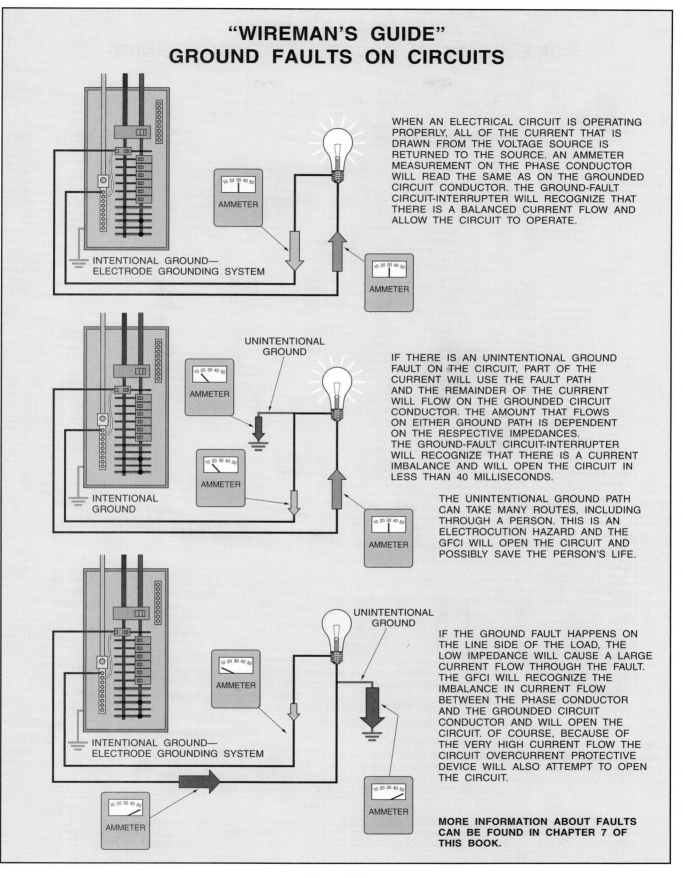

"WIREMAN'S GUIDE"
GROUND FAULTS ON CIRCUITS

WHEN AN ELECTRICAL CIRCUIT IS OPERATING PROPERLY, ALL OF THE CURRENT THAT IS DRAWN FROM THE VOLTAGE SOURCE IS RETURNED TO THE SOURCE. AN AMMETER MEASUREMENT ON THE PHASE CONDUCTOR WILL READ THE SAME AS ON THE GROUNDED CIRCUIT CONDUCTOR. THE GROUND-FAULT CIRCUIT-INTERRUPTER WILL RECOGNIZE THAT THERE IS A BALANCED CURRENT FLOW AND ALLOW THE CIRCUIT TO OPERATE.

IF THERE IS AN UNINTENTIONAL GROUND FAULT ON THE CIRCUIT, PART OF THE CURRENT WILL USE THE FAULT PATH AND THE REMAINDER OF THE CURRENT WILL FLOW ON THE GROUNDED CIRCUIT CONDUCTOR. THE AMOUNT THAT FLOWS ON EITHER GROUND PATH IS DEPENDENT ON THE RESPECTIVE IMPEDANCES. THE GROUND-FAULT CIRCUIT-INTERRUPTER WILL RECOGNIZE THAT THERE IS A CURRENT IMBALANCE AND WILL OPEN THE CIRCUIT IN LESS THAN 40 MILLISECONDS.

THE UNINTENTIONAL GROUND PATH CAN TAKE MANY ROUTES, INCLUDING THROUGH A PERSON. THIS IS AN ELECTROCUTION HAZARD AND THE GFCI WILL OPEN THE CIRCUIT AND POSSIBLY SAVE THE PERSON'S LIFE.

IF THE GROUND FAULT HAPPENS ON THE LINE SIDE OF THE LOAD, THE LOW IMPEDANCE WILL CAUSE A LARGE CURRENT FLOW THROUGH THE FAULT. THE GFCI WILL RECOGNIZE THE IMBALANCE IN CURRENT FLOW BETWEEN THE PHASE CONDUCTOR AND THE GROUNDED CIRCUIT CONDUCTOR AND WILL OPEN THE CIRCUIT. OF COURSE, BECAUSE OF THE VERY HIGH CURRENT FLOW THE CIRCUIT OVERCURRENT PROTECTIVE DEVICE WILL ALSO ATTEMPT TO OPEN THE CIRCUIT.

MORE INFORMATION ABOUT FAULTS CAN BE FOUND IN CHAPTER 7 OF THIS BOOK.

INTENTIONAL GROUND—
ELECTRODE GROUNDING SYSTEM

AMMETER

AMMETER

UNINTENTIONAL GROUND

AMMETER

AMMETER

INTENTIONAL GROUND

AMMETER

INTENTIONAL GROUND—
ELECTRODE GROUNDING SYSTEM

UNINTENTIONAL GROUND

AMMETER

AMMETER

AMMETER

Figure 2-5

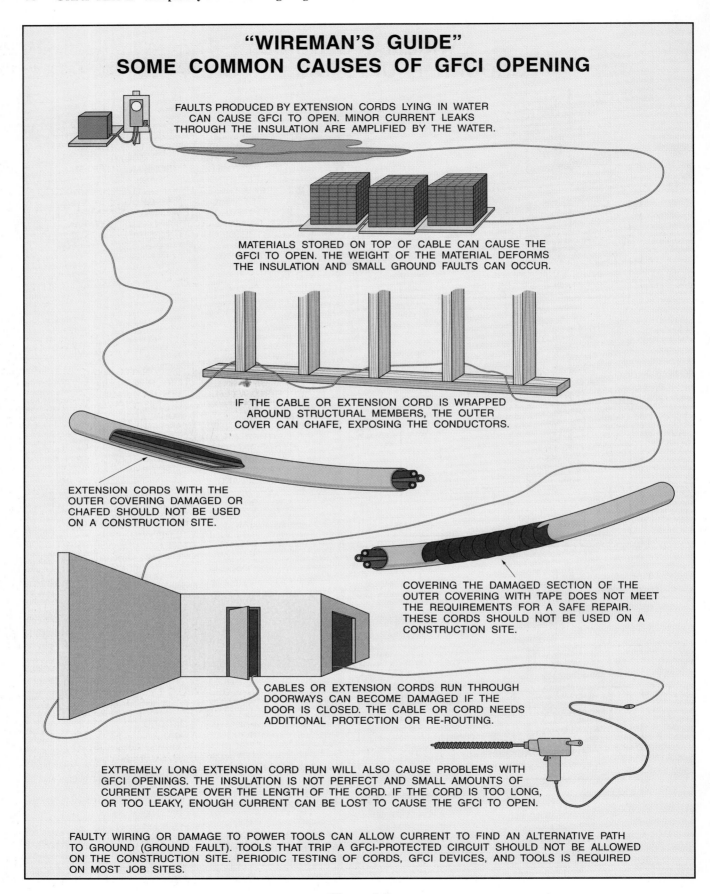

"WIREMAN'S GUIDE"
SOME COMMON CAUSES OF GFCI OPENING

FAULTS PRODUCED BY EXTENSION CORDS LYING IN WATER CAN CAUSE GFCI TO OPEN. MINOR CURRENT LEAKS THROUGH THE INSULATION ARE AMPLIFIED BY THE WATER.

MATERIALS STORED ON TOP OF CABLE CAN CAUSE THE GFCI TO OPEN. THE WEIGHT OF THE MATERIAL DEFORMS THE INSULATION AND SMALL GROUND FAULTS CAN OCCUR.

IF THE CABLE OR EXTENSION CORD IS WRAPPED AROUND STRUCTURAL MEMBERS, THE OUTER COVER CAN CHAFE, EXPOSING THE CONDUCTORS.

EXTENSION CORDS WITH THE OUTER COVERING DAMAGED OR CHAFED SHOULD NOT BE USED ON A CONSTRUCTION SITE.

COVERING THE DAMAGED SECTION OF THE OUTER COVERING WITH TAPE DOES NOT MEET THE REQUIREMENTS FOR A SAFE REPAIR. THESE CORDS SHOULD NOT BE USED ON A CONSTRUCTION SITE.

CABLES OR EXTENSION CORDS RUN THROUGH DOORWAYS CAN BECOME DAMAGED IF THE DOOR IS CLOSED. THE CABLE OR CORD NEEDS ADDITIONAL PROTECTION OR RE-ROUTING.

EXTREMELY LONG EXTENSION CORD RUN WILL ALSO CAUSE PROBLEMS WITH GFCI OPENINGS. THE INSULATION IS NOT PERFECT AND SMALL AMOUNTS OF CURRENT ESCAPE OVER THE LENGTH OF THE CORD. IF THE CORD IS TOO LONG, OR TOO LEAKY, ENOUGH CURRENT CAN BE LOST TO CAUSE THE GFCI TO OPEN.

FAULTY WIRING OR DAMAGE TO POWER TOOLS CAN ALLOW CURRENT TO FIND AN ALTERNATIVE PATH TO GROUND (GROUND FAULT). TOOLS THAT TRIP A GFCI-PROTECTED CIRCUIT SHOULD NOT BE ALLOWED ON THE CONSTRUCTION SITE. PERIODIC TESTING OF CORDS, GFCI DEVICES, AND TOOLS IS REQUIRED ON MOST JOB SITES.

Figure 2-6

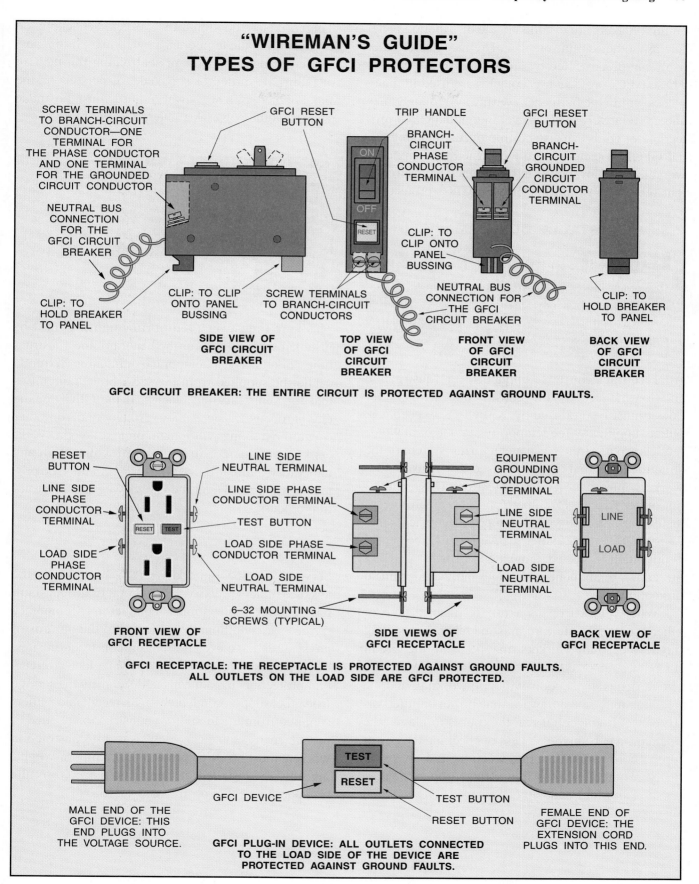

"WIREMAN'S GUIDE"
TYPES OF GFCI PROTECTORS

SCREW TERMINALS TO BRANCH-CIRCUIT CONDUCTOR—ONE TERMINAL FOR THE PHASE CONDUCTOR AND ONE TERMINAL FOR THE GROUNDED CIRCUIT CONDUCTOR

NEUTRAL BUS CONNECTION FOR THE GFCI CIRCUIT BREAKER

CLIP: TO HOLD BREAKER TO PANEL

GFCI RESET BUTTON

CLIP: TO CLIP ONTO PANEL BUSSING

SCREW TERMINALS TO BRANCH-CIRCUIT CONDUCTORS

TRIP HANDLE

ON

OFF

RESET

BRANCH-CIRCUIT PHASE CONDUCTOR TERMINAL

CLIP: TO CLIP ONTO PANEL BUSSING

NEUTRAL BUS CONNECTION FOR THE GFCI CIRCUIT BREAKER

GFCI RESET BUTTON

BRANCH-CIRCUIT GROUNDED CIRCUIT CONDUCTOR TERMINAL

CLIP: TO HOLD BREAKER TO PANEL

SIDE VIEW OF GFCI CIRCUIT BREAKER

TOP VIEW OF GFCI CIRCUIT BREAKER

FRONT VIEW OF GFCI CIRCUIT BREAKER

BACK VIEW OF GFCI CIRCUIT BREAKER

GFCI CIRCUIT BREAKER: THE ENTIRE CIRCUIT IS PROTECTED AGAINST GROUND FAULTS.

RESET BUTTON

LINE SIDE PHASE CONDUCTOR TERMINAL

LOAD SIDE PHASE CONDUCTOR TERMINAL

RESET   TEST

LINE SIDE NEUTRAL TERMINAL

LINE SIDE PHASE CONDUCTOR TERMINAL

TEST BUTTON

LOAD SIDE PHASE CONDUCTOR TERMINAL

LOAD SIDE NEUTRAL TERMINAL

6–32 MOUNTING SCREWS (TYPICAL)

EQUIPMENT GROUNDING CONDUCTOR TERMINAL

LINE SIDE NEUTRAL TERMINAL

LOAD SIDE NEUTRAL TERMINAL

LINE

LOAD

FRONT VIEW OF GFCI RECEPTACLE

SIDE VIEWS OF GFCI RECEPTACLE

BACK VIEW OF GFCI RECEPTACLE

GFCI RECEPTACLE: THE RECEPTACLE IS PROTECTED AGAINST GROUND FAULTS. ALL OUTLETS ON THE LOAD SIDE ARE GFCI PROTECTED.

TEST

RESET

GFCI DEVICE

TEST BUTTON

RESET BUTTON

MALE END OF THE GFCI DEVICE: THIS END PLUGS INTO THE VOLTAGE SOURCE.

FEMALE END OF GFCI DEVICE: THE EXTENSION CORD PLUGS INTO THIS END.

GFCI PLUG-IN DEVICE: ALL OUTLETS CONNECTED TO THE LOAD SIDE OF THE DEVICE ARE PROTECTED AGAINST GROUND FAULTS.

Figure 2-7

circuit breakers are installed in the distribution panel like any other circuit breaker. The receptacle outlet itself can be protected by an internal GFCI system. These GFCI-protected receptacles are used in many places within the permanent wiring system in homes and are relatively inexpensive. Individual GFCI protection for cords can also be employed. Such individual GFCI protectors are often owned by individual workers and allow power to be obtained from circuits that otherwise would not provide GFCI protection.

2.3: RECEPTACLE OUTLETS

KEY TERMS

Receptacle outlet   (See *NEC® Article 100*): An outlet in the electrical system that has a receptacle installed to provide access to the system for cord-and-plug connected loads.

All **receptacle outlets** supplied on a construction site for the use by the construction workers for power must be protected by a GFCI. Any receptacle outlet supplied by the temporary service or any outlet that is part of the permanent wiring system of the structure and that is being used by construction workers must have GFCI protection. This includes not only 125-volt 15-ampere and 20-ampere receptacles but also 125-volt 30-ampere receptacles and 240-volt receptacle outlets as well. If the receptacle is not GFCI protected, any extension cords or other utilization equipment must have individual built-in protection.

When a GFCI opens as the result of a fault or other event, extreme care must be taken in resetting the device. Many workers may be using the circuit for power for electrically operated tools. When the circuit opens, these tools will stop working, and the operator may not be aware that power failure was the cause. In the belief that the tool is to blame, the operator may be in the process of checking the power tool or cord for malfunction when the power is suddenly restored. Serious injury to the operator may result if the tool suddenly restarts. There is a real danger of electrocution if the worker is checking the connection to the temporary power. In addition, persons working on ladders or scaffolding run the risk of falling if startled or shocked. All persons on the site affected by the power failure must be made aware that the power is about to be restored.

2.4: TEMPORARY LIGHTING

KEY TERMS

Temporary lighting   Lighting that is installed in some buildings under construction to provide access and egress lighting for the workers but usually does not include specialized task lighting that may be required by some trades.

There are times when **temporary lighting** must be installed in dwellings for the benefit of the construction workers. These temporary luminaires (fixtures) cannot be supplied from the same branch circuit as that for temporary receptacles. If the GFCI for the circuit opens because of a ground fault or other event, the lighting will remain on to allow the workers safe entry and exit from the structure. With any temporary luminaire (fixture), the lamp must be protected from accidental damage during the construction process.

SUMMARY

Services can be quite dangerous because of the absence of any effective short-circuit or ground-fault protection on the supply conductors. Temporary services are additionally subjected to the hazards of physical damage ever present on construction sites. Temporary services must provide overload overcurrent protection and a means to disconnect the service from the supply conductors. All receptacle outlets supplied from the temporary construction service or permanent receptacles being used by construction workers must be GFCI protected. If temporary lighting must be installed, it cannot be powered by the same circuit as that for the receptacles.

REVIEW

1. Services can be particularly dangerous because they_____.

   a. have no effective fault overcurrent protection

   b. serve so many outlets

   c. are usually 240 volts

   d. carry high current flows

2. Temporary service usually uses _____ for an electrode to connect the service to the ground (earth).

   a. nails

   b. water pipes

   c. steel rebar

   d. rods

3. Temporary service must provide all of the following except _____.

   a. a disconnecting means

   b. line side fault overcurrent protection

   c. an electric meter

   d. load side fault and overload overcurrent protection

4. Ground-fault circuit-interrupters (GFCI) measure_____.

   a. the current flow on the phase conductor only

   b. the current flow on the grounded circuit conductor only

   c. the current flow on both the phase conductor and the grounded circuit conductor and compares them for a balanced current

   d. voltage changes that can be the result of current leaving the circuit

5. Which of the following is *not* a major cause of GFCI nuisance tripping?

   a. damage to the insulation on an extension cord

   b. a very long run of an extension cord

   c. an extension cord run through a puddle or standing water

   d. overload on the circuit

6. GFCI protection is commonly found in all of the following except _____.

   a. GFCI circuit breakers

   b. GFCI-protected receptacles

   c. GFCI-protected plug-in devices

   d. GFCI-protected luminaires (fixtures)

"WIREMAN'S GUIDE" REVIEW

1. Explain the advantages of a manufactured temporary construction panel over those constructed on the site using waterproof panels, boxes, and covers. What are some of the disadvantages?

2. Explain how a GFCI works. Will a GFCI also protect against overcurrents? Why or why not?

CHAPTER 3

Location of Receptacle Outlets

OBJECTIVES

On completion of this chapter, the student will be able to:

☑ Define the terms *outlet*, *switch point,* and *device*, as used in the Code.

☑ Identify those locations in a dwelling where receptacle outlets are required.

☑ Identify the locations of receptacles in the sample house that are installed at the request of the owner (nonrequired receptacles).

☑ Define the terms *line* and *load*.

☑ Describe arc-fault circuit-interrupter (AFCI) protection and identify the locations where arc-fault protection is required in a dwelling.

☑ Identify the locations where ground-fault circuit-interrupter (GFCI) protection is required.

INTRODUCTION

This chapter describes the locations of the various receptacles to be installed in the sample dwelling. Some of the receptacle locations are required by the *NEC®*, whereas others are installed for convenience of the owner. In this chapter, only receptacle outlets are considered. The study of outlets for utilization equipment other than receptacle outlets is the topic of Chapter 4.

3.1: OUTLETS, SWITCH POINTS, AND DEVICES

> **KEY TERMS**
>
> **Outlet** (See *NEC® Article 100*): A place in the electrical system where access is provided for the use of utilization equipment, such as appliances and luminaires (fixtures). Places of access to the electrical system intended for control, such as switch points, are not considered outlets.

In order to understand the *NEC®* requirements for lighting and receptacle outlets, it is necessary first to understand exactly what constitutes an outlet. An **outlet** is defined in *Article 100* as a place in which the wiring system can be accessed in order to obtain electrical energy to power luminaires (fixtures), receptacles, or other electrically operated appliances or equipment. A duplex receptacle such as that installed in the living room of the sample house is an example of an outlet. At places where access to the wiring is required, as in the case of a receptacle, a box is screwed or nailed to a structural member of the dwelling during rough-in so that the opening in the front of the box is flush with the finished surface of the wall or ceiling. The wiring is installed into a box, and the open face of the box allows access to the conductors. Chapter 6 of this book provides more detailed information about boxes.

A **receptacle** is connected to the supply conductors inside of the box, and the receptacle is then inserted into and screwed to the box. A cover plate that fits flat against the wall closes the opening between the receptacle and the box. A receptacle allows access to the electrical system by simply inserting the plug attached to the electrical equipment or appliance, as shown in Figure 3-1. There are many different sizes and configurations of receptacles. Each receptacle type has a particular configuration of pins, blades, or prongs that, when properly employed, makes it impossible to connect the equipment to an electrical system of the wrong voltage, phasing, or current rating. Receptacles are covered in more detail in Chapter 16 of this book.

A **luminaire (fixture)** installed in the wall or ceiling of a house must be connected to the electrical system in order get the power to operate. A ceiling-mounted luminaire (fixture) is shown in Figure 3-2. This luminaire (fixture) is not provided with an attachment plug and receptacle; instead, it is connected directly to the supply conductors at the outlet box (i.e., it is "hard-wired"). The luminaire (fixture) is then attached to the box, effectively closing the opening at the front of the box. Of course, some luminaires (fixtures), such as desk and swag lamps, are connected to the electrical system with attachment plugs, but the luminaires (fixtures) installed as part of the building are usually connected directly to the circuit conductors at the outlet box.

The electrical system wiring can be accessed from locations other than outlets. Conductors must be available for the connection of switches in order to control—to energize or de-energize—the utilization equipment. A place at which access to the electrical systems is gained in order to control the flow of current, to turn it on or off, is called a **switch point**. Switches are not classified as outlets because the switch itself is not utilization equipment. Chapters 4 and 16 of this book contain more detailed information about switches and switching systems.

A switch is an example of a type of electrical equipment called a **device**. *NEC® 100* defines a device as a piece of electrical equipment that carries current but that does not use any of the current. This definition is careful to declare that a device is not intended to utilize electrical energy, whereas the definition of an outlet is careful to note that an outlet supplies utilization equipment. Technically, a receptacle is also a device because it does carry current but uses no power on its own. Figure 3-3 provides additional information about switch points, outlets, and devices.

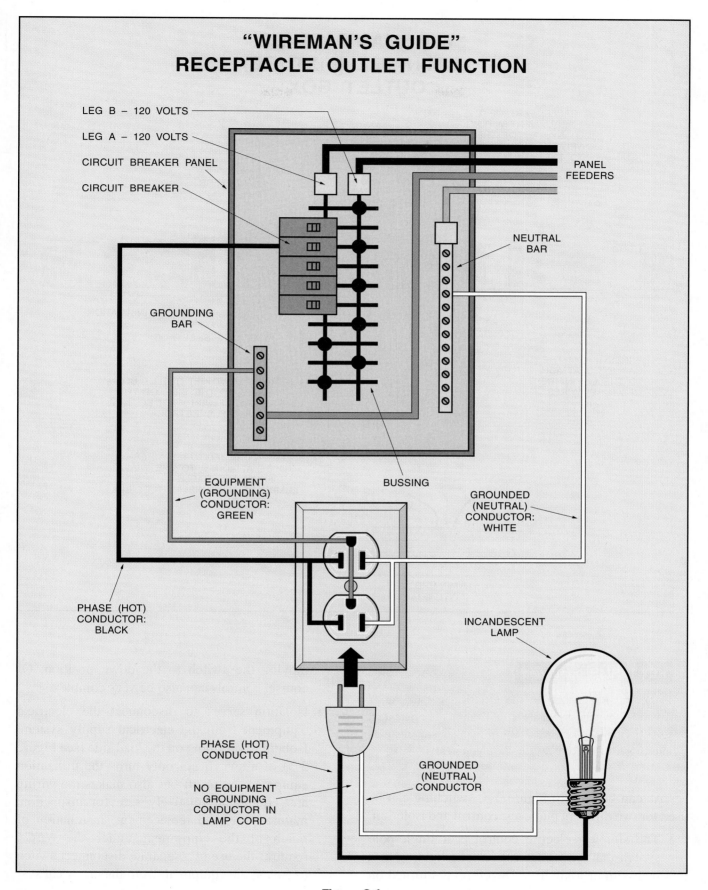

"WIREMAN'S GUIDE"
RECEPTACLE OUTLET FUNCTION

LEG B – 120 VOLTS

LEG A – 120 VOLTS

CIRCUIT BREAKER PANEL

CIRCUIT BREAKER

PANEL FEEDERS

NEUTRAL BAR

GROUNDING BAR

EQUIPMENT (GROUNDING) CONDUCTOR: GREEN

BUSSING

GROUNDED (NEUTRAL) CONDUCTOR: WHITE

PHASE (HOT) CONDUCTOR: BLACK

INCANDESCENT LAMP

PHASE (HOT) CONDUCTOR

NO EQUIPMENT GROUNDING CONDUCTOR IN LAMP CORD

GROUNDED (NEUTRAL) CONDUCTOR

Figure 3-1

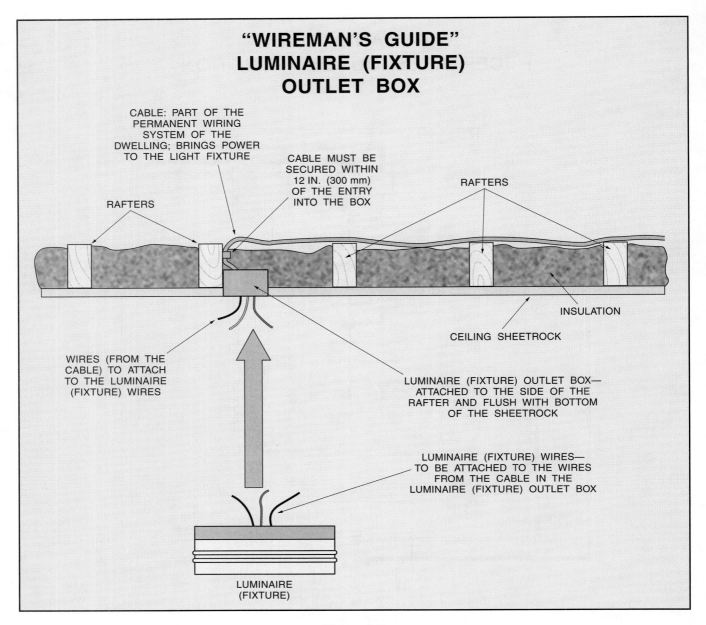

"WIREMAN'S GUIDE"
LUMINAIRE (FIXTURE)
OUTLET BOX

CABLE: PART OF THE PERMANENT WIRING SYSTEM OF THE DWELLING; BRINGS POWER TO THE LIGHT FIXTURE

CABLE MUST BE SECURED WITHIN 12 IN. (300 mm) OF THE ENTRY INTO THE BOX

RAFTERS

RAFTERS

INSULATION

CEILING SHEETROCK

WIRES (FROM THE CABLE) TO ATTACH TO THE LUMINAIRE (FIXTURE) WIRES

LUMINAIRE (FIXTURE) OUTLET BOX—ATTACHED TO THE SIDE OF THE RAFTER AND FLUSH WITH BOTTOM OF THE SHEETROCK

LUMINAIRE (FIXTURE) WIRES—TO BE ATTACHED TO THE WIRES FROM THE CABLE IN THE LUMINAIRE (FIXTURE) OUTLET BOX

LUMINAIRE (FIXTURE)

Figure 3-2

Isolation   Removal from the electrical circuit. An isolation switch disconnects utilization equipment from a circuit to ensure safety in maintenance or to interrupt the flow of power to the equipment.

As can be seen in Figure 3-4, switching can be used for two different purposes: **control** and **isolation**.

- The simplest electric control is a single-pole switch that controls the overhead lighting in a bedroom or a kitchen. The current is turned on by manually moving the switch to the up position. The current is turned off by manually

moving the switch to the down position. Of course, controls can also be very complex.

- Isolation serves to disconnect the electrical equipment from the electrical supply system. Isolating everything on the load side (see Figure 3-5) of a switch not only turns the utilization equipment on or off but also makes the wiring and equipment relatively safe for inspection, maintenance, and replacement. Installation of much of the equipment under the *NEC*® requires the use of a separate disconnect switch to isolate the equipment from the electrical system—notably, motors, appliances, and electric heaters.

"WIREMAN'S GUIDE"
COMPONENTS OF AN ELECTRICAL SYSTEM:
SWITCHES (CONTROLS) AND OUTLETS (LOADS)

AS SHOWN IN THE DIAGRAM, THERE ARE ONLY FIVE
MAJOR COMPONENTS OF ANY ELECTRICAL SYSTEM:
1. A POWER SUPPLY
2. CONDUCTORS
3. OVERCURRENT PROTECTION
4. CONTROLS
5. LOADS

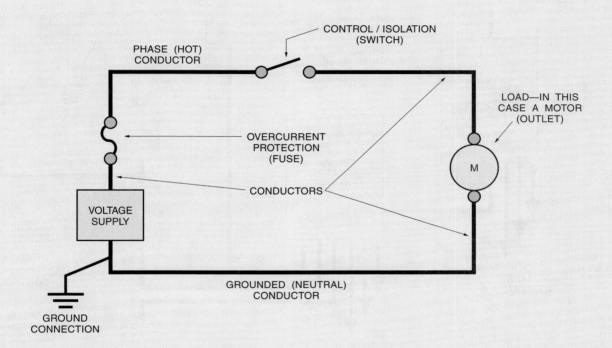

CONTROLS (AS SWITCHES) AND LOADS (AS OUTLETS) REPRESENT
TWO OF THE FIVE MAJOR ITEMS:
- **CONTROLS:** ALL CONTROLS ARE DEVICES. CONTROLS TURN THE
  UTILIZATION EQUIPMENT ON AND OFF. UTILIZATION EQUIPMENT
  IS APPLIANCES AND LUMINAIRES (FIXTURES) THAT CONSUME THE
  ELECTRICAL ENERGY. SWITCHES ALSO DISCONNECT. DISCONNECT
  MEANS TO ISOLATE. DISCONNECT SWITCHES ARE SWITCHES THAT
  ARE INTENDED TO ISOLATE RATHER THAN CONTROL.
- **LOADS:** ALL LOADS ARE OUTLETS. MORE SPECIFICALLY, ALL
  LOADS CONNECT TO AND DERIVE THEIR POWER FROM OUTLETS.
  RECEPTACLES ARE A COMMON FORM OF OUTLET, BUT ANY
  ARRANGEMENT WHEREIN UTILIZATION EQUIPMENT CONNECTS
  TO THE POWER SYSTEM IS BY DEFINITION AN OUTLET. THE
  RECEPTACLE ITSELF IS A DEVICE. A RECEPTACLE CARRIES
  THE ELECTRICAL CURRENT BUT USES NO ENERGY ITSELF
  (IT PRODUCES NO VOLTAGE DROP). THE RECEPTACLE IS A
  DEVICE THAT PROVIDES AN OUTLET FOR THE
  ELECTRICAL ENERGY.

Figure 3-3

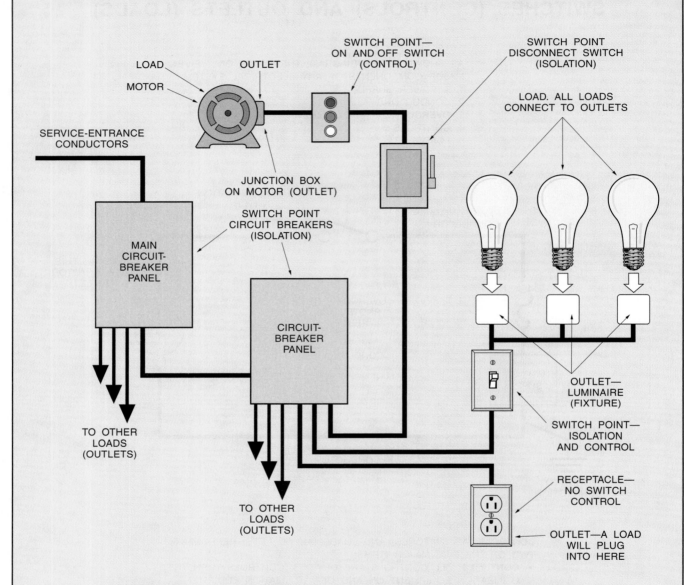

"WIREMAN'S GUIDE"
SWITCH POINTS AND OUTLETS: ONE-LINE DIAGRAM

FOR EACH OUTLET THERE CAN BE MULTIPLE SWITCH POINTS. THE MOTOR HAS A SWITCH POINT FOR CONTROL AND ANOTHER SWITCH POINT FOR ISOLATION. A SWITCH POINT CAN CONTROL SEVERAL OUTLETS AS SHOWN WITH THE LUMINAIRES (FIXTURES). THE ONE SWITCH CONTROLS THREE LAMPS. OUTLETS DO NOT HAVE TO HAVE CONTROLS, AS SHOWN WITH THE RECEPTACLE OUTLET. OF COURSE, THE CIRCUIT BREAKERS IN THE DISTRIBUTION PANEL CONSTITUTE SWITCHES FOR ISOLATION FOR ALL OF THE CIRCUITS.

THIS TYPE OF DIAGRAM IS CALLED A *ONE-LINE DIAGRAM* BECAUSE THE CIRCUITING IS REPRESENTED BY ONLY ONE LINE. THE NECESSARY PHASE CONDUCTORS, GROUNDED CIRCUIT CONDUCTORS, AND EQUIPMENT GROUNDING CONDUCTORS ARE ASSUMED TO BE INCLUDED WITHIN THE ONE LINE. ONE-LINE DIAGRAMS ARE USUALLY USED FOR SERVICE DIAGRAMS AND FOR SOME SYSTEM DIAGRAMS, SUCH AS FIRE DETECTION SYSTEMS OR SECURITY SYSTEMS, BUT ARE NOT USUALLY USED FOR WIRING DIAGRAMS. ONE-LINE DIAGRAMS SHOW THE ELECTRICAL SYSTEM IN A SCHEMATIC WAY THAT HAS NO RELATIONSHIP TO THE ACTUAL LOCATION OF THE COMPONENT PARTS. MANY ONE-LINE DIAGRAMS INCLUDE NOTES THAT SHOW THE CONDUCTOR COUNT AND SIZE AND CONDUIT TYPE AND SIZE NEEDED TO COMPLETE THE INSTALLATION.

Figure 3-4

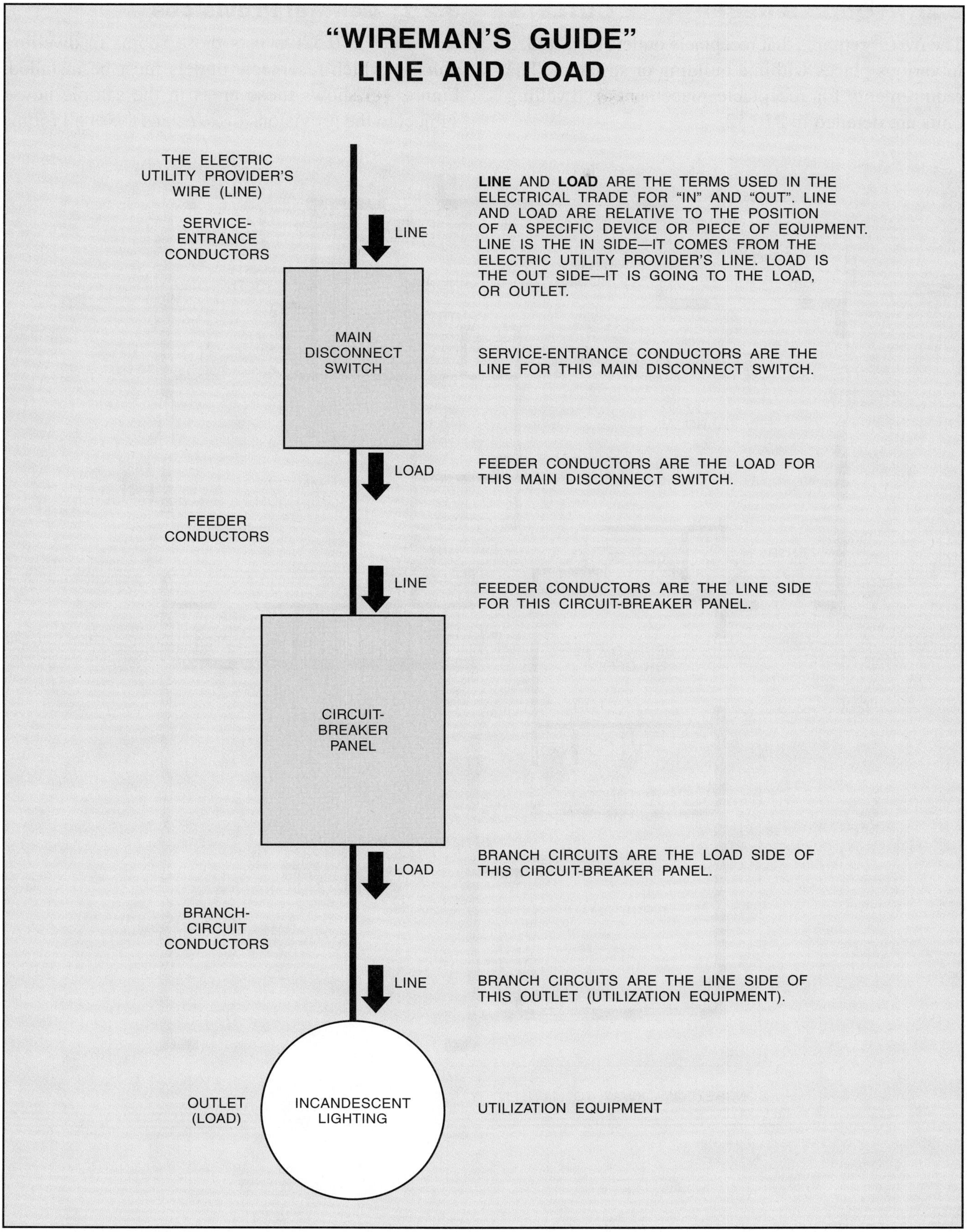

"WIREMAN'S GUIDE"
LINE AND LOAD

THE ELECTRIC
UTILITY PROVIDER'S
WIRE (LINE)

LINE

SERVICE-
ENTRANCE
CONDUCTORS

MAIN
DISCONNECT
SWITCH

LOAD

FEEDER
CONDUCTORS

LINE

CIRCUIT-
BREAKER
PANEL

LOAD

BRANCH-
CIRCUIT
CONDUCTORS

LINE

OUTLET
(LOAD)

INCANDESCENT
LIGHTING

LINE AND **LOAD** ARE THE TERMS USED IN THE
ELECTRICAL TRADE FOR "IN" AND "OUT". LINE
AND LOAD ARE RELATIVE TO THE POSITION
OF A SPECIFIC DEVICE OR PIECE OF EQUIPMENT.
LINE IS THE IN SIDE—IT COMES FROM THE
ELECTRIC UTILITY PROVIDER'S LINE. LOAD IS
THE OUT SIDE—IT IS GOING TO THE LOAD,
OR OUTLET.

SERVICE-ENTRANCE CONDUCTORS ARE THE
LINE FOR THIS MAIN DISCONNECT SWITCH.

FEEDER CONDUCTORS ARE THE LOAD FOR
THIS MAIN DISCONNECT SWITCH.

FEEDER CONDUCTORS ARE THE LINE SIDE
FOR THIS CIRCUIT-BREAKER PANEL.

BRANCH CIRCUITS ARE THE LOAD SIDE OF
THIS CIRCUIT-BREAKER PANEL.

BRANCH CIRCUITS ARE THE LINE SIDE OF
THIS OUTLET (UTILIZATION EQUIPMENT).

UTILIZATION EQUIPMENT

Figure 3-5

3.2: REQUIRED RECEPTACLE OUTLETS

The *NEC*® requires that receptacle outlets be installed in various places within a building or structure. The requirements for receptacle placement in dwelling units are detailed in *210.52*.

3.2.1: General Provisions

NEC® *210.52(A)* lists those rooms in dwelling units in which receptacle outlets must be installed. Figure 3-6 shows those areas in the sample house subject to the provisions of *201.52(A)*. Not all rooms

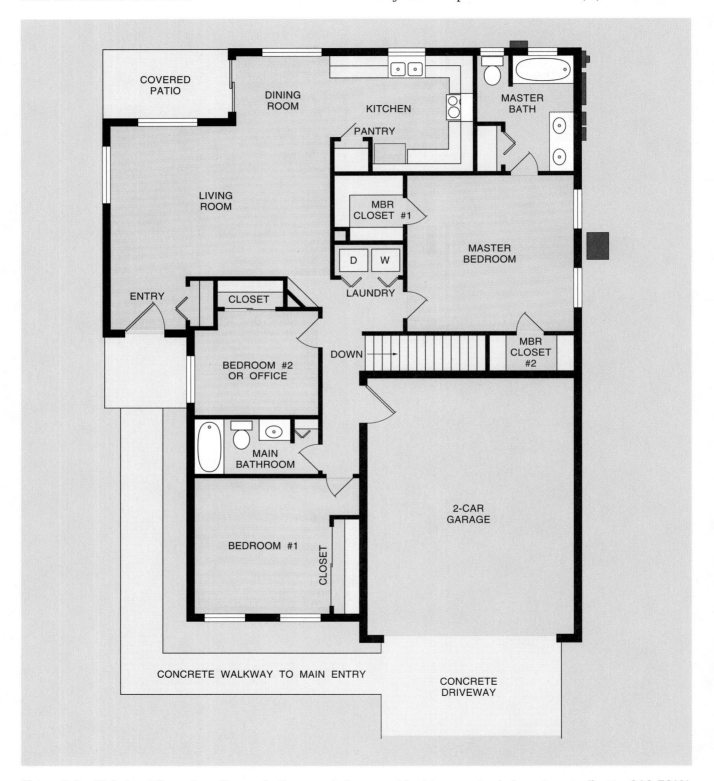

Figure 3-6 **Main level floor plan: Rooms in the sample house subject to receptacle layout according to *210.52(A)*.**

or areas are included. Each of the excluded areas is covered in a specific section of *210.52*: kitchen countertops in *(C)*, bathrooms in *(D)*, outdoor outlets in *(E)*, laundry rooms or areas in *(F)*, basements and garages in *(G)*, and hallways in *(H)*.

The rooms subject to the general provisions of *210.52(A)* must have receptacles installed in accordance with *210.52(A)(1)* through *210.52(A)(3)*. *NEC®* *210.52(A)(1)* requires that receptacles be installed so that no place along the wall is more than 6 ft (1.8 m) from an outlet. Receptacle outlets also should be spaced equal distances apart when practical. The point of this requirement is not placement of receptacle outlets at precisely 12 ft (3.7 m) apart but location of the receptacle outlets to serve the entire room—thus, they should not be grouped along one wall line. See Construction Drawing E-3 in Chapter 1 of this book for the electrical plan for the sample house, or Figures 1-6 and 1-7 for the electrical plans with either a square foot or square meter grid.

KEY TERMS

Wall space Space measured horizontally (linear space) along a wall in a dwelling or other structure. Doorways, fireplace openings, and sliding panels of glass doors are not considered wall space, although fixed panels of glass doors are considered wall space.

What is considered **wall space** is detailed in *210.52(A)(2)*. Wall space is defined as follows:

- Any space 2 ft (600 mm) or more in width along the floor line unbroken by doorways, fireplaces, sliding panels of glass doors, and similar openings. The space occupied by a fixed panel in a sliding glass door must also be counted as wall space.

- Freestanding bar type counters that project into an area, or that divide an area, must be counted as wall space.

- The railings that protect an exposed stairway or the railings that separate rooms on different levels must be included as wall space. Figure 3-7 shows a partial floor plan for a dwelling with an open stairway. The stairway has a freestanding railing protecting one side. This railing is considered wall space. The required outlet must be installed as a floor-mounted receptacle outlet. *NEC®* *210.52(A)(3)* states that the floor

receptacle must be within 18 in. (450 mm) of the railing to be counted as part of the required outlets. Figure 3-8 shows a floor receptacle installed adjacent to an open railing.

With a bedroom layout as shown in Figure 3-9 (page 62), no receptacle outlet is required on the wall between the two closets because that wall is only 18 in. (450 mm) wide. A receptacle may be installed in that wall if desired, but the Code does not require one. On progressing around the bedroom shown in Figure 3-9, it is clear that a receptacle must be installed within 6 ft (1.8 m) of the closet door. The next outlet, on progressing clockwise around the room, must be located within 12 ft (3.7 m) of the previous outlet. This placement ensures that no location on the wall line is more than 6 ft (1.8 m) from a receptacle outlet. Likewise, the next receptacle outlet must be no more than 12 ft (3.7 m) from the last receptacle. The next receptacle must be located no more that 12 ft (3.7 m) from the preceding outlet, but also no more than 6 ft (1.8 m) away from the edge of the bedroom entry doorway.

Figure 3-10 (page 63) shows wall space behind a bedroom door. Such wall space requires special consideration in the location of receptacle outlets. In this case, the door opens against the sidewall of the closet. This space is sometimes called "dead space" and is not specifically mentioned in the *NEC®*. There is no possibility of anyone's placing a dresser or table in that location because it would block the door from fully opening. However, the wall is longer than 2 ft (600 mm), and the Code requires a receptacle outlet to be installed on the wall on that size regardless of whether the space is behind the door. If there is a question about receptacles in dead space, the local AHJ should be consulted to determine how this provision of the Code should be addressed.

It should be noted that certain receptacles may not be counted as among the required number of outlets.

- The required receptacle outlets must be within 5½ ft (1.7 m) of the finished floor, as shown in Figure 3-11 (page 64). An outlet installed any higher (as requested by the owner, for example) cannot be included as one of the required outlets.

- Figure 3-11 also shows that receptacle outlets located within cabinets or cupboards are also not included as fulfilling Code requirements.

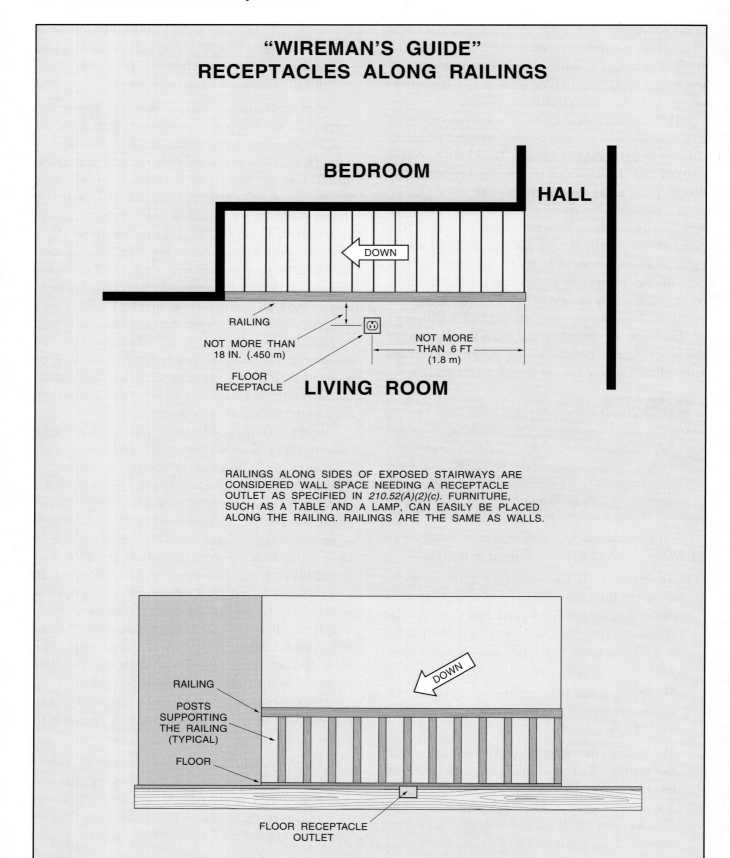

"WIREMAN'S GUIDE"
RECEPTACLES ALONG RAILINGS

BEDROOM

HALL

DOWN

RAILING

NOT MORE THAN
18 IN. (.450 m)

NOT MORE
THAN 6 FT
(1.8 m)

FLOOR
RECEPTACLE

LIVING ROOM

RAILINGS ALONG SIDES OF EXPOSED STAIRWAYS ARE
CONSIDERED WALL SPACE NEEDING A RECEPTACLE
OUTLET AS SPECIFIED IN *210.52(A)(2)(c)*. FURNITURE,
SUCH AS A TABLE AND A LAMP, CAN EASILY BE PLACED
ALONG THE RAILING. RAILINGS ARE THE SAME AS WALLS.

RAILING

POSTS
SUPPORTING
THE RAILING
(TYPICAL)

FLOOR

DOWN

FLOOR RECEPTACLE
OUTLET

Figure 3-7

Figure 3-8   A floor receptacle is installed because the railing is considered floor space.

- Many bathroom luminaires (fixtures) include a receptacle outlet as part of their design. Such outlets are considered to be there for convenience only and are not included as receptacle outlets as required by the *NEC®*. See Figure 3-12 (page 65).

- Receptacle outlets that are installed in electric baseboard heaters can be included as satisfying the provisions of required outlets under certain conditions listed in *210.52*. The FPN below *210.52* states that electric baseboard heaters may not be listed for installation below a receptacle outlet. This prohibition ensures that the supply cord of any electric equipment that may be plugged into the receptacle is not exposed to the electric baseboard heating elements. With some heaters, these elements can get hot enough to melt the insulation on the cord and cause a fire.

- According to *210.52(C)(5)*, if access to a kitchen countertop receptacle outlet is impeded by an appliance that is fastened in place or by an appliance occupying dedicated space, the coun-

tertop outlet shall not be considered as one of the required receptacle outlets. The *NEC®* considers any appliance that is fastened in place or that occupies a dedicated space to be permanently located, and the electrical service to these appliances must be provided by other than the required countertop small appliance receptacle outlets.

3.2.2: Kitchen Countertop Receptacles

KEY TERMS

Kitchen countertop   The top surface of a kitchen counter. Access to the electrical system must be available for electrical appliances that are used on kitchen countertops.

NEC® 210.52(C) deals with receptacle outlets above **kitchen countertops**, including peninsula countertops, and island countertops. Figure 3-13 (page 66) shows the areas of the sample house that are subject to the requirements of *210.52(C)*.

"WIREMAN'S GUIDE"
"DEAD SPACE" BEHIND DOOR

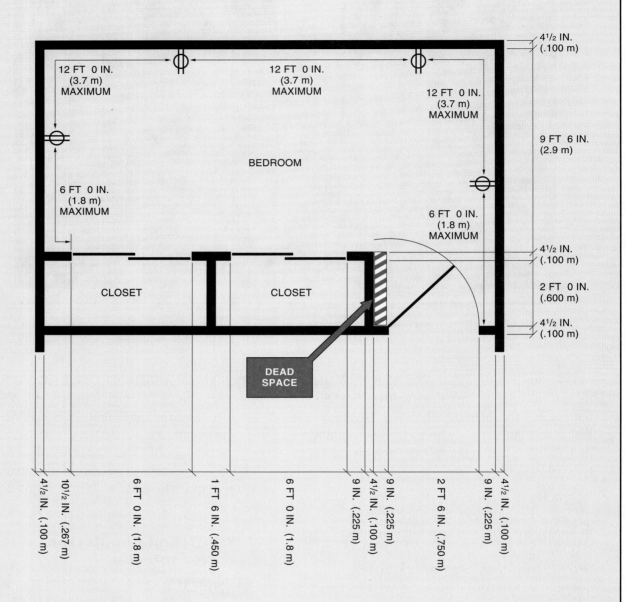

THE WALL SPACE BEHIND A DOOR IS SOMETIMES CALLED "DEAD SPACE."
THE *NEC*® MAKES NO MENTION OF DEAD SPACE; THEREFORE, THIS WALL
SPACE BEHIND THE DOOR NEEDS A RECEPTACLE OUTLET BECAUSE THE
WALL SPACE IS OVER 2 FT (.600 m) WIDE.

Figure 3-9

Figure 3-10 An example of "dead space" behind a bedroom entry door.

KEY TERMS

Wall countertop space Wall space that is located above countertops. Countertop wall space is measured horizontally along the wall where it meets the countertop.

The *NEC®* refers to **wall countertop space**, which is space above kitchen counters that are against a wall, as opposed to island counters, which have no border on a wall. *NEC® 210.52(C)(1)* says that a receptacle outlet must be installed above any wall countertop space that is 12 in. (300 m) in width or greater. If the counter is less than 1 ft (300 mm) wide, no receptacle outlet is required, but one may be installed there if desired. *NEC® 210.52(C)(1)* also says that no point on a wall countertop space can be more than 24 in. (600 mm) from another countertop receptacle outlet or more than 24 in. (600 mm) from the end of the countertop. This distance is measured horizontally along the wall line of the countertop.

KEY TERMS

Island countertop space Countertop space located above an island counter, which has no countertop wall space.

Peninsula countertop space Countertop space located above a peninsula, which has no countertop wall space except where the peninsula portion of the countertop is adjacent to a countertop that is not part of the peninsula space and that has countertop wall space.

Figure 3-14 (page 67) shows a typical layout of kitchen wall countertop space. This kitchen, in addition to the wall countertop, also has **island countertop space** and **peninsula countertop space**. In this layout, the pantry and the sink interrupt the wall countertop runs. The receptacle layout shown in Figure 3-14 meets all of the Code requirements for outlet placement.

NEC® 210.52(C)(5) specifies that receptacle outlets are to be installed above the countertop, but no higher than 20 in. (500 mm) from the countertop surface. Kitchen countertops usually have a backsplash where the counter meets the wall, as shown in Figure 3-15 (page 68). It is very important that the installing electrician be aware of possible problems related to placing the receptacle outlet boxes too low. If the boxes are installed too close to countertop height, they will interfere with the backsplash. A 6-in. (150-mm) clearance from the top of the counter to the bottom of the receptacle box is usually enough to avoid the backsplash, but the installing electrician should consult with the owner or builder about this detail before wiring the kitchen area. The requirement that the receptacle outlets can be no more than 20 in. (.500 m) above the countertop places an upper limit on the height of the outlets. Care must be taken in installing the boxes for the countertop receptacle outlets to ensure that the outlets are not placed so high that they will be covered by the upper sections of cabinets. Figure 3-16 (page 69) shows a typical kitchen countertop, backsplash, and upper cabinet configuration.

Also included in the same kitchen are a peninsula countertop and an island countertop. The receptacle requirements for the island countertop spaces are detailed in *210.52(C)(2)*. The Code states that at

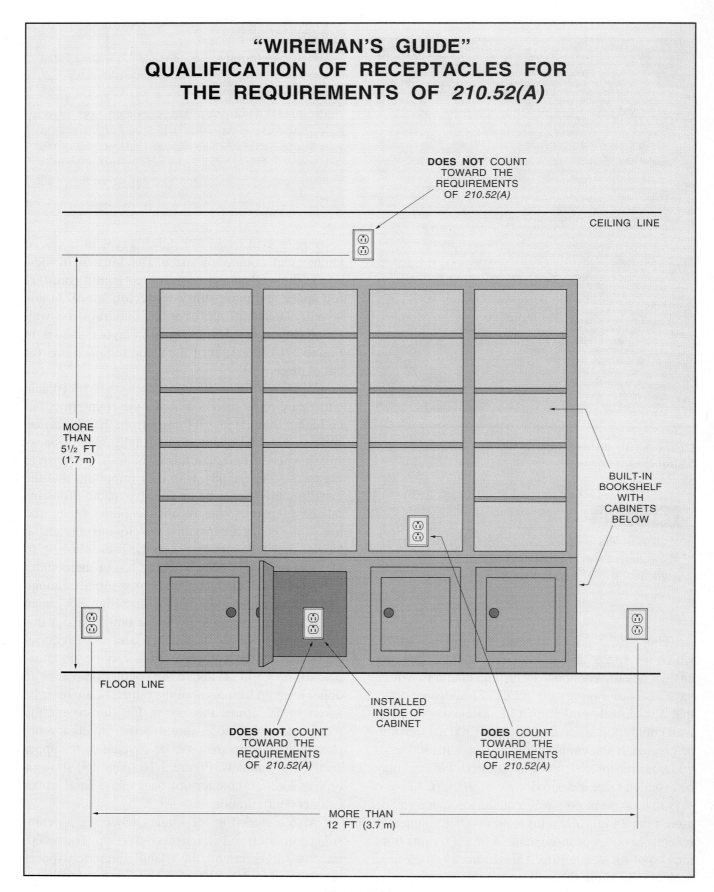

"WIREMAN'S GUIDE"
QUALIFICATION OF RECEPTACLES FOR
THE REQUIREMENTS OF *210.52(A)*

DOES NOT COUNT
TOWARD THE
REQUIREMENTS
OF *210.52(A)*

CEILING LINE

MORE
THAN
5¹/₂ FT
(1.7 m)

BUILT-IN
BOOKSHELF
WITH
CABINETS
BELOW

FLOOR LINE

INSTALLED
INSIDE OF
CABINET

DOES NOT COUNT
TOWARD THE
REQUIREMENTS
OF *210.52(A)*

DOES COUNT
TOWARD THE
REQUIREMENTS
OF *210.52(A)*

MORE THAN
12 FT (3.7 m)

Figure 3-11

"WIREMAN'S GUIDE"
RECEPTACLES IN RANGES,
LUMINAIRES (FIXTURES), AND
ELECTRIC BASEBOARD HEATERS

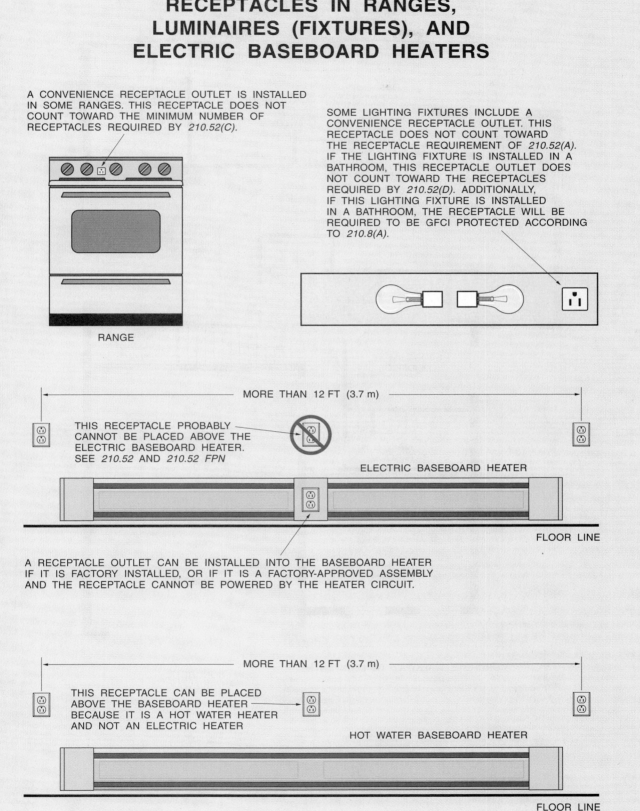

A CONVENIENCE RECEPTACLE OUTLET IS INSTALLED IN SOME RANGES. THIS RECEPTACLE DOES NOT COUNT TOWARD THE MINIMUM NUMBER OF RECEPTACLES REQUIRED BY *210.52(C)*.

SOME LIGHTING FIXTURES INCLUDE A CONVENIENCE RECEPTACLE OUTLET. THIS RECEPTACLE DOES NOT COUNT TOWARD THE RECEPTACLE REQUIREMENT OF *210.52(A)*. IF THE LIGHTING FIXTURE IS INSTALLED IN A BATHROOM, THIS RECEPTACLE OUTLET DOES NOT COUNT TOWARD THE RECEPTACLES REQUIRED BY *210.52(D)*. ADDITIONALLY, IF THIS LIGHTING FIXTURE IS INSTALLED IN A BATHROOM, THE RECEPTACLE WILL BE REQUIRED TO BE GFCI PROTECTED ACCORDING TO *210.8(A)*.

RANGE

MORE THAN 12 FT (3.7 m)

THIS RECEPTACLE PROBABLY CANNOT BE PLACED ABOVE THE ELECTRIC BASEBOARD HEATER. SEE *210.52* AND *210.52 FPN*

ELECTRIC BASEBOARD HEATER

FLOOR LINE

A RECEPTACLE OUTLET CAN BE INSTALLED INTO THE BASEBOARD HEATER IF IT IS FACTORY INSTALLED, OR IF IT IS A FACTORY-APPROVED ASSEMBLY AND THE RECEPTACLE CANNOT BE POWERED BY THE HEATER CIRCUIT.

MORE THAN 12 FT (3.7 m)

THIS RECEPTACLE CAN BE PLACED ABOVE THE BASEBOARD HEATER BECAUSE IT IS A HOT WATER HEATER AND NOT AN ELECTRIC HEATER

HOT WATER BASEBOARD HEATER

FLOOR LINE

Figure 3-12

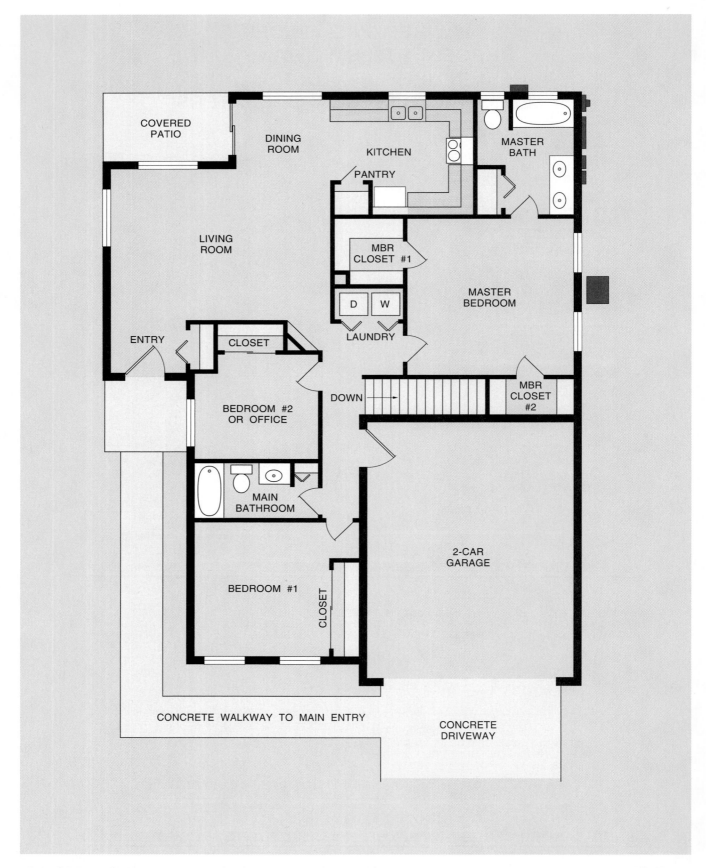

Figure 3-13 Main level floor plan: Areas of the sample house subject to receptacle layout according to *210.52(C)*.

"WIREMAN'S GUIDE"
SUGGESTED KITCHEN WALL
COUNTERTOP RECEPTACLE LOCATIONS
FOR THE SAMPLE HOUSE

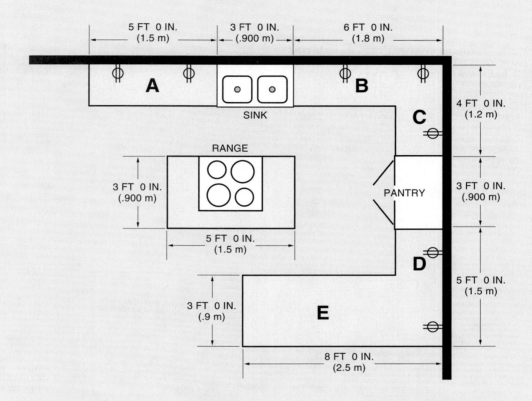

COUNTERTOP A: COUNTERTOP A IS 5 FT (1.5 m) WIDE. THE *NEC*® REQUIRES A RECEPTACLE OUTLET WITHIN 2 FT (.600 m) OF THE END OF THE COUNTERTOP AND NOT MORE THAN 4 FT (1.2 m) BETWEEN RECEPTACLE OUTLETS. THIS WILL REQUIRE TWO RECEPTACLE OUTLETS INSTALLED, AS SHOWN, TO MEET CODE REQUIREMENTS.

COUNTERTOP B: COUNTERTOP B IS 6 FT (1.8 m) WIDE. NO MATTER WHERE THE RECEPTACLE OUTLET IS INSTALLED AT THE LEFT SIDE OF THE COUNTERTOP, A SECOND RECEPTACLE OUTLET MUST BE INSTALLED ON THAT WALL. THE INSTALLATION SHOWN, AT 2 FT (.600 m) AND AT 5 FT (1.5 m) FROM THE LEFT SIDE WILL MEET CODE REQUIREMENTS AND PROVIDE FOR CONVENIENT LOCATIONS.

COUNTERTOP C: COUNTERTOP C IS 4 FT (1.2 m) WIDE, FROM THE CORNER WITH COUNTERTOP B TO THE LEFT SIDE OF THE PANTRY. THE LAST RECEPTACLE OUTLET ON COUNTERTOP B IS 1 FT (.300 m) FROM THE CORNER WITH COUNTERTOP C. THIS MEANS THAT THE RECEPTACLE OUTLET ON THE WALL FOR COUNTERTOP C MUST BE WITHIN 3 FT (.915 m) OF THE CORNER. THIS IS A CONVENIENT SPOT, LEAVING 1 FT (.300 m) TO THE LEFT SIDE OF THE PANTRY.

COUNTERTOP D: COUNTERTOP D IS 5 FT (1.5 m) WIDE. THERE MUST BE A RECEPTACLE OUTLET WITHIN 2 FT (.600 m) OF THE RIGHT SIDE OF THE PANTRY.

COUNTERTOP E: COUNTERTOP E IS A PENINSULA COUNTERTOP, WITH WALL ONLY ON ITS RIGHT END. THE CODE REQUIRES A RECEPTACLE OUTLET ALONG THE WALL ON COUNTERTOP D WITHIN 2 FT (.600 m) OF THE END OF THE COUNTERTOP AND NO MORE THAN 4 FT (1.2 m) FROM THE OTHER RECEPTACLE OUTLET. PLACING THE RECEPTACLES 1 FT (.300 m) FROM EACH END WILL MAKE A CONVENIENT INSTALLATION THAT MEETS *NEC*® REQUIREMENTS.

Figure 3-14

"WIREMAN'S GUIDE"
WALL COUNTERTOP BACKSPLASH
AND UPPER CABINETS

THERE IS USUALLY A BACKSPLASH AT THE BACK OF THE WALL COUNTERTOP SPACE TO KEEP WATER FROM GETTING BETWEEN THE COUNTERTOP AND THE WALL. THIS BACKSPLASH IS TYPICALLY 4 IN. (.100 m) HIGH, BUT THIS CAN VARY CONSIDERABLY FROM ONE HOUSE TO ANOTHER. IT IS BEST TO CHECK WITH THE BUILDER OR GENERAL CONTRACTOR BEFORE THE INSTALLATION OF RECEPTACLE BOXES ABOVE THE COUNTERTOPS.

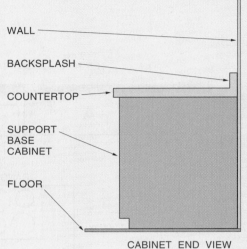

CABINET END VIEW

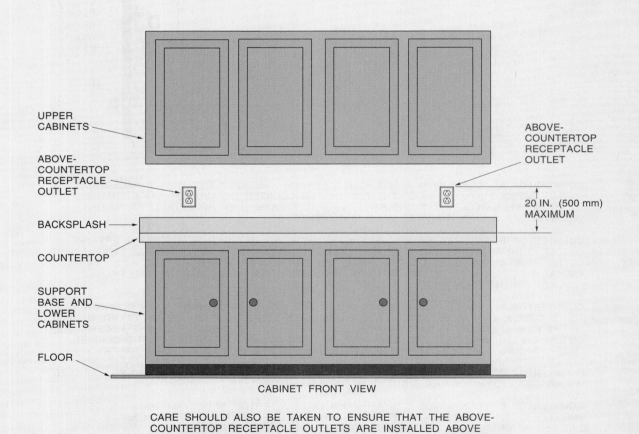

CABINET FRONT VIEW

CARE SHOULD ALSO BE TAKEN TO ENSURE THAT THE ABOVE-COUNTERTOP RECEPTACLE OUTLETS ARE INSTALLED ABOVE THE BACKSPLASH BUT NO HIGHER THAN 20 IN. (500 mm) ABOVE THE COUNTERTOP. THE UPPER CABINETS CREATE A POTENTIAL PROBLEM IF THEY ARE INSTALLED SO LOW THAT THEY COVER THE INTENDED RECEPTACLE LOCATIONS.

Figure 3-15

Figure 3-16   Kitchen countertop wall space with receptacle outlet.

least one receptacle outlet must be installed at an island counter space larger than 12 in. (300 mm) by 24 in. (600 mm). One receptacle outlet on an island countertop is all that is ever required, although additional outlets may be installed if desired.

The location of the receptacle outlet for the island countertop can present problems. Receptacle outlets are not to be installed in a face-up position on the countertop. A face-up position would allow water and other materials to enter the interior of the receptacle should a spill occur. However, the island has no wall in which a receptacle outlet could be installed. An exception to the installation requirements of *210.52(C)(5)* states that the receptacle outlet may be installed on the side of the island support base, but not more than 12 in. (300 mm) below the countertop surface, and only if one of the conditions stated in *(a)* and *(b)* is satisfied. Part *(a)* of *210.52(C)(5)* exception concerns kitchens for persons with physical impairments. Part *(b)* of the exception says that for island countertops, when the countertop is flat across its entire surface and no location for installation of a receptacle exists, the receptacle outlet may

be installed on the side of the supporting cabinet. Figures 3-17 and 3-18 show a typical kitchen island countertop and a receptacle outlet installed in the side of the island's supporting cabinet.

NEC® 210.52(C)(5) Exception also says that if a receptacle outlet needs to be installed in the side of a support cabinet and the countertop extends more than 6 in. (150 mm) over the support cabinet, the receptacle outlet cannot be installed, as also shown in Figure 3-17. Such a wide overhang would allow for too much appliance cord hanging over the edge of the countertop and would require the appliance to sit too close to the edge of the countertop for safe operation.

The peninsula countertop receptacle requirements are detailed in *210.52(C)(3)* and are much the same as for island countertops. At least one receptacle outlet must be installed to service the peninsula countertop if the countertop dimensions are 12 in. (.300 m) by 24 in. (.600 m) or greater. This measurement is from the edge of the peninsula countertop to where it meets the wall countertop. Figure 3-19 (page 72) shows receptacle outlet installations for

"WIREMAN'S GUIDE"
RECEPTACLE OUTLETS IN ISLAND COUNTERTOP SUPPORT CABINETS

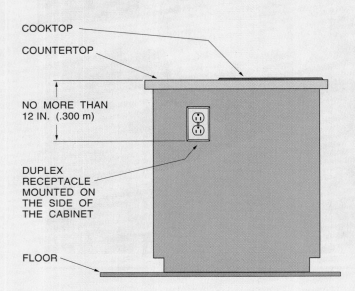

COOKTOP

COUNTERTOP

NO MORE THAN 12 IN. (.300 m)

DUPLEX RECEPTACLE MOUNTED ON THE SIDE OF THE CABINET

FLOOR

THE ISLAND COUNTERTOP HAS NO BACKSPLASH OR OTHER PLACE TO INSTALL A RECEPTACLE OUTLET ABOVE THE COUNTERTOP. IN THIS EVENT THE *NEC*® ALLOWS THE RECEPTACLE OUTLET TO BE INSTALLED IN THE SIDE OF THE SUPPORT CABINET. THE RECEPTACLE OUTLET CANNOT BE INSTALLED LOWER THAN 12 IN. (.300 m) BELOW THE COUNTERTOP. IT CAN BE INSTALLED IN A HORIZONTAL OR VERTICAL ORIENTATION. CARE MUST BE TAKEN IN CUTTING SO THAT THE SUPPORT CABINET IS NOT DAMAGED. THE WIRE MAY NEED PROTECTION INSIDE OF THE CABINET.

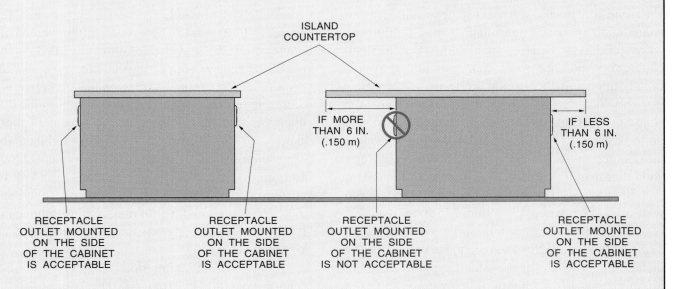

ISLAND COUNTERTOP

IF MORE THAN 6 IN. (.150 m)

IF LESS THAN 6 IN. (.150 m)

RECEPTACLE OUTLET MOUNTED ON THE SIDE OF THE CABINET IS ACCEPTABLE

RECEPTACLE OUTLET MOUNTED ON THE SIDE OF THE CABINET IS ACCEPTABLE

RECEPTACLE OUTLET MOUNTED ON THE SIDE OF THE CABINET IS NOT ACCEPTABLE

RECEPTACLE OUTLET MOUNTED ON THE SIDE OF THE CABINET IS ACCEPTABLE

RECEPTACLE OUTLETS CAN BE INSTALLED ON THE SIDE OF ISLAND SUPPORT CABINETS IF THE OVERHANG FROM THE COUNTERTOP IS 6 IN. (150 mm) OR LESS.

Figure 3-17

Figure 3-18   A receptacle outlet installed on the side of an island countertop support cabinet.

peninsula countertops. The same requirements that apply to the installation of a receptacle outlet on the side of the support base or cabinet of a kitchen island also apply with the peninsula.

If cables need to be installed inside cabinets, a wiring plan detailing how the wire is to be installed in the cabinet should also be developed. *NEC® 300.4* requires conductors to be protected against physical damage, and many inspectors require some method of protection for the power cable as it passes through the inside of a cabinet, as shown in Figure 3-20.

Another receptacle outlet needed in the kitchen —that for the refrigerator—is not usually placed above the countertop but is located on a wall that borders a countertop. This receptacle outlet is usually placed directly behind the refrigerator and can be at any height below 5½ ft (1.7 m) but most often is set at the same height as that for the kitchen countertop receptacle outlets. This placement allows easier access to the receptacle for disconnection of the refrigerator as needed.

3.2.3: Bathroom Receptacle Outlets

KEY TERMS

Bathroom   (See *NEC® Article 100*): A nonhabitable room that contains a washbasin and at least one other bathroom fixture such as a toilet, shower, or tub.

Figure 3-21 (page 74) shows that portion of the sample house subject to the requirements of *210.52(D)*, which specifies the receptacle outlet locations for a **bathroom**. At least one receptacle outlet must be located on the wall adjacent to the washbasin, or to the washbasins if there is more than one. The outlet must be within 36 in. (900 mm) of each basin. The distance is measured from the outside edge of the basin to the receptacle outlet. Figure 3-22 (page 75) shows a typical bathroom floor plan. In this design, there are two basins in the countertop. Three different methods of installing receptacle outlets are explored in Figure 3-22; either arrangement B or arrangement C is acceptable so long as the

"WIREMAN'S GUIDE"
RECEPTACLE OUTLETS INSTALLED ON THE SIDES OF PENINSULA AND ISLAND COUNTERTOP SUPPORT CABINETS

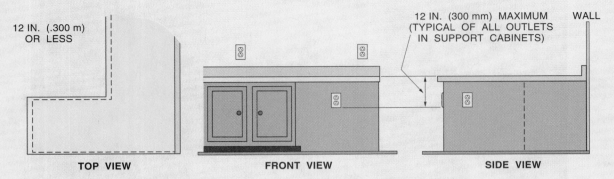

WALL COUNTERTOP AND SUPPORT CABINET WITH PENINSULA
SUPPORT CABINET AND COUNTERTOP WITH NO OVERHANG

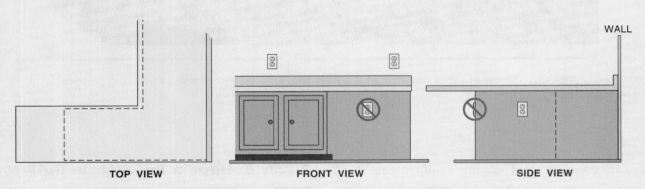

WALL COUNTERTOP AND SUPPORT CABINET WITH PENINSULA SUPPORT CABINET
AND COUNTERTOP WITH AN OVERHANG GREATER THAN 6 IN. (150 mm) ON THE END

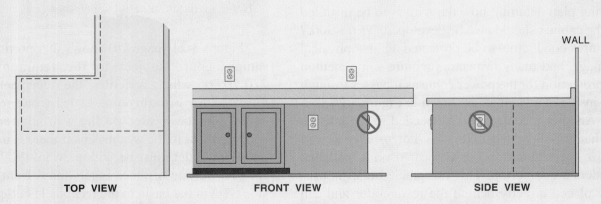

WALL COUNTERTOP AND SUPPORT CABINET WITH PENINSULA SUPPORT CABINET
AND COUNTERTOP WITH AN OVERHANG GREATER THAN 6 IN. (150 mm) ON THE SIDE

Figure 3-19

"WIREMAN'S GUIDE"
CABLE INSTALLED INSIDE
OF KITCHEN CABINET

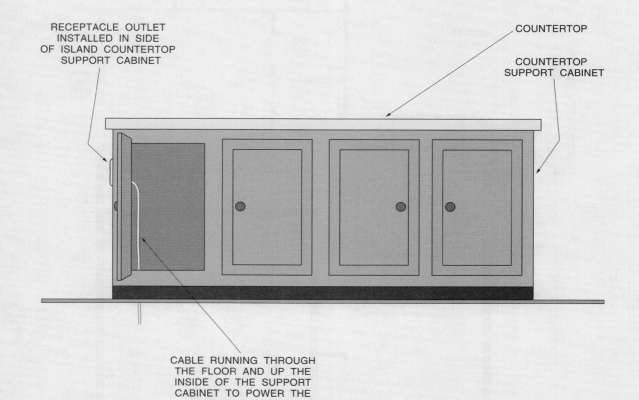

RECEPTACLE OUTLET
INSTALLED IN SIDE
OF ISLAND COUNTERTOP
SUPPORT CABINET

COUNTERTOP

COUNTERTOP
SUPPORT CABINET

CABLE RUNNING THROUGH
THE FLOOR AND UP THE
INSIDE OF THE SUPPORT
CABINET TO POWER THE
RECEPTACLE OUTLET

CABLE INSTALLED INSIDE OF KITCHEN ISLAND OR PENINSULA
SUPPORT CABINET THAT PROVIDES POWER TO THE RECEPTACLE
OUTLETS IS ALLOWED BY NEC®. IN MOST CASES THE AHJ
REQUIRES PROTECTION FOR THE CABLE FROM PHYSICAL DAMAGE.

Figure 3-20

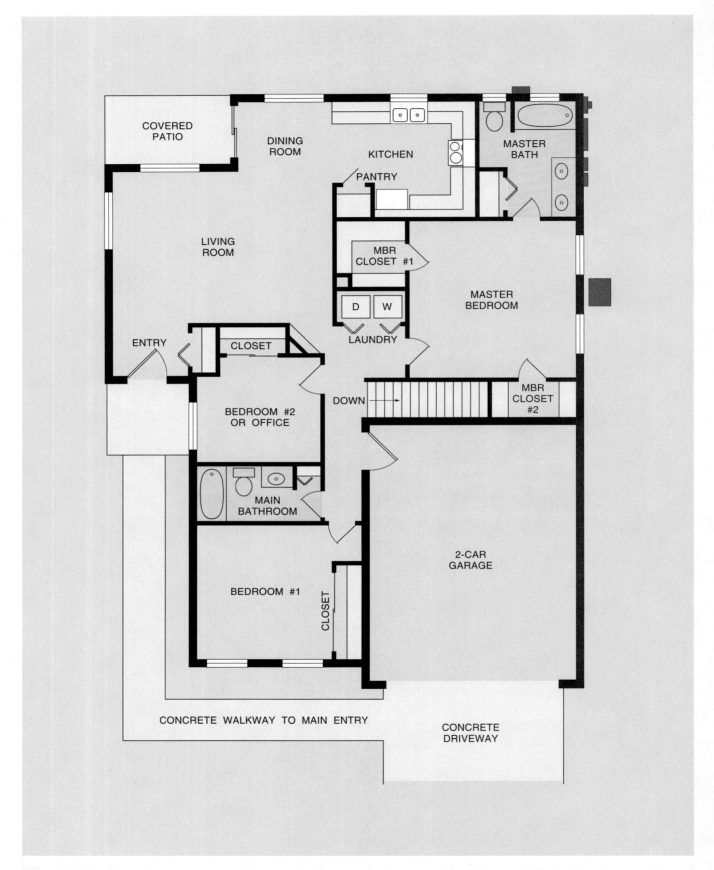

Figure 3-21 Main level floor plan: Areas of the sample house subject to receptacle layout according to *210.52(D).*

"WIREMAN'S GUIDE"
ABOVE-COUNTERTOP RECEPTACLE
OUTLETS IN BATHROOMS

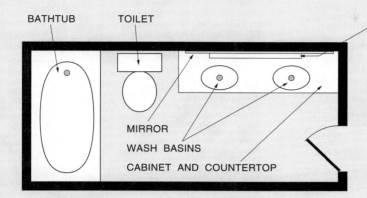

BATHTUB   TOILET

LUMINAIRE (FIXTURE) ON
WALL ABOVE MIRROR

MIRROR

WASH BASINS

CABINET AND COUNTERTOP

AT THE LEFT IS A FLOOR
PLAN OF A TYPICAL BATHROOM.
NOTICE THAT THERE IS A
LARGE MIRROR ON THE WALL
AT THE BACK OF THE CABINET
AND THAT THERE ARE TWO
WASHBASINS.

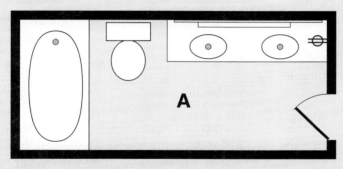

IN BATHROOM A THERE IS ONE
RECEPTACLE OUTLET, ON THE
WALL THAT IS TO THE RIGHT OF
THE CABINET. THERE IS ONLY
ONE RECEPTACLE INSTALLED.
THIS DOES NOT MEET CODE
REQUIREMENTS BECAUSE ONLY
ONE OF THE WASHBASINS IS
SERVED.

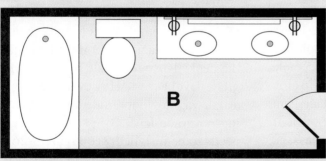

IN BATHROOM B THERE ARE TWO
RECEPTACLE OUTLETS INSTALLED
ABOVE THE COUNTERTOP. IF THE
MEASUREMENT FROM THE RECEP-
TACLE TO THE OUTSIDE EDGE OF
THE ADJACENT BASIN IS LESS THAN
36 IN. (900 mm), THIS INSTALLATION
MEETS THE CODE REQUIREMENTS
OF 210.52(D).

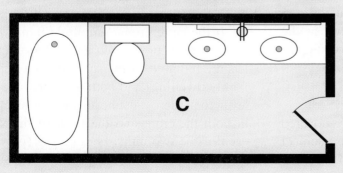

IN BATHROOM C THERE IS
ONLY ONE RECEPTACLE OUTLET
ABOVE THE COUNTERTOP. IF THE
MEASUREMENT FROM THE
RECEPTACLE TO THE OUTSIDE
EDGE OF EACH BASIN IS LESS
THAN 36 IN. (900 mm), THIS
INSTALLATION MEETS THE CODE
REQUIREMENTS OF 210.52(D).

Figure 3-22

receptacles are within 36 in. (900 mm) of the edge of the basin. Figure 3-23 shows a typical bathroom countertop with mirror and receptacle outlet.

Placement of a mirror on the wall directly behind the basin can cause installation problems because a hole would have to be cut in the mirror to accommodate the receptacle. Whenever possible, the receptacle outlets should be installed on the side walls to the bathroom cabinet to avoid potential problems. The bathroom countertop receptacle outlets do not have the same height requirement as that for the kitchen receptacle locations. In the kitchen, the receptacle outlet cannot be placed higher than 20 in. (500 mm) above the countertop and it must be installed above the countertop. In bathrooms, however, these requirements do not apply. The receptacle must be installed in a wall adjacent to the basin, but the location of the receptacle outlet does not have to be above the countertop and can be more than 20 in. (500 mm) above the counter, so long as the location is within 36 in. (900 mm) of the basin edge. Figure 3-24 shows several possible locations for the receptacle outlets that are legal according to the *NEC®*, although somewhat unconventional in everyday practice. *NEC® 210.52(D)* also has the same restriction that applies to kitchen countertops—that forbidding the installation in a face-up position on a countertop.

It is allowable to install bathroom receptacle outlets that are not associated with the basin. These receptacles can be installed anywhere on a wall line, at any height, but according to *406.8* a receptacle cannot be installed within a bathtub enclosure. Furthermore, there are other very strict placement requirements for the location of receptacle outlets if the bathtub is a hydromassage tub. A hydromassage bathtub is defined in *680.2* and has a pump, circulating system, and associated equipment that serve to circulate the water in the tub. The bathtub in the sample house is a standard bathtub with no pump or circulating system.

3.2.4: Outdoor Receptacle Outlets

Receptacle outlets for use outdoors are covered in *210.52(E)*. The areas of the sample house subject to these rules are shown in Figure 3-25. This Code section states that for single-family dwellings, and for each unit of two-family dwellings (duplex units)

Figure 3-23   A receptacle outlet on the side wall of a bathroom.

that are at grade level, at least one receptacle outlet must be installed in the front and one receptacle outlet must be installed in the back of the structure. The receptacle outlets must be accessible from grade level and not more than 6½ ft (2.0 m) above grade level. For example, for a deck off a second-floor master bedroom that has a stairway leading to grade level, a receptacle outlet must be installed to service the deck. This receptacle outlet is defined as having grade-level access. However, the same deck without a stairway down to grade level is considered as having no grade-level access, and no receptacle outlet needs to be installed; see Figure 3-26.

Figure 3-27 (page 80) shows a typical outside luminaire (fixture), mounted on a brick wall.

If the structure is a multiple-family dwelling of three or more units, then no outdoor receptacle outlets are required. If the structure is a duplex (a two-family dwelling) and both units are located at grade level, then there must be an outdoor receptacle outlet in both the front and the back of the structure for each of the two units. Figure 3-28 (page 81) shows the required outdoor receptacles for a single-family home, a duplex, and a multiple-family dwelling of more than two units.

"WIREMAN'S GUIDE"
POTENTIAL PROBLEMS IN BATHROOM
RECEPTACLE INSTALLATIONS

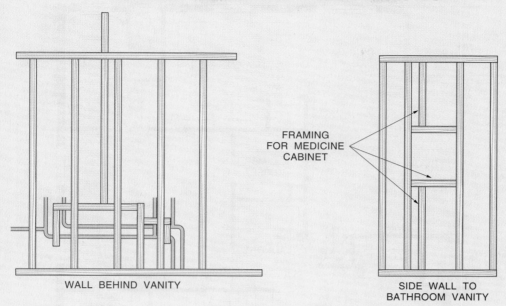

WALL BEHIND VANITY

FRAMING
FOR MEDICINE
CABINET

SIDE WALL TO
BATHROOM VANITY

THE PLUMBING PIPES CAN GET IN THE WAY
OF INSTALLING OUTLET BOXES FOR LUMINAIRES
(FIXTURES) AND RECEPTACLE OUTLETS.
COORDINATE WITH THE PLUMBER TO MINIMIZE
THESE PROBLEMS DURING ROUGH-IN.

IN MANY HOUSES THE MEDICINE CABINET IS
INSTALLED IN THE SIDE WALL OF THE BATHROOM
VANITY. LOOK FOR FRAMING THAT WILL GIVE
CLUES IF A MEDICINE CABINET IS TO BE INSTALLED
AND TAKE CARE TO NOT RUN WIRES OR MOUNT
BOXES IN THESE SPACES.

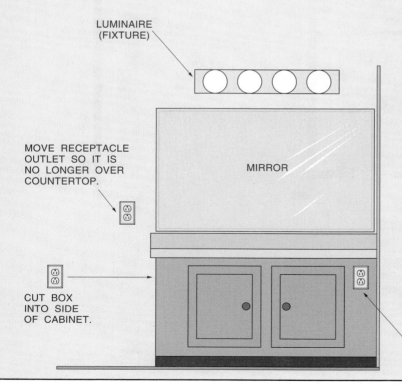

LUMINAIRE
(FIXTURE)

MOVE RECEPTACLE
OUTLET SO IT IS
NO LONGER OVER
COUNTERTOP.

MIRROR

CUT BOX
INTO SIDE
OF CABINET.

CUT BOX
INTO FRONT
OF CABINET.

THE BATHROOM RECEPTACLE
OUTLET IS REQUIRED BY *210.52(D)*
TO BE NO MORE THAN 36 IN.
(900 mm) FROM THE INSIDE EDGE
OF THE BASIN. THE CODE,
HOWEVER, DOES NOT HAVE THE
SAME RESTRICTIONS ON MOUNTING
RECEPTACLES INTO SUPPORT
CABINETS AS FOR KITCHENS.
CONSIDER MOUNTING THE RECEP-
TACLE OUTLET IN THE SIDE OF
THE CABINET, OR THE FRONT OF
THE CABINET. THE CABLE MAY
HAVE TO BE PROTECTED IF RUN
INSIDE OF A CABINET.

IF THERE IS A MIRROR INSTALLED
ABOVE THE COUNTERTOP, CONSIDER
MOVING THE RECEPTACLE OUTLET
TO ONE SIDE FAR ENOUGH TO
CLEAR THE MIRROR. *NEC® 210.52(D)*
DOES NOT REQUIRE THE RECEPTACLE
OUTLET TO BE ABOVE THE COUNTER-
TOP AS IN KITCHENS. IT STILL MUST
BE WITHIN 36 IN. (.900 m) OF THE
INSIDE EDGE OF THE BASIN,
HOWEVER.

Figure 3-24

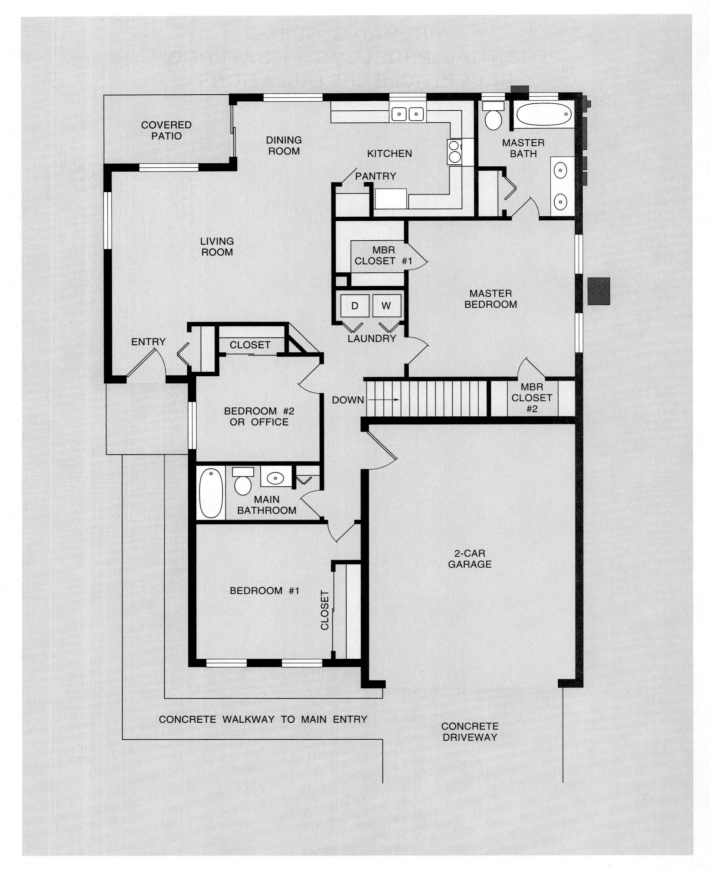

Figure 3-25 Main level floor plan: Areas of the sample house subject to receptacle layout according to *210.52(E).*

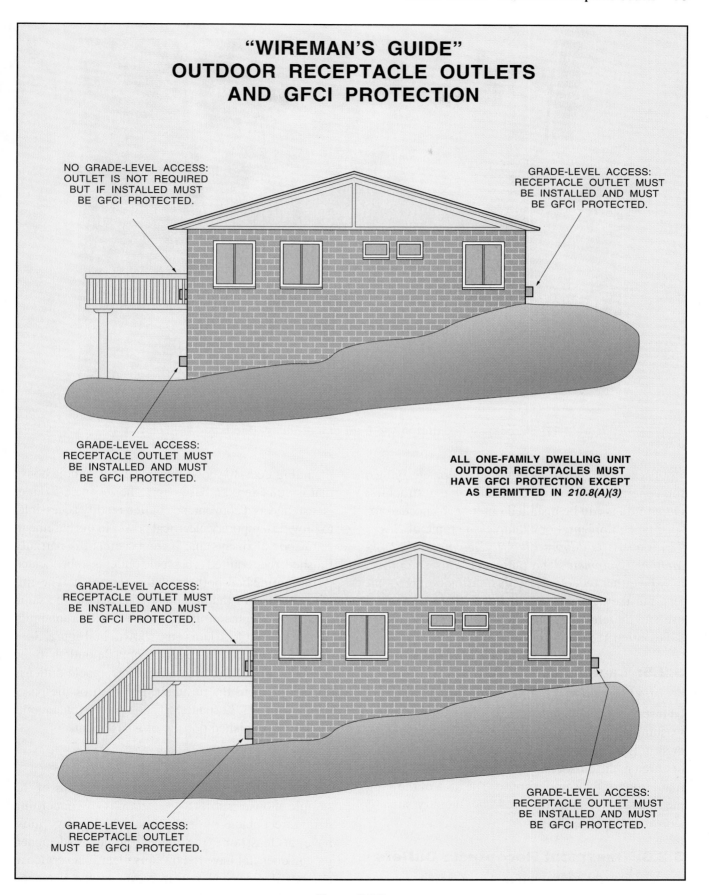

"WIREMAN'S GUIDE"
OUTDOOR RECEPTACLE OUTLETS AND GFCI PROTECTION

NO GRADE-LEVEL ACCESS: OUTLET IS NOT REQUIRED BUT IF INSTALLED MUST BE GFCI PROTECTED.

GRADE-LEVEL ACCESS: RECEPTACLE OUTLET MUST BE INSTALLED AND MUST BE GFCI PROTECTED.

GRADE-LEVEL ACCESS: RECEPTACLE OUTLET MUST BE INSTALLED AND MUST BE GFCI PROTECTED.

ALL ONE-FAMILY DWELLING UNIT OUTDOOR RECEPTACLES MUST HAVE GFCI PROTECTION EXCEPT AS PERMITTED IN *210.8(A)(3)*

GRADE-LEVEL ACCESS: RECEPTACLE OUTLET MUST BE INSTALLED AND MUST BE GFCI PROTECTED.

GRADE-LEVEL ACCESS: RECEPTACLE OUTLET MUST BE GFCI PROTECTED.

GRADE-LEVEL ACCESS: RECEPTACLE OUTLET MUST BE INSTALLED AND MUST BE GFCI PROTECTED.

Figure 3-26

Figure 3-27 A receptacle outlet at the front of a dwelling. One is required by the *NEC*®.

These rules apply to outdoor receptacle outlets for general use by the occupants of the structure. Other outlets can be installed outdoors as needed for specific equipment. For example, the topic of *NEC*® *426* is *Fixed Outdoors Electric Deicing and Snow-Melting Equipment.* The outlets for this equipment may be either receptacle outlets or junction box outlets, but they are not considered outlets for general use and therefore are not counted among the receptacle outlets required by *210.52(E)*.

3.2.5: Laundry Receptacle Outlet

NEC® *210.52(F)* requires a receptacle outlet to be installed in the laundry area for use by the clothes-washing machine. This area may be a separate room or simply a designated area—for example, the laundry area in the closet that opens into a hallway on the main floor of the sample house. The area of the sample house that is considered the laundry area is shown in Figure 3-29.

3.2.6: Basement Receptacle Outlets

NEC® *210.52(G)* specifies that at least one receptacle outlet is to be installed in an unfinished base-ment. Figure 3-30 shows the area in the sample house that is considered a basement. The receptacle outlet is installed as a convenience outlet, readily accessible, to provide temporary electrical power in the basement. The Code also says that if the basement is partially finished, the required receptacle outlet is to be located in the unfinished portion of the basement. The finished portion of the basement is subject to the same rules for receptacle placement as for the remainder of the habitable rooms, as shown in Figure 3-31 (page 84). Even if there are receptacle outlets in the finished area of the basement, a receptacle outlet is still required in the unfinished portion of the basement. Figure 3-32 (page 85) shows a receptacle outlet in the unfinished portion of a basement.

3.2.7: Garage Receptacle Outlets

Figure 3-33 (page 86) shows those areas of the sample house subject to *210.52(G)* concerning garage receptacle outlets. An attached garage must have at least one receptacle outlet. A detached garage does not have to have any electrical service at all, but if electrical power is run to the detached garage, then a receptacle outlet must be installed.

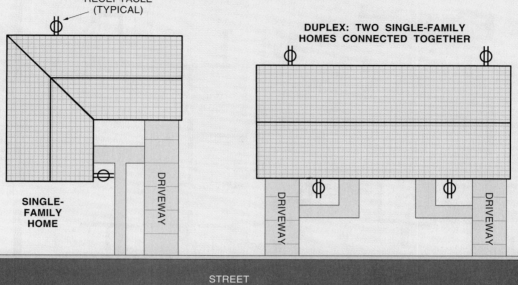

"WIREMAN'S GUIDE"
OUTDOOR RECEPTACLE OUTLETS
AND TYPE OF OCCUPANCY

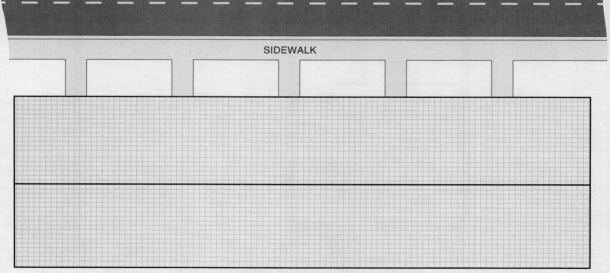

APARTMENT BUILDING—MULTIFAMILY HOME

NO OUTDOOR RECEPTACLE IS REQUIRED FOR
MULTIFAMILY DWELLING ACCORDING TO *210.52(E)*.

NEC® 210.52(E) REQUIRES OUTDOOR RECEPTACLE OUTLETS TO BE INSTALLED IN THE
FRONT AND IN THE BACK OF ONE-FAMILY DWELLINGS AND TWO-FAMILY DWELLINGS
(DUPLEX UNITS) WITH GRADE-LEVEL ACCESS. THESE RECEPTACLES MUST HAVE GFCI
PROTECTION. THE CODE DOES NOT REQUIRE ANY OUTDOOR RECEPTACLE OUTLETS
FOR THREE-FAMILY AND LARGER DWELLINGS, INCLUDING LARGE APARTMENT BUILDINGS.

Figure 3-28

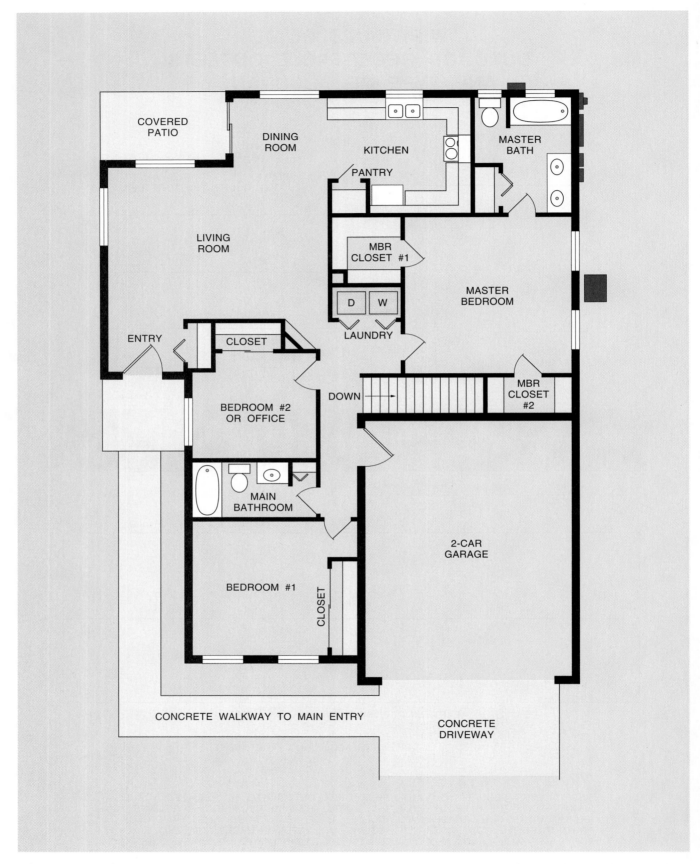

Figure 3-29 Main level floor plan: Area of the sample house subject to receptacle layout according to *210.52(F)*.

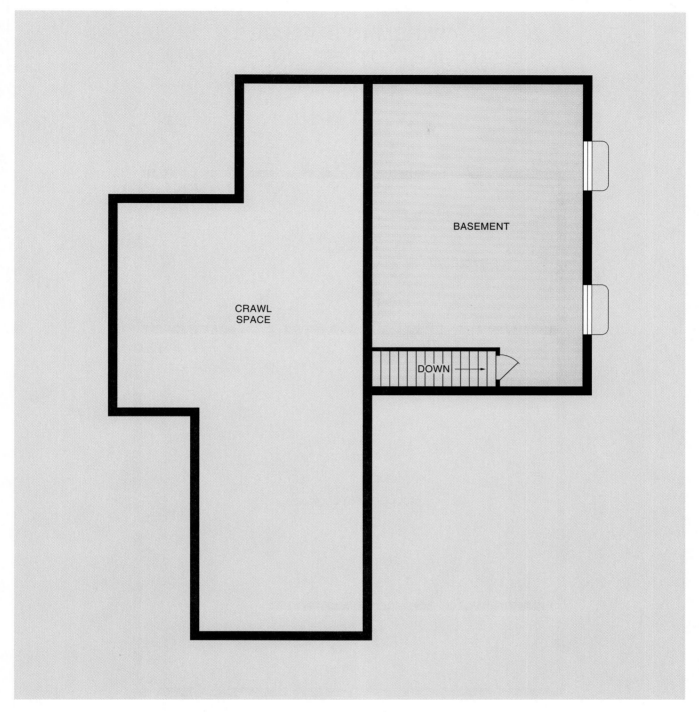

Figure 3-30 Basement area of the sample house subject to receptacle layout according to *210.52(G)*.

The garage outlet is installed for use by the dwelling occupant and should be easily accessible. Other receptacles can be installed if so desired. In order to comply with the requirements of *210.52*, the receptacle cannot be mounted higher than 5½ ft (1.7 m) above the floor.

Other receptacle outlets are routinely installed in garages that are not for the convenience of the occupants. A receptacle outlet to supply power to a freezer or refrigerator is sometimes installed in garages, and a receptacle outlet is very often installed in the ceiling to supply power to an electric garage door opener. It is acceptable to install these receptacles but they are not considered as required outlets.

"WIREMAN'S GUIDE"
RECEPTACLE OUTLETS IN PARTIALLY
FINISHED BASEMENTS

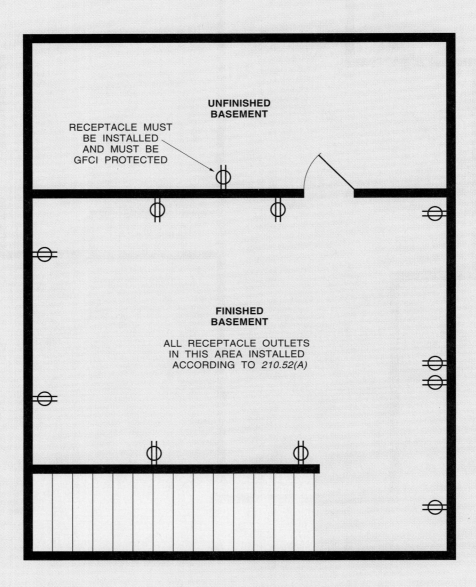

UNFINISHED BASEMENT

RECEPTACLE MUST BE INSTALLED AND MUST BE GFCI PROTECTED

FINISHED BASEMENT

ALL RECEPTACLE OUTLETS IN THIS AREA INSTALLED ACCORDING TO *210.52(A)*

NEC® 210.52(G) REQUIRES THE INSTALLATION OF AT LEAST ONE RECEPTACLE OUTLET IN AN UNFINISHED BASEMENT. IF PART OF THE BASEMENT IS FINISHED AND PART OF THE BASEMENT IS UNFINISHED, THE CODE STILL REQUIRES THE INSTALLATION OF AT LEAST ONE GFCI-PROTECTED RECEPTACLE OUTLET IN THE UNFINISHED PORTION.

Figure 3-31

Figure 3-32   A receptacle outlet installed in the unfinished portion of a dwelling basement.

3.2.8: Hallway Receptacle Outlets

The requirements for receptacle outlets in a dwelling unit hallway can be found in *210.52(H)*. In dwelling units, any hallway that is over 10 ft long must have at least one receptacle outlet installed. Figure 3-34 shows the areas of the sample house that are considered *hallway*. There is no Code guideline as to where in the hallway the receptacle outlet has to be installed, but common sense suggests that the receptacle should be somewhere near the center of the hallway to be the most convenient for the occupant. The centerline measurement of the hallway is to be used in determining the length, as shown in Figure 3-35 (page 88). The hallway in the figure has a turn to the right that makes one wall of the hallway longer. If the hallway happens to pass through a

doorway, a new hallway is considered to have begun, and a new 10-ft (3.0-m) measurement also begins. Figure 3-36 (page 89) shows a receptacle outlet installed in an upstairs hallway, immediately adjacent to a bedroom door.

3.2.9: Storage or Equipment Space Receptacle Outlets

NEC® *210.63* requires that a receptacle outlet be installed near any heating, air-conditioning, or refrigeration equipment that is installed in the attic or crawl space of a dwelling unit. This receptacle outlet is intended for use during servicing of the equipment, and the receptacle outlet must be installed on the same level as, and within 25 ft (7.5 m) of, the equipment needing servicing. Furthermore, the receptacle outlet cannot be supplied from the load side of the disconnecting switch that powers the equipment. In the sample house, no mechanical equipment is installed in the crawl space or the attic. Therefore, no receptacles are required.

3.3: RECEPTACLE OUTLETS OTHER THAN REQUIRED OUTLETS

Not all of the receptacle outlets installed in the sample house are required by the Code. Several additional receptacle outlets are installed to supply kitchen appliances and the clothes dryer. The location of these outlets is determined by the location of the equipment because *210.50(C)* requires that the outlets be within 6 ft (1.8 m) of the intended location.

3.3.1: Dishwasher Receptacle Outlet

In the sample house, the dishwasher is installed just to the left of the kitchen sink, as can be seen in Construction Drawing E-3 in Chapter 1 of this book. The dishwasher is to be powered by half of a duplex receptacle (shared with the garbage disposal) installed under the sink. A hole must be cut between the sink support cabinet and the dishwasher for the cord to go through.

3.3.2: Garbage Disposal Receptacle Outlet

The garbage disposal is installed in the cabinet directly below the kitchen sink. It is powered by half of a duplex receptacle (shared with the dishwasher) installed in the cabinet under the sink.

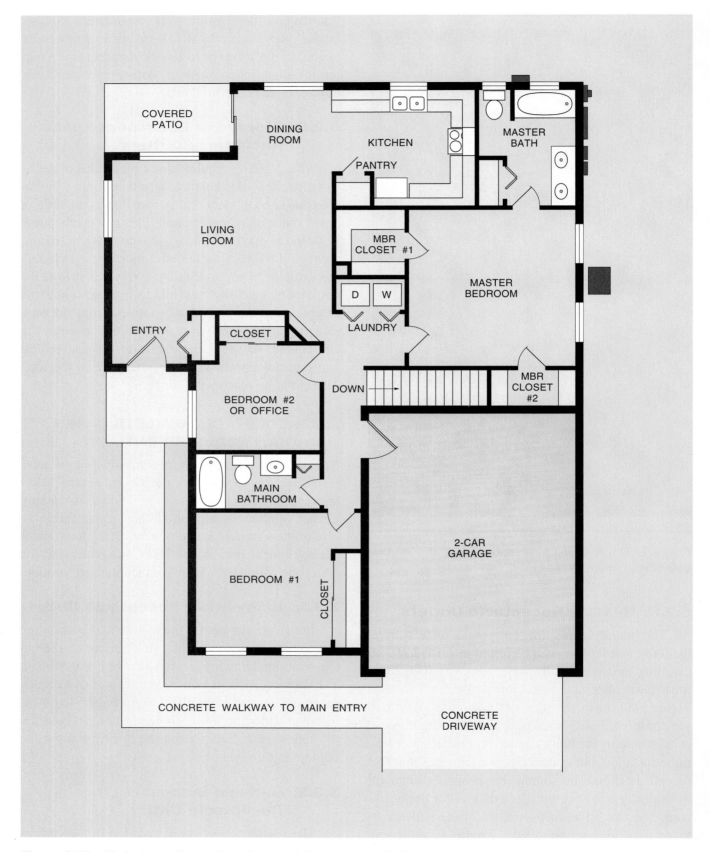

Figure 3-33   Main level floor plan: Areas of the garage of the sample house subject to receptacle layout according to *210.52(G)*.

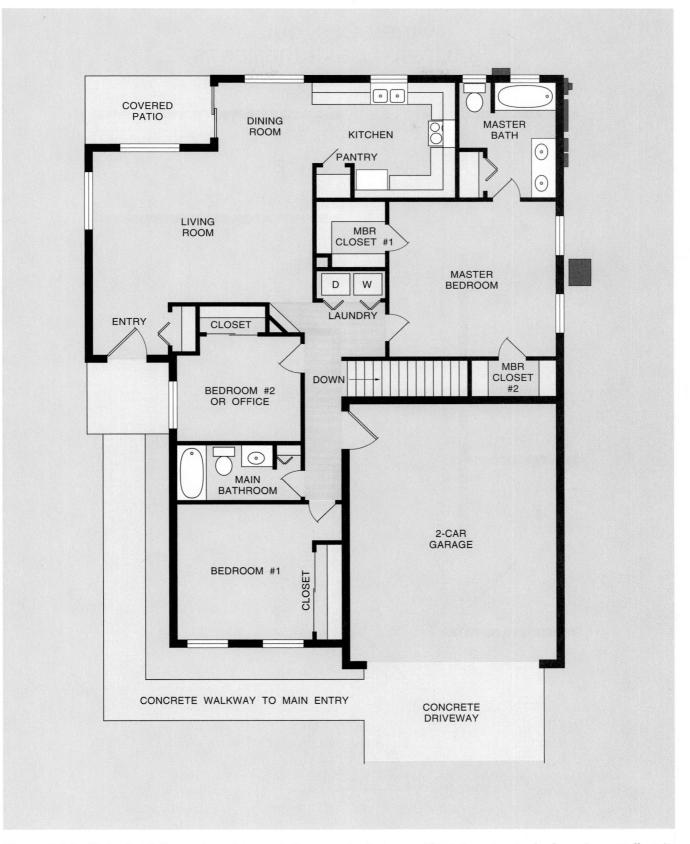

Figure 3-34 Main level floor plan: Areas of the sample house subject to receptacle layout according to 210.52(H).

"WIREMAN'S GUIDE"
HALLWAY MEASUREMENTS

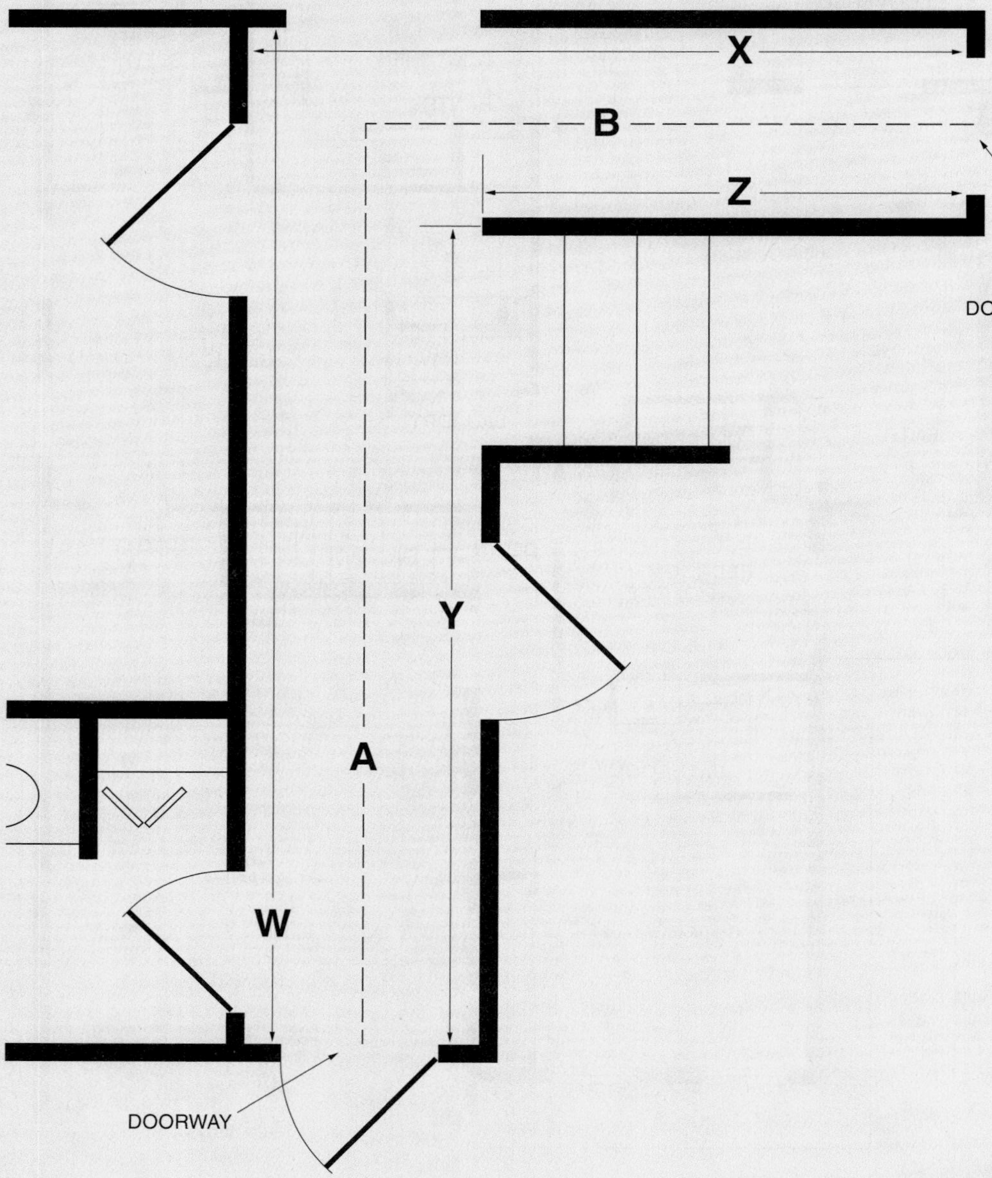

THE THREE WAYS TO MEASURE THE HALLWAY:

1. MEASURE THE WALL LINE WITH THE INSIDE CORNER. LENGTH = **W** + **X**.
   THIS WILL BE THE LONGEST DISTANCE.

2. MEASURE THE WALL LINE WITH THE OUTSIDE CORNER. LENGTH = **Y** + **Z**.
   THIS WILL BE THE SHORTEST DISTANCE.

3. MEASURE THE CENTERLINE OF THE HALLWAY (DOTTED LINE). LENGTH = **A** + **B**.
   THIS IS THE DISTANCE TO USE IN APPLYING *NEC® 210.52(H)*.

Figure 3-35

Figure 3-36 A receptacle outlet installed in an upstairs hallway in a dwelling.

3.3.3: Electric Range Receptacle Outlet

The electric range is also connected with a receptacle, a cord, and an attachment plug. Of course, this receptacle will not be of the same size and configuration as those for other additional kitchen receptacles because of the load requirements of the range. The receptacle for the range will be installed low (at 12 in. [300 mm] above the finished floor at the highest) on the wall directly behind the range.

3.3.4: Microwave Oven Receptacle Outlet

The microwave oven is installed immediately below the upper kitchen cabinet above the electric range. Power is supplied by a surface-mounted receptacle outlet installed in the upper cabinet. The supply cord for the microwave runs from below the cabinet, through a hole drilled in the bottom of the upper cabinet, to the receptacle.

3.3.5: Electric Clothes Dryer Receptacle Outlet

The electric clothes dryer is installed on the left side of the clothes-washing machine. The dryer is also connected with a receptacle, a cord, and an attachment plug, but as with the range, the receptacle must be of a specific size and plug configuration. The dryer receptacle box is mounted at 48 in. (1.2 m) above the finished floor to allow for ease of disconnection of the dryer, as necessary.

3.3.6: Garage Door Opener Receptacle Outlet

The receptacle outlet for the garage door opener is installed in the garage ceiling. The opener comes with a cord and an attachment plug, which also serve as the disconnecting means for the motor. Care must be taken in the location of the receptacle so that it is far enough back in the garage to allow travel of the door when it is opened.

3.3.7: Sump Pump Receptacle Outlet

The sump pump is receptacle-, attachment cord-and-plug connected. In the sample house, the sump pump is installed in the basement close to the east foundation wall. The receptacle is mounted into a box that is mounted on the surface of the foundation wall.

3.4: REQUIRED GROUND-FAULT CIRCUIT-INTERRUPTER PROTECTION

The use of electrical equipment or appliances can present a danger of electrocution. If an appliance is damaged or worn, some of the current may escape from the circuit. Leakage of current in this manner is called a *ground fault* and is illustrated in Figure 3-37. *Fault* is the word used in the electrical industry to define a condition in which current is able to find a path to ground other than the path it is supposed to take on the grounded circuit conductor. Obviously, this aberration may be very dangerous to anyone using the appliance.

A special device known as a *ground-fault circuit-interrupter* (GFCI) can detect the loss of current in the circuit caused by a ground fault of approximately 4 milliamperes (.004 ampere) or more. When a current loss is detected, the GFCI device immediately opens the circuit. The actual ground fault may be caused by long runs of wire. All types of insulation leak a small amount of current (there is no perfect insulator), and small amounts of current leakage over a long run may cause a loss of more than .004 ampere. Pinched insulation, insulation that was damaged at installation, or wires damaged from screws or nails over the life of the building can also cause a current leakage. The ground fault may also be caused by current running through an individual, from a damaged appliance, for example, to ground, thereby electrocuting the person. If the circuit is GFCI protected, the GFCI device will open the circuit within $\frac{1}{40}$ of a second after detection of the ground fault.

GFCI protection can be provided by either of two different devices. The protection can be built into the circuiting of a circuit breaker and installed within a circuit breaker panel. Such a device protects the entire circuit against ground faults. The second type of GFCI protection comes in the form of a receptacle with GFCI protection built in. When a GFCI receptacle is connected to a circuit, not only the GFCI receptacle itself but also all of the outlets on the load side of the GFCI receptacle can be protected against ground fault if desired. Figure 3-38 presents more detail on the connections to a GFCI receptacle and on the operation of the GFCI device.

Several receptacle locations in the sample house are required to have GFCI protection according to *210.8* of the Code, as seen in Figure 3-39. The Code requires GFCI protection virtually everywhere in a dwelling where electrical equipment can encounter a low-impedance ground path. This rule is not universal; the clothes washer does not have to be GFCI protected, for example. Following is a list of those locations in a dwelling that are required to have GFCI protection for general use receptacle outlets.

- Any receptacle outlet in a bathroom
- Any receptacle in a garage
- Any receptacle outlet installed outdoors
- Any receptacle installed in a crawl space
- Any receptacle installed in an unfinished basement
- Any receptacle installed above the countertop in kitchens, including any receptacles installed on the sides of island or peninsula countertops
- Any receptacle outlet located within 6 ft (1.8 m) of a wet bar sink

Outlets that are installed for specific appliances in the garage or basement—to supply a freezer, for example—are not required to be GFCI protected if the space is a dedicated space, if the appliance or equipment cannot be easily moved, and if there are no unprotected outlets installed other than the one necessary to supply the appliance or equipment. Other receptacle outlets in these areas that are not readily accessible, such as the receptacle in a garage ceiling to supply power to the garage door opener, or a receptacle installed to power snow-melting equipment on the roof, do not have to be GFCI protected.

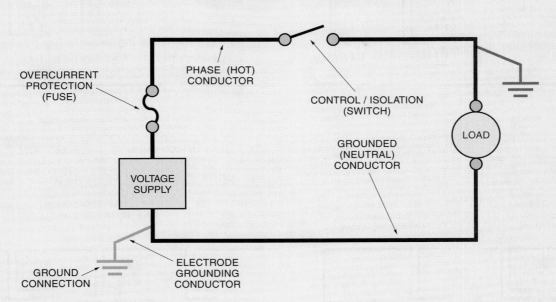

"WIREMAN'S GUIDE"
LINE SIDE AND LOAD SIDE
GROUND FAULTS

LINE SIDE GROUND FAULT: IF THERE IS A LINE SIDE GROUND FAULT ON THE CIRCUIT, WHEN THE SWITCH IS CLOSED THERE WILL BE A VERY LARGE CURRENT FLOW. THERE IS NO LOAD ON THE CIRCUIT (EXCEPT THE IMPEDANCE OF THE WIRE AND THE FAULT) TO CONTROL THE CURRENT FLOW, BUT THE OVERCURRENT PROTECTIVE DEVICE WILL OPEN THE CIRCUIT.

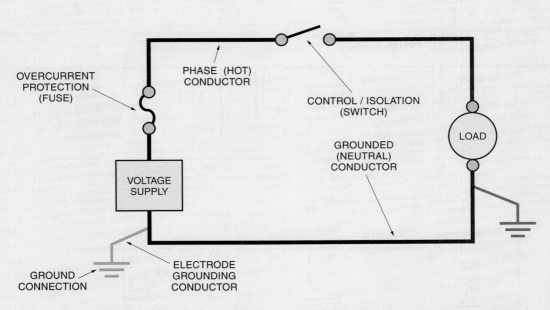

LOAD SIDE GROUND FAULT: IF THERE IS A GROUND FAULT ON THE LOAD SIDE OF THE CIRCUIT WHEN THE SWITCH IS CLOSED, THE CIRCUIT WILL APPEAR TO OPERATE PROPERLY. THE LOAD WILL CONTROL THE CURRENT FLOW TO LEVELS BELOW THE RATING OF THE OVERCURRENT DEVICE. IF THE FAULT IS CAUSED BY CURRENT TRAVELING THROUGH AN INDIVIDUAL, DEATH BY ELECTROCUTION MAY RESULT.

Figure 3-37

"WIREMAN'S GUIDE" GROUND-FAULT CIRCUIT-INTERRUPTERS

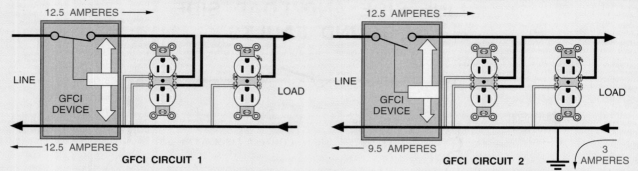

GFCI CIRCUIT 1   GFCI CIRCUIT 2

AN ELECTRICAL CIRCUIT THAT IS OPERATING PROPERLY WILL DRAW EXACTLY AS MUCH CURRENT FROM THE LINE AS IS RETURNED FROM THE LOAD. ALL OF THE CURRENT IS CONFINED TO THE ELECTRICAL CIRCUIT AS SHOWN IN CIRCUIT 1. IF THERE IS A GROUND FAULT, HOWEVER, AS SHOWN IN CIRCUIT 2, SOME OF THE CURRENT LEAKS FROM THE CIRCUIT AND THE CURRENT FLOW OUT DOES NOT EQUAL THE RETURN CURRENT FLOW. THE GFCI IS A DEVICE THAT MEASURES THE CURRENT DRAWN FROM THE LINE AND COMPARES IT TO THE CURRENT RETURNED FROM THE LOAD. IF THE TWO CURRENT FLOWS ARE NOT THE SAME (WITHIN 6 MILLIAMPERES), THE GFCI WILL IMMEDIATELY (WITHIN 40 MILLISECONDS) OPEN THE CIRCUIT. OBVIOUSLY, IF THE CURRENT IMBALANCE IS CAUSED BY CURRENT FLOWING THROUGH AN INDIVIDUAL, THIS DEVICE CAN BE A LIFESAVER.

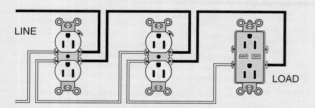

A GROUND-FAULT CIRCUIT-INTERRUPTER CAN BE INSTALLED AS A RECEPTACLE. THE RECEPTACLE HAS THE GFCI PROTECTION BUILT INTO THE DEVICE ITSELF AND AUTOMATICALLY PROTECTS ANY APPLIANCE THAT IS PLUGGED INTO THE DEVICE. THE DEVICE CAN BE INSTALLED ON A NORMAL GENERAL LIGHTING CIRCUIT WITH OTHER NONPROTECTED RECEPTACLES. GFCI-PROTECTED DEVICES ARE SHOWN IN BLUE.

GROUND-FAULT PROTECTION CAN BE OBTAINED FOR OUTLETS ON THE LOAD SIDE OF THE GFCI RECEPTACLE BY CONNECTING THE CIRCUIT TO THE LOAD TERMINALS OF THE RECEPTACLE. THE LOAD SIDE TERMINALS PROVIDE THE SAME PRO-TECTION AS CONFERRED BY THE GFCI RECEPTACLE ITSELF, AND ANY NUMBER OF RECEPTACLE CAN BE ATTACHED TO THE LOAD SIDE.

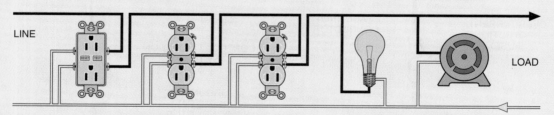

RECEPTACLES ARE NOT THE ONLY OUTLETS THAT CAN BE SUPPLIED BY A GFCI-PROTECTED CIRCUIT. LIGHTING LOADS AS WELL AS MOTOR LOADS CAN BE PROTECTED WITH GFCI DEVICES. SOMETIMES, AS WITH OUTLETS FOR FREEZERS IN A GARAGE OR BASEMENT, IT IS ADVISABLE NOT TO HAVE GFCI PROTECTION AS ALLOWED BY *210.8(C)* BECAUSE OF NUISANCE TRIPPING.

TO PROVIDE AN ENTIRE CIRCUIT WITH GFCI PROTECTIONS IT IS SOMETIMES PREFERABLE TO USE A GFCI CIRCUIT BREAKER THAT IS INSTALLED IN THE ELECTRICAL DISTRIBUTION PANEL.

Figure 3-38

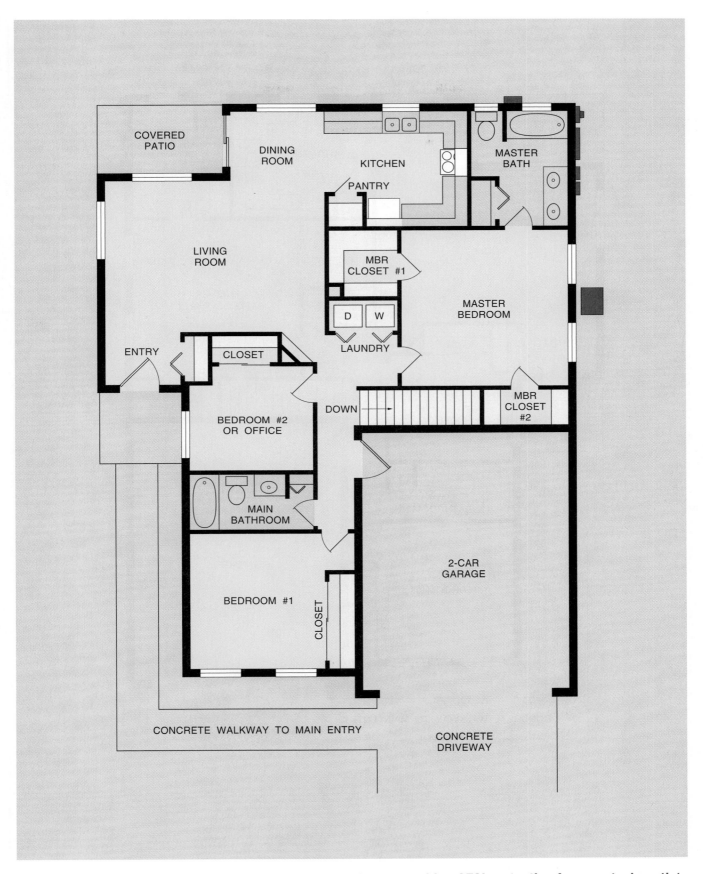

Figure 3-39   Main level floor plan: Areas of the sample house requiring GFCI protection for receptacle outlets according to *210.8*.

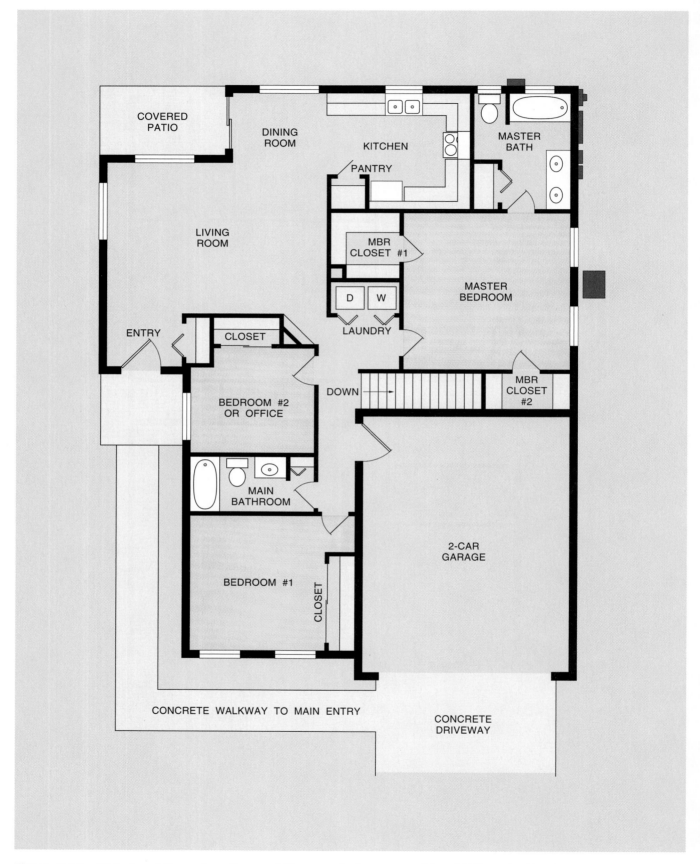

Figure 3-40 Main level floor plan: Areas of the sample house requiring AFCI protection for receptacle outlets according to *210.12*.

3.5: REQUIRED ARC-FAULT CIRCUIT-INTERRUPTER PROTECTION

KEY TERMS

Arc-fault circuit-interrupter (AFCI) A device capable of detecting arc faults that will shut off the flow of current to the circuit when an arc fault is detected.

NEC® *210.12* requires that all receptacle outlets in the bedrooms of dwelling units have **arc-fault circuit-interrupter (AFCI)** protection. Figure 3-40 shows those areas of the sample house that require AFCI protection. An arc-fault detector is a device that detects arcs in the electrical circuit. Arc faults are most often caused by loose terminations or splices. The arc created by a loose connection can generate intense heat, with the potential for extreme damage to the circuit that creates a fire hazard. Because no current is escaping from the electrical system, GFCI protectors do not detect arc faults. The circuit may apparently be operating normally, and the dwelling's occupants may not be aware of the potential danger.

As with GFCI protection, AFCI protection can be installed using an AFCI circuit breaker in the distribution panel. This device protects every outlet on the circuit, not just receptacle outlets, from the hazards of arcs. The device can also be installed as part of a receptacle outlet. Any outlet that is installed on the load side of the AFCI device can also be protected against arc faults by the receptacle. The same processes explored in Figure 3-38 for GFCI protectors also apply to AFCI protectors.

SUMMARY

Outlets provide access to the electrical system to power luminaires (fixtures), appliances, and other utilization equipment. Switch points provide access to the electrical system for control but not for utilization. The line side is the side of the load, device, cable, conductor, and so on that is located toward the voltage source. The load side is the side away from the voltage source. Many receptacles installed in a dwelling unit are required by the *NEC®*. There are also nonrequired receptacle outlets that may also be installed. In general, no location along a wall in a habitable room can be more than 6 ft (1.8 m) from a general use receptacle outlet. In kitchens, no location above the countertop can be more than 2 ft (600 mm) from a small appliance receptacle outlet. Receptacle outlets must be installed according to the Code in bathrooms, garages, basements, and many hallways, as well as outdoors and where mechanical equipment is located. A laundry area must have a receptacle outlet to power the clothes washer. Some of these outlets require GFCI and AFCI protection. Arc-fault protection is required on the circuits that supply dwelling unit bedrooms.

REVIEW

1. An outlet is _____ .
    a. a method for turning off (or out) several luminaires (fixtures) at the same time
    b. a place at which current can be controlled
    c. a place at which utilization equipment accesses the electrical system
    d. the short side opening of an LB

2. Most permanently installed luminaires (fixtures) in dwelling are _____ .
    a. cord-and-attachment plug–connected
    b. supported by the outlet box
    c. held in place by a device
    d. terminated at the switch point

3. The line side is _____.
   a. a marking on a conductor to identify it as neutral
   b. the righthand side of a switch
   c. the shortest side of a load
   d. the side toward the voltage source

4. The two main functions of a switch are to _____.
   a. control and isolate
   b. duck and cover
   c. control and cover
   d. isolate and cover

5. The Code requires receptacle outlets in all of the following areas except _____.
   a. the garage
   b. the crawl space
   c. the unfinished basement
   d. outdoors in the front of the dwelling

6. In a living room, no place on the wall line can be more than _____ from a general use receptacle outlet.
   a. 3 ft (900 mm)
   b. 6 ft (1.8 m)
   c. 9 ft (2.7 m)
   d. 12 ft (3.7 m)

7. Kitchen countertop receptacle outlets _____.
   a. can be placed face up on top of the counter
   b. must be at least 24 in. (600 mm) apart
   c. must be no more than 4 ft (1.2 m) apart
   d. must be at least 24 in. (600 mm) apart but not more than 4 ft (1.2 m) apart

8. The best location for installation of a bathroom receptacle outlet is _____.
   a. on the wall to the back of the basin where the mirror will be installed
   b. on a side wall adjacent to the basin
   c. on the side of the support base cabinet
   d. in the luminaire (fixture) above the basin

9. A receptacle outlet must be installed on any wall space _____ wide or greater.
   a. 12 in. (300 mm)
   b. 18 in. (450 mm)
   c. 24 in. (600 mm)
   d. 36 in. (900 mm)

"WIREMAN'S GUIDE" REVIEW

1. Which of the five major components of an electrical circuit can be eliminated without affecting the operation of the circuit? Explain why this is possible, and describe the potential problems associated with such alterations in the circuit.

2. Where in the *NEC*® are found the requirements for location of receptacle outlets?

3. Where in the *NEC*® are found the requirements for location of luminaires (fixtures)?

4. Where in the *NEC*® are found the requirements for switch points?

5. Explain why it is important to have more than one switch point for some luminaires (fixtures). Are multiple switch points required for any luminaires (fixtures) by the *NEC*®?

CHAPTER 4

Switches, Switch Points, and Lighting Outlets

OBJECTIVES

On completion of this chapter, the student will be able to:

☑ Name the switching system components.

☑ Explain the function and terminals of single-pole, 3-way, 4-way, and double-pole switches.

☑ Design single-pole, 3-way, 4-way, and double-pole switching systems.

☑ Describe poles and throws in relation to switches.

☑ Identify those locations in a dwelling where lighting outlets are required.

☑ Identify those locations in a dwelling where a switch control for luminaires (fixtures) is required.

INTRODUCTION

This chapter is about switches, switch points, and switching systems. It also includes those Code requirements for placement of luminaires (fixtures) and the requirements for installing switches to control the luminaires (fixtures) in a dwelling. The switching system in the sample house is representative of the most difficult controls that a residential electrician will encounter when wiring the average dwelling.

4.1: SWITCHES AND SWITCHING SYSTEMS

Switches are control and isolation devices. The access to the electrical system for a switch is called a switch point. There are four types of switches in the typical dwelling unit: single-pole, 3-way, 4-way, and double-pole.

The switches discussed here are switches intended for the control of the electrical system, specifically luminaires (fixtures), but, in some cases, receptacle outlets also, in the dwelling unit. The disconnect switches for the service are considered with the service installation in Chapter 13 of this book. The actual installation of the switches is covered in Chapter 16, in which "trimming" of the sample house is discussed.

4.1.1: Single-Pole Switch

KEY TERMS

Single-pole switch A switch containing terminals for only one phase conductor and one switch leg conductor.

Yoke Also called a *strap*. The metal or nonmetallic assembly that contains one or more devices, such as switches or receptacles. The device is attached to the yoke, or strap, which is screwed to the device box. The yoke or strap holds the device in place.

Toggle handle The switch handle of a standard single-pole, 3-way, 4-way, or double-pole (2-pole) switch that is the means for actuating the switch from an on to an off position.

except that now the phase conductor (power) is terminated to the common terminal of the 3-way switch in the lefthand box. The white wire and the red wire of the 3-wire cable that runs between the switch boxes are terminated to the traveler terminals of both switches. The common terminal of the right-hand 3-way switch is connected to the black wire of the 3-wire cable that contains the travelers, and this conductor is spliced to the switch leg phase conductor (black) in the 2-wire cable that is the switch leg for the lamp. In this case, the installing electrician ran the power into the lefthand 3-way switch and brought the switch leg all the way back from the righthand 3-way switch. As with the wiring diagram in Figure 4-6 (B), the grounded circuit conductor is spliced in the lefthand box to the white conductor in the 2-wire switch leg cable.

NEC® 200.7 requires that a white or natural gray conductor, or a conductor with three white stripes along its entire length, be used as the grounded circuit conductor only. However, *200.7(C)(2)* allows the use of a white or natural gray conductor, or a conductor with three white stripes along its entire length, to serve as a switch loop for single-pole, 3-way, and 4-way switching systems if it is part of a cable assembly. The white conductor must be used for wiring between switches (as travelers), or as the supply conductor to the switch (phase conductor), but not as the switch leg from the switch to the switched outlet. When a white conductor is used as a switch loop conductor, it must be re-identified at its terminals and at other places where it is accessible or visible to keep it from being confused with a grounded conductor. This re-identification can be achieved by painting or affixing colored tape to the wire or by other permanent means.

4.1.3: Double-Pole Switch

> **KEY TERMS**
>
> **Double-pole switch**   A switch that is capable of controlling the current from two separate circuits at the same time. The switch has terminals for two separate phase conductors and two separate switch leg conductors and resembles two single-pole switches connected together.

A **double-pole switch**—more properly, a double-pole, single-throw (DPST) switch—is a switch much

like two single-pole switches connected together. A DPST switch is used when the switch must control two separate circuits at the same time. The double-pole switch is also called a *2-pole switch*.

Figure 4-7 shows the configuration of a DPST switch. Note that there are line and load terminals in both the lefthand and righthand sides. When the toggle handle is moved from the down position to the up position, the switch closes the contacts on both sides at the same time. Moving the handle down likewise opens both sides at the same time. Because the double-pole switch operates as two single-pole switches connected together, the switch has definite on and off positions.

Figure 4-8 is a schematic diagram for a 2-pole switch. In this diagram, the circuit voltage is 120 volts to ground on both of the conductors and 240 volts between the conductors (rather than 120 volts as in Figure 4-5). The load is an electric heater that operates at 240 volts. The switch is located on both sides of the load. One half of the 2-pole switch opens or closes the lefthand 120-volt supply, and one half opens or closes the righthand 120-volt supply. The two halves of the switch are shown connected together by a dotted line, to show that the switch is a 2-pole switch and that both halves of the switch operate in unison.

4.1.4: Four-Way Switch

> **KEY TERMS**
>
> **Four-way switch**   A type of switch used in conjunction with two 3-way switches to allow for control of a switched outlet or outlets from more than two switching locations. Control can be obtained by adding as many 4-way switching locations as desired, but the system must include a 3-way switch at either end.

Figure 4-9 (page 110) shows a **four-way switch**—a double-pole, double-throw (DPDT) switch—in front, side, and back views. Initially, the 4-way switch looks very similar to the double-pole switch because they both have two terminals on each side plus an equipment grounding conductor terminal. However, the 4-way switch does not have open and closed positions, as do the single-pole and double-pole switches; rather, like the 3-way switch, it alternates between allowing current flow on one traveler

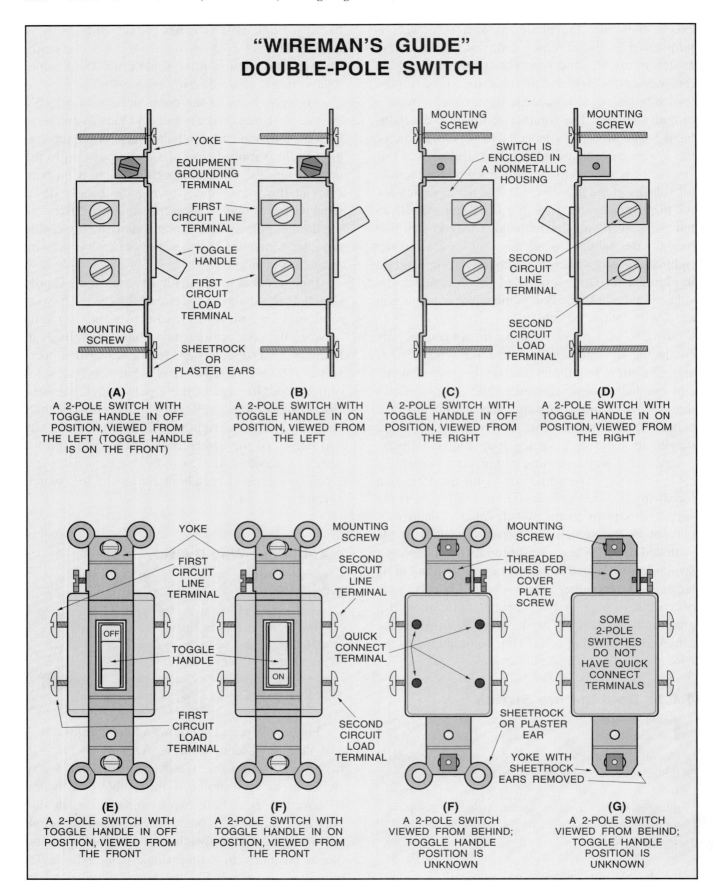

"WIREMAN'S GUIDE"
DOUBLE-POLE SWITCH

(A)
A 2-POLE SWITCH WITH TOGGLE HANDLE IN OFF POSITION, VIEWED FROM THE LEFT (TOGGLE HANDLE IS ON THE FRONT)

(B)
A 2-POLE SWITCH WITH TOGGLE HANDLE IN ON POSITION, VIEWED FROM THE LEFT

(C)
A 2-POLE SWITCH WITH TOGGLE HANDLE IN OFF POSITION, VIEWED FROM THE RIGHT

(D)
A 2-POLE SWITCH WITH TOGGLE HANDLE IN ON POSITION, VIEWED FROM THE RIGHT

(E)
A 2-POLE SWITCH WITH TOGGLE HANDLE IN OFF POSITION, VIEWED FROM THE FRONT

(F)
A 2-POLE SWITCH WITH TOGGLE HANDLE IN ON POSITION, VIEWED FROM THE FRONT

(F)
A 2-POLE SWITCH VIEWED FROM BEHIND; TOGGLE HANDLE POSITION IS UNKNOWN

(G)
A 2-POLE SWITCH VIEWED FROM BEHIND; TOGGLE HANDLE POSITION IS UNKNOWN

Figure 4-7

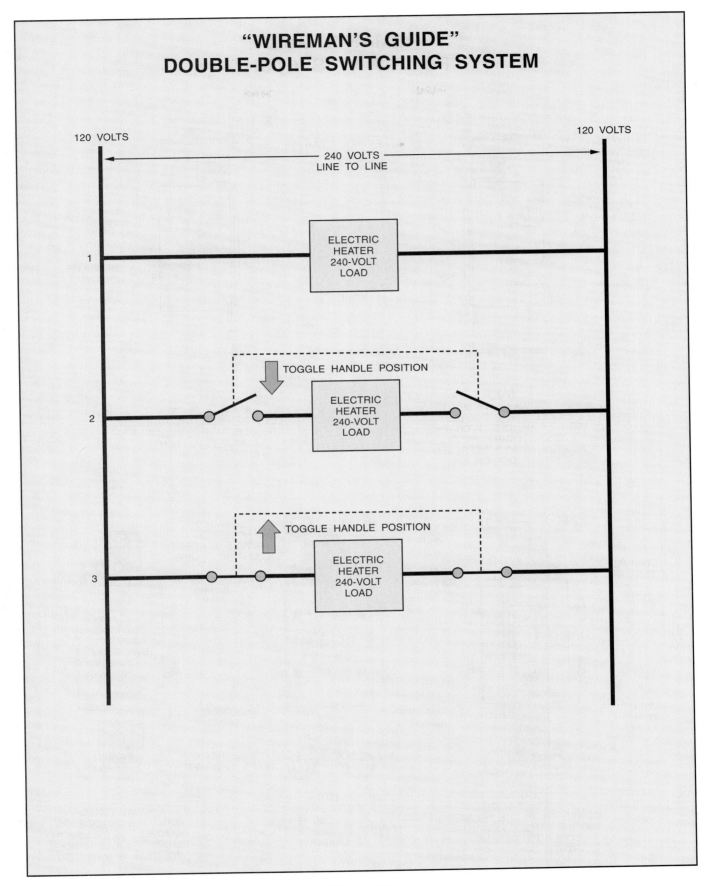

"WIREMAN'S GUIDE"
DOUBLE-POLE SWITCHING SYSTEM

Figure 4-8

"WIREMAN'S GUIDE" FOUR-WAY SWITCH

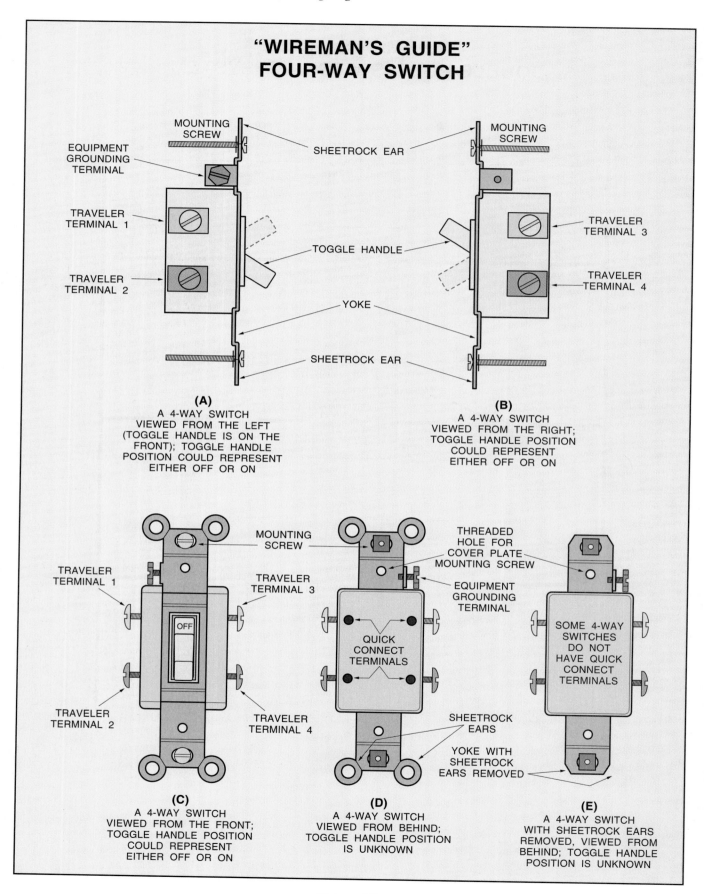

(A)
A 4-WAY SWITCH
VIEWED FROM THE LEFT
(TOGGLE HANDLE IS ON THE
FRONT); TOGGLE HANDLE
POSITION COULD REPRESENT
EITHER OFF OR ON

(B)
A 4-WAY SWITCH
VIEWED FROM THE RIGHT;
TOGGLE HANDLE POSITION
COULD REPRESENT
EITHER OFF OR ON

(C)
A 4-WAY SWITCH
VIEWED FROM THE FRONT;
TOGGLE HANDLE POSITION
COULD REPRESENT
EITHER OFF OR ON

(D)
A 4-WAY SWITCH
VIEWED FROM BEHIND;
TOGGLE HANDLE POSITION
IS UNKNOWN

(E)
A 4-WAY SWITCH
WITH SHEETROCK EARS
REMOVED, VIEWED FROM
BEHIND; TOGGLE HANDLE
POSITION IS UNKNOWN

Figure 4-9

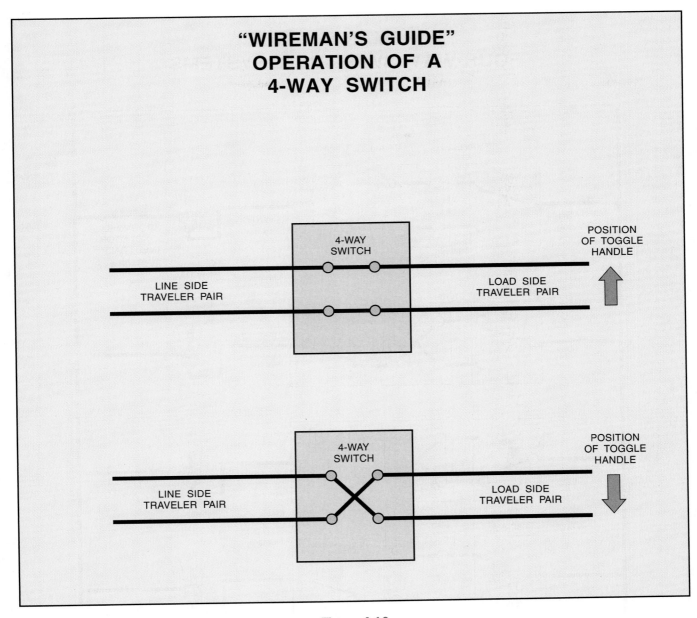

Figure 4-10

and then the other. Therefore, the 4-way switch does not have off and on positions stamped or printed on the toggle handle. Absence of such off and on markings is the easiest means of distinguishing the 4-way switch from the 2-pole switch.

Figure 4-10 shows the operation of a 4-way switch when the toggle handle is flipped. In one position, the switch directs the current from the incoming line traveler conductor to the adjacent outgoing load traveler conductor. Operation is the same for both sides of the switch. When the toggle handle is in the other position, the current is directed from the incoming line traveler conductor to the opposite side outgoing load traveler conductor.

The crossing and uncrossing function of the traveler conductors allows the 4-way switch to work, in conjunction with two 3-way switches, as another independent control of a switched circuit. This switching system is commonly referred to as a *4-way switching system*. Figure 4-11 shows the 4-way switching system in a schematic diagram. The 4-way switching system requires 3-way switches *at both ends* in order to operate properly. Careful study of the figure will show the independent operation

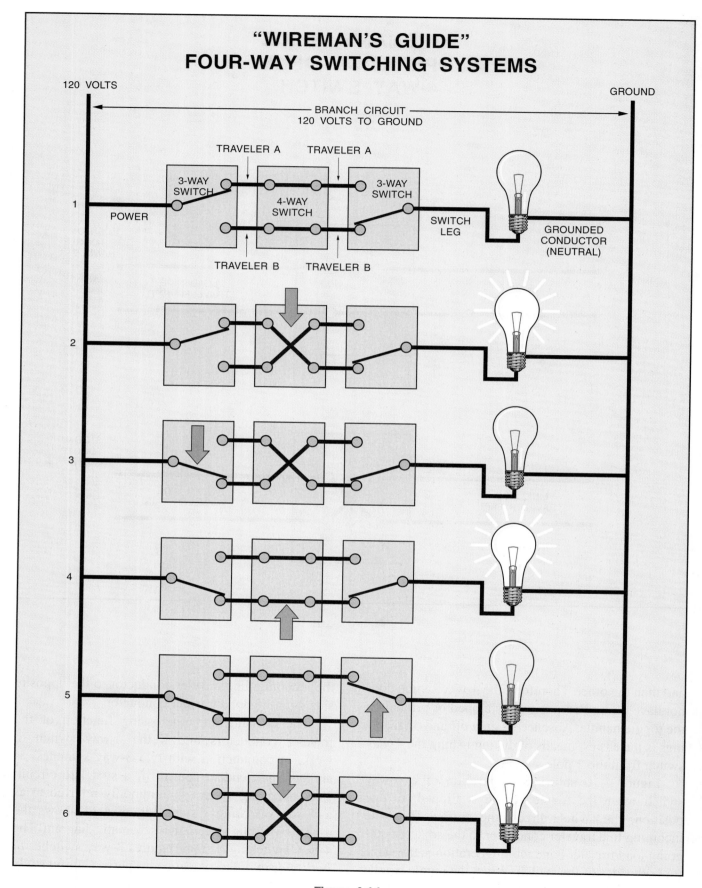

Figure 4-11

of the 4-way switch when it is inserted between two 3-way switches. There is no limit to the number of 4-way switches that can be installed between two 3-way switches, but each 4-way presents another location for independent control of the switched outlet.

As with 3-way switching systems, 4-way switching systems must have 3-wire cable installed between the switches. In fact, the 4-way switching system is no different from the 3-way switching system except that there is a 4-way (DPDT) switch inserted between two 3-way (SPDT) switches. Figure 4-12 shows three of the ways in which a 4-way switching system can be installed. When these systems are compared with the systems in Figure 4-6 for the 3-way switching systems, they are seen to be identical except for the inclusion of a 4-way switch. There are wiring configurations for a 4-way switching system other than the three shown in Figure 4-12.

4.2: REQUIRED LIGHTING OUTLETS AND SWITCH POINTS

The *NEC®* details specific locations in dwellings that must have switch-controlled lighting outlets. *NEC® 210-70(A)* lists three general groupings for these locations. Other than to specify that luminaires (fixtures) must be rated for the location in which they are installed, such as in wet locations, the Code does not require any particular type of luminaire (fixture). In some locations, however, certain restrictions are placed on the clearances between a particular type of luminaire (fixture) and the surroundings, such as in clothes closets or bathtub enclosures.

4.2.1: Habitable Room Lighting Outlets

> **KEY TERMS**
>
> **Habitable** Used for normal living functions such as eating, sleeping, and general living, but excluding unfinished portions of a dwelling, bathrooms, hallways, and closets.

A **habitable** room can be defined, for the purpose of determining the locations of required lighting outlets and switch points, as a room that is used for general day-to-day living. Habitable rooms include but are not limited to bedrooms, kitchens, living rooms, studies, libraries, sunrooms, family

rooms, and other rooms with finished walls, flooring, and ceilings. Habitable space does not include unfinished basements, garages, hallways, stairways, and outdoor areas such as porches or patios. Figure 4-13 shows those areas of the sample house considered habitable rooms by the *NEC®*. *NEC® 210.70(A)(1)* requires that at least one lighting outlet be installed in each habitable room and in all bathrooms.

> **KEY TERMS**
>
> **General use snap switch** (See *NEC® Article 100* [Switch — General Use Snap]): The term used by the *NEC®* to describe a switch used for single-pole, double-pole (2-pole), 3-way, or 4-way outlet control in a dwelling or other structure. General use snap switches come in various grades of quality, such as standard grade, specification grade, and hospital grade.

Lighting outlets can be installed in the ceiling or walls of the habitable room and may be of virtually any color, size, design, style, or brightness. The one feature that all the lighting outlets in a dwelling have in common is that they must be switch-controlled in order to satisfy the requirements of *210.70(A)(1)*. The standard type of switch that is installed in dwelling units is called a **general use snap switch** by the *NEC®*. *Article 404* requires that all switches be readily accessible and that the grip or toggle handle be no more than 6 ft 7 in. (2 m) above the floor. The most common height for installing switches for room illumination in dwellings is 48 in. (1.2 m) to either the top or the bottom of the device box. As shown in Figures 4-14 and 4-15 (page 116 and page 117), the switch must be readily accessible, and because the intent of the Code is to allow illumination of a room from the entry, it is also necessary to place the switch at a convenient location, usually just inside the room that it serves.

> **KEY TERMS**
>
> **Automatic** (See *NEC® Article 100*): Capable of operating without direct human action through a controller. Automatically operated equipment is inherently dangerous to maintain because of the possibility of its suddenly starting at any time.

There are two exceptions to *210.70(A)(1)*. The first exception states that in any habitable rooms other

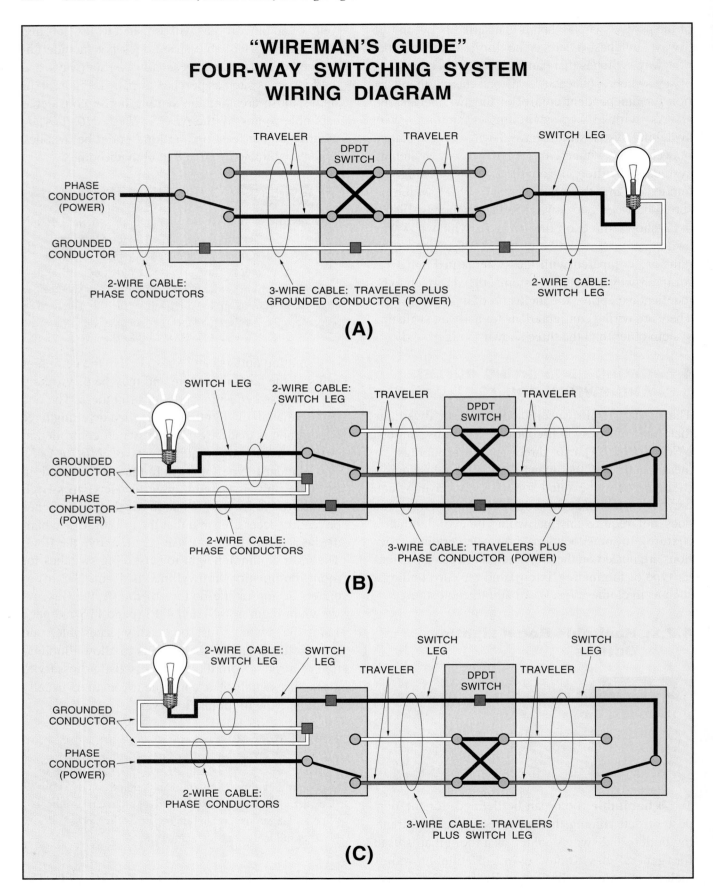

"WIREMAN'S GUIDE"
FOUR-WAY SWITCHING SYSTEM
WIRING DIAGRAM

Figure 4-12

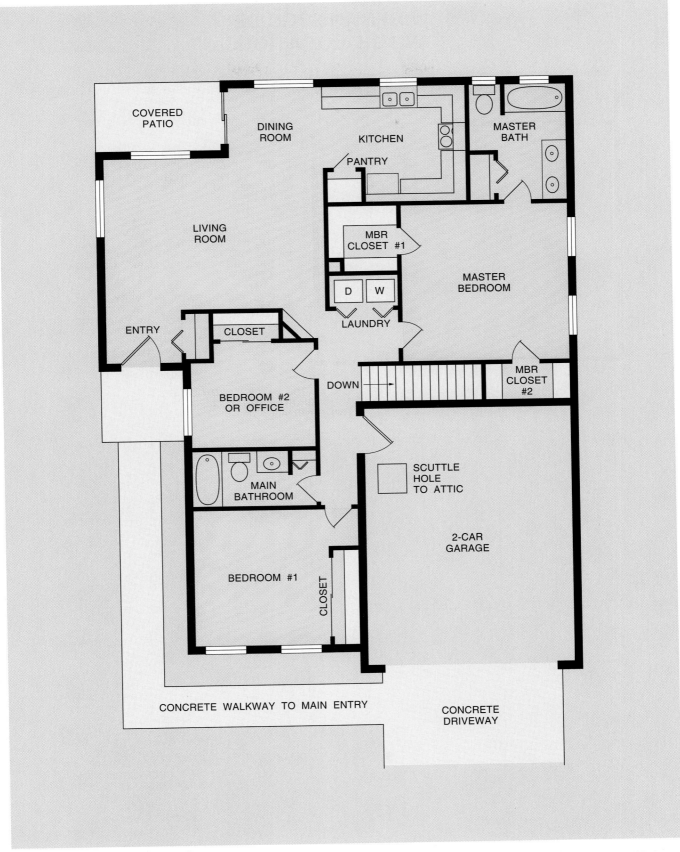

Figure 4-13 Main level floor plan: Areas of the sample house requiring lighting outlets according to 210.70(A)(1).

"WIREMAN'S GUIDE" SWITCH LOCATIONS

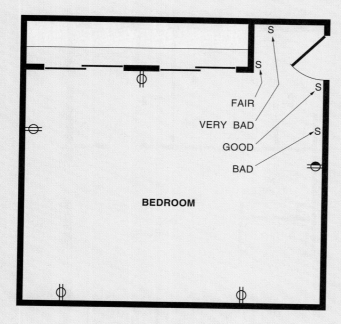

BEDROOM

GOOD: SWITCH IS MOUNTED CLOSE TO THE DOOR AND WHERE SWITCHES ARE TRADITIONALLY LOCATED. THE SWITCH IS EASY TO REACH AND EASY TO FIND IN THE DARK WITHOUT HAVING TO ENTER THE ROOM.

FAIR: SWITCH IS MOUNTED CLOSE TO THE DOOR. THE SWITCH IS IN AN UNUSUAL LOCATION. THE SWITCH IS RELATIVELY EASY TO REACH, BUT PARTIAL ENTRY INTO A DARK ROOM WILL PROBABLY BE NECESSARY.

BAD: ALTHOUGH THE SWITCH IS APPROXIMATELY WHERE SWITCHES ARE USUALLY LOCATED, IT IS TOO FAR DOWN THE WALL TO REACH EASILY. SWITCH IS HARD TO FIND AND WILL REQUIRE ENTRY INTO A DARK ROOM IN ORDER TO OPERATE.

VERY BAD: THE SWITCH IS VERY POORLY LOCATED. THE SWITCH IS LOCATED FOR THE OPPOSITE DOOR SWING. THE SWITCH IS VERY DIFFICULT TO LOCATE AND DIFFICULT TO REACH. ACCESS AND OPERATION WILL REQUIRE COMPLETE ENTRY INTO A DARK ROOM AND THE PARTIAL CLOSING OF THE DOOR.

ALL WALL SWITCHES USED FOR THE CONTROL OF LIGHTING OUTLETS SHOULD BE AT THE SAME HEIGHT THROUGHOUT THE DWELLING. THIS MAKES IT EASIER TO LOCATE THE SWITCH IN A DARK ROOM.

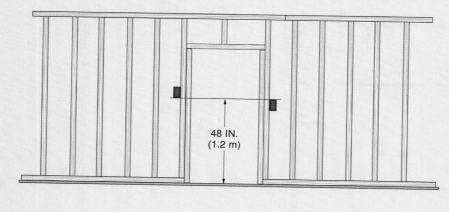

48 IN. (1.2 m)

SWITCH BOX HEIGHTS

THE *NEC*® HAS LITTLE TO SAY ABOUT THE MOUNTING HEIGHTS OF SWITCHES. SWITCHES MUST BE READILY ACCESSIBLE AND MOUNTED NO MORE THAN 6 FT 7 IN. (2 m) TO THE CENTER OF THE HANDLE THROW FROM THE FINISHED FLOOR.

A SWITCH MOUNTING HEIGHT THAT IS OFTEN USED IS EITHER 48 IN. (1.2 m) TO THE TOP OF THE BOX OR 48 IN. (1.2 m) TO THE BOTTOM OF THE BOX. SHEETROCK IS SOLD IN STANDARD 4 FT X 8 FT (1.2 X 2.5 m) OR 4 FT X 12 FT (1.2 X 3.7 m) SIZES, AND IS USUALLY MOUNTED WITH THE LONG SIDE OF THE SHEETROCK PARALLEL TO THE FLOOR. MOUNTING THE SWITCH AT 48 IN. (1.2 m) EITHER TO THE TOP OR TO THE BOTTOM LOCATES THE BOX ON A SHEETROCK SEAM. THIS MAKES FOR EASIER CUTS FOR THE SHEETROCK INSTALLERS AND AN EASILY REACHABLE STANDARD HEIGHT.

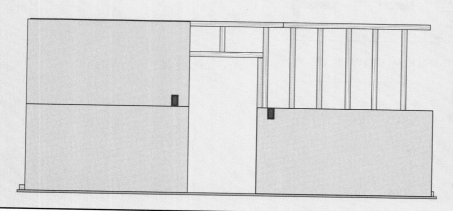

Figure 4-14

Figure 4-15 Switch should be easily reachable on entering a room.

than kitchens and bathrooms, a switch-controlled receptacle outlet can be used instead of the required lighting outlet. This is common practice, particularly in bedrooms and living rooms, because it saves the cost of an overhead luminaire (fixture). The second is for **automatic** control of lighting. Lighting outlets can be controlled by occupancy sensors, or motion detectors. Use of these automatic controls to turn lighting on and off in a habitable room is allowed if the sensor is in addition to the required wall switch and if the sensor is equipped with a manual control for both on and off positions. This Code requirement is based on the possibility of failure in the open posi-

tion, which would eliminate the wall-switched control of the luminaire (fixture).

4.2.2: Additional Locations for Lighting Outlets

Other locations in which a switch-controlled luminaire (fixture) must be installed in a residence are detailed in *210.70(A)(2)*. Figure 4-16 shows those areas of the sample house subject to *210.70(A)(2)*.

- **Hallways:** All hallways in a dwelling unit must have a switch-controlled luminaire (fixture) installed. There is no length definition for hallways, however, as there is for receptacle outlets (the 10-ft rule). Moreover, there is no requirement for the switching of the luminaire (fixture) from any particular place, or places, along the hallway.

- **Stairways:** All stairways in dwellings must have a switch-controlled luminaire (fixture) installed to illuminate the stairs. *NEC® 210.70(A)(2)* does require control of the stairway luminaire (fixture) at both the top and the bottom of the stairs; use of a 3-way switching system is required if the stairway contains six or more steps. See Figure 4-17.

- **Attached garages and detached garages with electric power**: All attached garages and all detached garages with electric power must have a switch-controlled lighting outlet installed. As with all switches, the switch control of the lighting outlet in the garage is required to be no more than 6 ft 7 in. (2 m) above the floor.

If an entry into the dwelling has grade-level access, it must have a switch-controlled lighting outlet on the exterior of the door, as shown in Figure 4-18 (page 120). There is no requirement to illuminate a walkway or sidewalk that is used to access the door, and there is no requirement to illuminate the street side of the car entrance doors in a garage, because the garage opening for motor vehicles is not considered an entry into the dwelling. Figure 4-19 (page 121) shows a luminaire (fixture) installed outdoors.

The one exception to *210.70(A)(2)* says that luminaires (fixtures) in hallways, stairways, and outdoors can be controlled automatically or remotely. As with the habitable rooms referred to in *210.70(A)(1)*, these lighting outlets can be controlled by automatic or sensory means.

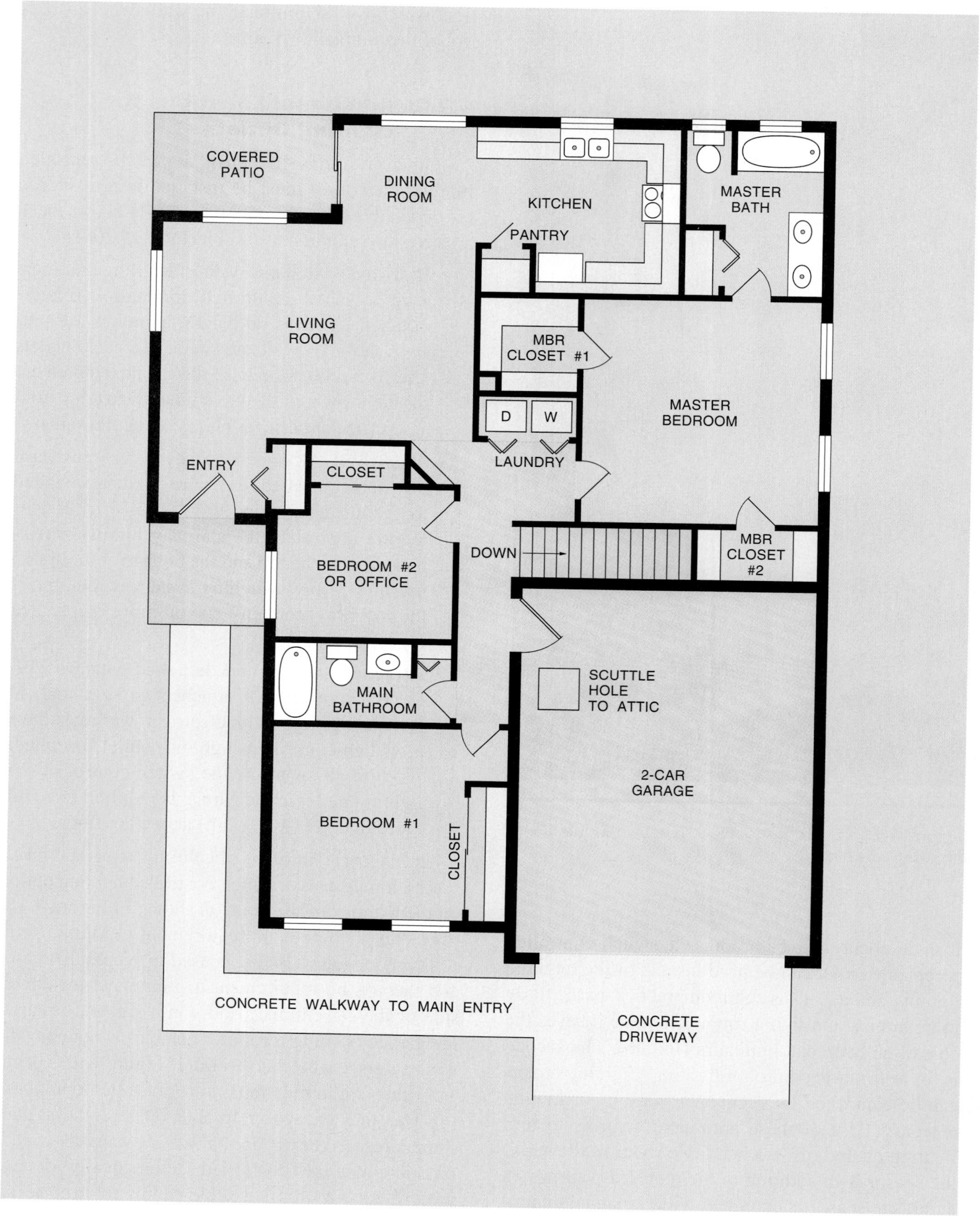

Figure 4-16 Main level floor plan: Areas of the sample house requiring lighting outlets according to 210.70(A)(2).

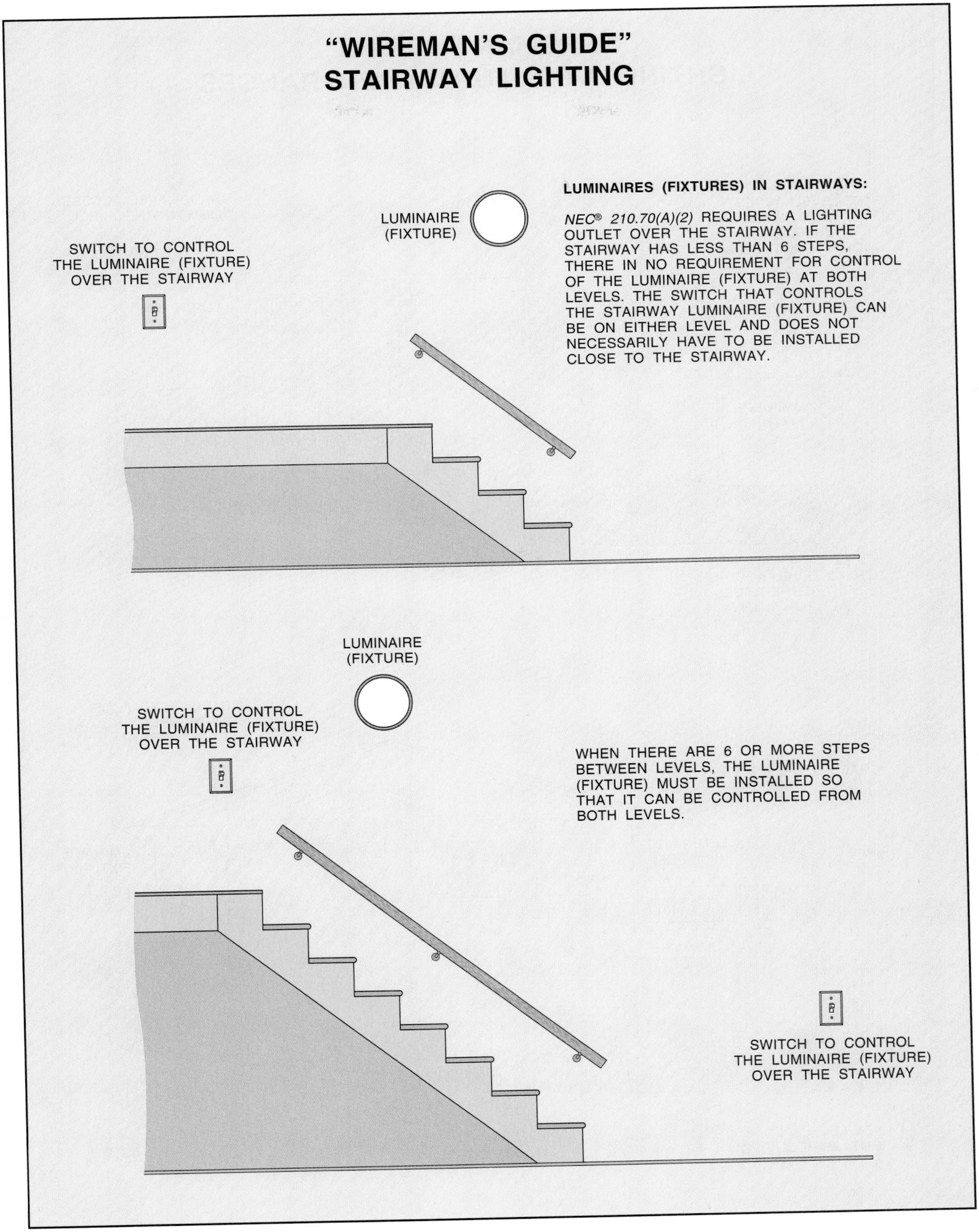

"WIREMAN'S GUIDE" STAIRWAY LIGHTING

LUMINAIRE
(FIXTURE)

SWITCH TO CONTROL
THE LUMINAIRE (FIXTURE)
OVER THE STAIRWAY

LUMINAIRES (FIXTURES) IN STAIRWAYS:

NEC® 210.70(A)(2) REQUIRES A LIGHTING
OUTLET OVER THE STAIRWAY. IF THE
STAIRWAY HAS LESS THAN 6 STEPS,
THERE IN NO REQUIREMENT FOR CONTROL
OF THE LUMINAIRE (FIXTURE) AT BOTH
LEVELS. THE SWITCH THAT CONTROLS
THE STAIRWAY LUMINAIRE (FIXTURE) CAN
BE ON EITHER LEVEL AND DOES NOT
NECESSARILY HAVE TO BE INSTALLED
CLOSE TO THE STAIRWAY.

LUMINAIRE
(FIXTURE)

SWITCH TO CONTROL
THE LUMINAIRE (FIXTURE)
OVER THE STAIRWAY

WHEN THERE ARE 6 OR MORE STEPS
BETWEEN LEVELS, THE LUMINAIRE
(FIXTURE) MUST BE INSTALLED SO
THAT IT CAN BE CONTROLLED FROM
BOTH LEVELS.

SWITCH TO CONTROL
THE LUMINAIRE (FIXTURE)
OVER THE STAIRWAY

Figure 4-17

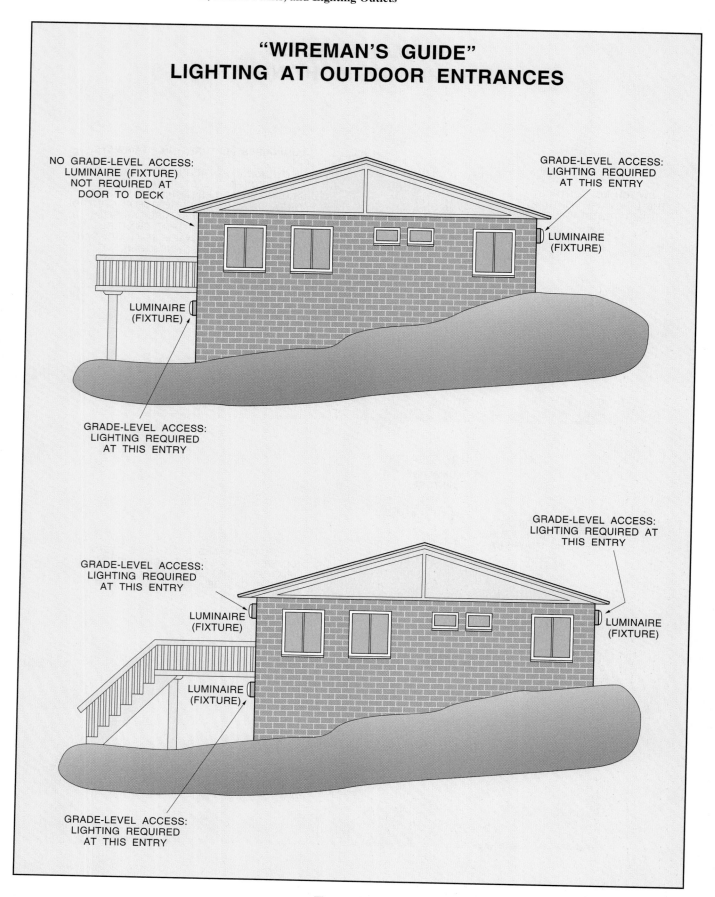

"WIREMAN'S GUIDE"
LIGHTING AT OUTDOOR ENTRANCES

NO GRADE-LEVEL ACCESS:
LUMINAIRE (FIXTURE)
NOT REQUIRED AT
DOOR TO DECK

GRADE-LEVEL ACCESS:
LIGHTING REQUIRED
AT THIS ENTRY

LUMINAIRE
(FIXTURE)

LUMINAIRE
(FIXTURE)

GRADE-LEVEL ACCESS:
LIGHTING REQUIRED
AT THIS ENTRY

GRADE-LEVEL ACCESS:
LIGHTING REQUIRED AT
THIS ENTRY

GRADE-LEVEL ACCESS:
LIGHTING REQUIRED
AT THIS ENTRY

LUMINAIRE
(FIXTURE)

LUMINAIRE
(FIXTURE)

LUMINAIRE
(FIXTURE)

GRADE-LEVEL ACCESS:
LIGHTING REQUIRED
AT THIS ENTRY

Figure 4-18

Figure 4-19   A typical outdoor luminaire (fixture).

4.2.3: Storage or Equipment Space Lighting Outlets

NEC® 210.70(A)(3) requires that a lighting outlet be installed at every unfinished and uninhabitable space in the dwelling, such as the attic or crawl space or an unfinished basement, that can be used for storage or that contains equipment that may require servicing or repair. Figure 4-20 shows those areas of the sample house subject to the requirements of *210.70(A)(3)*. The lighting outlet's switch control must be at the usual place of entry into the area. In addition to the required switched lighting outlet, other luminaires (fixtures) may be installed in the crawl space, or other storage area. These luminaires (fixtures) can provide additional illumination in the farthest reaches of the crawl space, but these lighting outlets do not have to be controlled with the same switch at the entry. A luminaire (fixture) in the basement of the sample house, for example, is not switch-controlled from the entry. Figure 4-21 shows an entry to a crawl space. A luminaire (fixture) is installed in the crawl space, and a switch to control the luminaire (fixture) is located just to the right of the entry door.

The last requirement of *210.70(A)* says that a lighting outlet must be installed close to any electrical or mechanical equipment that may require servicing or repair and that is located in an uninhabitable location in the dwelling. This requirement applies to basements and attic spaces if they enclose equipment that may need servicing or if these spaces can be used for storage. Almost any attic can be used for storage of some kind. For this reason, any dwelling unit crawl space or attic space must have a switch-controlled lighting outlet controlled by a switch point at the usual point of entry to the space; this requirement is widely enforced by the local AHJ. Basements are usually illuminated because a washer or some mechanical equipment, such as a water heater, is located there, but obviously they can also be used for storage.

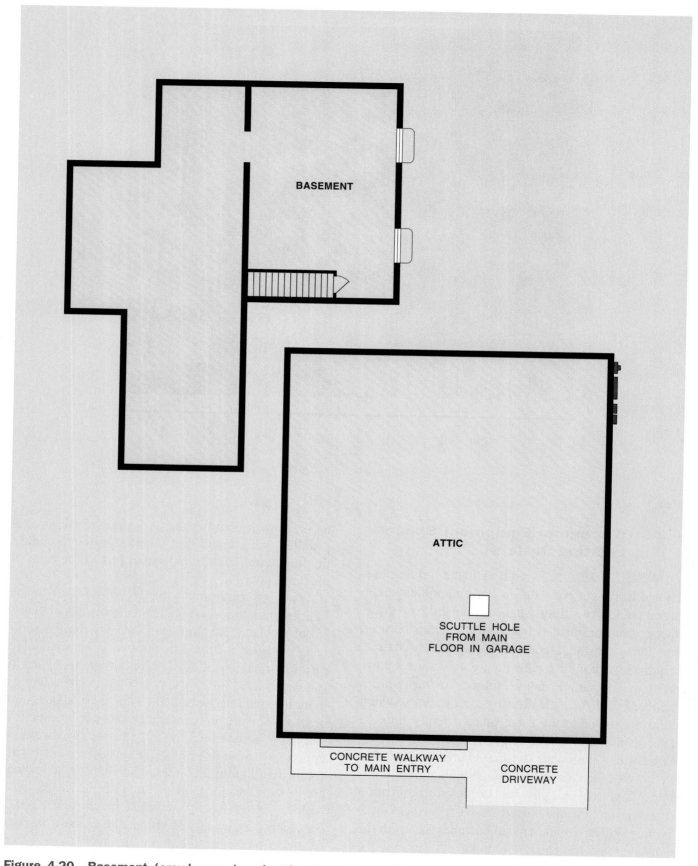

Figure 4-20   Basement (crawl space) and attic plans: Areas of the sample house requiring lighting outlets according to *210.70(A)(3)*.

Figure 4-21 The luminaire (fixture) inside of this crawl space is controlled by the switch to the right of the entry door.

4.3: LIGHTING OUTLETS AND SWITCH POINTS OTHER THAN REQUIRED

The sample house has lighting outlets other than those required by the *NEC*®. Figure 4-22 shows these additional lighting outlets and switch points. These outlets and switch points are installed to make the dwelling more salable or living more convenient. For example, the hallway requires one luminaire (fixture), controlled by just a single-pole switch system in the hallway. According to Construction Plan E-3, two luminaires (fixtures) controlled by a 4-way switching system instead of a single-pole system, are installed in the hallway. These outlets and switch points are not required but make the electrical sys-

tem more convenient to use. Figure 4-23 shows non-required lighting outlets installed in the basement and crawl space of the sample house.

4.4: EQUIPMENT OUTLETS

In addition to receptacle outlets, lighting outlets, and switch points, there are also outlets to service equipment fastened in place that do not connect with a receptacle, cord, and attachment plug. In the sample house, the furnace, located in the basement, and the air conditioner compressor, located on the north side of the house, are examples of this kind of equipment. The connection of the furnace and the air conditioner is covered in Chapter 18 of this book.

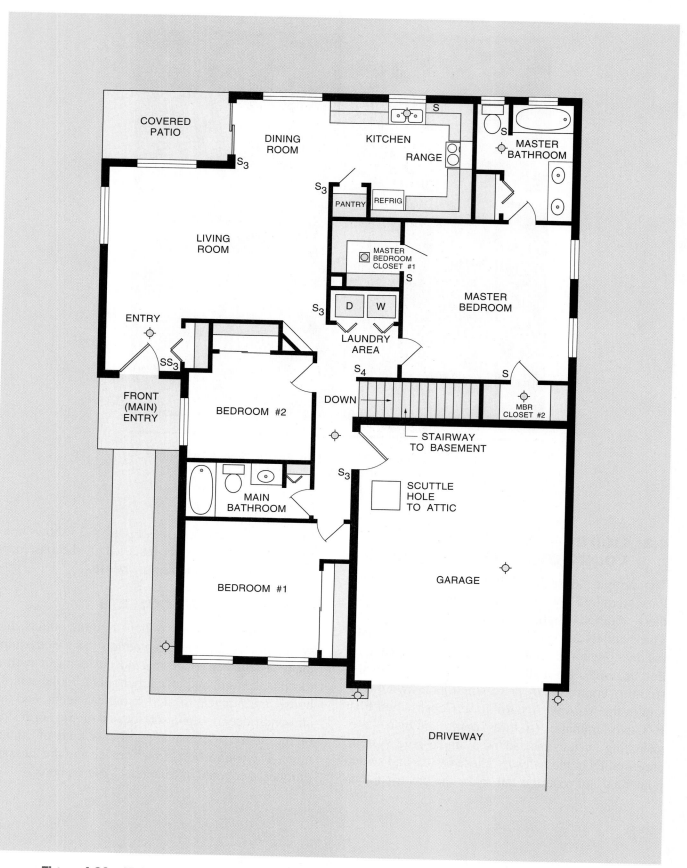

Figure 4-22 Main floor electrical plan: Lighting outlets and switch points other than required.

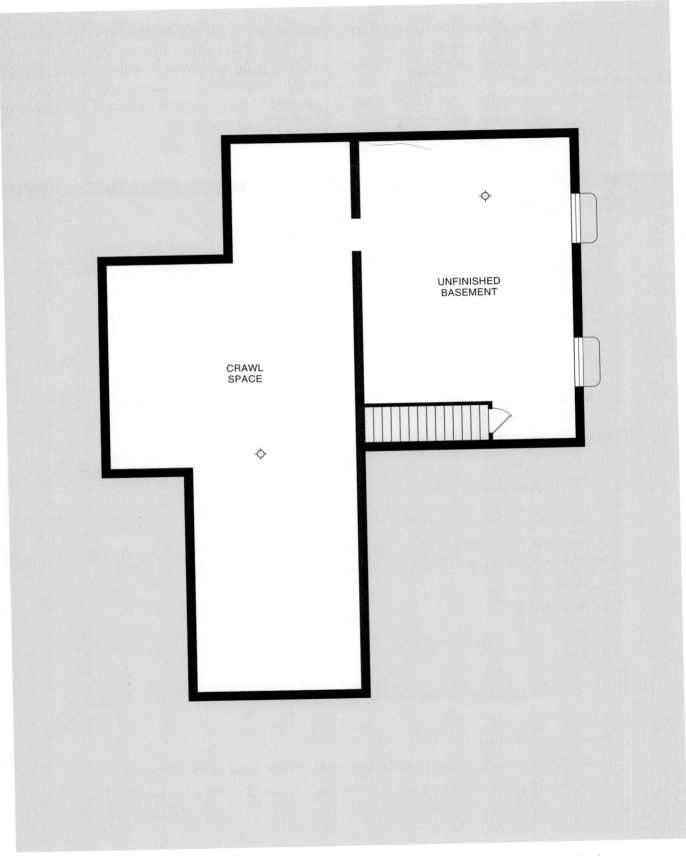

Figure 4-23 Basement and crawl space plans: Lighting outlets other than required.

SUMMARY

Switching systems can be designed to operate from any number of different locations. Single-pole switches are the simplest type of control system. There are also 3-way and 4-way switching systems for more complex control. The Code requires that for the purposes of illumination, habitable rooms, as well as other areas of a dwelling that occupants may use, be provided with a switched outlet at the point of entry to the area. Most switches in a dwelling are installed to control luminaires (fixtures), but they can also be used to isolate or control mechanical or other types of equipment.

REVIEW

1. Of the following paired terms, only the pair _____ and _____ are associated with 3-way switching systems.

    a. traveler, stationary

    b. common, separate

    c. stationary, separate

    d. traveler, common

2. The _____ and _____ types of switches have on and off positions marked on the toggle handle.

    a. single-pole, 4-way

    b. 3-way, 4-way

    c. single-pole, double-pole

    d. 3-way, double-pole

3. Among the following switch types, only the _____switch has a common terminal.

    a. single-pole

    b. double-pole

    c. 3-way

    d. 4-way

4. A _____ can replace a switch-controlled luminaire (fixture) in bedrooms.

    a. window

    b. switch-controlled wall receptacle

    c. hallway luminaire (fixture) located immediately outside the entry door to the bedroom

    d. nightlight plugged into a general use receptacle outlet without switch control

5. Stairways with _____ or more steps must have a 3-way or 4-way switch control of the required luminaire (fixture).

    a. 3

    b. 4

    c. 5

    d. 6

6. All of the following locations require switch-controlled illumination except _____ .

    a. an unfinished basement with only a furnace (space heating unit) installed

    b. the walkway to the front door of the dwelling

    c. an outdoor entry to the dwelling

    d. an attic that can be used for storage

7. A _____ switching system can control loads from two different locations only.

    a. single-pole

    b. double-pole

    c. 3-way

    d. 4-way

8. The _____ switch is a single-pole, double-throw switch.

    a. single-pole

    b. double-pole

    c. 3-way

    d. 4-way

9. The maximum allowable number of 4-way switches on a switching system is _____ .

    a. one

    b. one for each 3-way switch on the system

    c. one for each luminaire (fixture) being switched

    d. any number, so long as two 3-way switches are also installed

10. A detached garage requires a switch-controlled luminaire (fixture) if _____ .

    a. there is electricity run to the detached garage

    b. it is within 50 ft (15 m) from the dwelling

    c. it is located more than 50 ft (15 m) from the house

    d. the garage can be used for other purposes besides housing an automobile

"WIREMAN'S GUIDE" REVIEW

1. Draw a wiring plan for a 3-way switching system with the power run to the lighting outlet. Name and identify the color of the conductors including any that may need re-identification.

2. Discuss some of the considerations to be taken into account in locating the overhead luminaires (fixtures) in a residential kitchen.

3. Discuss some of the advantages of and drawbacks to switched receptacle outlets rather than overhead luminaires (fixtures).

4. Detail the location of an outdoor entrance to a dwelling that is not required to have a switched luminaire (fixture). Do not include the automobile door to the garage as an outdoor entrance to the dwelling.

5. Draw a wiring plan for a 4-way switching system with two 4-way switches, in which the phase conductor cable is run to one of the 4-way switch boxes and the switch leg cable is run to the other 4-way switch box. Name and identify the color of the conductors including any that may need re-identification.

6. What section of the *NEC*® details the requirements for switched luminaires (fixtures) in a dwelling?

Conductor Ampacity and Insulation

INTRODUCTION

This chapter is concerned with problems of current flow in a conductor. Considerations related to current flow include insulation type, conductor type, and allowable ampacity of the conductor. Another consideration is the effect of temperature on the ability of conductors to safely carry current over a long period. The material in this chapter provides a background for use in the study of overcurrent in Chapter 7 and wiring methods in Chapter 8 of this book.

5.1: TYPES OF CONDUCTORS

KEY TERMS

Conductor (See *NEC® Article 100*): A metallic bar or wire designed and intended to carry electrical current. It can be insulated, covered, or bare. Other components of an electrical circuit, such as metallic raceways, may also carry current in the event of a fault, but such components are not considered conductors because they are not intended to carry current during normal operations.

Impedance The total opposition to current flow in a circuit, consisting of resistance, inductive reactance, and capacitive reactance.

Current flow—the movement of electrons from one place to another in an electrical circuit—uses conductors as the pathway for flow. A **conductor** is a wire or bar (short for bus bar) that allows the electrons to move relatively easily because the material of the wire or bar has low electrical **impedance**. Bus bars are usually found within electrical switchgear, panelboards, and other electrical equipment that must carry a relatively large amount of current and are usually uninsulated. Residential electricians typically encounter bus bars used as the distribution system within a circuit-breaker panel. Circuit-breaker panels use bussing to distribute the electricity between the many branch circuits. Circuit breakers are connected to this bussing when they are installed, and the power is then routed through the circuit breaker to where it is required. A typical residential distribution panel is shown in Figure 5-1.

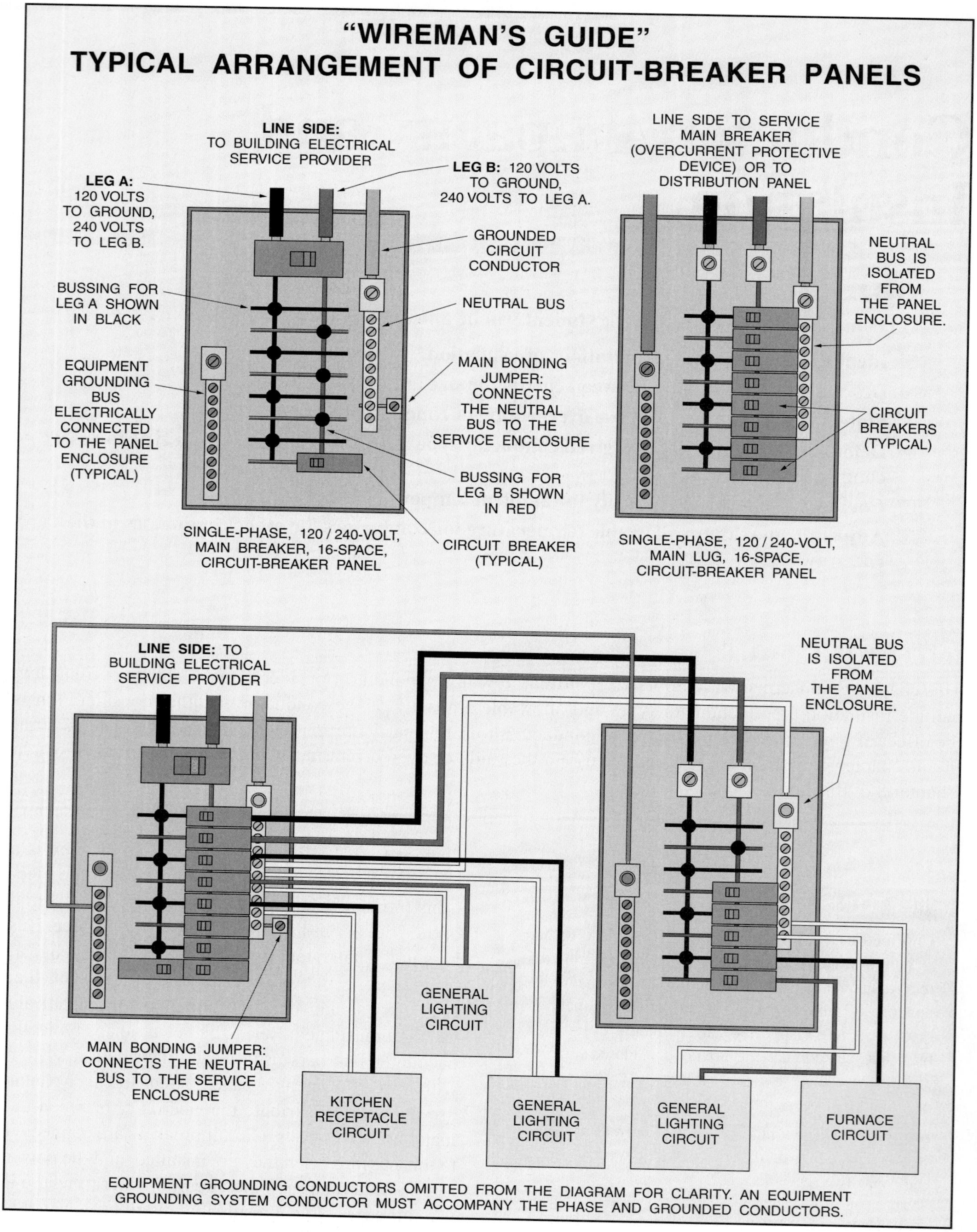

"WIREMAN'S GUIDE"
TYPICAL ARRANGEMENT OF CIRCUIT-BREAKER PANELS

LINE SIDE:
TO BUILDING ELECTRICAL
SERVICE PROVIDER

LEG A:
120 VOLTS
TO GROUND,
240 VOLTS
TO LEG B.

LEG B: 120 VOLTS
TO GROUND,
240 VOLTS TO LEG A.

GROUNDED
CIRCUIT
CONDUCTOR

BUSSING FOR
LEG A SHOWN
IN BLACK

NEUTRAL BUS

EQUIPMENT
GROUNDING
BUS
ELECTRICALLY
CONNECTED
TO THE PANEL
ENCLOSURE
(TYPICAL)

MAIN BONDING
JUMPER:
CONNECTS
THE NEUTRAL
BUS TO THE
SERVICE ENCLOSURE

BUSSING FOR
LEG B SHOWN
IN RED

SINGLE-PHASE, 120 / 240-VOLT,
MAIN BREAKER, 16-SPACE,
CIRCUIT-BREAKER PANEL

CIRCUIT BREAKER
(TYPICAL)

LINE SIDE TO SERVICE
MAIN BREAKER
(OVERCURRENT PROTECTIVE
DEVICE) OR TO
DISTRIBUTION PANEL

NEUTRAL
BUS IS
ISOLATED
FROM
THE PANEL
ENCLOSURE.

CIRCUIT
BREAKERS
(TYPICAL)

SINGLE-PHASE, 120 / 240-VOLT,
MAIN LUG, 16-SPACE,
CIRCUIT-BREAKER PANEL

LINE SIDE: TO
BUILDING ELECTRICAL
SERVICE PROVIDER

NEUTRAL BUS
IS ISOLATED
FROM
THE PANEL
ENCLOSURE.

MAIN BONDING JUMPER:
CONNECTS THE NEUTRAL
BUS TO THE SERVICE
ENCLOSURE

GENERAL
LIGHTING
CIRCUIT

KITCHEN
RECEPTACLE
CIRCUIT

GENERAL
LIGHTING
CIRCUIT

GENERAL
LIGHTING
CIRCUIT

FURNACE
CIRCUIT

EQUIPMENT GROUNDING CONDUCTORS OMITTED FROM THE DIAGRAM FOR CLARITY. AN EQUIPMENT
GROUNDING SYSTEM CONDUCTOR MUST ACCOMPANY THE PHASE AND GROUNDED CONDUCTORS.

Figure 5-1

Loose connection A flaw in a connection between two conductors, or between a conductor and a terminal, that allows for series arc faults. Tight connections prevent series arcs and the associated problem of generation of high levels of heat.

Wires are generally smaller and more flexible than bus bars and are referred to as *conductors.* *NEC® 310.2(B)* states that *conductors shall be constructed of aluminum, copper-clad aluminum, or copper unless otherwise specified.* Copper is the preferred conductor but has the disadvantage of being relatively heavy and relatively expensive. It is about 30% better at conducting current than aluminum. Aluminum is lightweight and relatively inexpensive but has been associated with problems at terminations in the circuit. A **loose connection** can cause an arc that can lead to fires. Aluminum terminations, if not installed properly, can become loose with use; therefore, in many areas of the country, the use of aluminum wire is restricted. The *NEC®* also says that if a conductor type is not specified, the use of copper conductors is to be assumed. The same principle applies in this book. If reference is not made to a particular conductor type, copper or aluminum, the use of copper conductors is to be assumed. If aluminum is to be considered, it is specifically identified.

The most widely used wires are termed *Conductors for General Wiring* by the *NEC®*, and these are the conductors examined most closely in this chapter. Other conductors are listed in the *NEC®*, however. Flexible cords and cables, such as extension cords and lamp cords, are covered in *400.* Fixture wires, like the ones that are internal to luminaires (fixtures), are covered in *402.* Elsewhere in the Code, ampacities are given for conductors other than conductors considered as wires for general wiring, but they are usually employed for specialized functions and are not usually found in dwellings.

5.2: TEMPERATURE LIMITATIONS

Insulated Describing a conductor covered with nonconductive material that is thick enough to be recognized by the *NEC®* as insulation. Insulation keeps the electrons confined to the conductor, preventing ground faults.

Insulation type There are many different types of insulations designed for specific locations and load types. Insulation type is usually designated by a letter code—for example, XHHW.

Wires are usually **insulated**—that is, covered with some type of material that (1) can easily bend without damage to the covering, (2) can survive many years of continuous high temperatures, in the range of 60°C (about 140°F), for 100 years or more, and (3) present a very high resistance to the current flow. Many different substances are used as insulation, many of them with specialized characteristics. Detailed information about the insulation for conductors for general wiring is available in *Table 310.13* of the *NEC®*. The **insulation type** is identified by a letter code. For example, type XHHW is listed in *Table 310.13* as moisture-resistant Thermoset insulation. As set forth in the Code, the use of certain types of conductor insulation may be required in cable assemblies.

Maximum operating temperature Each insulation type has a maximum operating temperature rating that should not be exceeded. If the maximum operating temperature of the conductor insulation is exceeded for a long period, the insulation can be damaged, usually by drying out and cracking, allowing electrons to escape the circuit.

One of the most important pieces of information available in *Table 310.13* is the conductor insulation's **maximum operating temperature**. The constant high operating temperature of conductors can cause severe problems in the electrical system. The heat causes the insulation materials to become brittle and to crack and break down over time, thus allowing electrons to escape from the circuiting. When electrons escape from the conductor system, fire is a very likely result. Restricting the maximum temperature to some level at which the insulation will not be damaged makes the entire electrical system more secure from fires caused by the thermal breakdown of the insulation. All conductors have a maximum allowable temperature rating that must not be exceeded if the electrical system is to remain functional over the life of the structure.

"WIREMAN'S GUIDE"
OPERATING TEMPERATURE OF CONDUCTORS

THE TOTAL OPERATING TEMPERATURE OF A CONDUCTOR IS THE SUM OF THE AMBIENT TEMPERATURE AND THE TEMPERATURE INCREASE FROM THE HEAT CREATED BY THE CURRENT FLOWING IN THE CIRCUIT. THE MAXIMUM ALLOWABLE OPERATING TEMPERATURE FOR CONDUCTOR INSULATION IS ESTABLISHED BY THE *NEC*® AFTER TESTING BY A RECOGNIZED TESTING LABORATORY.

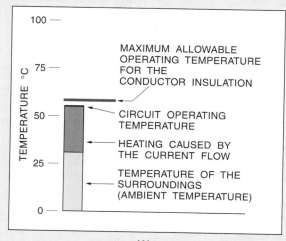

(A)

A 60°C MAXIMUM ALLOWABLE CONDUCTOR INSULATION TEMPERATURE. THE CURRENT FLOW IS LIMITED TO THE FLOW THAT WILL ALLOW SAFE OPERATION OF THE CONDUCTOR FOR THE LIFE OF THE STRUCTURE.

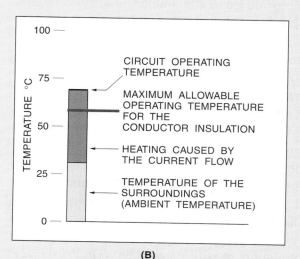

(B)

A 60°C MAXIMUM ALLOWABLE CONDUCTOR INSULATION TEMPERATURE. THE CURRENT FLOW IS TOO HIGH, AND THE CIRCUIT OPERATION TEMPERATURE IS ABOVE THE MAXIMUM ALLOWABLE BY CODE.

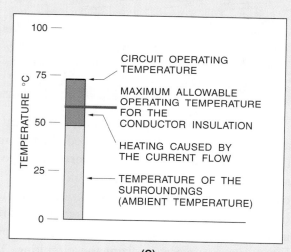

(C)

A 60°C MAXIMUM ALLOWABLE CONDUCTOR INSULATION TEMPERATURE. THE AMBIENT TEMPERATURE IS TOO HIGH, CAUSING THE OPERATING TEMPERATURE TO EXCEED THE INSULATION RATING. THE CURRENT FLOW IN THE CIRCUIT MUST BE REDUCED IN ORDER TO COMPLY WITH THE TEMPERATURE LIMITATIONS IN THE CODE.

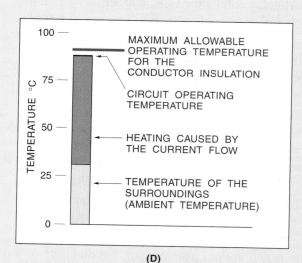

(D)

A 90°C MAXIMUM ALLOWABLE CONDUCTOR INSULATION TEMPERATURE. THIS CONDUCTOR IS ALLOWED TO CARRY MORE CURRENT AT THE SAME AMBIENT TEMPERATURE AS IN (A) BECAUSE OF THE HIGHER ALLOWABLE OPERATION TEMPERATURE.

Figure 5-2

Heat is introduced into the electrical system in two different ways, as shown in Figure 5-2. Heat is generated internally by the current flow. Each time an electron is conducted from one atom to the next, a small amount of the electrical energy is converted into heat. Second, external sources such as sunlight, or heat-producing equipment in close proximity to the wire, will heat the surroundings, thereby heating the conductors. The important point is that the temperature of a conductor must be controlled, either by restricting the environment where the conductor is to be installed or by limiting the level of current that the conductor is allowed to carry. Another possible method of controlling the temperature of a conductor is shown in Figure 5-3. As the conductor size increases, the allowable current also increases while the same operating temperature is maintained. Conductor size is the prevalent method used by the *NEC®* to control heating in a conductor.

Conductor insulation is not the only item in the electrical system that has temperature limitations. A fully loaded conductor can be expected to operate at close to the maximum allowable temperature rating. This means that the termination at which the conductors connect to controls or loads must be also able to withstand the constant high temperature. If any termination is rated for less than the conductor temperature rating, the ampacity of the conductor must be limited so that the temperature does not exceed that of the termination. See *110.14* for the *NEC®* requirements for temperature limitations of terminals.

5.3: CONDUCTOR SIZING

KEY TERMS

Circular mil   The unit of measure used for defining wire size by the *NEC®*. A circular mil is equal to the diameter of a conductor in thousandths of inches, squared. One circular mil equals approximately .0007 square inch.

American Wire Gauge   One of the two wire sizing systems employed by the *NEC®* to define the size of a conductor. The AWG system assigns a gauge number, such as 12 or 00, to standard sizes of conductors. The largest conductor defined under the AWG system is 0000 (4/0).

The fundamental unit of measure for the size of conductors is the **circular mil**. Mil is a term that means 1/1000 (one one-thousandth). In measuring conductor size, the mil is 1/1000 of an inch. **The American Wire Gauge** (AWG) is a system developed to identify conductor size using a gauge number to represent a particular cross-sectional area, measured in circular mils. The conductors are known by their AWG number, such as AWG 6 or AWG 000 (3/0). When conductor size reaches 250,000 circular mils, the sizing system changes from the gauge system to expression of the conductor size an the actual circular mil area. A conductor of 250,000 circular mils is called a 250-kcmil conductor (k = 1000, c = circular, mil = mil). *Table 8* from *Chapter 9* of the *NEC®* presents extensive information about conductors, including the size expressed in circular mils and in square inches.

5.4: CONDUCTOR AMPACITY

KEY TERMS

Ampacity   *(See NEC® Article 100)*: The amount of current that a conductor is allowed to carry continuously without heating to a point at which damage occurs to the conductor or the conductor's insulation.

Ampacity is the current that a conductor can carry continuously without damage to the insulation from excessively high temperatures. The Code uses a series of tables to provide information about ampacity for the electrician. The most widely used ampacity table is *Table 310.16* because it lists the most widely used insulation types including those employed in residential wiring. Other ampacity tables for conductors for general wiring are *Table 310.17*, through *310.21*. The application of the table and the insulation types involved are specified in the title of each of the tables. Figure 5-4 shows the details of determining rated ampacity for the various conductor sizes, insulation types, and temperature ratings.

Because of the variable nature of the products—conductors, boxes, circuit breakers, and other devices and equipment—used to wire dwelling units, the allowable ampacity of conductors and cables used in dwellings also varies, as follows:

- AWG 14 copper conductors are used for 15-ampere branch circuits.

"WIREMAN'S GUIDE"
CROSS-SECTIONAL AREA AND ALLOWABLE CURRENT FLOW

A ONE-LANE HIGHWAY WILL BE ABLE TO HANDLE ONLY A CERTAIN NUMBER OF CARS FOR ANY UNIT OF TIME.

CROSS-SECTIONAL AREA (END)

LARGER WIRE RESPONDS LIKE ANOTHER LANE ON A HIGHWAY. THE MORE LANES (THE LARGER THE CONDUCTOR), THE EASIER IT IS FOR ALL MOTORISTS.

THE CROSS-SECTIONAL AREA IS THE SURFACE EXPOSED AT THE END OF THE WIRE. THE CROSS-SECTION IS AN AREA MEASUREMENT; IN THE ELECTRICAL INDUSTRY THE UNITS OF AREA ARE CIRCULAR MILS RATHER THAN SQUARE INCHES. *TABLE 8* IN *NEC® CHAPTER 9* SHOWS THE MEASUREMENTS FOR EACH SIZE OF CONDUCTOR. FOR ALL SIZES OF WIRE, APPROXIMATE AREAS ARE SHOWN, BOTH IN SQUARE INCHES AND IN CIRCULAR MILS. EVERY CALCULATION IN THE CODE DEALING WITH CROSS-SECTIONAL AREA OF CONDUCTORS USES CIRCULAR MIL UNITS INSTEAD OF SQUARE INCH UNITS.

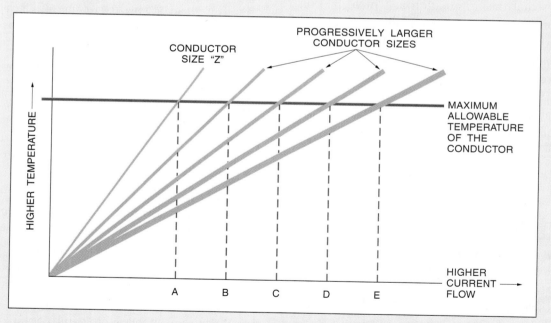

ALL FIVE CONDUCTORS SHOWN ON THE GRAPH ARE OF THE SAME INSULATION TYPE AND OF THE SAME CONDUCTOR TYPE (COPPER OR ALUMINUM), ARRANGED IN INCREASING SIZE. IN ORDER TO PROVIDE A LARGER CURRENT FLOW WITHOUT EXCEEDING THE MAXIMUM OPERATING TEMPERATURE OF THE CONDUCTOR, A LARGER SIZE IS NEEDED. FOR CONDUCTOR SIZE "Z" (FIRST CONDUCTOR ON THE LEFT), THE MAXIMUM ALLOWABLE AMPACITY IS CURRENT FLOW A. WITH THE NEXT LARGER WIRE SIZE, THE ALLOWABLE AMPACITY IS NOW AT CURRENT FLOW B. EACH CONDUCTOR TYPE (COPPER OR ALUMINUM), EACH INSULATION TYPE, AND EACH CONDUCTOR SIZE CAN BE PLOTTED AND WILL SHOW A LINE ON A GRAPH VERY MUCH LIKE THESE.

Figure 5-3

"WIREMAN'S GUIDE"
CONDUCTOR AMPACITY DETERMINATION

Table 310.16 Allowable Ampacities of Insulated Conductors Rated 0 Through 2000 Volts, 60°C Through 90°C (140°F Through 194°F), Not More Than Three Current-Carrying Conductors in Raceway, Cable, or Earth (Directly Buried), Based on Ambient Temperature of 30°C (86°F)

Size AWG or kcmil	60°C (140°F) Types TW, UF	75°C (167°F) Types RHW, THHW, THW, THWN, XHHW, USE, ZW	90°C (194°F) Types TBS, SA, SIS, FEP, FEPB, MI, RHH, RHW-2, THHN, THHW, THW-2, THWN-2, USE-2, XHH, XHHW, XHHW-2, ZW-2	60°C (140°F) Types TW, UF	75°C (167°F) Types RHW, THHW, THW, THWN, XHHW, USE	90°C (194°F) Types TBS, SA, SIS, THHN, THHW, THW-2, THWN-2, RHH, RHW-2, USE-2, XHH, XHHW, XHHW-2, ZW-2	Size AWG or kcmil
	COPPER			ALUMINUM OR COPPER-CLAD ALUMINUM			
18	—	—	14	—	—	—	—
16	—	—	18	—	—	—	—
14*	20	20	25	—	—	—	—
12*	25	25	30	20	20	25	12*
10*	30	35	40	25	30	35	10*
8	40	50	55	30	40	45	8
6	55	65	75	40	50	60	6
4	70	85	95	55	65	75	4
3	85	100	110	65	75	85	3
2	95	115	130	75	90	100	2
1	110	130	150	85	100	115	1
1/0	125	150	170	100	120	135	1/0
2/0	145	175	195	115	135	150	2/0
3/0	165	200	225	130	155	175	3/0
4/0	195	230	260	150	180	205	4/0
250	215	255	290	170	205	230	250
300	240	285	320	190	230	255	300
350	260	310	350	210	250	280	350
400	280	335	380	225	270	305	400
500	320	380	430	260	310	350	500
600	355	420	475	285	340	385	600
700	385	460	520	310	375	420	700
750	400	475	535	320	385	435	750
800	410	490	555	330	395	450	800
900	435	520	585	355	425	480	900
1000	455	545	615	375	445	500	1000
1250	495	590	665	405	485	545	1250
1500	520	625	705	435	520	585	1500
1750	545	650	735	455	545	615	1750
2000	560	665	750	470	560	630	2000

COPPER SIDE OF TABLE ⇦ ⇨ ALUMINUM SIDE OF TABLE

CORRECTION FACTORS

Ambient Temp. (°C)	For ambient temperatures other than 30°C (86°F), multiply the allowable ampacities shown above by the appropriate factor shown below.						Ambient Temp. (°F)
21–25	1.08	1.05	1.04	1.08	1.05	1.04	70–77
26–30	1.00	1.00	1.00	1.00	1.00	1.00	78–86
31–35	0.91	0.94	0.96	0.91	0.94	0.96	87–95
36–40	0.82	0.88	0.91	0.82	0.88	0.91	96–104
41–45	0.71	0.82	0.87	0.71	0.82	0.87	105–113
46–50	0.58	0.75	0.82	0.58	0.75	0.82	114–122
51–55	0.41	0.67	0.76	0.41	0.67	0.76	123–131
56–60	—	0.58	0.71	—	0.58	0.71	132–140
61–70	—	0.33	0.58	—	0.33	0.58	141–158
71–80	—	—	0.41	—	—	0.41	159–176

* See 240.4(D).

REPRINTED WITH PERMISSION FROM NFPA 70-2002.

I. THE *NEC®* ASSIGNS EACH CONDUCTOR TYPE (COPPER OR ALUMINUM) AND EACH CONDUCTOR SIZE (AWG OR kcmil) AN AMPACITY RATING AND LISTS THEM IN *TABLE 310.16*, REPRODUCED AT LEFT. THIS IS THE MOST COMMONLY USED AMPACITY TABLE AND IS HERE TO DEMONSTRATE HOW TO DETERMINE THE AMPACITY OF A CONDUCTOR.

THESE AMPERE RATINGS APPLY UNDER THE CONDITIONS STATED IN THE TITLE OF THE TABLE. OTHER AMPACITY TABLES APPLY FOR DIFFERENT CONDITIONS, SUCH AS CONDUCTORS SUSPENDED IN FREE AIR INSTEAD OF IN CONDUIT OR CABLE, OR DIFFERENT INSULATION TYPES USED FOR TEMPERATURES HIGHER THAN 90°C. (SEE *TABLES 310.17, 310.18, 310.19, 310.20,* AND *310.21*.) THE TITLE OF *TABLE 310.16* STIPULATES THAT THE AMPACITIES ARE FOR INSTALLATIONS WITH NO MORE THAN THREE CURRENT-CARRYING CONDUCTORS IN CONDUIT OR CABLE, OR DIRECTLY BURIED, AND WITH AN AMBIENT TEMPERATURE OF 30°C.

NOTICE THAT THE TABLE IS DIVIDED INTO HALVES: THE LEFTHAND SIDE IS FOR COPPER CONDUCTORS AND THE RIGHTHAND SIDE IS FOR ALUMINUM OR COPPER-CLAD ALUMINUM CONDUCTORS. DOWN THE SIDES OF THE TABLE ARE CONDUCTOR SIZES IN AWG AND kcmil.

II. ON EACH SIDE OF THE TABLE, SHOWN AGAIN TO THE RIGHT, THERE ARE THREE COLUMNS THAT GROUP TOGETHER CONDUCTORS WITH THE SAME MAXIMUM TEMPERATURE RATING. THESE TEMPERATURE GROUPINGS ARE ACCORDING TO *TABLE 310.13*, WHICH PROVIDES DETAILED INFORMATION ABOUT ALL INSULATION TYPES FOR CONDUCTORS FOR GENERAL WIRING. THERE ARE COLUMNS FOR 60°C, 75°C, AND 90°C CONDUCTOR INSULATION, DESIGNATING THE MAXIMUM ALLOWABLE CONDUCTOR TEMPERATURE.

THE EXAMPLE HERE USES **3/0 COPPER, TYPE THWN INSULATION**.

STEP 1: DETERMINE WHICH SIDE OF THE TABLE TO USE. IS THIS A COPPER OR AN ALUMINUM CONDUCTOR?

STEP 2: LOCATE THE COLUMN FOR THE SUBJECT INSULATION TYPE (THWN): THE 75°C COLUMN.

STEP 3: LOCATE THE ROW WITH THE SUBJECT CONDUCTOR SIZE: THE ROW FOR 3/0 CONDUCTORS.

STEP 4: LOCATE THE AMPACITY FROM THE INTERSECTION OF THE ROW CONTAINING THE CONDUCTOR SIZE WITH THE COLUMN FOR THWN INSULATION TYPE. THE NUMBER AT THAT INTERSECTION IS THE CONDUCTOR MAXIMUM ALLOWABLE AMPERE RATING. IN THIS EXAMPLE, THE AMPERE RATING OF THE 3/0 COPPER, THWN INSULATION CONDUCTOR IS 200 AMPERES.

Table 310.16 Allowable Ampacities of Insulated Conductors Rated 0 Through 2000 Volts, 60°C Through 90°C (140°F Through 194°F), Not More Than Three Current-Carrying Conductors in Raceway, Cable, or Earth (Directly Buried), Based on Ambient Temperature of 30°C (86°F)

Size AWG or kcmil	60°C (140°F) Types TW, UF	75°C (167°F) Types RHW, THHW, THW, THWN, XHHW, USE, ZW	90°C (194°F) Types TBS, SA, SIS, FEP, FEPB, MI, RHH, RHW-2, THHN, THHW, THW-2, THWN-2, USE-2, XHH, XHHW, XHHW-2, ZW-2	60°C (140°F) Types TW, UF	75°C (167°F) Types RHW, THHW, THW, THWN, XHHW, USE	90°C (194°F) Types TBS, SA, SIS, THHN, THHW, THW-2, THWN-2, RHH, RHW-2, USE-2, XHH, XHHW, XHHW-2, ZW-2	Size AWG or kcmil
	COPPER			ALUMINUM OR COPPER-CLAD ALUMINUM			
18	—		14	—	—	—	—
16	—		18	—	—	—	—
14*	20		25	—	—	—	—
12*	25		30	20	20	25	12*
10*	30		40	25	30	35	10*
8	40		55	30	40	45	8
6	55		75	40	50	60	6
4	70		95	55	65	75	4
3	85		110	65	75	85	3
2	95		130	75	90	100	2
1	110		150	85	100	115	1
1/0	125	150	170	100	120	135	1/0
2/0	145	175	195	115	135	150	2/0
3/0	165	200	225	130	155	175	3/0
4/0	195	230	260	150	180	205	4/0
250	215	255	290	170	205	230	250
300	240	285	320	190	230	255	300
350	260	310	350	210	250	280	350
400	280	335	380	225	270	305	400
500	320	380	430	260	310	350	500
600	355	420	475	285	340	385	600
700	385	460	520	310	375	420	700
750	400	475	535	320	385	435	750
800	410	490	555	330	395	450	800
900	435	520	585	355	425	480	900
1000	455	545	615	375	445	500	1000
1250	495	590	665	405	485	545	1250
1500	520	625	705	435	520	585	1500
1750	545	650	735	455	545	615	1750
2000	560	665	750	470	560	630	2000

CORRECTION FACTORS

Ambient Temp. (°C)	For ambient temperatures other than 30°C (86°F), multiply the allowable ampacities shown below by the appropriate factor shown below.						Ambient Temp. (°F)
21–25	1.08	1.05	1.04	1.08	1.05	1.04	70–77
26–30	1.00	1.00	1.00	1.00	1.00	1.00	78–86
31–35	0.91	0.94	0.96	0.91	0.94	0.96	87–95
36–40	0.82	0.88	0.91	0.82	0.88	0.91	96–104
41–45	0.71	0.82	0.87	0.71	0.82	0.87	105–113
46–50	0.58	0.75	0.82	0.58	0.75	0.82	114–122
51–55	0.41	0.67	0.76	0.41	0.67	0.76	123–131
56–60	—	0.58	0.71	—	0.58	0.71	132–140
61–70	—	0.33	0.58	—	0.33	0.58	141–158
71–80	—	—	0.41	—	—	0.41	159–176

* See 240.4(D).

REPRINTED WITH PERMISSION FROM NFPA 70-2002.

Figure 5-4

- AWG 12 copper conductors are used for 20-ampere branch circuits.

- AWG 10 copper conductors are used for 30-ampere branch circuits.

- AWG 8 copper conductors are used for 40-ampere branch circuits.

- AWG 6 copper conductors are used for 50-ampere branch circuits.

Although these conductors are subject to the reductions in allowable ampacity required by the *NEC®* and outlined in Figure 5-5 for ambient temperature and number of conductors in a conduit or cable, they rarely have to be employed because of the characteristics of the conductors used to construct the cable assemblies. The cables inside of these cable assemblies are 90°C rated, but the ampacity must be applied as if the conductors were actually 60°C rated. This is a Code requirement for nonmetallic-sheathed (Type NM) and armored cable (Type AC) cable assemblies and is usually applied with use of metal-clad (Type MC) cable assemblies, although not specifically required.

KEY TERMS

Nonmetallic-sheathed cable Electrical cable with 90°C-rated conductors installed in an outer sheath of nonmetallic material. Use of this type of cable is the most popular and widely used wiring method for dwellings.

These wiring methods are the ones most commonly employed for dwelling-unit branch circuits and are readily available in both two-wire and three-wire designs with equipment grounding conductor cable assemblies. **Nonmetallic-sheathed cable** (NM cable) is used to install the branch-circuit conductors in an overwhelming percentage of the dwellings constructed in the United States each year. See Chapter 8 of this book for more information on these wiring methods.

The ampacity ratings for service-entrance conductors and for main power feeders for dwellings are designated according to *310.15(B)(6)* and its accompanying table. These ampacities are not subject to the reduction in allowable ampacities required for ambient temperature and for the number of conductors in a cable. This Code section applies to 120/240-volt, three-wire, single-phase, dwelling-unit services only.

"WIREMAN'S GUIDE"
AMPACITY CORRECTION FACTORS
AND ADJUSTMENT FACTORS

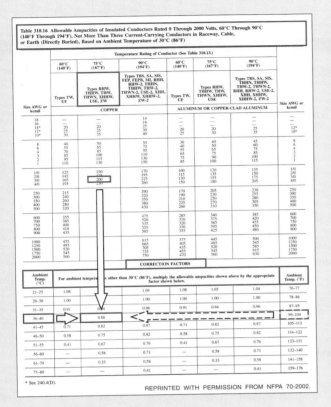

Table 310.16 Allowable Ampacities of Insulated Conductors Rated 0 Through 2000 Volts, 60°C Through 90°C (140°F Through 194°F), Not More Than Three Current-Carrying Conductors in Raceway, Cable, or Earth (Directly Buried), Based on Ambient Temperature of 30°C (86°F)

REPRINTED WITH PERMISSION FROM NFPA 70-2002.

* See 240.4(D).

ONCE THE ALLOWABLE AMPACITY FOR CONDUCTORS IS DETERMINED, TWO OUTSIDE FACTORS THAT ALSO CONTRIBUTE TO THE OPERATING TEMPERATURE OF CONDUCTORS MUST BE CONSIDERED: (1) AMBIENT TEMPERATURE AND (2) PROXIMITY OF OTHER CURRENT-CARRYING CONDUCTORS. IF THE AMBIENT TEMPERATURE IS OTHER THAN THAT LISTED IN THE TITLE, CORRECTION FACTORS MUST BE APPLIED. DIRECTLY BELOW THE LISTED AMPACITIES IN *TABLE 310.16* ARE *CORRECTION FACTORS*. THE SAME COLUMN THAT PROVIDED THE ALLOWABLE AMPACITY IS USED FOR TEMPERATURE CORRECTION. THE CORRECTION FACTOR CAN BE FOUND WHERE THE COLUMN USED IN THE AMPACITY TABLE INTERSECTS WITH THE ROW CONTAINING THE AMBIENT TEMPERATURE IN THE CORRECTION FACTOR TABLE. FOR EXAMPLE, WITH AN AMBIENT TEMPERATURE OF 38°C AND EIGHT 3/0 COPPER, THWN CONDUCTORS IN A CONDUIT, THE CORRECTION FACTOR IS 0.88, MEANING THAT THE CONDUCTORS CAN CARRY ONLY 88% OF THE CURRENT LISTED IN THE AMPACITY TABLE. THEREFORE:

200 AMPERES X .88 = 176 AMPERES

176 AMPERES IS THE NEW MAXIMUM ALLOWABLE AMPACITY FOR THESE CONDUCTORS.

THERE ARE ALSO EIGHT CURRENT-CARRYING CONDUCTORS IN THE CONDUIT. BECAUSE THE TITLE OF THE TABLE STIPULATES NO MORE THAN THREE CURRENT-CARRYING CONDUCTORS, THE ALLOWABLE AMPACITY MUST BE ADJUSTED FOR THE ADDITIONAL HEAT CREATED BY THE CURRENT FLOW IN THESE CONDUCTORS. *TABLE 310.15(B)(2)(a)* IS USED FOR THESE ADJUSTMENTS.

DOWN THE LEFT SIDE OF *TABLE 310.15(B)(2)(a)*, REPRODUCED BELOW, ARE GROUPINGS OF CONDUCTORS, AND DOWN THE RIGHT COLUMN ARE LISTED ADJUSTMENT FACTORS. TO FIND THE ADJUSTMENT FACTOR FOR ANY GIVEN NUMBER OF CURRENT-CARRYING CONDUCTORS, TRACE DOWN THE LEFT COLUMN FOR THE PROPER NUMBER—IN THIS EXAMPLE, EIGHT—AND OBTAIN THE CORRESPONDING ADJUSTMENT FROM THE RIGHT COLUMN. THE ADJUSTMENT FACTOR FOR THIS EXAMPLE IS 70%. THIS MEANS THAT THE CONDUCTORS CAN CARRY ONLY 70% OF THE CURRENT AS CORRECTED FOR AMBIENT TEMPERATURE. THUS:

176 AMPERES X .70 = 123.2 AMPERES—THE ALLOWABLE AMPACITY OF THE CONDUCTORS.

ONE OTHER PROCEDURE MUST BE FOLLOWED FOR DETERMINING CONDUCTOR AMPACITY. *NEC® 110.14(C)* REQUIRES THAT FOR CONDUCTORS RATED 100 AMPERES OR LESS, THE ALLOWABLE AMPACITY CANNOT EXCEED THAT GIVEN IN THE 60°C COLUMN ON THE AMPACITY TABLES. FOR ABOVE 100 AMPERES, THE 75°C COLUMN AMPACITY MUST NOT BE EXCEEDED. THESE REQUIREMENTS APPLY AFTER ANY CORRECTIONS FOR AMBIENT TEMPERATURE OR ADJUSTMENTS FOR THE NUMBER OF CONDUCTORS HAVE BEEN COMPLETED. THE ALLOWABLE AMPACITY FROM THE FOREGOING CALCULATIONS ARE COMPARED WITH THE ALLOWABLE AMPACITY DIRECTLY FROM THE 60°C OR THE 75°C COLUMN, AND THE SMALLER AMPACITY IS THE ONE THAT IS USED.

Table 310.15(B)(2)(a) Adjustment Factors for More Than Three Current-Carrying Conductors in a Raceway or Cable	
Number of Current-Carrying Conductors	Percent of Values in Tables 310.16 through 310.19 as Adjusted for Ambient Temperature if Necessary
4–6	80
7–9	70
10–20	50
21–30	45
31–40	40
41 and above	35

GROUPING OF NUMBER OF CURRENT CARRYING CONDUCTORS IN CABLE OR RACEWAY

ADJUSTMENT FACTOR

Figure 5-5

SUMMARY

Insulation is damaged by high temperatures. The current flow in a circuit must be restricted to keep the conductors from overheating and damaging the insulation. The three major factors affecting conduc-tor ampacity are the conductor size, the insulation type, and the ambient temperature. The *NEC®* describes procedures for correcting and adjusting conductor allowable ampacity that must be employed when necessary.

REVIEW

1. The following is not a standard insulation temperature rating.
   a. 30°C
   b. 60°C
   c. 75°C
   d. 90°C

2. If all other factors are kept constant, increasing the size (AWG or kcmil) of a conductor will _____ the allowable ampacity.
   a. increase
   b. not affect
   c. decrease
   d. cannot be determined from the information given

3. The temperature rating of the conductor insulation must be _____ the termi-nal temperature rating.
   a. greater than
   b. the same as
   c. less than
   d. not applicable—the two ratings have nothing to do with each other

4. What is the maximum allowable ampacity for six current-carrying 12 AWG copper conductors installed in a conduit in a dwelling unit?
   a. 15 amperes
   b. 20 amperes
   c. 25 amperes
   d. 30 amperes

5. Unless otherwise marked, what is the maximum temperature rating of a termination for a 70-ampere load?

   a. 30°C

   b. 60°C

   c. 75°C

   d. 90°C

6. When the allowable ampacity is corrected for ambient temperature and number of conductors in a conduit, which must be applied first, the correction factors or the adjustments?

   a. the correction factor

   b. the adjustments

   c. either one—the order makes no difference

   d. The adjustment is divided by the correction factor before being applied to the allowable ampacity.

"WIREMAN'S GUIDE" REVIEW

1. The cost of delivering amperes can be measured by the amount of copper needed to carry the load. The smaller the amount of copper needed for each ampere delivered, the more economical the installation. For conductor sizes 10 AWG through 4/0 AWG, calculate the circular mils of conductor needed for each ampere allowed (the area in circular mils divided by the allowable ampacity). Use *Table 8, Chapter 9,* for the circular mil area and *Table 310.16* and the 60°C copper column for the allowable ampacity. What do the calculations tell you about the larger wire sizes?

2. Explain the relationship among conductor size, allowable ampacity, and conductor temperature.

3. A circular mil is defined as the area of a circle with a diameter of 1/1000 of an in. (.0254 mm) squared. How does this relate to the area measured in square inches (square millimeters)?

4. What is the allowable ampacity of a 6 AWG XHHW copper conductor installed in a conduit with three other XHHW copper conductors, in an ambient environment of 45°C in a dry location?

CHAPTER 6

Boxes

INTRODUCTION

This chapter deals with boxes and box installation. Boxes are required at virtually all junction points and splice points, at all outlets, and at all switch points. Chapter 3 describes the locations of boxes for receptacle outlets; Chapter 4 details the locations of boxes for lighting outlets and switch points. The various types of boxes, as well as their uses and installation, are discussed in this chapter.

6.1: FUNCTIONS OF BOXES

KEY TERMS

Boxes   Housings in the electrical circuit that contain splices and terminations. Boxes can be metallic or nonmetallic and may house devices or simply contain conductors, but they provide a barrier between the electrical system of the structure and the living or working space of that structure.

The electrical system must be easily accessible to electricians to install devices, receptacles and switches, and to effect repairs or additions. Access to the electrical system is provided by the use of **boxes.** All splices, taps, and terminations (with very few exceptions) of the wiring system must be accomplished using an approved electrical box or equipment wiring enclosure. Boxes also serve an important function in the event of a fire by containing the fire inside the housing for some time. In many cases, the fire is discovered before it can spread to other parts of the structure.

Chapter 5 of this book emphasizes that conductor insulation must be protected against excessive heating due to the level of current flow and ambient temperature influences. Boxes are another potential source of heating problems for the electrical system. Current-carrying conductors and ambient temperature influences contribute heat to the inside of the box. Wiring for a device such as a dimmer switch not only takes up space in the box, thus allowing less available air to cool the conductors, but also contributes to the heating of the conductors. Obviously, the number of conductors allowed in the box must be limited in order to control the conductor operating temperature within the box.

6.2: TYPES OF BOXES

The *NEC®* divides boxes into two distinct groups according to the size of the largest conductor that enters the box. This two-group system is used because the *NEC®* is concerned with the heating of the conductors inside the device and outlet boxes. Air cannot circulate easily when the box is crowded with conductors and devices, and very high temperatures can be attained in an overloaded box. When conductors become larger, the concern changes to the amount of bending space available within the box. Larger conductors take up a considerable amount of room. Boxes that contain no conductor larger than 6 AWG are called device, outlet, and junction boxes, and boxes with conductors 4 AWG and larger are called junction and pull boxes.

6.2.1: Boxes for Conductors 6 AWG and Smaller

KEY TERMS

Device boxes Electrical boxes intended to house and make available devices such as switches and receptacles.

Lighting outlet boxes Boxes that are designed to supply outlets for luminaires (fixtures). These boxes are usually round in shape, can be metallic or nonmetallic, and have 8-32 threaded holes for the connection of the luminaire (fixture) support hardware.

Square boxes Electrical boxes that are square in shape and can be metallic or nonmetallic. Square boxes can be used for flush and surface installations by employing plaster rings or industrial covers and are the most common type of box used in commercial electrical work.

Waterproof boxes Electrical boxes designed to be used in wet and damp areas, such as outdoors, where moisture can enter the raceway system or the boxes themselves, thereby causing faulting problems.

Electrical outlet and device boxes come in many different sizes, configurations, and materials. Metal boxes are very commonly used, but boxes constructed from plastic and fiber compounds (nonmetallic) are most prevalent in residential construction today. The boxes, designed for use with receptacle outlets and switching devices, can be divided into four major groupings: (1) **device boxes**, (2) **lighting outlet boxes**, (3) **square boxes**, and (4) **waterproof boxes**. Table 6-1 presents a brief look at the various boxes for small conductors and some of their uses and limitations.

Device Boxes

This group of boxes contains a variety of boxes, some for surface mounting and others for flush mounting in walls or ceilings. The boxes can be metal or nonmetallic. Metal boxes can be single-gang or multigang, or they can be gangable (allowing connection of two or more boxes together to make a multigang box). Nonmetallic boxes can be single-gang or multigang boxes. Some of these device boxes are designed for remodeling work, in which the boxes will be cut into existing walls, and others are designed strictly for installation onto structural framing members during construction. Figure 6-1 shows a selection of typical metal device boxes, and Figure 6-2 (page 144) shows a selection of nonmetallic device boxes. Figure 6-3 (page 145) shows a typical box rough-in in a dwelling.

One feature that all device boxes have in common is the shape of the opening in the front of the box. The opening is rectangular, nominally 3 in. (75 mm) by 2 in. (50 mm) for each gang. A three-gang device box (or three single-gang metal device boxes ganged together) has a nominal opening of 3 in. (75 mm) by 6 in. (150 mm). Receptacle devices and switching devices are designed to fit into a space of that size at trim.

Another feature that all device boxes have in common is that they all provide threaded size 6-32 holes to facilitate the installation of receptacles and switches (devices). The 6-32 screw is used to mount these devices, and the devices usually fit inside the box for a flush finish.

Device boxes are not intended to support luminaires (fixtures). Lighting outlet boxes are provided with size 8-32 threaded holes. The extra size of the screws is needed to support the weight of the luminaire (fixture). Size 6-32 screws have not been listed as a supporting means for luminaires (fixtures), and the Code does not allow it, with one exception, as shown in Figure 6-4 (page 146).

FEATURE	METAL OCTAGON BOX	METAL SQUARE BOX	METAL DEVICE BOX	METAL MASONRY BOX	METAL HANDY BOX	WATERPROOF DEVICE BOXES	WATERPROOF ROUND BOX	NONMETAL SQUARE BOX	NONMETAL ROUND BOX	NONMETAL DEVICE BOX
Surface mounted										
Junction or pull box	X	X			X	X	X	X	X	X
Support luminaire (fixture)	X	X(1)				X	X	X(1)	X	X
Support device	X	X(1)			X	X	X	X(1)	X	X
Mounted flush inside of wall or ceiling										
Junction or pull box	X	X(2)	X	X				X(2)		
Support luminaire (fixture)	X	X(2)	X	X				X(2)		
Support devices	X	X(2)	X	X				X(2)		
Available with										
Built-in AC, MC, or NM cable clamps	X	X								
Without built-in AC, MC, or NM cable clamps	X	X	X	X	X		X	X	X	X(3)
Additional exterior cable clamp	X	X	X	X	X			X		X(4)
Side mounting bracket	X	X	X		X			X	X	X
Face mounting bracket	X	X	X		X			X	X	X
Extension box	X	X	X		X	X		X		
Gangable or multigang		X(5)(6)	X	X		X		X(5)		X
Threaded hole size for mounting to box	8-32	8-32 (7)	6-32	6-32	6-32	6-32	8-32	8-32(7)	8-32	6-32

(1) With proper industrial cover.
(2) With proper plaster ring installed.
(3) Most single-gang boxes do not have clamps.
(4) Single gang only.
(5) One-gang or two-gang plastic rings available.
(6) Multigang box available with multigang plaster ring.
(7) The box itself has 8-32 holes. Plaster ring can have 6-32 or 8-32 threaded holes.

Table 6-1   Selected boxes: Their uses and limitations.

"WIREMAN'S GUIDE"
SELECTED METAL DEVICE BOXES

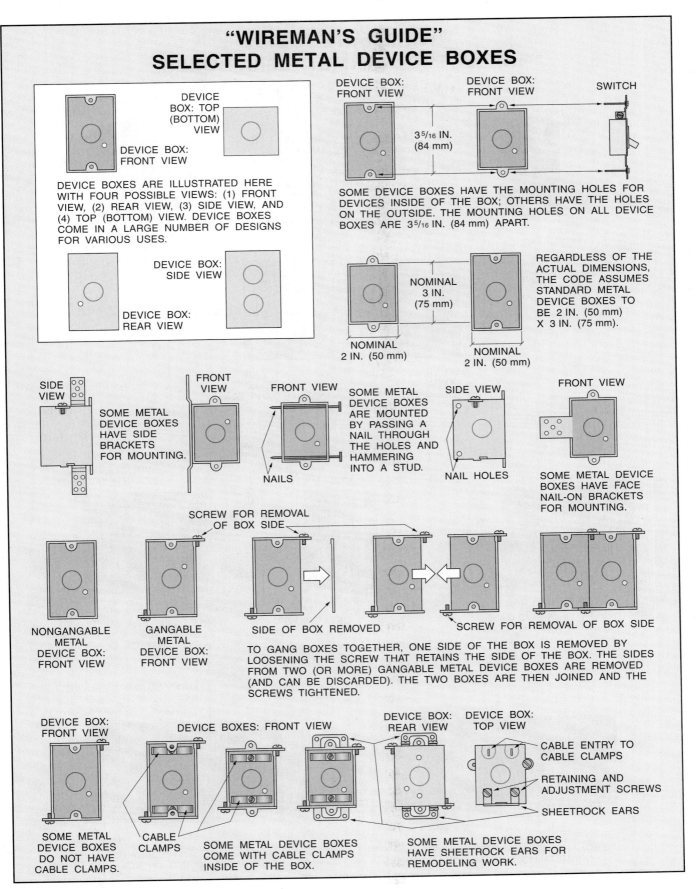

DEVICE BOX: TOP (BOTTOM) VIEW

DEVICE BOX: FRONT VIEW

DEVICE BOXES ARE ILLUSTRATED HERE WITH FOUR POSSIBLE VIEWS: (1) FRONT VIEW, (2) REAR VIEW, (3) SIDE VIEW, AND (4) TOP (BOTTOM) VIEW. DEVICE BOXES COME IN A LARGE NUMBER OF DESIGNS FOR VARIOUS USES.

DEVICE BOX: SIDE VIEW

DEVICE BOX: REAR VIEW

DEVICE BOX: FRONT VIEW

DEVICE BOX: FRONT VIEW

SWITCH

$3^5/_{16}$ IN. (84 mm)

SOME DEVICE BOXES HAVE THE MOUNTING HOLES FOR DEVICES INSIDE OF THE BOX; OTHERS HAVE THE HOLES ON THE OUTSIDE. THE MOUNTING HOLES ON ALL DEVICE BOXES ARE $3^5/_{16}$ IN. (84 mm) APART.

NOMINAL 3 IN. (75 mm)

NOMINAL 2 IN. (50 mm)

NOMINAL 2 IN. (50 mm)

REGARDLESS OF THE ACTUAL DIMENSIONS, THE CODE ASSUMES STANDARD METAL DEVICE BOXES TO BE 2 IN. (50 mm) X 3 IN. (75 mm).

SIDE VIEW

FRONT VIEW

SOME METAL DEVICE BOXES HAVE SIDE BRACKETS FOR MOUNTING.

FRONT VIEW

NAILS

SOME METAL DEVICE BOXES ARE MOUNTED BY PASSING A NAIL THROUGH THE HOLES AND HAMMERING INTO A STUD.

SIDE VIEW

NAIL HOLES

FRONT VIEW

SOME METAL DEVICE BOXES HAVE FACE NAIL-ON BRACKETS FOR MOUNTING.

NONGANGABLE METAL DEVICE BOX: FRONT VIEW

GANGABLE METAL DEVICE BOX: FRONT VIEW

SCREW FOR REMOVAL OF BOX SIDE

SIDE OF BOX REMOVED

SCREW FOR REMOVAL OF BOX SIDE

TO GANG BOXES TOGETHER, ONE SIDE OF THE BOX IS REMOVED BY LOOSENING THE SCREW THAT RETAINS THE SIDE OF THE BOX. THE SIDES FROM TWO (OR MORE) GANGABLE METAL DEVICE BOXES ARE REMOVED (AND CAN BE DISCARDED). THE TWO BOXES ARE THEN JOINED AND THE SCREWS TIGHTENED.

DEVICE BOX: FRONT VIEW

SOME METAL DEVICE BOXES DO NOT HAVE CABLE CLAMPS.

DEVICE BOXES: FRONT VIEW

CABLE CLAMPS

SOME METAL DEVICE BOXES COME WITH CABLE CLAMPS INSIDE OF THE BOX.

DEVICE BOX: REAR VIEW

DEVICE BOX: TOP VIEW

CABLE ENTRY TO CABLE CLAMPS

RETAINING AND ADJUSTMENT SCREWS

SHEETROCK EARS

SOME METAL DEVICE BOXES HAVE SHEETROCK EARS FOR REMODELING WORK.

Figure 6-1

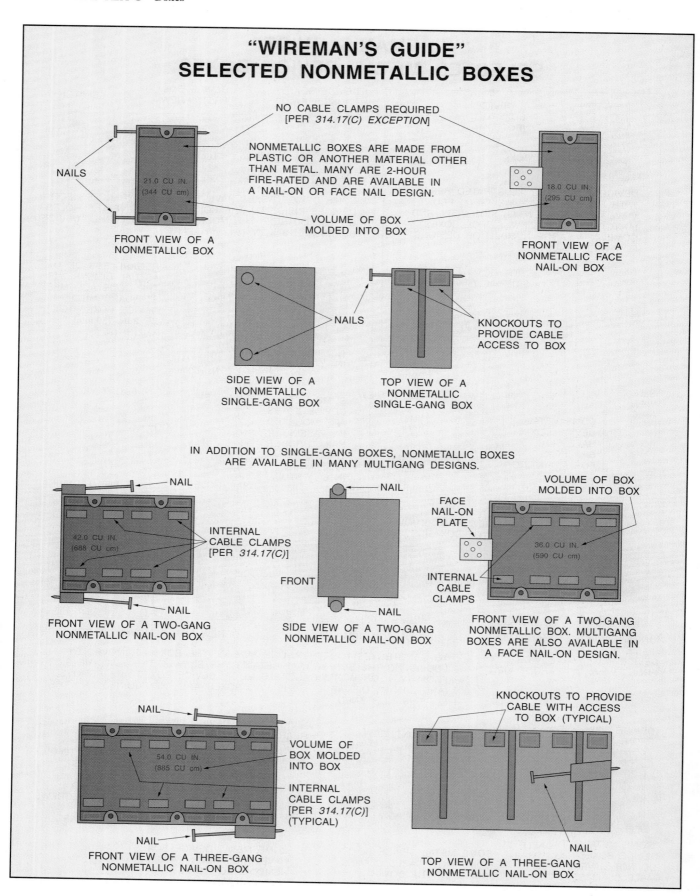

Figure 6-2

Figure 6-3 A typical nail-on box rough installation.

Lighting Outlet Boxes

Like device boxes, boxes for lighting outlets are available in metal and nonmetal designs, and they are available with special brackets for remodeling installations or in a nail-on design intended for new construction. Figure 6-5 shows examples of lighting outlet boxes. Figure 6-6 (page 148) shows a typical nail-on lighting outlet box, and Figure 6-7 (page 148) shows a standard lighting outlet box supported by a hanger bar for exact positioning. Lighting outlet boxes differ from device boxes in several significant ways:

- The openings in the boxes, and the boxes themselves, are round or octagonal instead of rectangular.

- The round and octagonal boxes are available in a 3-in. (.075-m)-diameter and 4-in. (.100-m)-diameter sizes.

- Lighting outlet boxes have threaded holes for the mounting of the luminaire (fixture) with 8-32 screws.

- Lighting outlet boxes are not designed to enclose the luminaire (fixture), in contrast to device boxes as used with receptacles and switches.

- There are no gangs with lighting outlet boxes. Each box is intended for only one luminaire (fixture).

Metal lighting outlet boxes are available with face nail-on and side nail-on bracket designs. Lighting outlet boxes can sometimes be installed on the bottom of a ceiling joist or truss. The luminaire (fixture) mounts over the box using a canopy or domed cover. A luminaire (fixture) with a domed cover can be used with a ½-in.-deep (.013-m-deep) lighting outlet box under certain conditions, as shown in Figures 6-8 and 6-9 (page 149 and page 150).

There are many different ways in which luminaires (fixtures) are connected to a box. Sometimes the luminaire (fixture) screws directly into the 8-32 threaded holes in the box. In many cases, room must be allowed in the box for a luminaire (fixture) stud or nipple that attaches to the box with a special

"WIREMAN'S GUIDE"
LUMINAIRES (FIXTURES) INSTALLED TO DEVICE BOXES

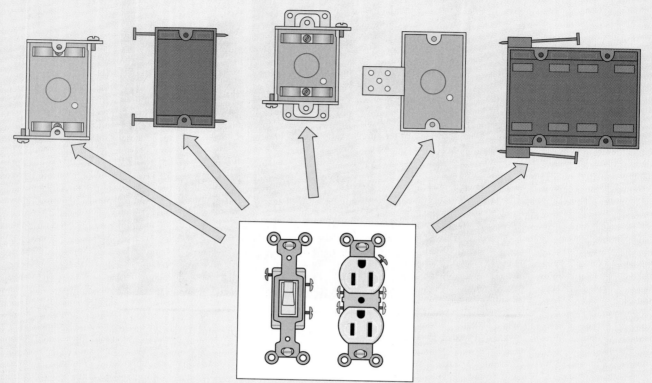

ALL STANDARD DEVICES ARE MOUNTED TO THE BOXES WITH 6-32 SCREWS, PROVIDED ON THE DEVICE OUT OF THE PACKAGE.

DEVICE BOXES HAVE THREADED OPENINGS FOR THE 6-32 SCREWS. THEY WILL NOT ACCEPT AN 8-32 SCREW.

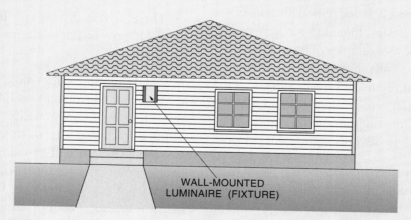

WALL-MOUNTED
LUMINAIRE (FIXTURE)

ACCORDING TO *314.27*, LUMINAIRES (FIXTURES) MUST BE MOUNTED TO BOXES DESIGNED FOR THIS USE—THAT IS, LIGHTING OUTLET BOXES WITH 8-32 THREADED HOLES. HOWEVER, THE EXCEPTION TO *314.27* ALSO ALLOWS MOUNTING OF A WALL-MOUNTED LUMINAIRE (FIXTURE), WEIGHING LESS THAN 6 LB (3 KG) TO A DEVICE BOX OR DEVICE PLASTER RING AS LONG AS THE LUMINAIRE (FIXTURE) IS MOUNTED USING AT LEAST TWO 6-32 SCREWS. IN THE DRAWING, THIS INSTALLATION IS OUTDOORS. THE CODE DOES NOT REQUIRE THAT THE LUMINAIRE (FIXTURE) BE OUTDOORS FOR THIS EXCEPTION. INTERIOR INSTALLATIONS ARE ALSO ACCEPTABLE.

Figure 6-4

"WIREMAN'S GUIDE"
SELECTED METAL AND NONMETALLIC LIGHTING OUTLET BOXES

METAL OCTAGON LIGHTING BOXES

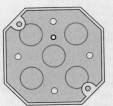

FRONT VIEW

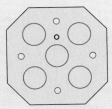

REAR VIEW

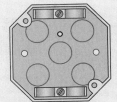

BOX WITH INTERNAL CABLE CLAMPS
FRONT VIEW

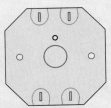

REAR VIEW

SIDE VIEW

METAL LIGHTING OUTLET BOXES ARE USUALLY OCTAGONAL IN SHAPE. THEY ARE AVAILABLE IN SEVERAL DEPTHS AND IN 3-IN. (75-mm) AND 4-IN. (100-mm) DIAMETER SIZES. THEY ARE ALSO AVAILABLE WITH OR WITHOUT INTERNAL CABLE CLAMPS. A BOX WITHOUT CABLE CLAMPS CAN BE USED WITH CABLE IF AN EXTERNAL CLAMP IS INSTALLED TO THE BOX.

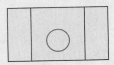

SIDE VIEW:
4-IN.-DEEP BOX

PAN BOXES

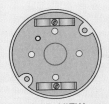

TOP VIEW
NOTICE THAT THE PAN BOX HAS INTERNAL CABLE CLAMPS.

SIDE VIEW

REAR VIEW

PAN BOXES ARE ONLY ½ IN. (13 mm) DEEP. THIS DEPTH IS ALLOWED BY THE CODE WITH CERTAIN EXCLUSIONS IN *314.16.* PAN BOXES ARE AVAILABLE IN 3-IN. (75-mm) AND 4-IN. (100-mm) DIAMETERS.

EXTENSION BOX

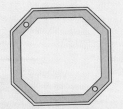

OCTAGON EXTENSION BOXES ARE MADE FROM STANDARD BOXES WITH THE BACK REMOVED.

MOUNTING METHODS

BAR HANGERS ALLOW EXACT PLACEMENT OF THE LUMINAIRE (FIXTURE) BETWEEN CEILING RAFTERS OR JOISTS. BAR HANGERS ARE ALSO AVAILABLE FOR OTHER BOXES, INCLUDING DEVICE BOXES.

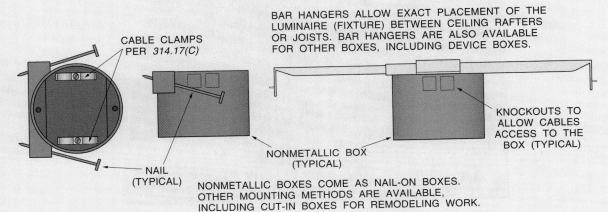

CABLE CLAMPS PER *314.17(C)*

NAIL (TYPICAL)

NONMETALLIC BOX (TYPICAL)

KNOCKOUTS TO ALLOW CABLES ACCESS TO THE BOX (TYPICAL)

NONMETALLIC BOXES COME AS NAIL-ON BOXES. OTHER MOUNTING METHODS ARE AVAILABLE, INCLUDING CUT-IN BOXES FOR REMODELING WORK.

Figure 6-5

Figure 6-6   A standard nail-on lighting box installation.

Figure 6-7   A bar hanger lighting box installation is used when the exact location of the luminaire (fixture) is critical.

"WIREMAN'S GUIDE"
ATTACHING A PAN BOX TO THE BOTTOM
OF A FLOOR JOIST

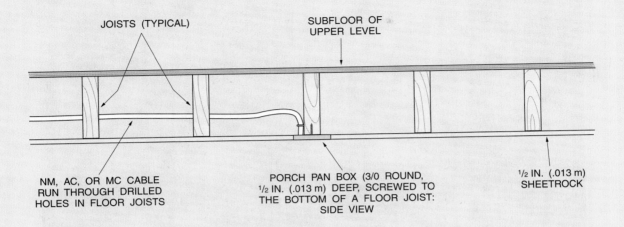

JOISTS (TYPICAL)

SUBFLOOR OF
UPPER LEVEL

NM, AC, OR MC CABLE
RUN THROUGH DRILLED
HOLES IN FLOOR JOISTS

PORCH PAN BOX (3/0 ROUND,
1/2 IN. (.013 m) DEEP, SCREWED TO
THE BOTTOM OF A FLOOR JOIST:
SIDE VIEW

1/2 IN. (.013 m)
SHEETROCK

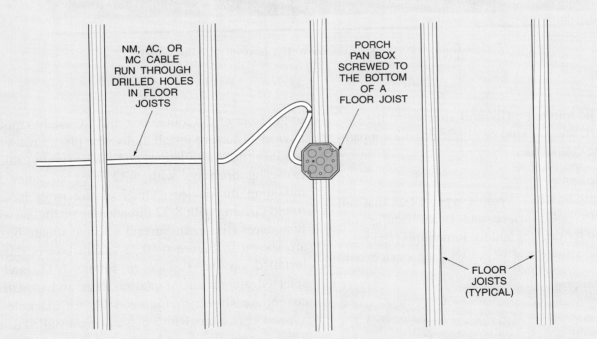

NM, AC, OR
MC CABLE
RUN THROUGH
DRILLED HOLES
IN FLOOR
JOISTS

PORCH
PAN BOX
SCREWED TO
THE BOTTOM
OF A
FLOOR JOIST

FLOOR
JOISTS
(TYPICAL)

VIEW OF CEILING ABOVE

ACCORDING TO 314.24, BOXES CANNOT BE LESS THAN 1/2 IN. (12.7 MM) DEEP. WITH PAN BOXES, HOWEVER, THERE IS NOT ENOUGH ROOM FOR THE CONDUCTORS IN THE BOX IF ALL CONDUCTORS ARE COUNTED ACCORDING TO 314.16. THE EXCEPTION TO 314.16(B)(1) ALLOWS THE ELIMINATION OF SOME CONDUCTORS FROM THE CONDUCTOR COUNT FOR INSTALLING A LUMINAIRE (FIXTURE) WITH A CANOPY OR DOMED COVER. USE OF A PAN BOX IS OFTEN ACCEPTABLE TO THE AHJ BECAUSE OF THIS EXCEPTION. WHEN INSTALLED IN THIS MANNER, THE PAN BOX DOES NOT EXTEND BEYOND THE LOWER EDGE OF THE SHEETROCK, AND THE BOTTOM OF THE JOIST PROVIDES A VERY SECURE MOUNTING.

Figure 6-8

Figure 6-9   A pan box attached to the bottom of a floor joist.

bracket. The luminaire (fixture) attaches to the special bracket, and the stud or nipple provides a pathway for the conductors.

Square Boxes

A square box is a versatile type of box that can be used for both surface and flush installations, can be used for devices and for luminaires (fixtures), and can be a one-gang or a two-gang device box.

KEY TERMS

Plaster ring   An accessory that is used with square boxes to allow them to be used in flush installations.

Square boxes are available in both metal and non-metal designs and can be readily obtained with side mounting brackets or face mounting brackets or with no bracket at all.

Square boxes owe their versatility to the use of special rings, called **plaster rings** or mud rings, and special covers, called industrial covers, in addition to a simple blank cover. These special covers install to the front of the square box and are used to mount devices in surface installations. The plaster rings are used for flush installations and can be single-gang or two-gang openings with 6-32 threaded holes for mounting the devices. Other plaster rings have a round opening with 8-32 threaded holes for use with luminaires (fixtures). Several selected square boxes are shown in Figure 6-10, a square box is shown installed on the surface in Figure 6-11, and a selected assortment of plaster rings and industrial covers are shown in Figure 6-12. An example of a metal square box with a plaster ring installed using a bar hanger for exact location is shown in Figure 6-13 (page 154), and a nonmetallic square box with a single-gang plaster ring is shown in Figure 6-14 (page 154).

Waterproof Boxes

Weatherproof boxes are designed to be installed on the exterior surface of dwellings and other structures. If a receptacle outlet is to be installed flush with the building surface, a normal single-gang device box is installed to house the device. A special

"WIREMAN'S GUIDE"
4-SQUARE AND 4¹¹/₁₆-SQUARE BOXES

A LARGE SELECTION OF SQUARE BOXES IS AVAILABLE. SQUARE BOXES ARE VERY VERSATILE; THEY CAN BE USED AS SURFACE- OR FLUSH-MOUNTED JUNCTION BOXES, SURFACE- OR FLUSH-MOUNTED DEVICE BOXES, OR SURFACE- OR FLUSH-MOUNTED LIGHTING OUTLET BOXES, DEPENDING ON THE PLASTER RING, INDUSTRIAL COVER, OR BLANK COVER USED. SQUARE BOXES ARE ALSO AVAILABLE WITH MOUNTING DESIGNS SUCH AS SIDE AND FACE BRACKETS. THEY ARE AVAILABLE WITH OR WITHOUT CABLE CLAMPS, IN THREE DIFFERENT DEPTHS—1¼ IN. (32 mm), 1½ IN. (38 mm), AND 2⅛ IN. (54 mm)—AND IN SEVERAL DIFFERENT KNOCKOUT PATTERNS AND SIZES.

4-SQUARE BOX

A 4-SQUARE BOX WITH ½ IN. TRADE SIZE (13 mm) KNOCKOUTS. THIS SIZE SQUARE BOX IS CALLED A 4-SQUARE BOX BECAUSE IT MEASURES 4 IN. (100 mm) ON EACH SIDE. THIS BOX IS A 1¼-IN. (32 mm)-DEEP BOX. NOTICE THE 8-32 THREADED HOLES IN OPPOSITE CORNERS FOR SECURING THE PLASTER RING OR COVER.

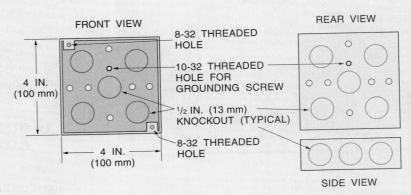

4-SQUARE COMBINATION BOX

FRONT VIEW

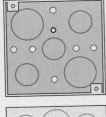

A 4-SQUARE BOX WITH ½ IN. AND ¾ IN. TRADE SIZE (13 mm AND 19 mm) KNOCKOUTS. IT IS ALSO CALLED A 4-SQUARE COMBINATION (COMBO) BOX.

SIDE VIEW   SIDE VIEW

4-SQUARE COMBINATION BOXES ARE AVAILABLE IN 1½ IN. (38 mm) AND 2⅛ IN. (54 mm) DEPTHS.

A 2⅛-IN. (54-mm)-DEEP BOX ALLOWS UP TO 1¼ IN. TRADE SIZE (32 mm) KNOCKOUTS.

A 4-SQUARE BOX WITH CABLE CLAMPS. THERE ARE STANDARD KNOCKOUTS ON THE BOX AS WELL AS CABLE CLAMPS.

FRONT VIEW

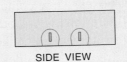

SIDE VIEW

A FRONT VIEW OF A 4-SQUARE BOX WITH A FACE NAIL-ON BRACKET AND A SIDE VIEW OF A 4-SQUARE BOX WITH A SIDE NAIL-ON MOUNTING BRACKET.

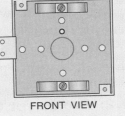

FRONT VIEW

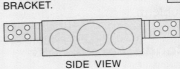

SIDE VIEW

4¹¹/₁₆ SQUARE BOX

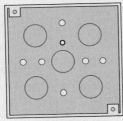

FRONT VIEW

A SQUARE BOX THAT MEASURES 4¹¹/₁₆ IN. (120 mm) ON EACH SIDE IS ALSO AVAILABLE IN THE SAME KNOCKOUT PATTERNS AND SIZES AS FOR THE 4-SQUARE BOX.

SIDE VIEW: 1½-IN. (38 mm)-DEEP, 4¹¹/₁₆-IN. (120 mm)-SQUARE BOX

SIDE VIEW: 2⅛ IN.-IN. (54 mm)-DEEP, 4¹¹/₁₆-IN. (120 mm)-SQUARE BOX

Figure 6-10

Figure 6-11 A typical Surface mounted 4-square box.

waterproof cover is used to keep out moisture. If the receptacle is to mount on the exterior surface of the building, a waterproof box is needed to house the device and to keep out moisture.

Weatherproof boxes are available as device boxes and lighting outlet boxes. They come with a variety of knockout configurations with both ½-in. (.013 m) and ¾-in. (.019-m) threaded knockouts on the ends and also, if desired, on the sides of the boxes. Weatherproof device boxes are readily available in one-gang and two-gang sizes with waterproof covers for receptacles and switches. Weatherproof boxes are usually made from a cast aluminum alloy so that rusting is not a problem.

6.2.2: Boxes for Conductors 4 AWG and Larger

No branch circuits are normally found in a dwelling that requires 4 AWG or larger conductors. They are sometimes used in service conductor installations or feeders, but boxes for these larger conductor sizes are also large. These larger boxes are available in certain common sizes, such as 8 in. × 8 in. × 6 in. (.200 m × .200 m × .150 m), or 12 in. × 12 in. × 8 in. (.300 m × .300 m × .200 m), but they can easily be manufactured in any size, thickness, configuration, or shape desired. These boxes can be

weatherproof (National Electrical Manufacturers Association [NEMA] 3R) or dry location (NEMA 1) boxes, or of any other type desired. Larger boxes are intended for conductors only, and no devices should be installed in this type of box.

6.3: SIZING BOXES

For sizing the boxes, the *NEC*® uses the same two-group system. The rules for sizing the smaller boxes are intended to limit the heat build-up inside the box. The rules for sizing the larger boxes are concerned with providing adequate bending space within the box.

6.3.1: Sizing Boxes for Conductors 6 AWG and Smaller

The rules for sizing boxes for 6 AWG and smaller conductors can be found in *314.16* of the Code. Determining the number of conductors allowed in a given box requires the following information:

- The volume of the box—that is, the space, measured in cubic inches or cubic centimeters, enclosed within the box. This volume can be obtained in either of two places. For boxes that the Code considers to be standard boxes, the volume can be obtained from *Table 314.16(A)* in the Minimum Volume column. For boxes that are not standard boxes, the volume is displayed somewhere on the box itself. It may be stamped into the metal or molded into the box when it is manufactured, or it may be printed or stamped on the box.

- Box fill—that is, the number and the size of the conductors that will be installed in the box. This information will come from the routing of the cables and the details of the circuiting.

- Conductor volume requirements—the volume that each conductor size is required to receive according to *Table 314.16(B)*. Obviously, a 6 AWG conductor will occupy more volume than a 14 AWG conductor of the same length.

The process for calculating the number of wires allowed in the box involves three steps: (1) Multiply the number of each of the conductor sizes by the required volume allowance from *Table 314.16(B)*. (2) Add the products. This number should represent

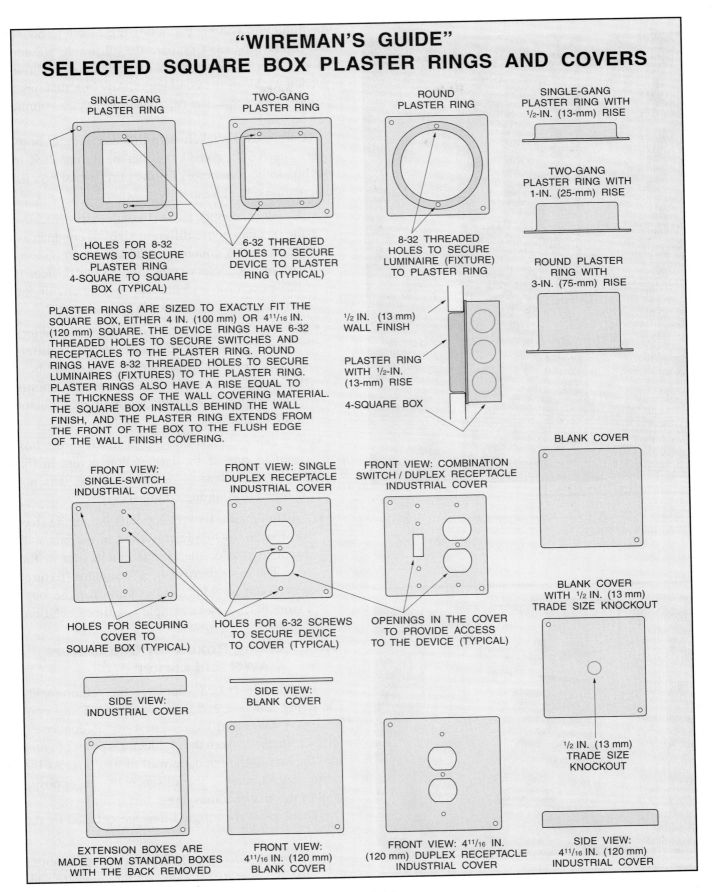

"WIREMAN'S GUIDE"
SELECTED SQUARE BOX PLASTER RINGS AND COVERS

SINGLE-GANG
PLASTER RING

TWO-GANG
PLASTER RING

ROUND
PLASTER RING

SINGLE-GANG
PLASTER RING WITH
1/2-IN. (13-mm) RISE

HOLES FOR 8-32
SCREWS TO SECURE
PLASTER RING
4-SQUARE TO SQUARE
BOX (TYPICAL)

6-32 THREADED
HOLES TO SECURE
DEVICE TO PLASTER
RING (TYPICAL)

8-32 THREADED
HOLES TO SECURE
LUMINAIRE (FIXTURE)
TO PLASTER RING

TWO-GANG
PLASTER RING WITH
1-IN. (25-mm) RISE

ROUND PLASTER
RING WITH
3-IN. (75-mm) RISE

PLASTER RINGS ARE SIZED TO EXACTLY FIT THE SQUARE BOX, EITHER 4 IN. (100 mm) OR 4 11/16 IN. (120 mm) SQUARE. THE DEVICE RINGS HAVE 6-32 THREADED HOLES TO SECURE SWITCHES AND RECEPTACLES TO THE PLASTER RING. ROUND RINGS HAVE 8-32 THREADED HOLES TO SECURE LUMINAIRES (FIXTURES) TO THE PLASTER RING. PLASTER RINGS ALSO HAVE A RISE EQUAL TO THE THICKNESS OF THE WALL COVERING MATERIAL. THE SQUARE BOX INSTALLS BEHIND THE WALL FINISH, AND THE PLASTER RING EXTENDS FROM THE FRONT OF THE BOX TO THE FLUSH EDGE OF THE WALL FINISH COVERING.

1/2 IN. (13 mm) WALL FINISH

PLASTER RING WITH 1/2-IN. (13-mm) RISE

4-SQUARE BOX

BLANK COVER

FRONT VIEW: SINGLE-SWITCH INDUSTRIAL COVER

FRONT VIEW: SINGLE DUPLEX RECEPTACLE INDUSTRIAL COVER

FRONT VIEW: COMBINATION SWITCH / DUPLEX RECEPTACLE INDUSTRIAL COVER

HOLES FOR SECURING COVER TO SQUARE BOX (TYPICAL)

HOLES FOR 6-32 SCREWS TO SECURE DEVICE TO COVER (TYPICAL)

OPENINGS IN THE COVER TO PROVIDE ACCESS TO THE DEVICE (TYPICAL)

BLANK COVER WITH 1/2 IN. (13 mm) TRADE SIZE KNOCKOUT

SIDE VIEW: INDUSTRIAL COVER

SIDE VIEW: BLANK COVER

1/2 IN. (13 mm) TRADE SIZE KNOCKOUT

EXTENSION BOXES ARE MADE FROM STANDARD BOXES WITH THE BACK REMOVED

FRONT VIEW: 4 11/16 IN. (120 mm) BLANK COVER

FRONT VIEW: 4 11/16 IN. (120 mm) DUPLEX RECEPTACLE INDUSTRIAL COVER

SIDE VIEW: 4 11/16 IN. (120 mm) INDUSTRIAL COVER

Figure 6-12

Figure 6-13 A metallic 4-square box with a single-gang plaster ring.

Figure 6-14 A nonmetallic 4-square box with a single-gang plaster ring.

the minimum volume that a box must have to house the conductors. (3) Compare the allowable volume of the various boxes with the calculated volume needed for all of the conductors. Any box that has a volume larger than the calculated necessary volume can be used.

Care must be taken with this calculation, however. The *NEC®* considers other items installed in the box to be conductors for the purposes of box fill calculations. In general,

- A device counts as two conductors. Because the box can house different sizes of conductors, the device is counted as the same wire size as that of conductors connected to the device. If the device is on a 15-ampere lighting circuit, the switch counts as two 14 AWG conductors.

- All of the equipment grounding conductors together count as a single conductor. If different sizes of conductors are present in the box, the equipment grounding conductor counts as the largest equipment grounding conductor in the box.

- Any cable clamps that are installed in the box count as one of the largest conductors in the box. If the clamp is outside the box, it is not included in counting conductors.

- Any luminaire (fixture) supports, luminaire (fixture) studs, or nipples installed in the box will count as one of the largest conductors in the box. The domed cover of a luminaire (fixture) may count as additional volume for the box. Figure 6-15 shows a sample box fill calculation.

6.3.2: Sizing Boxes for Conductors 4 AWG and Larger

In general, *314.28* stipulates the requirements for sizing junction and pull boxes containing conductors 4 AWG and larger. The Code is concerned with the distance from the conductor's point of entry into the box to the opposite wall of that box. As the cable size increases, the distance to the opposite wall of the box becomes larger.

Three general configurations are detailed by the Code:

1. *Straight pulls*, in which the conductors enter the box and exit the box on the opposite wall of the box, as shown in Figure 6-16. The rules are covered in detail in *314.28(A)(1)*.

"WIREMAN'S GUIDE"
SAMPLE DEVICE FILL CALCULATION

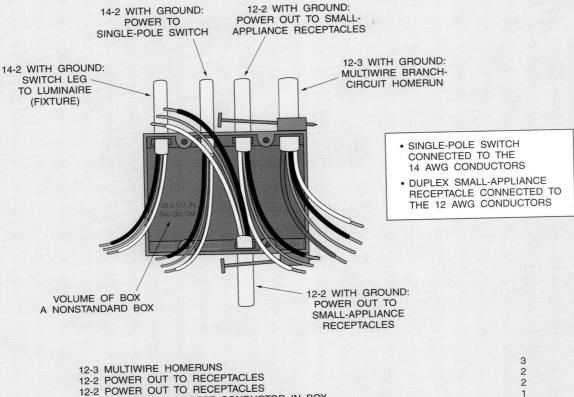

14-2 WITH GROUND:
POWER TO
SINGLE-POLE SWITCH

12-2 WITH GROUND:
POWER OUT TO SMALL-
APPLIANCE RECEPTACLES

14-2 WITH GROUND:
SWITCH LEG
TO LUMINAIRE
(FIXTURE)

12-3 WITH GROUND:
MULTIWIRE BRANCH-
CIRCUIT HOMERUN

- SINGLE-POLE SWITCH
  CONNECTED TO THE
  14 AWG CONDUCTORS
- DUPLEX SMALL-APPLIANCE
  RECEPTACLE CONNECTED TO
  THE 12 AWG CONDUCTORS

39.5 CU. IN.
648 CU. CM

VOLUME OF BOX
A NONSTANDARD BOX

12-2 WITH GROUND:
POWER OUT TO
SMALL-APPLIANCE
RECEPTACLES

12-3 MULTIWIRE HOMERUNS	3
12-2 POWER OUT TO RECEPTACLES	2
12-2 POWER OUT TO RECEPTACLES	2
CABLE CLAMPS—LARGEST CONDUCTOR IN BOX	1
GROUNDING CONDUCTORS—LARGEST CONDUCTOR IN BOX	1
DUPLEX RECEPTACLE—CONNECTED TO 12 AWG CONDUCTORS	2
	11

TOTAL: 12 AWG

14-2 POWER IN TO SWITCH	2
14-2 POWER OUT TO LUMINAIRE (FIXTURE)	2
SINGLE-POLE SWITCH—CONNECTED TO 14 AWG CONDUCTORS	2
	6

TOTAL: 14 AWG

11 (TOTAL 12 AWG) @ 2.25 IN.^{3} (36.9 CM^3) EACH = 24.75 IN.^{3} (405.9 CM^3)

6 (TOTAL 14 AWG) @ 2.00 IN.^{3} (32.8 CM^3) EACH = 12.00 IN.^{3} (393.6 CM^3)

TOTAL NECESSARY BOX VOLUME = 36.75 IN.^{3} (799.5 CM^3)

TO OBTAIN THE VOLUME OF THE BOX:
1. FOR STANDARD BOXES [THOSE BOXES LISTED IN *TABLE 314.16(A)*], THE VOLUME OF THE BOX CAN BE TAKEN FROM THE TABLE UNDER THE COLUMN HEADED MINIMUM VOLUME.
2. FOR NONSTANDARD BOXES, THE VOLUME WILL BE FOUND STAMPED, ETCHED, PRINTED, OR MOLDED INTO THE BOX SURFACE, USUALLY INSIDE OF THE BOX.

THE TOTAL VOLUME OF THE BOX INCLUDES THAT VOLUME ADDED BY EXTENSION BOXES, RAISED COVERS (SUCH AS INDUSTRIAL COVERS), AND RAISED PLASTER RINGS. THE ADDED VOLUME CAN BE INCLUDED ONLY IF THE ACTUAL VOLUME IS MARKED ON THE EXTENSION, COVER, OR RING OR IF THE EXTENSION, COVER, OR RING IS MADE FROM A STANDARD BOX. SQUARE AND OCTAGONAL EXTENSION BOXES ARE USUALLY MADE FROM STANDARD BOXES. INDUSTRIAL COVERS AND PLASTER RINGS ARE NOT MADE FROM STANDARD BOXES, AND THEREFORE THEIR VOLUME MUST BE MARKED ON THE COVER OR RING.

Figure 6-15

"WIREMAN'S GUIDE"
BOX SIZING FOR STRAIGHT PULLS FOR BOXES WITH CONDUCTORS 4 AWG OR LARGER

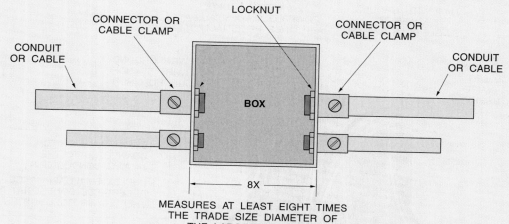

MEASURES AT LEAST EIGHT TIMES
THE TRADE SIZE DIAMETER OF
THE LARGEST RACEWAY

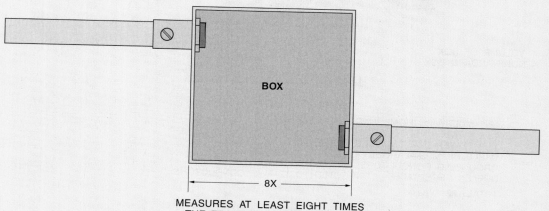

MEASURES AT LEAST EIGHT TIMES
THE TRADE SIZE DIAMETER OF
THE LARGEST RACEWAY

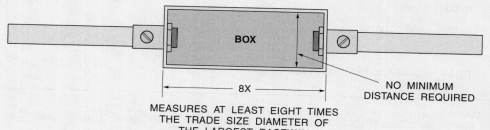

MEASURES AT LEAST EIGHT TIMES
THE TRADE SIZE DIAMETER OF
THE LARGEST RACEWAY

A STRAIGHT PULL IS A CONFIGURATION IN WHICH THE CABLE OR CONDUIT EXITS (LOAD SIDE) ON THE OPPOSITE WALL OF THE BOX FROM THAT OF THE ENTRY (LINE SIDE) CONDUIT OR CABLE. THE BOX IS SIZED TO THE LARGEST CABLE OR CONDUIT ENTERING THE BOX IF THERE IS MORE THAN ONE ENTRY PER SIDE. ALL OF THE EXAMPLES SHOWN ARE CONSIDERED STRAIGHT PULLS. NOTICE THAT THERE IS NO MINIMUM DISTANCE REQUIRED BETWEEN THE WALLS ADJACENT TO THE ENTRY AND EXIT WALLS OF THE BOX.

Figure 6-16

2. *Angle pulls*, including U-pulls, in which the conductors enter a box on one wall and exit the box on an adjacent wall (a wall that is other than the opposite wall). This configuration is shown in Figure 6-17. Also considered in this figure are rows of cable or conduit entries in the same box. If a box is used with two or more rows of openings, the box size is controlled by the row that requires the largest box. The rules for angle pulls are covered in *314.28(A)(2)*.

3. *Front-access angle pulls*, in which the conductors enter the back of a box that has a removable cover. The minimum depth of the box is the value given in *Table 312.6(A)*, in the 1 Wire Per Terminal column. This special type of angle pull is detailed in Figure 6-18.

6.4: INSTALLING BOXES

Boxes can be installed in one of two ways: (1) surface mounted or (2) flush mounted. Some device boxes are intended for use with only one mounting method, such as nail-on nonmetallic device boxes, and others, such as square boxes, can be installed either surface mounted with the use of an industrial cover or flush mounted with the use of a plaster ring. Large junction and pull boxes can also be surface mounted or flush mounted, depending on the covers employed.

6.4.1: Surface mounted Device Boxes

Surface mounted device boxes must be securely attached to the building surface. Several fastening methods can be employed in mounting boxes, as shown in Figure 6-19 (page 160) for hollow wall construction and in Figure 6-20 (page 161) for solid wall construction. A Surface mounted device box can be either a handy box or a square box. Handy boxes allow for the direct installation of the device, and the covers are designed for surface mounting. Surface mounted square boxes do not allow for the direct installation of a device but must employ industrial covers to secure the device in place.

When a box is surface mounted, the cable or conduit must enter from the rear of the box in order to be hidden from view. If the wiring method is surface mounted—for example, as with electrical metallic tubing (EMT), then the box must also be surface mounted (see "Wiring Methods" in Chapter 8 of this book).

6.4.2: Flush Mounted Device Boxes

Most types of device boxes are designed to be installed flush with the outer surface of a wall or ceiling. *NEC® 314.20* requires that the outer edge of the box be no more than ¼ in. (6.4 mm) from the finished wall surface if the wall is constructed of noncombustible materials, and that it be exactly flush with the outer wall surface if the surface of the wall is constructed of wood or other flammable materials. Obviously, accuracy in installing boxes is very important, as shown in Figure 6-21 (page 162). The electrician must be aware of the type of wall finish—sheetrock, plaster, or wood paneling—and the thickness of the wall covering in order to properly install the device box. If the box is recessed too far into the wall, it will be in violation of the *NEC®*. If the box is installed too far from the edge of the stud or framing member, it will pro-trude from the wall surface, causing problems when the device is installed. The box must also be mounted square to the stud or joist and plumb with the wall.

Many flush-mounted device boxes are available with the means for mounting, usually nails, included in the box from the manufacturer. Nails are acceptable for mounting the boxes on wooden studs or joists, but if nonmetallic framing methods are used, screws must be employed to mount the boxes. The box must be secure and rigidly attached. Any movement in the box will cause problems when the receptacle, switch, or other device is installed at the time of trim and will cause the box to eventually become separated from the mounting surface with use over time.

If there is any gap or opening between the wall surface and the edge of the box greater than ⅛ in. (3 mm), the gap must be filled in or repaired to eliminate the opening. The covers for flush device installations are constructed so that the cover plate is larger than the device box. This design allows the cover plate to hide the edge between the box and the wall surface, ensuring a clean appearance.

6.4.3: Surface mounted Junction or Pull Boxes

Surface mounted junction and pull boxes must also be rigidly attached to the building surface. The cover for the box must be of the same size as the box so that there is no overhang to the cover, as shown in

"WIREMAN'S GUIDE"
BOX SIZING FOR ANGLE PULLS AND U-PULLS FOR BOXES WITH 4 AWG CONDUCTORS OR LARGER

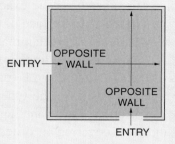

FRONT VIEW:
OPPOSITE WALL OF THE BOX

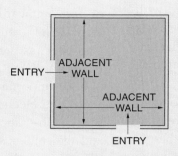

FRONT VIEW:
ADJACENT WALL OF THE BOX

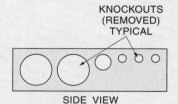

KNOCKOUTS (REMOVED) TYPICAL

SIDE VIEW

KNOCKOUTS CAN BE TAKEN IN ONE ROW OF HOLES. FOR A STRAIGHT PULL, THE MEASUREMENT FROM THIS SIDE OF THE BOX TO THE OPPOSITE SIDE OF THE BOX MUST BE AT LEAST **8 TIMES** THE TRADE DIAMETER OF THE LARGEST RACEWAY.

EXAMPLE:
THE KNOCKOUTS FOR THE BOX SHOWN ABOVE ARE, FROM LEFT TO RIGHT: 2 IN., 2 IN., 1 IN., 1/2 IN., 1/2 IN., AND 1/2 IN. (50 mm, 50 mm, 25 mm, 13 mm, 13 mm, AND 13 mm). THE DISTANCE FROM THIS WALL TO THE OPPOSITE SIDE WALL MUST BE NO LESS THAN:

2 IN. X 8 = 16 IN.
50 mm X 8 = 400 mm

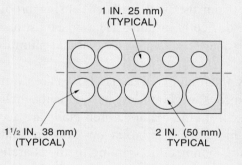

1 IN. 25 mm)
(TYPICAL)

1 1/2 IN. 38 mm)
(TYPICAL)

2 IN. (50 mm)
TYPICAL

KNOCKOUTS CAN BE PLACED ON DIFFERENT ROWS, THEREBY DIVIDING THE BOX INTO TWO HALVES. THE MINIMUM DISTANCE TO THE OPPOSITE SIDE IS BASED ON THE ROW THAT REQUIRES THE GREATEST LENGTH.

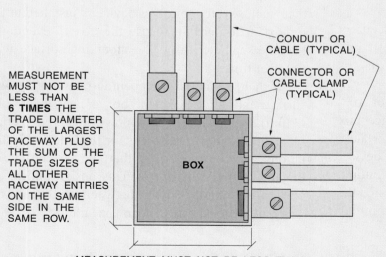

MEASUREMENT MUST NOT BE LESS THAN **6 TIMES** THE TRADE DIAMETER OF THE LARGEST RACEWAY PLUS THE SUM OF THE TRADE SIZES OF ALL OTHER RACEWAY ENTRIES ON THE SAME SIDE IN THE SAME ROW.

CONDUIT OR CABLE (TYPICAL)

CONNECTOR OR CABLE CLAMP (TYPICAL)

BOX

MEASUREMENT MUST NOT BE LESS THAN **6 TIMES** THE TRADE DIAMETER OF THE LARGEST RACEWAY PLUS THE SUM OF THE TRADE SIZES OF ALL OTHER RACEWAY ENTRIES ON THE SAME SIDE IN THE SAME ROW.

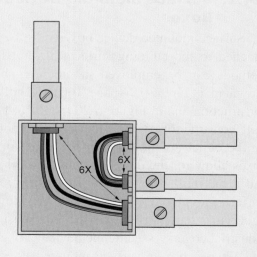

6X

THE DISTANCE BETWEEN THE RACEWAY ENTRIES CONTAINING THE SAME CONDUCTORS CANNOT BE LESS THAN **6 TIMES** THE TRADE DIAMETER OF THE RACEWAY.

Figure 6-17

"WIREMAN'S GUIDE"
BOX SIZING WITH ENTRY OPPOSITE
A REMOVABLE COVER FOR
BOXES WITH CONDUCTORS 4 AWG OR LARGER

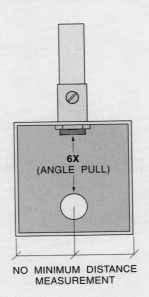

6X
(ANGLE PULL)

NO MINIMUM DISTANCE
MEASUREMENT

BOXES THAT HAVE A CONDUIT OR CABLE
THAT ENTERS A WALL OPPOSITE A REMOVABLE
COVER ARE SIZED BY THE "ONE WIRE PER
TERMINAL" COLUMN IN *TABLE 312.6(A)*. HOWEVER,
TABLE 312.6(A) IS BASED ON THE SIZE OF THE
CONDUCTORS ENTERING ON THE WALL OPPOSITE
THE REMOVABLE COVER. CONDUIT SIZE MUST
BE CONVERTED TO CONDUCTOR SIZE. THE
CONDUCTOR SIZE IS DETERMINED FROM THE
SMALLEST CONDUIT OR THE SMALLEST CABLE
CONNECTOR SIZE REQUIRED TO HOUSE THAT
NUMBER AND SIZE OF THE CONDUCTORS
INSTALLED IN THE CONDUIT OR CABLE.

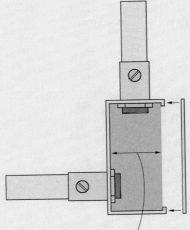

THE DISTANCE FROM THE ENTRY WALL
OF THE BOX TO THE REMOVABLE COVER
IS DETERMINED FROM *TABLE 312.6(A)*.

Figure 6-18

"WIREMAN'S GUIDE"
THREE COMMON TYPES OF
HOLLOW WALL FASTENERS

TOGGLE BOLT

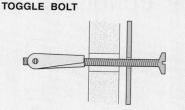

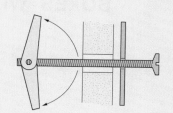

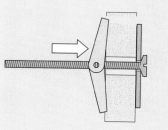

A TOGGLE BOLT CONSISTS OF A BOLT AND A SPRING-LOADED, HINGED WING NUT. A HOLE IS DRILLED IN THE HOLLOW WALL LARGE ENOUGH TO ACCOMMODATE THE WING NUT. THE BOLT IS INSERTED THROUGH THE MOUNTING HOLE IN THE ITEM TO BE ATTACHED (THE BOX STRAP), AND THE WING NUT IS THREADED ONTO THE BOLT. THE BOLT ASSEMBLY IS THEN INSERTED THROUGH THE HOLE IN THE HOLLOW WALL. INSIDE THE WALL, THE SPRING LOAD ON THE WING NUT CAUSES IT TO OPEN. THE BOLT IS THEN TIGHTENED, DRAWING THE WING NUT FAST AGAINST THE INSIDE OF THE HOLLOW WALL COVERING.

AUGER ANCHOR

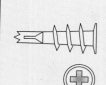
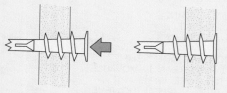
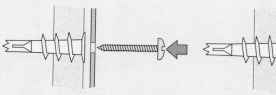

AN AUGER ANCHOR IS A ONE-PIECE METALLIC OR NONMETALLIC SLEEVE WITH A HOLE THROUGH THE MIDDLE FOR THREADING A TAPPING SCREW. A HOLE IS DRILLED IN THE HOLLOW WALL LARGE ENOUGH TO ACCOMMODATE THE SHANK OF THE ANCHOR, AND THE ANCHOR IS SCREWED INTO THE HOLE. THE THREADS ON THE ANCHOR DIG INTO THE WALLBOARD, THEREBY HOLDING THE ANCHOR FAST. THE ITEM TO BE ATTACHED IS THEN PLACED OVER THE ANCHOR, AND A SHEET METAL OR TAPPING SCREW IS SCREWED INTO THE ANCHOR. THIS TYPE OF ANCHOR CAN BE USED ONLY WHEN THE WALL IS CONSTRUCTED OF A SOFT MATERIAL, SUCH AS SHEETROCK, SO THAT THE AUGER THREADS CAN EFFECTIVELY DIG INTO THE WALL. THIS TYPE OF ANCHOR IS INTENDED FOR LIGHT-DUTY APPLICATIONS ONLY.

EXPANSION ANCHOR

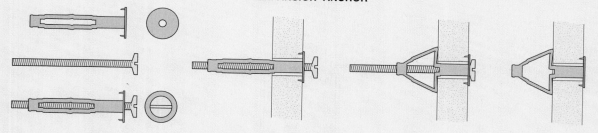

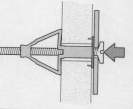

AN EXPANSION ANCHOR CONSISTS OF A BOLT AND A NUT ATTACHED TO A COLLAPSIBLE METAL FRAME, OR CAGE, WITH A RETAINING FLANGE ON THE OUTSIDE END. A HOLE IS DRILLED IN THE HOLLOW WALL LARGE ENOUGH TO ACCOMMODATE THE ANCHOR. THE ANCHOR IS THEN INSERTED INTO THE HOLE AND TIGHTENED. AS THE BOLT IS TIGHTENED, THE FRAME COLLAPSES, CONSTRICTING AGAINST THE INSIDE OF THE WALLBOARD. THE BOLT IS THEN REMOVED AND INSERTED INTO THE HOLE OF THE ITEM TO BE ATTACHED. THE BOLT IS THEN RE-INSERTED INTO THE HOLE, THREADED INTO THE NUT, AND TIGHTENED.

Figure 6-19

"WIREMAN'S GUIDE"
THREE COMMON TYPES OF SOLID WALL FASTENERS

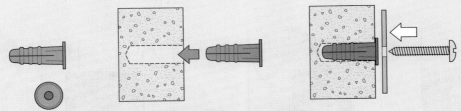

PLASTIC ANCHOR OR SHIELD

A HOLE IS DRILLED INTO THE SOLID WALL OF THE PROPER SIZE FOR THE ANCHOR TO BE INSTALLED. THE ANCHOR IS INSERTED INTO THE HOLE AND TAPPED INTO PLACE WITH A HAMMER. A TAPPING SCREW OF THE PROPER SIZE IS THEN INSERTED THROUGH THE HOLE IN THE ITEM TO BE FASTENED (BOX OR STRAP) AND SCREWED INTO THE ANCHOR. THIS TYPE OF ANCHOR IS INTENDED FOR LIGHT-DUTY APPLICATIONS ONLY, AND PLASTIC SHIELDS SHOULD NEVER BE USED IN SHEETROCK OR PLASTER WALLS. THESE ANCHORS COME IN VARIOUS SIZES, AND THE SCREW SHOULD ALWAYS BE SIZED ACCORDING TO THE INSTALLATION INSTRUCTIONS ACCOMPANYING THE ANCHORS. NO SPECIAL TOOLS ARE REQUIRED FOR THIS TYPE OF ANCHOR.

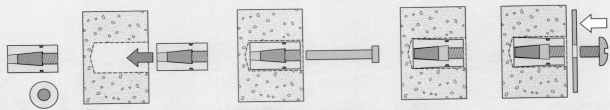

DROP-IN ANCHORS

A DROP-IN ANCHOR CONSISTS OF A METAL SLEEVE WITH A THREADED HOLE THROUGH THE CENTER. INSIDE OF THE THREADED HOLE IS A METAL WEDGE. A PROPERLY SIZED HOLE IS DRILLED INTO THE SOLID WALL AND THE DROP-IN ANCHOR IS INSERTED INTO THE HOLE. A SPECIAL SETTING TOOL IS THEN INSERTED INTO THE HOLE, AND A HAMMER IS USED TO STRIKE THE TOOL, DRIVING THE WEDGE FURTHER INTO THE HOLE. THE METAL SLEEVE IS CAUSED TO EXPAND BY THE WEDGE, THEREBY CONSTRICTING THE SLEEVE AGAINST THE SIDES OF THE SOLID WALL. THE ITEM TO BE FASTENED IS THEN PLACED OVER THE HOLE, AND A PROPERLY SIZED MACHINE SCREW IS INSERTED. THE SCREW IS THEN TIGHTENED. DROP-IN ANCHORS ARE AVAILABLE IN VARIOUS SIZES AND ARE DESIGNED FOR MEDIUM- TO HEAVY-DUTY APPLICATIONS. IT IS ALWAYS NECESSARY TO USE THE PROPER-SIZE ANCHOR FOR THE JOB AND THE PROPER-SIZE MACHINE SCREW FOR THE ANCHOR. A SETTING TOOL IS CONTAINED IN EACH BOX OF ANCHORS. THE PROPER SETTING TOOL MUST ALWAYS BE USED TO ENSURE SATISFACTORY RESULTS.

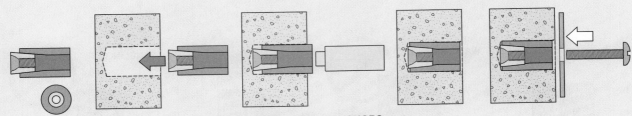

LEAD SHIELD ANCHORS

LEAD SHIELD ANCHORS CONSIST OF A LEAD OUTER SLEEVE SURROUNDING A STEEL CORE. THE STEEL CORE IS FLARED AT ONE END WITH A THREADED HOLE THROUGH THE CENTER. A PROPERLY SIZED HOLE IS DRILLED IN THE SOLID WALL, AND THE LEAD ANCHOR IS INSERTED INTO THE HOLE. IT IS VERY IMPORTANT THAT THE CORRECT END IS INSERTED, OR THE ANCHOR WILL NOT FASTEN PROPERLY. THE END OF THE ANCHOR THAT IS FLARED IS INSERTED INTO THE HOLE. A SPECIAL SETTING TOOL IS THEN USED WITH A HAMMER TO DRIVE THE LEAD INTO THE HOLE. THE LEAD CONSTRICTS AROUND THE FLARED END OF THE STEEL CORE AND THE WALL OF THE HOLE TO HOLD THE ANCHOR FAST. A PROPERLY SIZED MACHINE SCREW IS THEN USED TO FASTEN THE BOX OR STRAP TO THE WALL. LEAD ANCHORS ARE AVAILABLE IN SEVERAL SIZES AND ARE RATED FOR MEDIUM- TO HEAVY-DUTY FASTENING. WHEN INSTALLED PROPERLY, THEY WILL SUPPORT A CONSIDERABLE AMOUNT OF WEIGHT. A SETTING TOOL IS INCLUDED IN EACH BOX OF ANCHORS. BECAUSE OF THE RELATIVELY HIGH COST OF LEAD, THESE ANCHORS TEND TO BE MORE EXPENSIVE THAN OTHER TYPES.

Figure 6-20

"WIREMAN'S GUIDE"
INSTALLING DEVICE BOXES

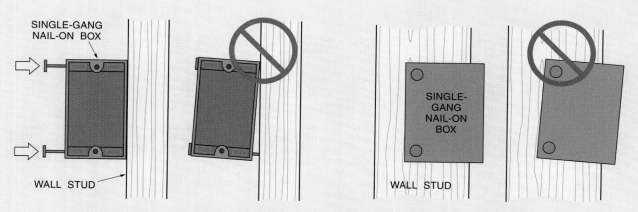

BOXES MUST BE MOUNTED SQUARE TO THE STUD OR JOIST AND PLUMB WITH THE WALL.

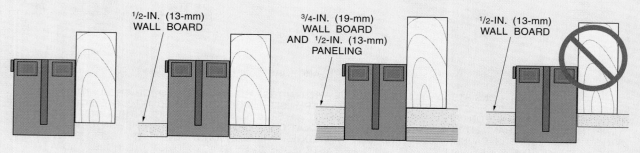

THE BOX SHOULD BE INSTALLED SO THAT THE FRONT EDGE OF THE BOX WILL BE FLUSH WITH THE FINISHED WALL SURFACE. THE ELECTRICIAN MUST BE AWARE OF WHAT TYPE OF WALL FINISHING IS TO BE INSTALLED AND THE THICKNESS OF THE FINISH. IF THE BOX IS INSTALLED SO THAT IT EXTENDS PAST THE FINISHED WALL, TRIM PROBLEMS WILL RESULT.

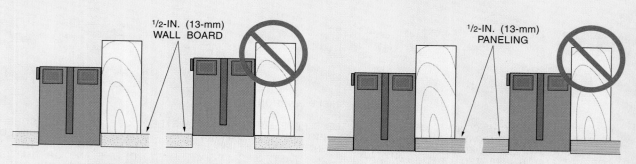

IN NONFLAMMABLE WALLS SUCH AS SHEETROCK AND PLASTER, THE FRONT EDGE OF THE BOX CAN BE NO MORE THAN 1/4 IN. (6.4 mm) FROM THE FINISHED WALL SURFACE. IN FLAMMABLE WALLS SUCH AS PANELING OR WOODEN SIDING, THE FRONT EDGE OF THE BOX MUST BE FLUSH WITH THE FINISHED WALL SURFACE. THE BOX IS DESIGNED TO RETARD THE SPREAD OF FIRE SHOULD A PROBLEM OCCUR WITHIN THE BOX. IF ANY FLAMMABLE WALL SURFACE IS EXPOSED TO THE INSIDE OF THE BOX, FIRE COULD SPREAD MORE EASILY.

Figure 6-21

Figure 6-22. As with Surface mounted device boxes, the cables or conduits must enter through the back of the box in order to be hidden from view. If the wiring method is surface mounted, the junction or pull box must also be surface mounted.

6.4.4: Flush-Mounted Junction or Pull Boxes

Many large junction and pull boxes can be flush mounted. Several considerations arise with such installations: (1) Can the box be rigidly attached to a stud or other framing members within the wall or ceiling? The box must be attached on opposite sides in order to be rigidly attached. (2) Is the wall deep enough to accommodate the depth of the box? (3) Is the box cover larger than the box itself? If the cover is of the same size as the box, as with Surface mounted boxes, the edge of the box and the edge of the wall covering will present an unfinished appearance that will be unacceptable in most installations. Figure 6-22 presents more information about flush-mounted junction and pull boxes.

6.5: CONDUIT BODIES

> **KEY TERMS**
>
> **Conduit body**   (See *NEC® Article 100*): A type of raceway fitting that allows for a rapid change in direction of wiring, or for the tapping of a raceway, without the use of a box. Conduit bodies have removable covers that allow access to the interior of the fitting to facilitate the installation of the conductors.

Conduit bodies are fittings that aid in running conduit around tight corners or for tapping off from a conduit run. Residential electricians occasionally have a need to install conduit in remodeling work or outside circuit installations. Conduit bodies are covered by *314.16(C)*.

> **KEY TERMS**
>
> **Short conduit body**   A type of conduit body designed for use with smaller sizes of raceways but with limited bending space of the conductors. The length of short conduit bodies is considerably less than the length of a regular conduit body of the same raceway trade size.

Conduit bodies are available as **short conduit bodies** as well as regular conduit bodies. Short conduit bodies are intended for the installation of conductors 6 AWG or smaller. No taps or splices are to be accomplished in a short conduit body. Regular conduit bodies can include splices and taps if the conduit body is durably and legibly marked with the volume. The number of conductors allowed in a conduit body is the same as the number of conductors allowed in the conduit to which it is attached, although this may not be the case with some larger (conductor size 300 kcmil and up) conduit bodies. Care should be used in sizing the conduit bodies for large conduit runs. Figure 6-23 shows some examples of conduit bodies.

"WIREMAN'S GUIDE"
COVERS AND LARGE JUNCTION OR
PULL BOX INSTALLATION

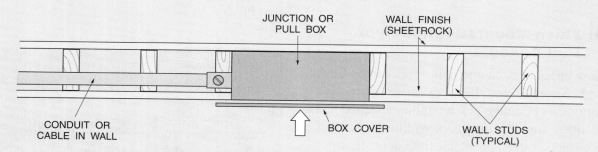

JUNCTION OR
PULL BOX

WALL FINISH
(SHEETROCK)

CONDUIT OR
CABLE IN WALL

BOX COVER

WALL STUDS
(TYPICAL)

FOR FLUSH MOUNTING OF LARGE JUNCTION OR PULL BOXES, THE BOX
COVER MUST BE LARGER THAN THE BOX TO PROVIDE A NEAT AND
PROFESSIONAL APPEARANCE.

IF THE BOX COVER IS THE SAME SIZE AS THE BOX, THE UNFINISHED EDGE OF
THE WALL COVERING WILL SHOW WHERE IT WAS CUT TO ACCEPT THE BOX. THIS
TYPE OF COVER IS DESIGNED FOR SURFACE MOUNTING OF THE BOX.

FOR FLUSH MOUNTING OF LARGE JUNCTION OR PULL BOXES, CARE MUST BE
TAKEN TO ENSURE THAT THE BOX IS NOT TOO DEEP FOR THE WALL WHERE IT
IS TO BE INSTALLED.

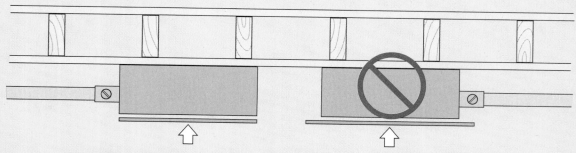

FOR SURFACE-MOUNTED LARGE JUNCTION OR PULL BOXES, THE COVER MUST
BE OF THE SAME SIZE AS THAT OF THE BOX, OR IT WILL OVERHANG THE BOX,
PRESENTING A POOR APPEARANCE, AND A HAZARD FOR SCRATCHING OR SNAGGING.

Figure 6-22

"WIREMAN'S GUIDE"
CONDUIT BODIES

CONDUIT BODIES ARE INSTALLED IN CONDUIT RUNS TO TURN SHARP CORNERS, TO PROVIDE WIRE PULLING POINTS, AND FOR MAKING SPLICES. CONDUIT BODIES CAN BE MANUFACTURED FROM CAST ALUMINUM, CAST IRON, STEEL, OR PVC AND ARE AVAILABLE IN ALL SIZES OF CONDUIT, FROM 1/2 IN. (13 mm) TO 6 IN. (150 mm).

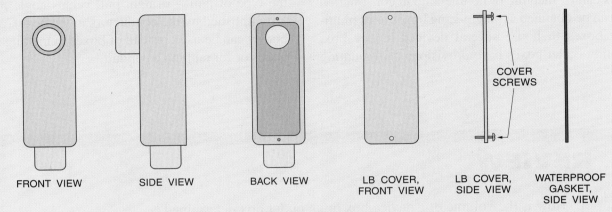

FRONT VIEW SIDE VIEW BACK VIEW LB COVER, FRONT VIEW LB COVER, SIDE VIEW COVER SCREWS WATERPROOF GASKET, SIDE VIEW

TYPE OF LB CONDUIT BODY

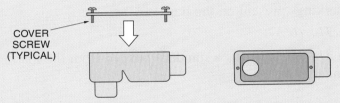

COVER SCREW (TYPICAL)

SIDE VIEW AND BACK VIEW OF A SHORT LB (SLB) CONDUIT BODY. THESE CONDUIT BODIES ARE USED FOR CONDUCTORS UP TO 6 AWG. CONDUCTORS LARGER THAN 6 AWG REQUIRE A FULL-SIZED LB CONDUIT BODY.

SOME OTHER AVAILABLE CONDUIT BODIES

IN ORDER TO SPLICE OR TERMINATE IN A CONDUIT BODY, IT MUST BE MARKED WITH THE VOLUME. CONDUCTOR FILL IS CALCULATED THE SAME AS FOR BOXES IN *314.16.*

18 CU. IN. (295 MM)

SIDE VIEW OF A TYPE LR CONDUIT BODY

SIDE VIEW OF A TYPE LL CONDUIT BODY

SIDE VIEW OF A TYPE C CONDUIT BODY

SIDE VIEW OF A TYPE T CONDUIT BODY

SIDE VIEW OF A TYPE E CONDUIT BODY

SOME OTHER AVAILABLE CONDUIT BODIES

Figure 6-23

SUMMARY

Boxes must be used for all devices, outlets, and switch points. Boxes for conductors 6 AWG and smaller are termed device boxes and are sized by the number of conductors, devices, and fittings to be housed by the box. Device boxes have provisions for the mounting of devices using 6-32 screws. Device boxes can be metallic or nonmetallic in construction and can be obtained as single-gang boxes or as multi-gang boxes to house several devices in one box. Lighting outlet boxes have provisions for mounting luminaires (fixtures) with 8-32 screws. Lighting outlet boxes can also be of metallic or nonmetallic construction. Square boxes are a special kind of box that is very versatile and can be used for surface and concealed installations and as device boxes or lighting outlet boxes. Boxes containing conductors larger than 6 AWG are termed pull boxes and are sized by the type of pull—straight pull versus angle pull—accomplished by the conductors in the box. Conduit bodies can be used instead of boxes in making sharp bends or for tapping a conduit.

REVIEW

1. How is the volume of a device box or an outlet box determined?
   a. by measuring the box
   b. from the markings present on the box
   c. from *Table 314.16(A)*
   d. from *Table 314.16(A)* if it is a standard box, or from the markings on the box if it is a nonstandard box

2. Boxes in which the largest conductor is _____ or smaller are device or outlet boxes.
   a. 4 AWG
   b. 6 AWG
   c. 8 AWG
   d. none of the above—the size is determined by the device installed in the box

3. Metallic device boxes are limited to _____ gangs.
   a. one
   b. two
   c. three
   d. none of the above—there is no physical limit to the number of gangs allowed

4. Metallic square boxes are limited to _____ gangs
   a. one
   b. two
   c. three
   d. none of the above—there is no physical limit to the number of gangs allowed

5. In concealed work, square boxes need to have _____ used with the box.

    a. industrial covers

    b. plaster rings

    c. external cable clamps

    d. blank covers

6. What is the maximum number of 14 AWG conductors allowed in a 3-in. × 2-in. × 3½-in. (75-mm × 50-mm × 90-mm) device box?

    a. 10 conductors

    b. 9 conductors

    c. 8 conductors

    d. none of the above—there is no physical limit to the number of conductors allowed in a device box

7. A device installed in a device box is counted as _____ conductor(s).

    a. one

    b. two

    c. three

    d. none of the above—devices are not conductors

8. For U-pulls or angle pulls in boxes with 250 kcmil conductor installed, the distance between raceway opens containing the same conductors must be _____ times the trade diameter of the raceway or cable.

    a. six

    b. seven

    c. eight

    d. none of the above—the distance is determined by the number of conductors, not by the size of the conduit or cable

9. For device boxes, the number of conductors allowed in the box is reduced by one conductor for each _____ cable clamp(s).

    a. one

    b. two

    c. three

    d. none of the above—the number of allowable conductors is reduced by one if there are any cable clamps present in the box, regardless of the actual number of clamps

10. Metallic device boxes are available _____.

    a. only without internal cable clamps

    b. with or without internal cable clamps

    c. only with internal cable clamps

    d. with internal cable clamps for AC and MC cable only

"WIREMAN'S GUIDE" REVIEW

1. What size device box (in cubic in. [cubic mm]) is needed for four 12 AWG conductors, five 14 AWG conductors, three cable clamps, one receptacle connected to the 12 AWG conductors, and one switch connected to the 14 AWG conductors? The cable used is NM cable with equipment grounding conductors not included in the foregoing count of conductors.

2. Explain when it is allowable to use a pan box (porch pan). Why do some AHJs forbid the use of pan boxes altogether?

3. All cables installed in a box must be secured to the box and stapled or strapped within 12 in. (300 mm) of the box, except for single-gang nonmetallic boxes. Explain why the Code excludes these boxes and what rules change if a clamp is not used with a single-gang nonmetallic box. Why does this not apply to single-gang metal boxes?

4. Explain the major drawback to using *Table 314.16(A)* to calculate box size, and why the method given in *Table 314.16(B)* is superior.

5. Explain when the volume of industrial covers, plaster rings, or extension boxes can be included in the volume of the device box.

CHAPTER 7

Overcurrent and Types of Circuits

OBJECTIVES

On completion of this chapter, the student will be able to:

☑ Identify the causes of overcurrent.

☑ Explain the function of fuses and circuit breakers, and state the advantages and disadvantages of each.

☑ Explain the difference between overload and circuit overutilization.

☑ Explain the difference between short circuits and ground faults.

☑ Describe what controls the current flow to a motor or other inductive load.

☑ Differentiate among motor starting overcurrent, motor running overcurrent, and locked-rotor current.

☑ Define and explain the differences between services, branch circuits, and feeder.

INTRODUCTION

This chapter is concerned with overcurrent and overcurrent protection. The causes of overcurrent, along with grounding, are among the most misunderstood subjects in the electrical trade. Overcurrent is the major issue that electricians and electrical system designers must deal with on an everyday basis, and overcurrent protection is fundamental to a safe and useful installation. This chapter explores the two main types of overcurrent, faults and overloads, and examines the different problems they present to the electrician. Also covered is the requirement for overcurrent protection with branch circuits, feeders, and service conductors. The material in this chapter provides a background for use in the study of the various wiring method installation requirements and restrictions presented in Chapter 8.

7.1: THE CAUSES AND NATURE OF OVERCURRENT

KEY TERMS

Overcurrent *(See NEC® Article 100)*: Too much current in a circuit. Overcurrent has four possible causes: short circuits, ground faults, overloads, and circuit overutilization.

Providing overcurrent protection for circuits and equipment is one of the primary responsibilities of the electrician. *NEC® 240.20* requires that an overcurrent-protective device be connected in series with each ungrounded conductor of a circuit. **Overcurrent** in an electrical circuit is a condition in which too much current is flowing through the circuit. The two causes of overcurrent are faults and overloads.

Two principal types of overcurrent-protective devices are installed in the circuit to protect against any faults and overloads: fuses and circuit breakers.

KEY TERMS

Fuse A device that is sensitive to heating caused by circuit current flow. If the current becomes too high, the heating will cause the fuse link to open, thus eliminating the current flow through the circuit.

Disconnect switch An enclosure that houses a switch mechanism and sometimes also fuses. The switch is designed to disconnect a portion of the circuit or a load from the electrical supply and when fuses are included, the switch can provide overcurrent protection.

- **Fuses.** A fuse is a removable (and replaceable) device that provides a link in series with the circuit load that is sensitive to changes in current flow, as shown in Figure 7-1. If the current exceeds the rating on the fuse, the fuse link will melt, thereby opening the circuit. Fuses are usually housed in **disconnect switches** that are manufactured to accept the installation of fuses. Disconnect switches are also available without provision for fuses, and fuses can also be installed in boxes rather than in disconnect switches. Fuses are available in many standard sizes, as listed in *240.6*. The physical size and design of fuses are such that any brand-name fuse can be installed into virtually any other brand-name disconnect switch. Fuses and disconnect switches are also grouped by ampacity rating and physical size so that it is not possible to install a larger ampere–rated fuse into a disconnect switch designed for less current, as shown in Figure 7-2. Fuses come in a large number of types for different applications, voltages, and fuse links. Some fuses have a dual link that allows for a time delay on low-level overloads but also quick response for faults. The fuses used most often in dwellings are Type S plug fuses and Type RK-1 and Type RK-5 cartridge fuses. More information about the *NEC®* requirements for fuses can be found in *Part V* of *240* beginning at *240.50* of the Code. References to fuses are found elsewhere in the *NEC®*, however, and care must be taken before installation of any circuits or equipment that the proper fusing is employed.

KEY TERMS

Circuit breaker (See *NEC® Article 100*): A device that is sensitive to both the heating caused by current flow and the magnetic field created by current flow. The device will open the circuit if the current flow

becomes too great. Circuit breakers are usually housed in circuit-breaker panels, although enclosures for individual circuit breakers are available.

- **Circuit breakers.** Circuit breakers, or "breakers" for short, are also devices that are sensitive to changes in current flow. A circuit breaker is also placed in series with the circuit load and will open the circuit if the current flow exceeds the ampere rating. The circuit breaker, however, opens a set of contacts within the breaker housing, or case, as shown in Figure 7-3. These contacts can be manually reset after the reason for the overcurrent has been eliminated from the circuit. The standard ampere ratings of circuit breakers are also listed in *240.6*. Circuit breakers either clip or screw onto bussing of electrical distribution panels. Each brand-name electrical distribution panel has specific bus and clip configurations, so that the circuit breakers and the distribution panel must be manufactured by the same company. Circuit breakers are discussed in the *NEC®* beginning with *240.80*. As with fuses, additional requirements for circuit-breaker use are scattered throughout the Code, and in using a circuit breaker to protect any particular load, care must be taken to ensure that it is approved for that piece of equipment, and that it is properly sized.

One or both of these overcurrent-protective devices can be used to protect circuits and equipment against faults and overloads. Usually for a specific load, multiple overcurrent-protective devices are installed on the supply conductors between the connection to the power utility and the actual load.

7.1.1: Faults

KEY TERMS

Fault The term used by the *NEC®* to define a circuit that is not operating properly, so that current flow is using an unintended pathway.

Fault is the term that the *NEC®* uses to describe a problem with an electrical circuit that is not operating properly. The problem may be caused by physical damage to the system, faulty installation, faulty equipment, or any number of other factors. In most

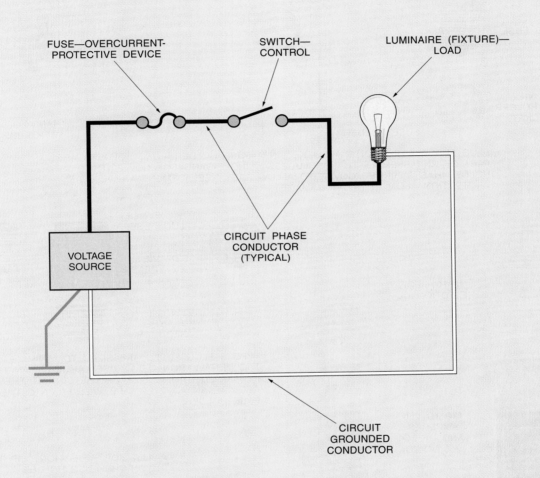

"WIREMAN'S GUIDE"
OVERCURRENT-PROTECTIVE DEVICE
IN SERIES WITH THE LOAD

FUSE—OVERCURRENT-
PROTECTIVE DEVICE

SWITCH—
CONTROL

LUMINAIRE (FIXTURE)—
LOAD

VOLTAGE
SOURCE

CIRCUIT PHASE
CONDUCTOR
(TYPICAL)

CIRCUIT
GROUNDED
CONDUCTOR

OVERCURRENT-PROTECTIVE DEVICES—FUSES AND
CIRCUIT BREAKERS—ARE INSTALLED IN SERIES
WITH THE CIRCUIT LOAD. IF THE CURRENT BECOMES
TOO LARGE, THE OVERCURRENT DEVICE WILL OPEN
THE CIRCUIT AND STOP THE CURRENT FLOW. A FUSE
IS SHOWN IN THE DRAWING; CIRCUIT BREAKERS
ARE ALSO INSTALLED IN SERIES WITH THE LOAD.

Figure 7-1

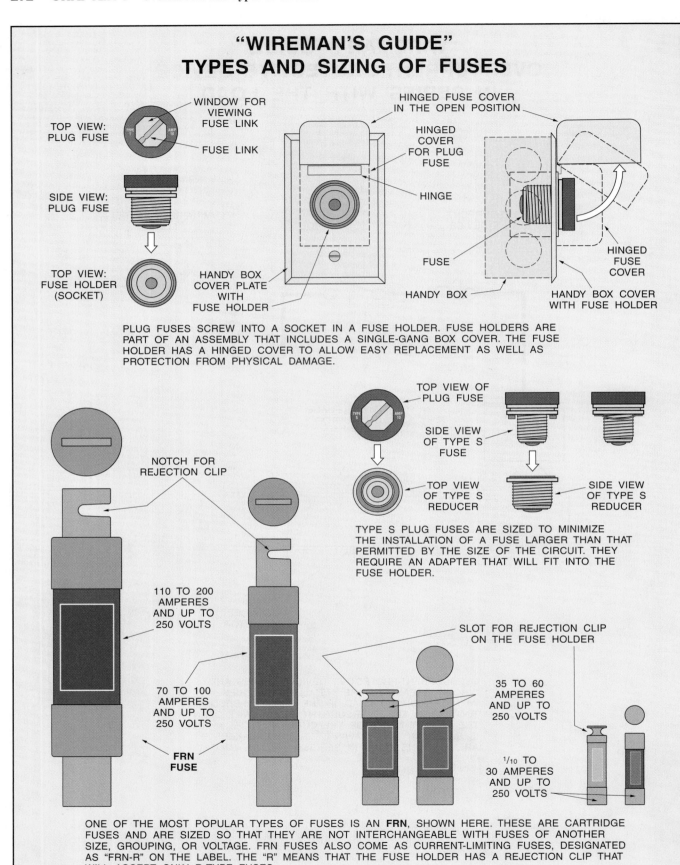

"WIREMAN'S GUIDE"
TYPES AND SIZING OF FUSES

TOP VIEW:
PLUG FUSE

WINDOW FOR
VIEWING
FUSE LINK

FUSE LINK

HINGED FUSE COVER
IN THE OPEN POSITION

HINGED
COVER
FOR PLUG
FUSE

HINGE

SIDE VIEW:
PLUG FUSE

TOP VIEW:
FUSE HOLDER
(SOCKET)

HANDY BOX
COVER PLATE
WITH
FUSE HOLDER

FUSE

HANDY BOX

HINGED
FUSE
COVER

HANDY BOX COVER
WITH FUSE HOLDER

PLUG FUSES SCREW INTO A SOCKET IN A FUSE HOLDER. FUSE HOLDERS ARE
PART OF AN ASSEMBLY THAT INCLUDES A SINGLE-GANG BOX COVER. THE FUSE
HOLDER HAS A HINGED COVER TO ALLOW EASY REPLACEMENT AS WELL AS
PROTECTION FROM PHYSICAL DAMAGE.

NOTCH FOR
REJECTION CLIP

TOP VIEW OF
PLUG FUSE

SIDE VIEW
OF TYPE S
FUSE

TOP VIEW
OF TYPE S
REDUCER

SIDE VIEW
OF TYPE S
REDUCER

TYPE S PLUG FUSES ARE SIZED TO MINIMIZE
THE INSTALLATION OF A FUSE LARGER THAN THAT
PERMITTED BY THE SIZE OF THE CIRCUIT. THEY
REQUIRE AN ADAPTER THAT WILL FIT INTO THE
FUSE HOLDER.

110 TO 200
AMPERES
AND UP TO
250 VOLTS

70 TO 100
AMPERES
AND UP TO
250 VOLTS

**FRN
FUSE**

SLOT FOR REJECTION CLIP
ON THE FUSE HOLDER

35 TO 60
AMPERES
AND UP TO
250 VOLTS

1/10 TO
30 AMPERES
AND UP TO
250 VOLTS

ONE OF THE MOST POPULAR TYPES OF FUSES IS AN **FRN**, SHOWN HERE. THESE ARE CARTRIDGE
FUSES AND ARE SIZED SO THAT THEY ARE NOT INTERCHANGEABLE WITH FUSES OF ANOTHER
SIZE, GROUPING, OR VOLTAGE. FRN FUSES ALSO COME AS CURRENT-LIMITING FUSES, DESIGNATED
AS "FRN-R" ON THE LABEL. THE "R" MEANS THAT THE FUSE HOLDER HAS A REJECTION CLIP THAT
WILL ACCEPT ONLY R-TYPE FUSES.

Figure 7-2

"WIREMAN'S GUIDE"
CIRCUIT BREAKERS

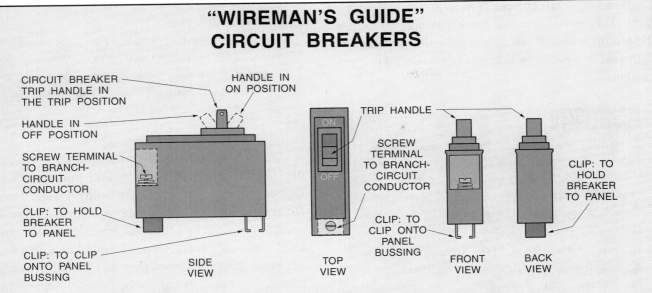

CIRCUIT BREAKER TRIP HANDLE IN THE TRIP POSITION

HANDLE IN ON POSITION

HANDLE IN OFF POSITION

SCREW TERMINAL TO BRANCH-CIRCUIT CONDUCTOR

CLIP: TO HOLD BREAKER TO PANEL

CLIP: TO CLIP ONTO PANEL BUSSING

SIDE VIEW

TRIP HANDLE

SCREW TERMINAL TO BRANCH-CIRCUIT CONDUCTOR

CLIP: TO CLIP ONTO PANEL BUSSING

TOP VIEW

FRONT VIEW

CLIP: TO HOLD BREAKER TO PANEL

BACK VIEW

CIRCUIT BREAKER BRAND X: NOTICE THAT THE METHOD OF ATTACHING THE CIRCUIT BREAKER TO THE PANEL AND THE BUSSING IS DIFFERENT THAN FOR BRAND Y BELOW.

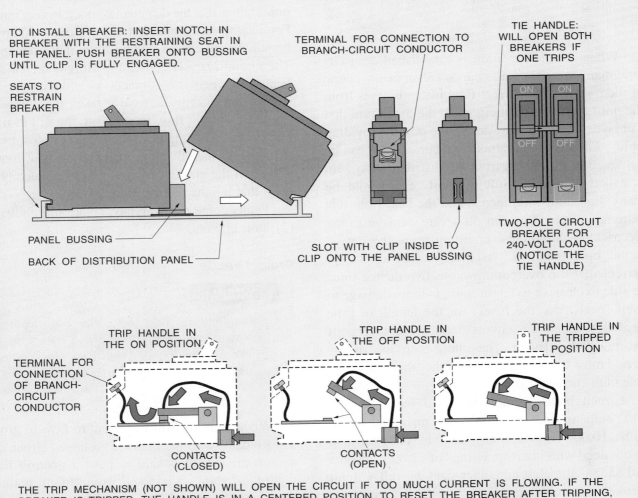

TO INSTALL BREAKER: INSERT NOTCH IN BREAKER WITH THE RESTRAINING SEAT IN THE PANEL. PUSH BREAKER ONTO BUSSING UNTIL CLIP IS FULLY ENGAGED.

SEATS TO RESTRAIN BREAKER

PANEL BUSSING

BACK OF DISTRIBUTION PANEL

TERMINAL FOR CONNECTION TO BRANCH-CIRCUIT CONDUCTOR

SLOT WITH CLIP INSIDE TO CLIP ONTO THE PANEL BUSSING

TIE HANDLE: WILL OPEN BOTH BREAKERS IF ONE TRIPS

TWO-POLE CIRCUIT BREAKER FOR 240-VOLT LOADS (NOTICE THE TIE HANDLE)

TRIP HANDLE IN THE ON POSITION

TRIP HANDLE IN THE OFF POSITION

TRIP HANDLE IN THE TRIPPED POSITION

TERMINAL FOR CONNECTION OF BRANCH-CIRCUIT CONDUCTOR

CONTACTS (CLOSED)

CONTACTS (OPEN)

THE TRIP MECHANISM (NOT SHOWN) WILL OPEN THE CIRCUIT IF TOO MUCH CURRENT IS FLOWING. IF THE BREAKER IS TRIPPED, THE HANDLE IS IN A CENTERED POSITION. TO RESET THE BREAKER AFTER TRIPPING, PUSH THE HANDLE TO THE FULLY OFF POSITION AND THEN PUSH IT TO THE FULL ON POSITION.

Figure 7-3

cases, a very high current flow is associated with a fault. This is because the current is able to bypass the circuit load and obtain ground using a shortened pathway, or a pathway that was not intended to carry current.

> ### KEY TERMS
>
> **Available fault current** The maximum amount of current available if a fault occurs. The available current is limited by the amount of current the power provider's generator can produce, the impedance in the delivery system circuiting, and the let-through current of the power provider's transformers.
>
> **Interrupt rating** (See *NEC® Article 100*): The rating of an overcurrent-protective device that defines the maximum amount of current the device can pass without failing. The overcurrent-protective device must have an interrupt rating that is at least as high as the available fault current.

When a fault condition is established on a circuit, many things begin to happen all at once—none of them good. The current flow instantly rises from the normal load on that circuit to the maximum that can be delivered to that spot by the power-providing utility. The maximum amount of current that can be delivered to the circuit from the distribution system is called **available fault current**. This could be many thousands of amperes, and the heat from this extreme current flow can do severe damage to the electrical system if the overcurrent-protective device cannot open the circuit fast enough. The amount of current that the overcurrent-protective device must be able to clear, or to let through, without damage to itself (except for the fuses) for the length of time needed to open the circuit defines the **interrupt rating** of the device. The interrupt rating of the device must be the same as or higher than the available fault current.

Circuit breakers must be rated for at least a 5000-ampere interrupt rating, and many are found with a 10,000-ampere interrupt rating. They can also be obtained with interrupt ratings of 22,000 amperes and 65,000 amperes, and other sizes as well. Circuit breakers typically must carry the fault current for approximately one cycle (1/60 of a second) because that is how long it takes the circuit breaker to recognize a fault situation and for the contacts inside the breaker to operate.

> ### KEY TERMS
>
> **Current-limiting fuses** A class of fuses that limit the maximum amount of current that can flow through a circuit regardless of the available fault current. These types of fuses are designed to operate very quickly in the event of a fault.

Fuses are routinely found with interrupt ratings of 100,000 amperes and 200,000 amperes. Some types of fuses, called **current-limiting fuses**, operate so quickly that the circuit is opened before the current has time to get to the available fault current level. Because events happen so quickly with a fault, the exact level of current flow at which the overcurrent device operates cannot be precisely determined and is not of any real importance as long as it does not exceed the interrupt current level. What is most important is that the device must operate very quickly. For example, a fault that occurs on a 20-ampere branch circuit causes current flow of 5000 amperes. The circuit breaker senses the current increase and begins to open the contacts as soon as the fault occurs, but during the time it takes the contacts to open, the current flow increases to the full available current of 5000 amperes. The breaker is constructed to be able to pass that much current, and the contacts will open the circuit before the heating becomes so great that it damages the breaker.

Faults come in two types: (1) ground faults and (2) short circuits.

Ground Faults

> ### KEY TERMS
>
> **Ground fault** A flaw in an electrical circuit in which some current is escaping and is flowing to ground using a pathway other than the one intended. Ground faults pose an electrocution hazard.

When a circuit allows current to flow to ground using a pathway other than the grounded circuit conductor, the circuit is said to have a **ground fault**. Figure 7-4 shows a line side ground fault with low impedance to ground. The low impedance allows the current in the circuit to instantly increase to tens of thousands of amperes when the ground fault occurs. This much current flow through a circuit that is not designed to handle it can cause many problems and dangers if it is not eliminated very quickly. The large current flow can superheat the air inside enclosures, causing them to explode, very much like a bomb. The current is also being conducted over equipment enclosures and raceways that are not supposed to carry current. This presents an electrocution hazard for anyone in contact with the energized equipment enclosure or raceway.

Figure 7-4 also shows a load side ground fault—a fault between the grounded circuit conductor and the grounding path. A load side ground fault allows current to use an unintended path to ground, which presents an electrocution hazard, but very high current flows do not occur because the load is still connected to the circuit. The current probably will not increase enough to cause the overcurrent-protective device to open the circuit and may not increase at all. Load side ground faults are not even defined by the *NEC*® as ground faults, but they constitute a potential hazard to the system.

Short Circuits

> **KEY TERMS**
>
> **Short circuit** A circuit that is not operating properly because the current is bypassing the load(s) and returning to ground using the circuit conductors. Short circuits do not pose an electrocution hazard.

The second type of fault is a **short circuit**. Short circuits differ from ground faults in one major way: they involve only the circuit conductors, and no current is escaping from the circuit. Therefore, short circuits do not present an electrocution hazard. However, with short circuits, the load is bypassed, and the circuit conductors themselves represent the only impedance on the circuit, as seen in Figure 7-5. This situation virtually ensures that a short circuit will have a small enough impedance to produce a very high current flow—which has the potential to cause major damage to equipment and personnel if it is not rapidly eliminated. If the overcurrent-protective device does not operate very quickly, fire and explosion can result.

7.1.2: Circuit Overloading

> **KEY TERMS**
>
> **Overload** (See *NEC*® *Article 100*): A condition in which too much current is flowing through a circuit

although the circuit is operating properly (as it should). This condition results during the normal operation of certain types of loads—motor loads, for example—that utilize reactance as the major component of their impedance.

> **Operating overcurrents** Overcurrents that can be present on a circuit as the result of the nature of the connected loads, such as motor loads. As the motor is placed under load, the amount of counter EMF is reduced because of the slower speed of the motor. Current flow to the motor from the source increases. This increased current flow, or operating overcurrent, is a temporary condition; when the motor speed returns to normal, the circuit current flow will also return to normal.

The other causes of overcurrent are related to circuit overloading. An **overload** is characterized by small overcurrents that can last for a few milliseconds to several seconds or longer. Overloads on circuits are not caused by damage to the circuit or system failure, but rather result from a load (or loads) causing the current to be drawn from the source. Overloads are termed **operating overcurrents**, meaning that the overload is taking place while the circuit is operating properly and no current is leaving the electrical system.

In contrast, when it is important for the fault overcurrent protection to operate very quickly, a time delay must be incorporated into the overload protection apparatus. This time delay puts off the operation of the device when the overcurrent is small. This feature is necessary to allow normal running overcurrents to clear themselves before the overcurrent device shuts down the circuit. The time delay is an inverse time delay, meaning that the higher the overcurrent, the faster the overcurrent device operates.

There are two basic types of overloads: circuit overutilization and overload.

Circuit Overutilization

> **KEY TERMS**
>
> **Circuit overutilization** A condition that causes overcurrent on the circuit caused by too many loads being connected in parallel, and operating at the same time, for one circuit to supply. The circuit is operating properly, but each load connected in parallel reduces the overall impedance of the circuit.

"WIREMAN'S GUIDE"
LINE SIDE AND LOAD SIDE GROUND FAULTS

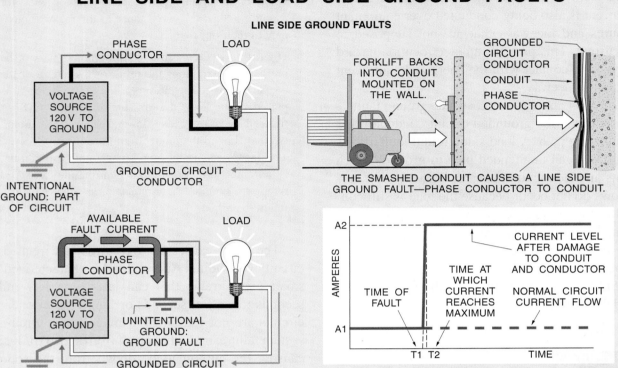

LINE SIDE GROUND FAULTS

THE SMASHED CONDUIT CAUSES A LINE SIDE GROUND FAULT—PHASE CONDUCTOR TO CONDUIT.

BEFORE THE FORKLIFT DAMAGES THE CONDUIT AND THE CONDUCTORS, THE CIRCUIT IN THE DIAGRAM IS OPERATING PROPERLY. THE CURRENT DRAW THROUGH THE CIRCUIT IS ONLY ENOUGH TO SERVE THE ONE LUMINAIRE (FIXTURE). WHEN THE FORKLIFT HITS THE CONDUIT AND THE GROUND FAULT IS CREATED, THE CURRENT FLOW IMMEDIATELY INCREASES TO THE AVAILABLE FAULT CURRENT. THIS IS A VERY DANGEROUS SITUATION THAT COULD RESULT IN FIRE, EXPLOSION, AND SEVERE DAMAGE TO THE SYSTEM.

LOAD SIDE GROUND FAULTS

THE SMASHED CONDUIT CAUSES A LOAD SIDE GROUND FAULT—GROUNDED CONDUCTOR TO CONDUIT.

THERE IS A MAJOR DIFFERENCE WITH LOAD SIDE GROUND FAULTS. WITH THESE FAULTS, THE FAULT IS BETWEEN THE GROUNDED CIRCUIT CONDUCTOR (NEUTRAL) AND THE UNINTENDED GROUND PATH (THE CONDUIT IN THIS EXAMPLE). BECAUSE THE LOAD IS STILL PART OF THE CIRCUIT, THE CURRENT FLOW THROUGH THE CIRCUIT IS STILL CONTROLLED, SO THE CURRENT FLOW DOES NOT INCREASE SIGNIFICANTLY. THERE MAY NOT BE ANY OUTWARD INDICATION THAT A FAULT EXISTS. THIS IS STILL A DANGEROUS SITUATION—ANY TIME CURRENT IS LEAVING THE CONDUCTORS AND IS FLOWING IN AN UNINTENDED PATH, DANGER EXISTS. THERE ARE UNDOUBTEDLY MANY LOAD SIDE FAULTS THAT GO UNCORRECTED FOR EXTENDED LENGTHS OF TIME.

Figure 7-4

"WIREMAN'S GUIDE"
SHORT CIRCUITS

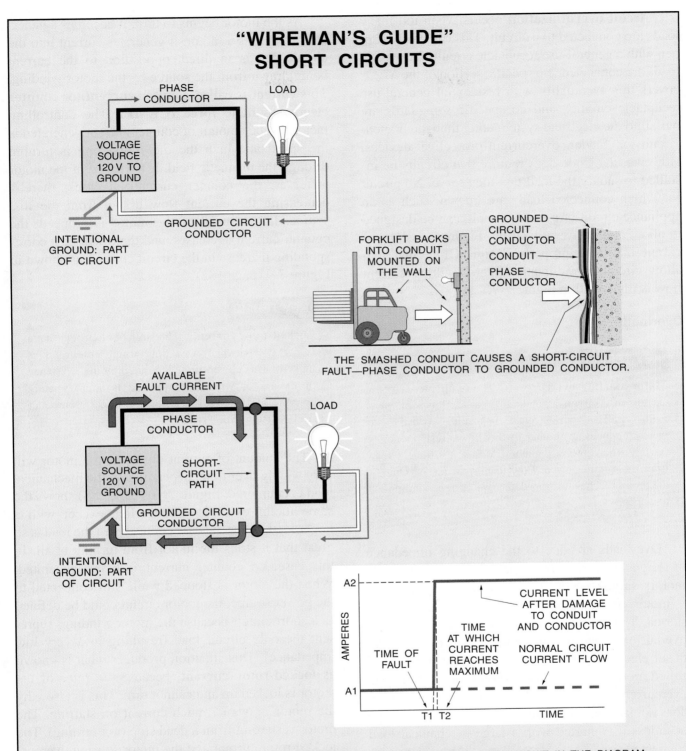

BEFORE THE FORKLIFT DAMAGES THE CONDUIT AND THE CONDUCTORS, THE CIRCUIT IN THE DIAGRAM IS OPERATING PROPERLY. THE CURRENT DRAW THROUGH THE CIRCUIT IS CONTROLLED BY THE LOAD. WHEN THE FAULT OCCURS, A SHORT CIRCUIT IN THIS CASE, THE CURRENT IS ALLOWED TO BYPASS THE LOAD, BUT IT IS STILL USING THE CIRCUIT CONDUCTORS AS THE PATH TO GROUND (NO CURRENT IS LEAVING THE SYSTEM, BUT IT IS NOT GOING THROUGH THE LOAD). THE CURRENT FLOW WILL IMMEDIATELY INCREASE TO THE AVAILABLE FAULT, AND WILL REMAIN THERE UNTIL THE OVERCURRENT-PROTECTIVE DEVICE (NOT SHOWN) OPENS THE CIRCUIT OR UNTIL MAJOR DAMAGE IS DONE TO THE EQUIPMENT OR CONDUCTORS THAT CLEAR THE FAULT. THIS IS A VERY DANGEROUS SITUATION BECAUSE FIRE OR EXPLOSION COULD OCCUR IF THE OVERCURRENT-PROTECTIVE DEVICE DOES NOT OPERATE VERY QUICKLY. A SHORT CIRCUIT IS NOT AN ELECTROCUTION HAZARD, HOWEVER.

Figure 7-5

Circuit overutilization occurs when too many loads are connected to a circuit. This can easily happen with a general-use receptacle circuit if too many loads are connected. No specific section of the *NEC®* covers this eventuality with residential general-use receptacle circuits, and no specific safeguards are built into the electrical system other than the branch-circuit or feeder overcurrent-protective devices. However, the Code does require that circuits be installed to satisfy the load that they serve. No circuit for which connected loads are known, such as an appliance circuit or a motor circuit, can be designed to allow circuit overutilization. Figure 7-6 shows a circuit that is improperly designed and that will allow circuit overutilization. The resulting current flow is illustrated in Figure 7-7.

Overloads

> ### KEY TERMS
>
> **Motor starting overcurrent** The temporary overcurrent in a circuit associated with starting a motor from rest. The high current flow results because the motor produces no counter EMF while at rest, and, on starting, the motor's counter EMF is insufficient to limit the circuit current flow to normal levels. Once the motor reaches full speed, the overcurrent condition disappears.

Overloads are due to the changing impedance of large inductive or capacitive loads, usually motors. Figure 7-8 details the overload created when a motor, representing a high inductive load, is started. The starting of the motor is classified as an overcurrent event that is a routine part of operation of an electrical circuit, and the current involved is called **motor starting overcurrent**, or motor inrush overcurrent. The motor starting overcurrent usually does not last very long unless the motor is very large or unless it is started with a large mechanical load connected.

> ### KEY TERMS
>
> **Counter current/counter electromagnetic force (CEMF)** The current generated within a turning motor that is in opposition to the current flow drawn from the source.

As the motor begins to turn, it becomes a generator as well as a motor. It generates current into the circuit that is in direct opposition to the current being drawn from the source by the motor winding. This current is called **counter current**, or **counter electromagnetic force (CEMF)**. The controlling factor for the amount of counter current generated is the speed at which the motor winding is turning through the magnetic field. If the speed of the motor increases, the counter current increases, thereby decreasing the current flow drawn from the line source. If the speed of the motor is reduced, the counter current decreases, and there will be a corresponding increase in the circuit current, as shown in Figure 7-9 (page 182).

> ### KEY TERMS
>
> **Locked-rotor current** The amount of current that a motor draws when it is not turning (the rotor is locked). This current flow is limited by the resistance of the motor windings only and is approximately equal to the amount of current the motor requires at start (starting overcurrent).

The amount of current drawn by the motor will vary widely over time, depending on the mechanical load encountered. Figure 7-10 (page 183) shows the same motor example used earlier, except with a larger load. A special case occurs when the load is so great that it stops the motor from turning at all. In this case, no counter current is being generated. When the motor is stopped while still connected to the power source, the motor circuit could be defined as a short circuit because the motor windings represent the only circuit load (resulting in a very low impedance). This situation produces what is known as **locked-rotor current**, because the rotor of the motor is locked up and cannot turn. This is also why the motor draws so much current on starting. The motor is starting from a dead stop (not turning). The locked-rotor current and the motor starting overcurrent ampere measurements should be the same, except on starting, that the motor may begin to turn before the full locked-rotor current is obtained. The locked rotor is also illustrated in Figure 7-10. Figure 7-11 (page 184) takes a more detailed look at overcurrent protection for overloads, and a summary of the types of overcurrent is presented in Figure 7-12 (page 185).

"WIREMAN'S GUIDE"
CIRCUIT OVERUTILIZATION

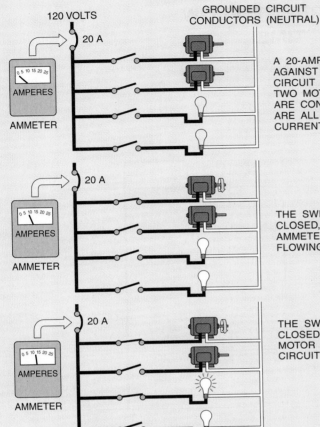

A 20-AMPERE OVERCURRENT DEVICE PROTECTS THE CIRCUIT AGAINST FAULTS AND ALSO AGAINST OVERLOADS DUE TO CIRCUIT OVERUTILIZATION. CONNECTED TO THE CIRCUIT ARE TWO MOTORS AND TWO LUMINAIRES (FIXTURES). ALL LOADS ARE CONTROLLED BY SEPARATE SINGLE-POLE SWITCHES, WHICH ARE ALL IN THE OPEN POSITION. AN AMMETER RECORDS THE CURRENT FLOW THROUGH THE CIRCUIT.

THE SWITCH CONTROLLING THE FIRST MOTOR IS CLOSED, AND THE MOTOR BEGINS TO RUN. THE AMMETER SHOWS APPROXIMATELY 8 AMPERES NOW FLOWING THROUGH THE CIRCUIT.

THE SWITCH CONTROLLING THE FIRST LUMINAIRE (FIXTURE) IS CLOSED, AND THE LAMP BEGINS TO PRODUCE LIGHT. THE FIRST MOTOR IS ALSO RUNNING AND THE AMMETER SHOWS A TOTAL CIRCUIT CURRENT OF APPROXIMATELY 11 (8 + 3) AMPERES.

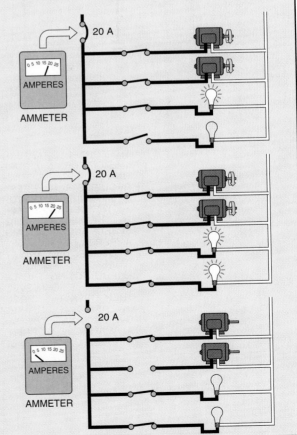

THE SECOND MOTOR IS STARTED, AND THE TOTAL CURRENT CARRIED BY THE CIRCUIT IS AGAIN INCREASED. THE AMMETER NOW RECORDS A CURRENT FLOW OF APPROXIMATELY 19 (8 + 3 + 8) AMPERES.

THE SWITCH FOR THE FINAL LOAD ON THE CIRCUIT, THE SECOND LUMINAIRE (FIXTURE), IS CLOSED, AND THE LUMINAIRE (FIXTURE) BEGINS TO PRODUCE LIGHT. THE AMMETER RECORDS THE CURRENT FLOW OF APPROXIMATELY 22 (8 + 8 + 3 + 3) AMPERES. BECAUSE THIS IS A 20-AMPERE CIRCUIT, THE AMOUNT OF CURRENT FLOW RESULTS IN AN OVERCURRENT EVENT. THE CIRCUIT WILL PROBABLY CONTINUE TO OPERATE FOR SOME PERIOD OF TIME UNDER THESE CONDITIONS.

THE 20-AMPERE OVERCURRENT-PROTECTIVE DEVICE OPENS AFTER SOME AMOUNT OF TIME OPERATING WITH THE OVERLOAD. THE LENGTH OF TIME FOR WHICH THE BREAKER WILL OPERATE DEPENDS ON THE MAGNITUDE OF THE OVERCURRENT, THE AMBIENT TEMPERATURE, THE LOCATION OF THE OVERCURRENT DEVICE, AND MANY OTHER FACTORS. DAMAGE TO THE CONDUCTOR INSULATION WILL BE VERY SMALL IN THIS EXAMPLE.

Figure 7-6

"WIREMAN'S GUIDE"
OVERUTILIZED
CIRCUIT CURRENT FLOW
OVER TIME

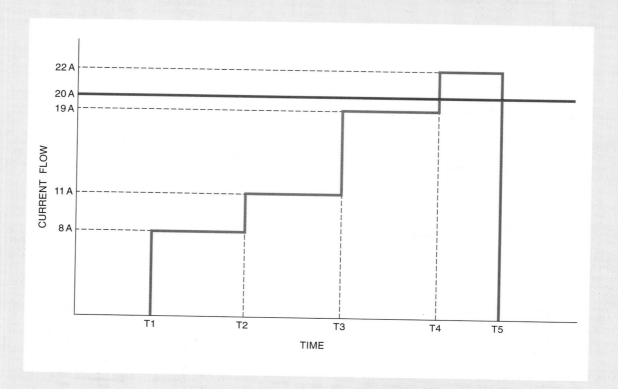

THIS GRAPH CHARTS THE CURRENT FLOW OVER TIME FOR THE CIRCUIT IN FIGURE 7-6. THE FIRST MOTOR IS SWITCHED ON AND OBTAINS FULL SPEED AT TIME T1. THE MOTOR REQUIRES 8 AMPERES TO RUN AT FULL SPEED. AT TIME T2, THE LUMINAIRE (FIXTURE) IS SWITCHED ON, AND THE CURRENT FLOW GOES UP 3 AMPERES, TO 11 AMPERES TOTAL. AT TIME T3, THE SECOND MOTOR IS SWITCHED ON AND OBTAINS FULL SPEED, AND THE CURRENT IN THE CIRCUIT GOES UP TO 19 AMPERES. AT TIME T4, THE REMAINING LUMINAIRE (FIXTURE) IS CONNECTED, AND THE CURRENT FLOW FOR THE CIRCUIT RISES TO 22 AMPERES. THE 20-AMPERE CIRCUIT BREAKER THAT IS PROTECTING THE CIRCUIT AGAINST OVERCURRENT WILL ALLOW THE LOW-LEVEL OVERLOAD FOR SOME TIME AND WILL THEN OPEN THE CIRCUIT. COMMON CIRCUIT BREAKERS ARE INVERSE TIME BREAKERS, MEANING THAT THE HIGHER THE OVERCURRENT, THE FASTER THE BREAKER WILL OPEN. WITH A SMALL OVERCURRENT AS IN THIS EXAMPLE, IT MAY TAKE MINUTES OR HOURS BEFORE THE BREAKER OPENS THE CIRCUIT.

Figure 7-7

"WIREMAN'S GUIDE"
STARTING OVERCURRENT OF A MOTOR

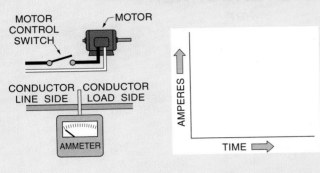

SHOWN IS A MOTOR, A CLAMP-ON AMMETER AROUND ONE OF THE CONDUCTORS THAT SUPPLIES THE MOTOR, AND A GRAPH OF THE AMMETER READINGS OVER TIME. THE MOTOR IS AT REST (NOTE THAT THE CONTROL SWITCH IS OPEN) AND THE AMMETER READS NO (ZERO) CURRENT FLOW. THE TWO AXES OF THE GRAPH ARE TIME, ON THE HORIZONTAL AXIS, AND CURRENT FLOW, MEASURED IN AMPERES, ON THE VERTICAL AXIS.

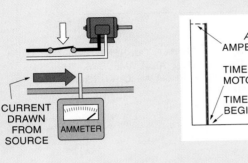

THE SWITCH IS CLOSED AT TIME T1, ALLOWING CURRENT TO FLOW THROUGH THE MOTOR CIRCUIT. IT APPEARS AS A SHORT CIRCUIT TO THE CURRENT SOURCE, AND THE CURRENT FLOW INCREASES UNTIL TIME T2 AT CURRENT FLOW *A* AMPERES WHEN THE MOTOR BEGINS TO TURN. THE LENGTH OF TIME BETWEEN T1 AND T2 AND THE VALUE OF *A* AMPERES DEPEND ON THE SIZE OF THE MOTOR, THE MOTOR DESIGN, THE APPLICATION OF THE MOTOR, AND THE LOAD THAT THE MOTOR EXPERIENCES WHEN STARTING. THIS MOTOR IS A RELATIVELY SMALL MOTOR, STARTING WITHOUT ANY LOAD. THE CURRENT FLOW REPRESENTED BY THE RED ARROW IS THE CURRENT THAT IS DRAWN BY THE RESISTANCE OF THE MOTOR WINDING.

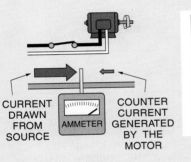
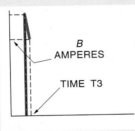

AS SOON AS THE MOTOR BEGINS TO TURN, IT BECOMES A GENERATOR. THE MOTOR WINDING TURNING INSIDE OF THE MOTOR'S MAGNETIC FIELD GENERATES CURRENT IN THE OPPOSITE DIRECTION TO THAT OF THE MOTOR'S CURRENT DRAW FROM THE SOURCE. THIS GENERATED CURRENT IS CALLED *COUNTER CURRENT*. THE CURRENT DRAW FROM THE SOURCE AS MEASURED BY THE AMMETER HAS DECREASED FROM *A* AMPERES TO *B* AMPERES AT TIME T3.

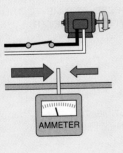
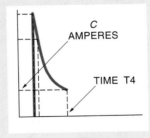

AS THE MOTOR CONTINUES TO INCREASE IN SPEED, THE COUNTER CURRENT CONTINUES TO INCREASE, MAKING THE CURRENT FLOW THROUGH THE CIRCUIT DROP FROM *B* AMPERES TO *C* AMPERES AT TIME T4. THE MOTOR IS APPROACHING FULL SPEED.

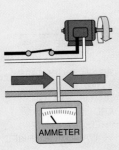
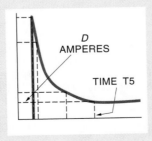

THE MOTOR IS NOW RUNNING AT FULL SPEED. THE CURRENT FLOW HAS STABILIZED AT *D* AMPERES, AND THE MOTOR WILL CONTINUE TO DRAW THIS CURRENT FROM TIME T5 UNTIL IT IS PUT UNDER LOAD OR UNTIL IT IS TURNED OFF. AT THIS POINT, THE COUNTER CURRENT IS AT MAXIMUM. THE MOTOR WILL ALWAYS DRAW SOME CURRENT FROM THE SOURCE BECAUSE THERE IS ALWAYS SOME ENERGY LOST IN GENERATING COUNTER CURRENT, BUT THE MORE EFFICIENT THE MOTOR, THE SMALLER WILL BE THE FULL-SPEED CURRENT DRAW FROM THE SOURCE.

Figure 7-8

"WIREMAN'S GUIDE"
MOTOR RUNNING OVERCURRENT

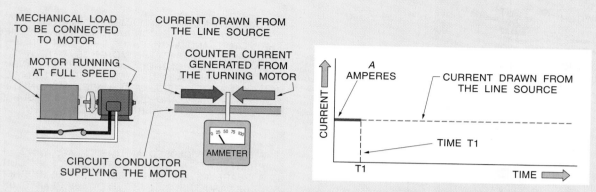

THE MOTOR IS RUNNING AT FULL SPEED. THE MAXIMUM AMOUNT OF COUNTER CURRENT IS BEING PRODUCED, AND THE CURRENT DRAW OF THE MOTOR IS AT THE MINIMUM AT *A* AMPERES. THE MOTOR HAS NOT YET BEEN REQUIRED TO DO ANY MECHANICAL WORK SUCH AS CUTTING WOOD. AT TIME T1, THE MOTOR WILL BE CONNECTED TO THE MECHANICAL LOAD.

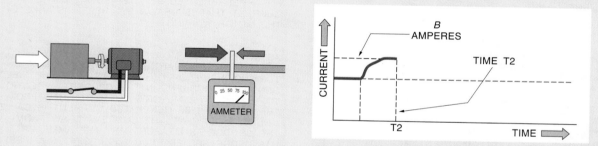

AT TIME T1, THE MECHANICAL LOAD IS CONNECTED TO THE MOTOR, AND THE SPEED OF THE MOTOR IS DECREASED. THIS DROP IN MOTOR SPEED ALSO PRODUCES A DROP IN THE COUNTER CURRENT BEING GENERATED BY THE MOTOR. THE DROP IN COUNTER CURRENT CAUSES THE CURRENT DRAW IN THE CIRCUIT TO INCREASE. THE MOTOR SPEED STABILIZES AT THE SLOWER SPEED AT TIME T2 AND AT *B* AMPERES. THE CURRENT DRAW AT THE NEW MOTOR SPEED IS SHOWN ON THE AMMETER.

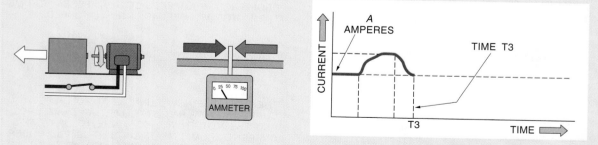

WHEN THE MECHANICAL LOAD IS REMOVED FROM THE MOTOR AT T3, THE SPEED AGAIN INCREASES, CAUSING A CORRESPONDING DROP IN CURRENT FLOW BACK TO *A* AMPERES. THE LOAD COULD BE REMOVED EITHER BY PHYSICALLY DISCONNECTING THE LOAD FROM THE MOTOR OR BY THE MOTOR HAVING ACCOMPLISHED THE NECESSARY WORK, SUCH AS CUTTING A BOARD IN HALF. THE LENGTH OF THE OVERLOAD IS DETERMINED BY THE LOAD TO BE ACCOMPLISHED, AND THE MAGNITUDE OF THE OVERLOAD IS DEPENDENT ON THE RATE AT WHICH THE WORK WAS PERFORMED.

Figure 7-9

"WIREMAN'S GUIDE"
LOCKED-ROTOR CURRENT

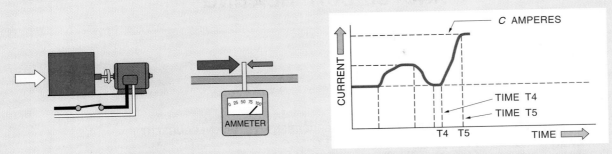

AT TIME T4, THE MOTOR IS AGAIN PUT UNDER LOAD, BUT THE LOAD IS CONSIDERABLY LARGER THAN IN FIGURE 7-9. THE MOTOR'S SPEED IS AGAIN REDUCED BECAUSE OF THE LOAD, BUT IT SLOWS DOWN MUCH MORE THAN BEFORE. THE CIRCUIT LINE CURRENT INCREASES WITH THE REDUCTION IN SPEED AND IN COUNTER CURRENT TO *C* AMPERES AT TIME T5.

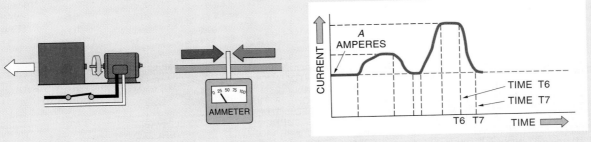

WHEN THE LOAD IS SATISFIED (OR WHEN THE LOAD IS REMOVED FROM THE MOTOR), AT TIME T6, THE SPEED INCREASES UNTIL THE MOTOR AGAIN REACHES FULL SPEED AT TIME T7. THE CIRCUIT CURRENT FLOW IS REDUCED BACK TO *A* AMPERES, AND THE COUNTER CURRENT IS AGAIN AT THE MAXIMUM.

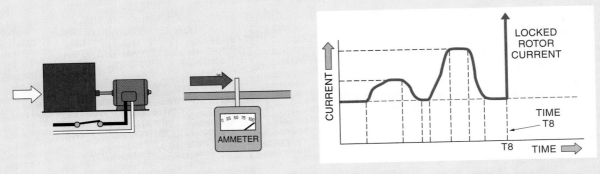

THE MOTOR IS AGAIN PUT UNDER LOAD AT TIME T8. THE LOAD NOW IS SO LARGE THAT THE MOTOR CAN NO LONGER TURN, AND IT STOPS COMPLETELY. THERE IS NO COUNTER CURRENT BEING GENERATED, AND THEREFORE THE ONLY CONTROL OVER THE CURRENT FLOW IS THE RESISTANCE OF THE MOTOR WINDING. THE CIRCUIT CURRENT FLOW INSTANTLY INCREASES TO APPROXIMATELY THE LEVEL OF THE STARTING OVERCURRENT. WHEN THE MOTOR IS NOT ALLOWED TO TURN, IT IS CALLED A *LOCKED ROTOR*, AND THE RESULTING LARGE CURRENT FLOW IS CALLED *LOCKED-ROTOR CURRENT*.

Figure 7-10

"WIREMAN'S GUIDE"
OVERLOAD OVERCURRENT PROTECTION AND CIRCUIT HEATING

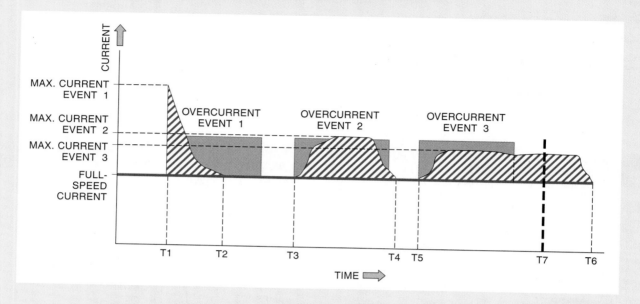

THE BLUE BOX REPRESENTS THE OVERLOAD OVERCURRENT-PROTECTIVE DEVICE. THE DEVICE IS SET AT SLIGHTLY ABOVE THE MOTOR'S FULL-SPEED CURRENT RATING (USUALLY 25% ABOVE) AND ALSO FOR SOME ALLOWABLE DURATION OF TIME. THE AREA UNDER THE CURVE (CROSS-HATCHING) FOR THE OVERLOAD REPRESENTS HEATING CAUSED BY THE CURRENT FLOW FOR THE DURATION OF THE OVERLOAD. THE AREA OF THE OVERCURRENT EVENT BOX REPRESENTS THE AMOUNT OF HEATING ALLOWABLE SHORT OF DAMAGING THE MOTOR AND THE CIRCUIT CONDUCTORS. AS LONG AS THE AREA UNDER THE CURVE FOR THE ACTUAL CURRENT FLOW THROUGH THE CIRCUIT IS SMALLER THAN THE AREA OF THE OVERCURRENT EVENT BOX, THE OVERCURRENT-PROTECTIVE DEVICE WILL NOT OPEN THE CIRCUIT. IF THE AREA OF THE ACTUAL OVERLOAD IS LARGER THAN THE AREA OF THE OVERCURRENT EVENT BOX, THE OVERLOAD OVERCURRENT DEVICE WILL OPEN THE CIRCUIT.

OVERCURRENT EVENT 1: THE OVERCURRENT IN THIS CASE APPROXIMATES THE STARTING OVERCURRENT OF A MOTOR. ALTHOUGH THE MAXIMUM CURRENT FLOW IS WELL ABOVE THE ALLOWABLE OVERLOAD MAGNITUDE, THE OVERLOAD DURATION FROM TIME T1 TO TIME T2 IS NOT VERY LONG. THE AREA UNDER THE OVERCURRENT CURVE IS LESS THAN THE AREA OF THE OVERCURRENT EVENT BOX; THEREFORE, THE OVERLOAD OVERCURRENT-PROTECTIVE DEVICE **SHOULD NOT** OPEN THE CIRCUIT.

OVERCURRENT EVENT 2: THE OVERCURRENT IN THIS CASE APPROXIMATES THE RUNNING OVERCURRENT OF A MOTOR. THE DURATION OF THE OVERLOAD FROM TIME T3 TO TIME T4 IS SLIGHTLY LONGER THAN THE ALLOWABLE TIME, AND THE MAXIMUM MAGNITUDE FROM THE FULL-SPEED CURRENT OF THE MOTOR TO THE MAXIMUM CURRENT IS SLIGHTLY ABOVE THE ALLOWABLE LEVEL, BUT THE AREA UNDER THE OVERCURRENT CURVE IS LESS THAN THE AREA OF THE OVERCURRENT EVENT BOX, SO THE OVERLOAD OVERCURRENT-PROTECTIVE DEVICE **SHOULD NOT** OPEN THE CIRCUIT.

OVERCURRENT EVENT 3: IN THIS CASE, THE MAXIMUM CURRENT LEVEL IS WELL BELOW THE ALLOWABLE LEVEL, BUT THE DURATION OF THE OVERLOAD FROM TIME T5 TO TIME T6 IS SUBSTANTIALLY LONGER THAN THE ALLOWABLE TIME. WITH THIS OVERLOAD, THE OVERLOAD OVERCURRENT-PROTECTIVE DEVICE **SHOULD** OPEN THE CIRCUIT TO PROTECT THE MOTOR AND THE CONDUCTORS AGAINST DAMAGE FROM EXCESSIVE HEATING. THE OVERLOAD OVERCURRENT-PROTECTIVE DEVICE SHOULD OPEN THE CIRCUIT AT APPROXIMATELY TIME T7.

Figure 7-11

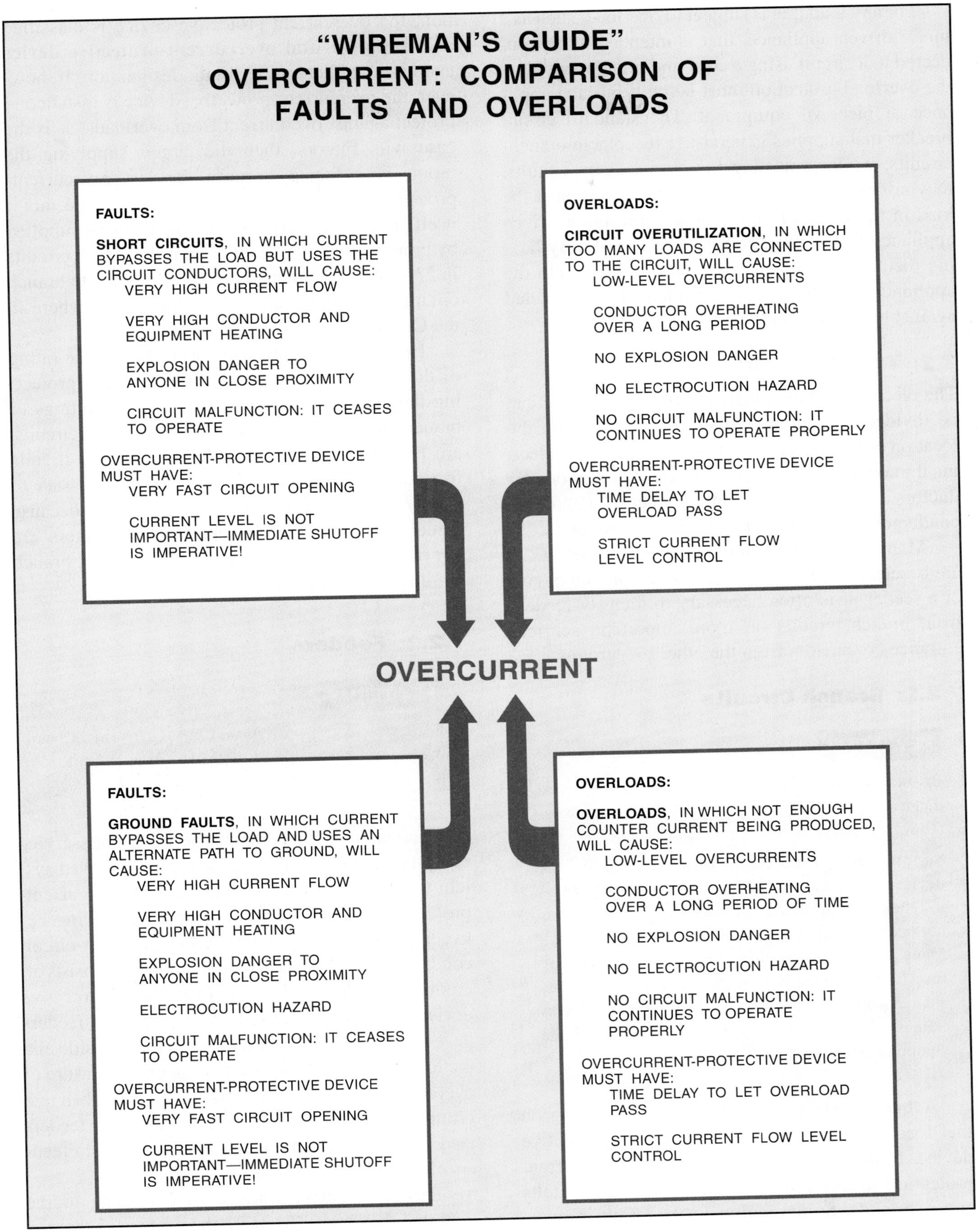

"WIREMAN'S GUIDE"
OVERCURRENT: COMPARISON OF
FAULTS AND OVERLOADS

FAULTS:

SHORT CIRCUITS, IN WHICH CURRENT BYPASSES THE LOAD BUT USES THE CIRCUIT CONDUCTORS, WILL CAUSE:
    VERY HIGH CURRENT FLOW

    VERY HIGH CONDUCTOR AND EQUIPMENT HEATING

    EXPLOSION DANGER TO ANYONE IN CLOSE PROXIMITY

    CIRCUIT MALFUNCTION: IT CEASES TO OPERATE

OVERCURRENT-PROTECTIVE DEVICE MUST HAVE:
    VERY FAST CIRCUIT OPENING

    CURRENT LEVEL IS NOT IMPORTANT—IMMEDIATE SHUTOFF IS IMPERATIVE!

OVERLOADS:

CIRCUIT OVERUTILIZATION, IN WHICH TOO MANY LOADS ARE CONNECTED TO THE CIRCUIT, WILL CAUSE:
    LOW-LEVEL OVERCURRENTS

    CONDUCTOR OVERHEATING OVER A LONG PERIOD

    NO EXPLOSION DANGER

    NO ELECTROCUTION HAZARD

    NO CIRCUIT MALFUNCTION: IT CONTINUES TO OPERATE PROPERLY

OVERCURRENT-PROTECTIVE DEVICE MUST HAVE:
    TIME DELAY TO LET OVERLOAD PASS

    STRICT CURRENT FLOW LEVEL CONTROL

OVERCURRENT

FAULTS:

GROUND FAULTS, IN WHICH CURRENT BYPASSES THE LOAD AND USES AN ALTERNATE PATH TO GROUND, WILL CAUSE:
    VERY HIGH CURRENT FLOW

    VERY HIGH CONDUCTOR AND EQUIPMENT HEATING

    EXPLOSION DANGER TO ANYONE IN CLOSE PROXIMITY

    ELECTROCUTION HAZARD

    CIRCUIT MALFUNCTION: IT CEASES TO OPERATE

OVERCURRENT-PROTECTIVE DEVICE MUST HAVE:
    VERY FAST CIRCUIT OPENING

    CURRENT LEVEL IS NOT IMPORTANT—IMMEDIATE SHUTOFF IS IMPERATIVE!

OVERLOADS:

OVERLOADS, IN WHICH NOT ENOUGH COUNTER CURRENT BEING PRODUCED, WILL CAUSE:
    LOW-LEVEL OVERCURRENTS

    CONDUCTOR OVERHEATING OVER A LONG PERIOD OF TIME

    NO EXPLOSION DANGER

    NO ELECTROCUTION HAZARD

    NO CIRCUIT MALFUNCTION: IT CONTINUES TO OPERATE PROPERLY

OVERCURRENT-PROTECTIVE DEVICE MUST HAVE:
    TIME DELAY TO LET OVERLOAD PASS

    STRICT CURRENT FLOW LEVEL CONTROL

Figure 7-12

For any load that is subject to overload, such as a motor-driven appliance, that is intended to be connected to a circuit using a cord-and-attachment plug, the overload protection must be built into the appliance or piece of equipment. The standard circuit breaker that supplies general-use receptacle branch-circuits, while protecting the circuit against faults, may not be of the proper size or type to protect the load in the event of an overload on a motor-driven appliance that is plugged into a receptacle. By having their own overload protection built in, both the appliance and the circuit are adequately protected against both faults and overloads.

7.2: TYPES OF CIRCUITS

The electrical circuits installed in any structure can be divided into three distinct groups based on the location of the overcurrent-protective devices protecting the conductor of a circuit. The three groups of conductors are (1) branch-circuit conductors, (2) feeder conductors, and (3) service-entrance conductors.

Many of the procedures required by the *NEC®* are based on whether a conductor is a branch circuit or a feeder. It is often necessary to identify feeders from branch circuits or, more important, service-entrance conductors from the other two groups.

7.2.1: Branch Circuits

> **KEY TERMS**
>
> **Branch circuit** (See *NEC® Article 100*): A circuit that has overcurrent protection, other than supplemental overcurrent protection, on the line side only.
>
> **Supplemental overcurrent-protective device** A device providing overcurrent protection that is installed to protect a circuit load or component but is not intended to protect the circuit against overcurrent. Many pieces of electronic equipment are protected by fuses internal to the equipment. These fuses are considered supplemental overcurrent-protective devices because they protect the equipment, or part of the equipment, but not the circuit.

A **branch circuit** is defined as that wiring on the load side of the last overcurrent-protective device. In dealing with branch circuits, overcurrent protection means protective devices to limit faults and overloads. Some equipment has built-in overload protection. If the protection is in addition to the required fault-current and overload-current protec-

tion, the overcurrent-protective device is classified as a **supplemental overcurrent-protective device** and has no effect on the circuit designation. If, however, the overcurrent-protective device is installed to protect against overcurrent from overloads, as is the case with motors, then the circuit supplying the motor is a feeder because the last overcurrent-protective device on the circuit is inside the motor itself. There is no requirement that loads be supplied by branch circuits. The *NEC®* covers branch circuits in *210*, but there are many other references to branch circuits and branch-circuit conductors elsewhere in the Code.

Branch circuits are sized according to the rating of the overcurrent-protective device that is protecting the circuit, as detailed in *210.3*. The ratings of branch circuits, other than individual branch circuits, are 15, 20, 30, 40, and 50 amperes. Branch circuits for individual loads can be of any size necessary to supply the load. This is an important point because feeder circuits of service-entrance conductors are not rated in this manner. Some examples of branch circuits are shown in Figure 7-13.

7.2.2: Feeders

> **KEY TERMS**
>
> **Feeder** (See *NEC® Article 100*): A circuit that has overcurrent protection on both the line side and the load side.

A **feeder** is that wiring between the service equipment, or the source of a separately derived system (as in a transformer), and the last overcurrent-protective device—in other words, any wiring between the service equipment and the branch-circuit conductors. Because the service equipment consists of, among other things, the main overcurrent-protective devices for the structure, it is obvious that feeders have overcurrent protection on both the line side and the load side. Thus, feeders have an extra measure of overcurrent protection not present in the other two groups of conductors. The requirements for feeders are covered in *215*, but there are references to feeders elsewhere in the Code.

Feeder conductors are sized according to the load that they serve. In *215.2* the Code says that the feeder will be sized according to the sum of all of the branch-circuit loads as determined by *220*,

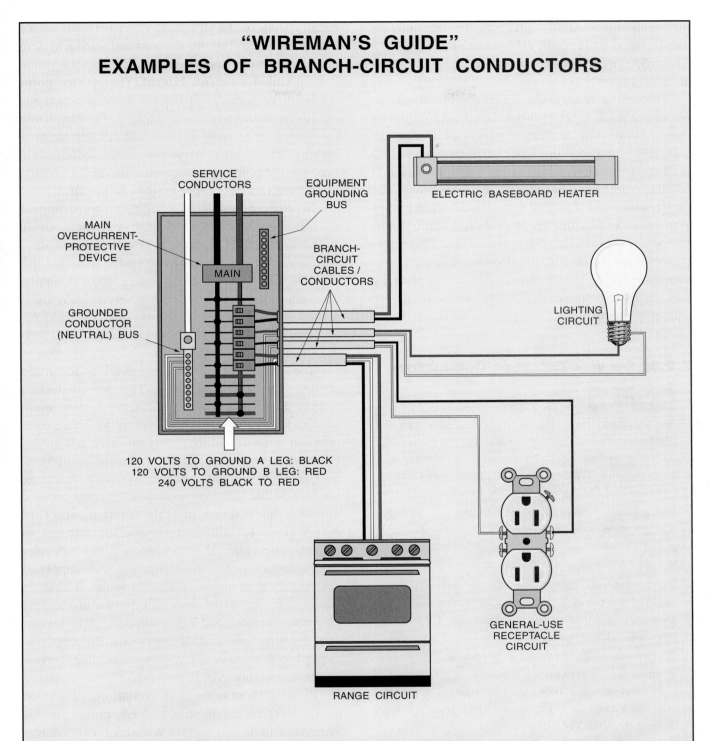

"WIREMAN'S GUIDE"
EXAMPLES OF BRANCH-CIRCUIT CONDUCTORS

BRANCH-CIRCUIT CONDUCTORS CARRY CURRENT FROM THE LAST OVERCURRENT-PROTECTIVE DEVICE TO THE LOAD. NO OVERCURRENT-PROTECTIVE DEVICES ARE INSTALLED BETWEEN THE PANEL AND THE LOADS. IT IS POSSIBLE TO PLUG SOME EQUIPMENT THAT HAS SEPARATE OVERLOAD PROTECTION BUILT IN (SUCH AS A THERMALLY PROTECTED MOTOR) INTO THE GENERAL-USE RECEPTACLE CIRCUIT, BUT THE CIRCUIT WOULD STILL BE DEFINED AS A BRANCH CIRCUIT. THE EQUIPMENT GROUNDING CONDUCTORS HAVE BEEN OMITTED FROM THE DIAGRAM FOR CLARITY.

Figure 7-13

including all allowable adjustments for demand factors detailed in *Parts II, III,* and *IV* of that article. The procedure for determining a feeder load is covered in Chapter 12 of this book.

According to *215.4(A)*, feeder conductors of different phases or legs are allowed to share a grounded circuit conductor, creating a multiwire feeder circuit. Also, in *215.9* the Code says that if ground-fault interrupter protection is provided for a feeder, any receptacle or other outlets that are required to be GFCI protected by *210.8* and *527* (Temporary Wiring) served by that feeder does not require individual GFCI protection. The ground-fault interrupter protector for the feeder will also satisfy any requirement for GFCI branch-circuit outlets supplied by that feeder. Some examples of feeder circuits are shown in Figure 7-14.

7.2.3: Service-Entrance Conductors

KEY TERMS

Service conductors (See *NEC® Article 100*): Those conductors from the service point (point of service) to the main, or first, overcurrent-protective device. For all intents and purposes, these conductors are not protected against faults.

Service drop (See *NEC® Article 100*): Overhead wiring system from the power utility's equipment to the point of service. The service-drop conductors connect to the service conductors (service-entrance conductors).

Service lateral (See *NEC® Article 100*): Underground wiring system from the power utility's equipment to the point of service. Lateral conductors often terminate in the meter enclosure.

Service point (See *NEC® Article 100*): The location at which the power utility's conductors (drop or lateral conductors) connect to the dwelling wiring. Overhead drop conductors usually terminate at the service conductors from the riser or mast. (See Chapter 13 for more about risers and masts.) Lateral conductors usually terminate in the meter enclosure.

A clear understanding of the nature and function of a service of **service conductors** is very important to comprehension of the various rules of the *NEC®* and the procedures for installing circuits. A *service* can be considered to include those conductors, and that equipment, employed to provide electrical energy from the power-providing utility to the electrical system of the structure. The electrical service can be supplied through an overhead wiring system called a **service drop** or through an underground system called a **service lateral**. The **service point** (or point of service) is the location at which the wiring from the power-providing utility physically connects to the wiring of the structure. Service conductors, also called *service-entrance conductors*, begin at the point of service and include the wiring up to the terminals of the service equipment.

Service equipment is equipment consisting of an overcurrent-protective device or devices and a switch or switches connected to the load end of service conductors and intended to be the main control and disconnection location of the electrical supply. The requirements for service and service equipment are detailed in Chapter 13 of this book, and in *NEC® 230* although it should be noted that services are referenced elsewhere in the Code.

The most important consideration regarding services, service equipment, and service conductors is that they are not protected against overcurrent caused by faults in any useful way. The power provider (electric utility) may or may not install overcurrent-protective devices for its equipment. The provider's overcurrent-protective devices are not accessible to the residential electrician and almost certainly do not meet the requirements of the *NEC®* (electric utilities follow a separate set of codes). Therefore, the electrician must consider those conductors and the equipment they supply as *not* protected in any way against faults. The main overcurrent-protective devices installed at the load end of the service conductors will protect the service and the structure against overloads, however. An example of a service and the associated service equipment is shown in Figure 7-15.

Because there is no fault overcurrent protection on the service conductors and equipment, in the event of a fault, the current will reach exceedingly high levels and may carry that excessive current for a considerable length of time. When a fault occurs on a feeder or a branch circuit, the equipment grounding system and the affected equipment must carry the fault current only for the time it takes to open the circuit breaker (or melt the fuse). It should be a matter of only a few milliseconds before the overcurrent-protective device opens the circuit. When there is no device to open the circuit, the fault may continue indefinitely (see Figure 7-16).

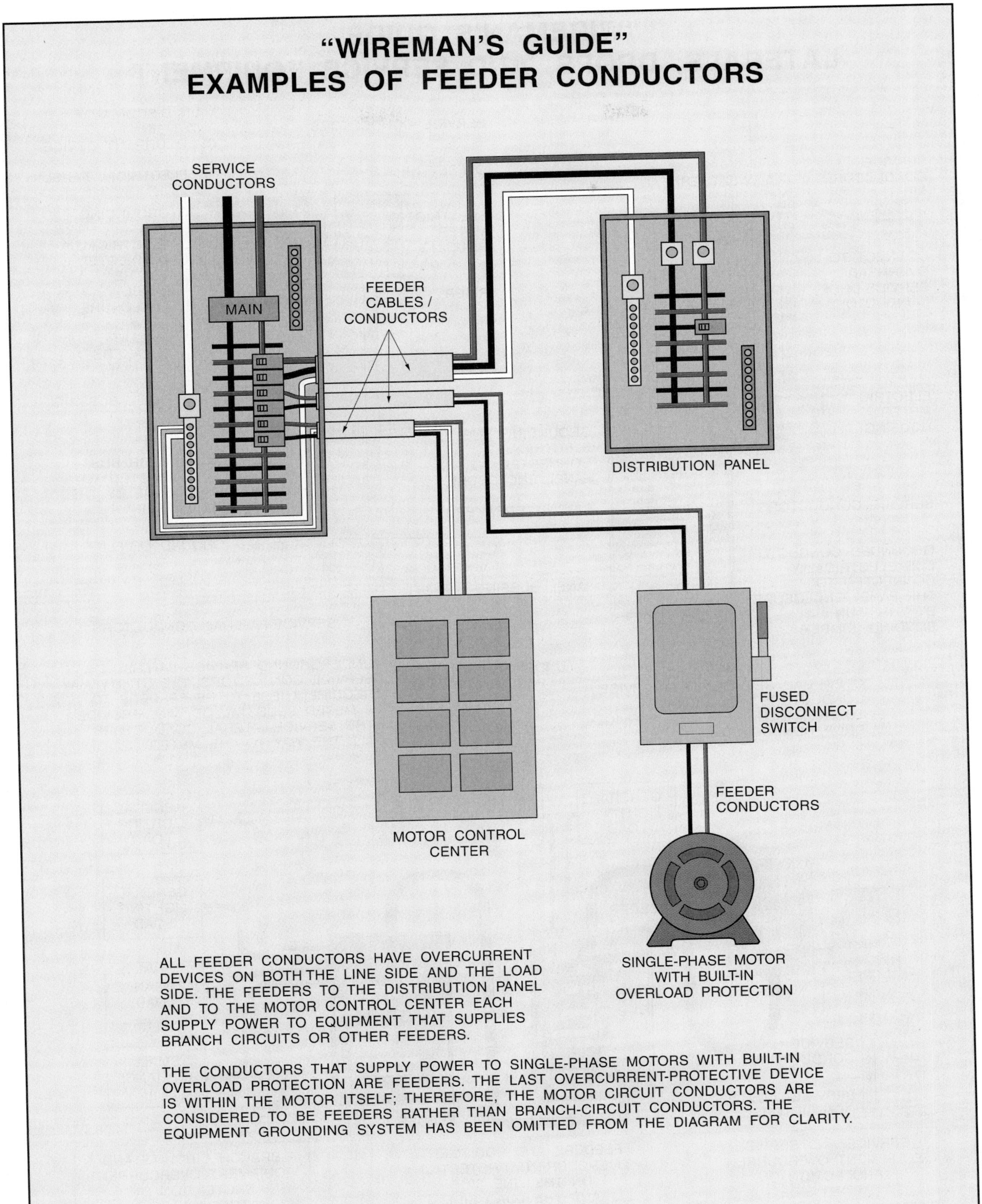

"WIREMAN'S GUIDE"
EXAMPLES OF FEEDER CONDUCTORS

SERVICE
CONDUCTORS

MAIN

FEEDER
CABLES /
CONDUCTORS

DISTRIBUTION PANEL

MOTOR CONTROL
CENTER

FUSED
DISCONNECT
SWITCH

FEEDER
CONDUCTORS

SINGLE-PHASE MOTOR
WITH BUILT-IN
OVERLOAD PROTECTION

ALL FEEDER CONDUCTORS HAVE OVERCURRENT DEVICES ON BOTH THE LINE SIDE AND THE LOAD SIDE. THE FEEDERS TO THE DISTRIBUTION PANEL AND TO THE MOTOR CONTROL CENTER EACH SUPPLY POWER TO EQUIPMENT THAT SUPPLIES BRANCH CIRCUITS OR OTHER FEEDERS.

THE CONDUCTORS THAT SUPPLY POWER TO SINGLE-PHASE MOTORS WITH BUILT-IN OVERLOAD PROTECTION ARE FEEDERS. THE LAST OVERCURRENT-PROTECTIVE DEVICE IS WITHIN THE MOTOR ITSELF; THEREFORE, THE MOTOR CIRCUIT CONDUCTORS ARE CONSIDERED TO BE FEEDERS RATHER THAN BRANCH-CIRCUIT CONDUCTORS. THE EQUIPMENT GROUNDING SYSTEM HAS BEEN OMITTED FROM THE DIAGRAM FOR CLARITY.

Figure 7-14

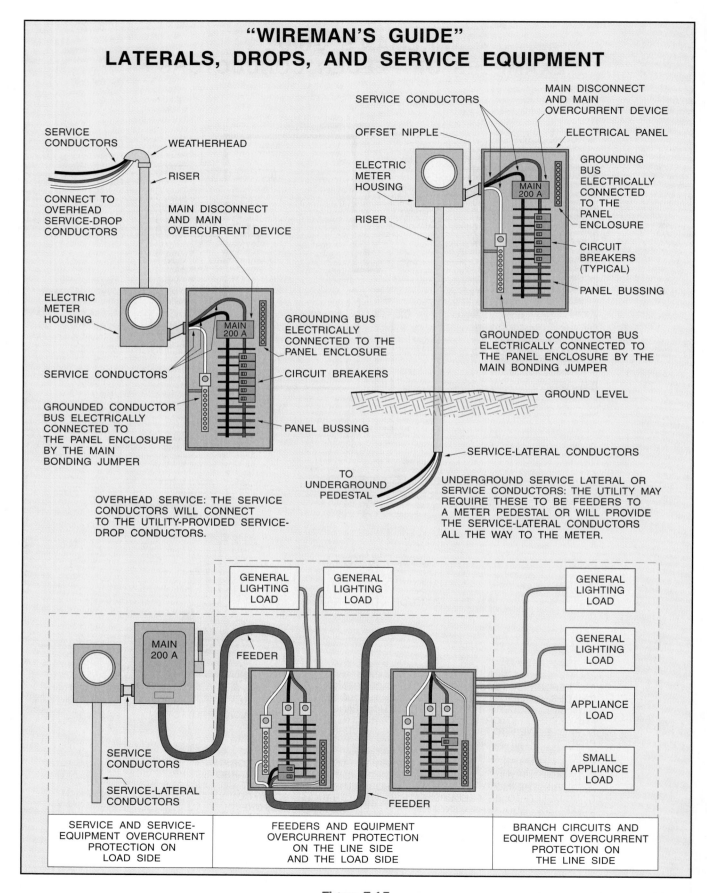

"WIREMAN'S GUIDE"
LATERALS, DROPS, AND SERVICE EQUIPMENT

SERVICE CONDUCTORS

WEATHERHEAD

RISER

CONNECT TO OVERHEAD SERVICE-DROP CONDUCTORS

MAIN DISCONNECT AND MAIN OVERCURRENT DEVICE

ELECTRIC METER HOUSING

MAIN 200 A

GROUNDING BUS ELECTRICALLY CONNECTED TO THE PANEL ENCLOSURE

SERVICE CONDUCTORS

CIRCUIT BREAKERS

GROUNDED CONDUCTOR BUS ELECTRICALLY CONNECTED TO THE PANEL ENCLOSURE BY THE MAIN BONDING JUMPER

PANEL BUSSING

OVERHEAD SERVICE: THE SERVICE CONDUCTORS WILL CONNECT TO THE UTILITY-PROVIDED SERVICE-DROP CONDUCTORS.

SERVICE CONDUCTORS

OFFSET NIPPLE

ELECTRIC METER HOUSING

RISER

MAIN DISCONNECT AND MAIN OVERCURRENT DEVICE

ELECTRICAL PANEL

GROUNDING BUS ELECTRICALLY CONNECTED TO THE PANEL ENCLOSURE

MAIN 200 A

CIRCUIT BREAKERS (TYPICAL)

PANEL BUSSING

GROUNDED CONDUCTOR BUS ELECTRICALLY CONNECTED TO THE PANEL ENCLOSURE BY THE MAIN BONDING JUMPER

GROUND LEVEL

SERVICE-LATERAL CONDUCTORS

TO UNDERGROUND PEDESTAL

UNDERGROUND SERVICE LATERAL OR SERVICE CONDUCTORS: THE UTILITY MAY REQUIRE THESE TO BE FEEDERS TO A METER PEDESTAL OR WILL PROVIDE THE SERVICE-LATERAL CONDUCTORS ALL THE WAY TO THE METER.

GENERAL LIGHTING LOAD

GENERAL LIGHTING LOAD

GENERAL LIGHTING LOAD

GENERAL LIGHTING LOAD

APPLIANCE LOAD

SMALL APPLIANCE LOAD

MAIN 200 A

FEEDER

FEEDER

SERVICE CONDUCTORS

SERVICE-LATERAL CONDUCTORS

| SERVICE AND SERVICE-EQUIPMENT OVERCURRENT PROTECTION ON LOAD SIDE | FEEDERS AND EQUIPMENT OVERCURRENT PROTECTION ON THE LINE SIDE AND THE LOAD SIDE | BRANCH CIRCUITS AND EQUIPMENT OVERCURRENT PROTECTION ON THE LINE SIDE |

Figure 7-15

"WIREMAN'S GUIDE"
GROUND FAULTS AND SERVICE EQUIPMENT

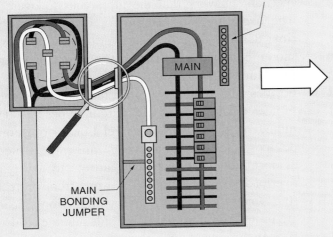

EQUIPMENT GROUNDING
BUS ELECTRICALLY CONNECTED
TO THE PANEL ENCLOSURE

MAIN

MAIN
BONDING
JUMPER

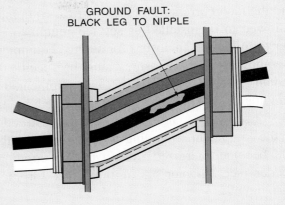

GROUND FAULT:
BLACK LEG TO NIPPLE

A GROUND FAULT HAS DEVELOPED BETWEEN THE METAL NIPPLE AND THE BLACK PHASE CONDUCTOR. BECAUSE THERE IS NO OVERCURRENT DEVICE AHEAD OF THE SERVICE CONDUCTORS, THERE IS NO FUSE OR CIRCUIT BREAKER TO STOP THE CURRENT FLOW. IF IT PERSISTS FOR SUFFICIENT TIME, THERE ARE THREE POSSIBILITIES FOR THE OUTCOME:

1 THE PHASE CONDUCTOR MAY BURN THROUGH OR BLOW ITSELF AWAY, CLEARING THE FAULT. THE HEAT AND THE MAGNETIC FIELDS DEVELOPED CAN EASILY PRODUCE ENOUGH MOVEMENT IN THE CONDUCTOR TO ELIMINATE THE FAULT. THIS IS POTENTIALLY A VERY DANGEROUS SITUATION. IF SOMEONE OPENS THE ENCLOSURE TO TROUBLESHOOT THE ELECTRICAL PROBLEM, THE MOVEMENT OF THE DOOR OR ENCLOSURE COVER MAY CAUSE THE CONDUCTOR TO AGAIN COME INTO CONTACT WITH THE NIPPLE, AND THE RESULTING FLASH AND BURNING MAY SERIOUSLY INJURE OR KILL THAT PERSON.

2 THE PHASE CONDUCTOR PRODUCING THE FAULT MAY CAUSE SO MUCH DAMAGE TO THE SERVICE CONDUIT OR EQUIPMENT THAT THE FAULT IS ELIMINATED. THE EXTENT OF THE DAMAGE WILL BE SUCH THAT IT WILL BE OBVIOUS TO ANYONE APPROACHING THE SITE, SO THAT THE DANGER CAN BE AVOIDED. THIS MAY SOUND LIKE A CRUDE METHOD FOR DETERMINING AN ELECTRICAL PROBLEM, BUT IT IS THE LEAST DANGEROUS OF THE THREE POSSIBILITIES.

3 WHEN THE FAULT OCCURS, THE CONDUCTOR IS NOT SEPARATED FROM THE NIPPLE, SO CURRENT CONTINUES TO FLOW. IF THE CONNECTION BETWEEN THE NIPPLE AND THE SERVICE EQUIPMENT IS A WEAK POINT, THE GROUNDING CONNECTION MAY OPEN. THE GROUND FAULT WILL APPEAR TO BE CLEARED BECAUSE THE CIRCUIT TO GROUND IS ELIMINATED, BUT THIS WILL LEAVE THE NIPPLE ENERGIZED BY THE FAULTING CONDUCTOR. ANYONE WHO APPROACHES THIS SERVICE CAN PROVIDE THE GROUND PATH THE CURRENT NEEDS, AND THE CONSEQUENCES ARE LIKELY TO BE DEADLY.

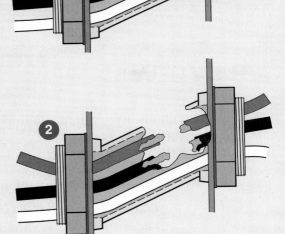

1

2

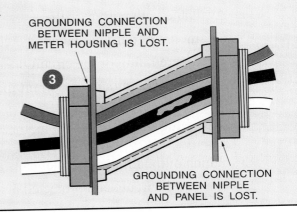

GROUNDING CONNECTION
BETWEEN NIPPLE AND
METER HOUSING IS LOST.

3

GROUNDING CONNECTION
BETWEEN NIPPLE
AND PANEL IS LOST.

Figure 7-16

The design and construction of service equipment and enclosures, and of the service grounding system, must be more exacting than elsewhere in the electrical system so that these components can withstand the fault currents. Not all electrical equipment is allowed to be installed as service equipment. In order to utilize an enclosure for service equipment, the equipment and enclosures must be specifically listed and labeled for use as service equipment. The grounding and bonding system mandated by the Code for services is larger and of better construction than the equipment grounding conductor system that could become energized from a branch circuit or a feeder (see Chapter 14 of this book for more on grounding).

SUMMARY

Overcurrent can be caused by damage or improper installation of equipment and circuits. Short-circuit faults and ground faults are the result. Overcurrent-protective devices must operate very quickly in the event of faults because of the tremendous amount of heat that the excessive current flow will produce. Overcurrents also occur in systems that are operating normally when there is a large inductive load, such as motor loads. The overload is caused by the mechanical work that the motor has to overcome. The overcurrent protection for overloads must be time delayed to allow the overload to clear as the motor completes its work. Therefore, overcurrent protection actually involves two separate protective devices, one to protect against faults and the other to protect against overloads. Circuit types are determined by their location in relation to overcurrent devices. Service conductors are not protected against faults, although they are protected against overloads by their load side overcurrent protection (the service main overcurrent protection). Feeders are protected on both the line side and the load side. Branch circuits are protected against overcurrent on the load side only.

REVIEW

1. When a fault occurs the most important function of the overcurrent-protective device protecting against faults is to _____.
   a. open the circuit very quickly
   b. delay operation to allow the fault to clear itself
   c. open when the current flow goes over 20 amperes
   d. transfer the load to a circuit that has no fault

2. Available fault current is the _____.
   a. amount of current that an overcurrent device will let through before it opens
   b. ampere rating of the circuit breaker or fuse
   c. maximum amount of current that the source can deliver to any given point in the circuit
   d. starting overcurrent of a motor

3. Which of the following statements is *not* correct?
   a. Branch circuits are protected against overcurrent on the load side only.
   b. Feeders have overcurrent protection on both the line side and the load side.

   c. Service conductors are protected against overcurrent on the line side only.

   d. Supplementary overcurrent protection is usually installed on individual pieces of equipment.

4. Which of the following statements is *not* correct?

   a. Circuit breakers from one manufacturer usually cannot be installed in another manufacturer's distribution panel.

   b. Fuses are standardized in size so that one manufacturer's fuses will fit into any fuseholder of the proper ampere rating.

   c. Circuit breakers usually have a higher current interrupt rating than that of fuses.

   d. Some fuses are current limiting.

5. What is the standard interrupt rating of a circuit breaker? (What is the rating if it is not marked on the breaker?)

   a. 5000 amperes

   b. 10,000 amperes

   c. 100,000 amperes

   d. 200,000 amperes

6. *True or False:* Short circuits do not pose an electrocution hazard.

7. *True or False:* Ground faults do not pose an electrocution hazard.

8. *True or False:* The power-providing utility protects each individual dwelling service against fault currents.

9. *True or False:* Many of the procedures required by the *NEC*® are designed to protect dwelling branch circuits against circuit overutilization.

10. *True or False:* Short circuits and ground faults are the same thing.

11. *True or False:* Circuit overutilization means that too many loads are connected to the circuit.

12. *True or False:* The type of load—for example, motor loads or lighting loads—has little effect on the type of overcurrent protection needed on the circuit.

13. *True or False:* Motor starting overcurrent is defined as 50% of the locked-rotor current rating.

14. *True or False:* Overcurrent protection must be installed on all circuits except small-appliance circuits.

15. *True or False:* When overloads occur, the circuit is operating normally.

"WIREMAN'S GUIDE" REVIEW

1. Explain why fault protection must operate very quickly.

2. Explain the method of protecting general-use receptacle circuits against motor overloads.

3. Explain the method used to ensure that only current-limiting fuses will be installed as replacement fuses for current-limiting fuses.

4. Explain what Type S fuses and fuse adapters are designed to prevent.

5. Can loads be served by any other circuit type except branch circuits? Explain.

6. Make two lists. On the first list include those loads that you can think of that are not subject to overloading. On the second list include those loads that you can think of that are subject to overloading. What do the loads in the first list have in common?

7. Read *210.19(A)(3)* and *Exception No. 1*. Why would the *NEC*® allow a 50-ampere circuit breaker to be used for overcurrent protection of a 20-ampere conductor?

CHAPTER 8

Wiring Methods

OBJECTIVES

On completion of this chapter, the student will be able to:

☑ Determine which wiring methods are acceptable for a given installation.

☑ Use the Code book for instructions concerning the requirements for any given wiring method.

☑ Identify *NEC*® wiring requirements, including the location of necessary detailed information, for service-entrance (SE) cable, underground service-entrance (USE) cable, underground feeder (UF) and branch-circuit cable, rigid nonmetallic conduit (RNC), nonmetallic surface raceway (NSR), metal surface raceway (MSR), flexible metal conduit (FMC), liquidtight flexible metal conduit (LFMC), liquidtight flexible nonmetallic conduit (LFNC), electrical nonmetallic tubing (ENT), and electrical metallic tubing (EMT).

☑ Explain the requirements and installation procedures for use of armored cable (AC cable) and metal-clad (MC) cable.

☑ Demonstrate competence in all of the requirements and installation procedures for use of nonmetallic-sheathed cable (NMC).

INTRODUCTION

KEY TERMS

Wiring method   A specific set of materials and installation procedures used to deliver electrical energy from one place to another. Conductors, boxes, raceways or cables, assemblies, straps, clamps, and many other fittings may be used.

This chapter is concerned with **wiring methods**—designating the materials and procedures used to install the electrical system in a building. Any of 14 different wiring methods may reasonably be used, at some location, in dwellings. The suitability of the wiring method to the surroundings, as well as the cost of the material and labor, usually drives the final choice of methods. Dwelling-unit branch circuits are typically wired using one of the three most popular and versatile wiring methods (the remaining 11 wiring methods are used in special situations and locations). Featured in one of the three most common methods used in dwellings, nonmetallic-sheathed cable (NM) is used in about 90% of all dwellings constructed today, and it is the method examined most closely in this chapter. The determination of the proper wiring method or methods to employ is covered first, followed by the actual installation of the wiring in the sample house.

8.1: OVERVIEW OF WIRING METHODS

Conductors are intended to carry electrical current to the loads. Each conductor has three properties that make it unique: (1) conductor type (copper or aluminum); (2) size (measured in AWG or circular mils); and (3) insulation type. However, for safety reasons and for engineering reasons, single conductors cannot be installed as they are; single conductors need to be installed in a raceway or within a cable assembly.

Raceway (See *NEC® Article 100*): A pathway specifically designed and installed to house conductors for protection against physical damage. Raceways may be metallic or nonmetallic and are usually in the shape of a pipe, or conduit, although some raceways are rectangular and have removable covers.

A **raceway** is a metal or nonmetallic channel designed expressly for holding wires or cables. Raceway systems are installed complete, and then the wires are pulled into or onto the raceway. The installation of raceways and the installation of the conductors into the raceways are two separate procedures. In many larger buildings, the conduit work, meaning the installation of the raceway, may be under way for many months before any wire is installed into the conduit.

Cable A group of conductors that are associated with each other by being twisted together or covered with an outer jacketing that provides electrical energy to the load. Cables may have a metallic or a nonmetallic outer covering, or sheathing, and come in many different sizes, types, and styles for use according to the intended installation environment and the nature of the load served.

A **cable** is an assembly of several conductors bundled together. The wire bundle usually has an outer covering that can be either a metallic or nonmetallic material. The size, number, and insulation type for the conductors included in the cable by the manufacturer, as well as the construction of the outer jacket or covering, vary widely. A cable assembly can be produced in almost unlimited combinations of conductor size and type and outer jacketing material. Even communication wires and also fiber-optic systems can be found together in a cable assembly, sometimes with power and lighting conductors included also.

The most popular and useful cable assemblies can be readily purchased at electrical supply companies nationwide. These cables usually have commonplace functions, can be used in many different types of environments and structures, and are capable of serving many possible loads. Other cable assemblies are custom produced, made to order, and good for only one particular function. Some cables are enclosed in a metal jacket that provides excellent protection against physical damage. Some cables are covered with thermoplastic, some with rubber, and some with polyvinyl chloride (PVC). Some cables have no outer jacket; instead, the individual conductors are twisted around each other, and they remain twisted together after installation.

The raceway or cable assembly provides protection against physical damage, keeps the circuit conductors in close proximity, and provides a method of getting the conductors to the loads. The conductors that are installed in the conduit or within the cable are selected to provide the necessary current to the loads without heating to a point at which damage will be done to the conductor insulation over a period of many years. Conductors that are properly selected will last indefinitely. The structure, materials, and installation requirements of these cable and conduit systems are collectively called *wiring methods*. *Chapter 3* of the *NEC®* is devoted to wiring methods.

Excluding wiring for boxes, there are 38 recognized wiring methods, which can be found in *320* to *398* in the Code. For the purposes of this book, these wiring methods are divided into four categories:

1. All 38 of the available wiring methods listed in *Chapter 3* of the Code, including those wiring methods not used in residential construction.

2. Wiring methods that may possibly be used in larger residential projects, such as in construction of high-rise apartment buildings but that are not usually found in residential construction. The residential electrician needs to be aware of these methods but extensive knowledge about them is not usually necessary.

3. Wiring methods that are widely used for residential construction. The residential electrician must be very familiar with these methods and have experience with their installation.

4. Wiring using nonmetallic-sheathed (NM) cable. This is the method used most often in residential construction. The residential electrician must be expert with the installation techniques and with the Code requirements for NM cable.

8.2: AVAILABLE WIRING METHODS

All 38 of the possible wiring methods detailed in the *NEC*® are listed in Table 8-1 (note that wiring methods covered in *338* include underground service entrance [USE] and service entrance [SE], both type R [SER] and type U [SEU], which are considered separately in this book). Of those possible wiring methods, 18 are not used in residential construction and are not discussed further in this book. These wiring methods are the ones that are *not* highlighted on Table 8-1.

8.3: "POSSIBLE" WIRING METHODS

Wiring methods that may be used in residential construction are shown in Table 8-1. Twenty-one wiring methods fit into this category. They are designated as either "possible" or "probable" in the Residential Installation column of the table. Although each of these is considered in this book, the methods more likely to be used are examined in greater detail (notice that the wiring method covered in *338* includes three specific methods: SER, SEU, and USE). The particular section in the Code in which each of the wiring methods is considered in detail is also listed. For more information about the wall thickness and dimensions of conduits, see *Table 4* in *Chapter 9* of the *NEC*®. Following is a list of "possible" wiring methods as designated in Table 8-1:

- *NEC*® *342*—Intermediate metal conduit (IMC): IMC is a round conduit that has extra-thick walls. The walls are not as thick as those of rigid conduit but are thick enough to be threaded at couplings and connections. IMC is used where conductors are subject to severe physical stresses, such as in stubs through concrete floors.

- *NEC*® *344*—Rigid metal conduit (RMC): RMC is virtually the same as IMC except that the conduit wall is considerably thicker.

- *NEC*® *354*—Nonmetallic underground conduit and conductors (NUCC): NUCC is a conduit and cable assembly with thin-walled nonmetallic outer covering. The cables are assembled in the factory to specific predetermined lengths but can also be obtained in bulk. NUCC is used for branch circuits and feeders in place of AC or MC cables when allowed.

- *NEC*® *360*—Flexible metal tubing (FMT): FMT is a metallic conduit that is flexible and liquid tight without being covered with a water-tight covering. There are special bending restrictions for FMT. It has very limited residential applications.

- *NEC*® *376*—Metal wireways: Metal wireways are metal channels and fittings with removable covers. The channel is usually square and available in several different lengths. Metal wireways are used for large conductors and services.

- *NEC*® *378*—Nonmetallic raceway: Nonmetallic raceways are the same as metal wireways except that they are constructed of nonmetallic material.

- *NEC*® *380*—Multioutlet assembly: Multioutlet assemblies are metallic surface-mounted raceways that house receptacle outlets with removable covers. The assembly serves as the raceway, the box, and the mounting support for the receptacles. The receptacle outlet spacing varies from a few inches to several feet apart.

8.4: "PROBABLE" WIRING METHODS

Wiring methods that are commonly used in residential construction are designated "probable" in Table 8-1. Excellent familiarity with these 14 wiring methods is essential to the professional competence of a residential electrician. Each of the methods is approved for specific installation environments and for serving specific wiring demands.

Table 8-2 provides some detail about the various wiring methods and their allowable installation locations. The wiring method highlighted in red is that using NM cable. It is the most widely used wiring method for dwellings and is considered separately from the other "probable" wiring methods. It is important that the electrician read the Code to obtain full requirements for installing and using these wiring methods. Table 8-2 summarizes the requirements but may not contain all of the necessary information for an acceptable installation. This table and those that follow in this chapter are to be used as guides for comparing one wiring method with another and as aids in selecting appropriate wiring methods in specific locations. In addition, it should be noted that the use of these wiring methods may not be allowed with certain types of construction.

NEC® ARTICLE	WIRING METHOD	TYPE	RACEWAY	SINGLE CONDUCTOR	CABLE ASSEMBLY	RESIDENTIAL INSTALLATION	NOTES
320	Armored cable	AC			X	Probable	Required dwelling wiring in some areas of the country
322	Flat cable assemblies	FC			X	No	
324	Flat conductor cable	FCC			X	No	
326	Integrated gas spacer cable	IGS			X	No	
326	Medium-voltage cable	MV			X	No	
330	Metal-clad cable	MC			X	Probable	Required dwelling wiring in some areas of the country
332	Mineral-insulated, metal-sheathed cable	MI			X	No	
334	Nonmetallic-sheathed cable	NM,NMC,NMS			X	Yes	Most common wiring method
336	Power and control tray cable	TC			X	No	
338	Service-entrance cable	SEU,SER,USE		X	X	Probable	Services, ranges, dryers
340	Underground feeder and branch-circuit cable	UF			X	Probable	Underground direct buried cable
342	Intermediate metal conduit (IMC)	IMC	X			Possible	To protect conductors only—not as raceway
344	Rigid metal conduit (RIGID)	RMC	X			Possible	To protect conductors only—not as raceway
348	Flexible metal conduit (FLEX)	FMC	X			Probable	Equipment connection
350	Liquidtight flexible metal conduit	LFMC	X			Probable	Outdoor equipment connection
352	Rigid nonmetallic conduit (PVC)	RNC	X			Probable	Underground raceway
354	Nonmetallic underground conduit and conductors	NUCC			X	Possible	Factory assembly
356	Liquidtight flexible nonmetallic conduit	LFNC	X			Probable	
358	Electrical metallic tubing (EMT)	EMT	X			Probable	To protect wire/outdoor surface; most popular conduit
360	Flexible metal tubing	FMT	X			Possible	Equipment connection
362	Electrical nonmetallic tubing (ENT)	ENT	X			Probable	For raceway if needed
366	Auxiliary gutters		X			No	
368	Busways		X			No	
370	Cable bus		X			No	
372	Cellular concrete floor raceways		X			No	
374	Cellular metal floor raceways		X			No	
376	Metal wireways		X			Possible	Services/multi-family applications
378	Nonmetallic raceway		X			Possible	Exposed wiring only
380	Multioutlet assembly		X			Possible	Plug mold—mainly for remodeling
382	Nonmetallic extensions				X	No	
384	Strut-type channel raceway		X			No	
386	Surface metal raceway		X			Probable	Exposed only
388	Surface nonmetallic raceway		X			Probable	Wire mold—mainly for remodeling
390	Underfloor raceway		X			No	Wire mold—mainly for remodeling
392	Cable tray		X		X	No	
394	Concealed knob-and-tube wiring			X		No	
396	Messenger-supported wiring			X		No	
398	Open wiring on insulators			X		No	

Table 8-1　"Possible" wiring methods and their application in residential wiring.

RESTRICTIONS

FEATURE/REQUIREMENT	320 AC	330 MC	334 NM	338 USE	338 SE	340 UF	348 FMC	350 LFMC	352 RNC	356 LFNC	358 EMT	362 ENT	386 MSR	388 NSR
System requirements														
Can be used as service conductors	N	Y	N	Y	Y	N	Y	Y	Y(4)(5)	N	Y	N	N	N
Can be used as feeder conductors	Y	Y	Y(6)	N(2)	Y(4)	Y	Y	Y	Y(4)(5)	Y	Y	Y(4)	Y	Y
Can be used as branch-circuit conductors	Y	Y	Y(6)	N(2)	Y(2)(4)	Y	Y	Y	Y(4)(5)	Y	Y	Y(4)	Y	Y
Location requirements														
Indoors	Y	Y	Y	N	Y(2)	Y	Y	Y	Y	Y	Y	Y	Y	Y
Outdoors	N	Y	N	Y	Y	Y	N	Y	Y	Y(3)	Y	Y	N	N
Wet location	N	N(2)	Y(2)	Y	Y	Y	N(2)	Y	Y(2)	Y	Y	Y	N	N
Damp location	N	N	Y	Y	Y	Y	Y	Y	Y	Y	Y	Y	Y	Y
Dry location	Y	Y	Y	Y	Y	Y	Y	Y	Y	Y	Y	Y	Y	Y
Exposed	Y(1)	Y	Y(1)	N	Y(1)	Y(1)	Y(1)	Y(1)	Y(1)	Y(1)	Y(9)	N(2)	Y(9)	Y(9)
Direct sunlight	Y	Y	Y(3)	Y	Y(3)	Y(3)	Y	Y	Y	Y(3)	Y	Y(3)	N	N
Direct buried (underground)	N	Y(3)	N	Y	Y	Y	N	Y(3)	Y	Y(3)	Y(8)	N(2)	N	N
Under slab on grade	N	Y(3)	N	Y	Y	Y	N	N	Y	Y(3)	Y(8)	Y	N	N
Encased in concrete	Y	Y	N	N	N	N	N	N	Y	N	Y(8)	Y	N	N
Embedded in plaster/masonry	Y	Y	Y	Y	Y	Y	Y	Y	Y	Y	Y	Y(7)	N	N
"Fish" into walls and masonry	Y	Y	Y	Y	Y	Y	Y	Y	Y	Y	Y	Y(7)	N	N
Concealed in floors	Y	Y	Y	N	Y	Y	Y	Y	Y	Y	Y	Y(7)	N	N
Concealed in walls	Y	Y	Y	N	Y	Y	Y	Y	Y	Y	Y	Y(7)	N	N
Concealed in ceilings	Y	Y	Y	N	Y	Y	Y	Y	Y	Y	Y	Y(7)	N	N
Above suspended ceiling	Y	Y	Y	N	Y	Y	Y	Y	Y	Y	Y	Y(7)	N	N
Grounding requirements														
A separate equipment grounding conductor is needed.	N	Y(11)	N	—	N	N	N(2)	N(2)	Y	Y	N	Y	N	Y
Conduit/sheath is the equipment grounding conductor.	Y	N(11)	N	N	N	N	Y(2)	Y(2)	N	N	Y	N	Y	N
Equipment grounding conductor is included in cable.	Y(10)	Y(11)	Y	—	Y	Y	—	—	—	—	—	—	—	—

Y, yes; N, no.

Special Condition Notes:
(1) Must be protected from physical damage.
(2) For detailed restrictions or requirements, see the *NEC*®.
(3) Must be listed for the use.
(4) There are thermal insulation limitations.
(5) Expansion joints are required.
(6) Multi-family dwelling must be of Type III, IV, or V construction.
(7) Can be installed above three floor if a 15-minute thermal barrier is installed. See *NEC*® 362.10 for more about thermal barriers.

(8) Corrosion protection is required.
(9) Must be protected from severe physical damage.
(10) Available with or without separate equipment grounding conductor.
(11) Required with interlocked metal tape sheath Type MC cable. Other types have sheathing approved as equipment grounding conductor.

Table 8-2 Restrictions on the use of various wiring methods in dwelling units.

For example, rigid nonmetallic conduit (RNC) is listed in Table 8-2 as usable above suspended ceilings. However, if the space above a suspended ceiling is used as an air plenum for the heating and cooling systems, then use of RNC is not allowed. Care must be taken to select the proper wiring method for the type of construction.

In Table 8-1, the Wiring Method column is followed by a column of letter codes designating the wiring method type (see Table 8-2). These codes are how wiring methods are referred to in the electrical trade—for example, "12-2 with ground steel AC cable." Following is a list of what these letter codes mean:

- M: Any wiring method with an M in the letter code is a metallic wiring method (for example, EMT).

- N: Any wiring method with an N in the letter code is a nonmetallic wiring method (for example, RNC). In the case of NM cable, the N stands for "non" and the M stands for "metallic (sheathed)"; thus, NM cable is nonmetallic-sheathed cable.

- U: Any wiring method with a U in the letter code is an underground wiring method (for example, UF).

- F: Any wiring method with an F in the letter code is a flexible wiring method (for example, FMC).

- L: Any wiring method with an L in the letter code is a liquidtight wiring method (for example, LFNC).

- S: Any wiring method with an S in the letter code is a wiring method for use with services (for example, SE).

- A: Any wiring method with an A in the letter code is an armored wiring method (for example, AC).

- E: The letter E in a letter code stands for *electrical* (for example, EMT).

- C: The letter C in a letter code can mean *conduit* or *cable* (for example, with AC, the C is for cable, and with FMC, the C is for conduit).

In order to more closely study the wiring methods that may be used in dwelling units, they are divided into three categories: cable assemblies, NM conduits and raceways, and metallic conduits and raceways.

8.4.1: Cable Assemblies

Of those wiring methods commonly used in dwellings, a total of six utilize cable assemblies, as shown in Table 8-3.

NEC® 320—Armored cable (AC): Armored cable has a metallic outer covering. The conductors enclosed in the metal jacket are 90°C rated, and their ampacity is determined from the 60°C column of *Table 310.16*. AC cable is manufactured with an internal bonding strip of either aluminum or copper that must be in contact with the sheathing for its entire length. This bonding strip can be cut off when the termination is made. The AC cable method is the wiring method of choice for residential wiring when NM cable cannot be used because of local codes or customs. With AC cable, bushings must be inserted into the termination of the armor to keep the conductors from chafing and faulting to the sheath. See Figure 8-1 for more information on AC cable.

NEC® 330—Metal-clad cable (MC): MC cable comes in three other varieties: (1) with a smooth metal outer sheath, (2) with a corrugated metal outer sheath, and (3) with an interlocking metal strip, as shown in Figure 8-2. The most common type of MC cable is the interlocking metal strip, which looks much the same as AC cable. MC cable can be manufactured using any wiring type approved for MC cable listed in *Table 310.13* and can include fiber-optic cables as well.

NEC® 334—Nonmetallic-sheathed cable (NM): Type NM cable is the wiring method of choice. The cable assembly includes two or three insulated conductors (black, white, and red) and a bare or green equipment grounding conductor. The conductors are laid side-by-side in two-wire plus equipment grounding conductor cables and wrapped around themselves and the equipment grounding conductor in three-wire cables. The equipment grounding conductor is assumed for NM cable. For example, 12-2 NM cable is a cable with two insulated circuit conductors (black and white), size 12 AWG, and one bare (or green, covered, or insulated) equipment grounding conductor, size 12 AWG. The conductors of type NM cable are separated from the outer sheathing and from each other by a wrapping of paper or thin cardboard. Chapter 9 of this book covers the installation of NM cable in a dwelling. Figure 8-3 (page 204) presents more information on type NM cables. Figure 8-4 (page 205) shows NM cable runs installed in a typical dwelling.

REQUIREMENTS

FEATURE	AC 320	MC 330	NM 334	USE 338	SE 338	UF 340
Sizes allowed			14 AWG—2 AWG [334.104]	14 AWG—2000 kcmil [Table 310.13]		14—4/0 AWG [340.104]
Bends—number of bends allowed in run	Unlimited	Unlimited	Unlimited	Unlimited	Unlimited	Unlimited
Bends—minimum radius of bends in cable	5 times the diameter [320.24]	Varies with type of MC [330.24]	5 times the diameter [334.24]	5 times the diameter [338.24]	5 times the diameter [338.24]	5 times the diameter [340.24]
Supports—at boxes—inches from box	12 in. (300 mm) [320.30]	12 in. (300 mm) [330.30(C)]	12 in. (300 mm) [334.30]	Interior wiring not allowed [338.10(B)(4)]	12 in. (300 mm) [338.10(B)(4)]	12 in. (300 mm) [340.10]
Supports—along cable run—inches from previous support	4½ ft (1.4 m) [320.30]	6 ft (1.8 m) [330.30(A)]	4½ ft (1.4 m) [334.30]	Interior wiring not allowed [338.10(B)(4)]	4½ ft (1.4 m) [338.10(B)(4)]	4½ ft (1.4 m) [340.10]
Bushings required at termination	Yes [320.40]					
Conductor insulation type	90°C from Table 310-13 [320.80(A)]	Per Table 310.13 [330.112]	90°C from Table 310.13 [334.112]	Use [Table 310.13]	Any in Table 310.13 for 90°C [338.10(B)(4)]	Any moisture-resistant type from Table 310.13 [340.112]
Temperature rating of conductors	60°C [320.80(A)]	Per Table 310.13 [330.112]	60°C [334.80]	75°C [Table 310.13]	90°C [338.10(B)(4)]	60°C [340.80]
In accessible attic—across top of joists—guarding	Guard strips [320.23]	Guard strips [330.23]	Guard strips [334.23]	Interior wiring not allowed [338.10(B)(4)]	Guard strips [338.10(B)(4)]	Guard strips [340.10]
Attic—no stairs or ladder—across top of joists—guarding	Not within 6 ft (1.8 m) of hole [320.23(A)]	Not within 6 ft (1.8 m) of hole [330.23]	Not within 6 ft (1.8 m) of hole [334.23]	Interior wiring not allowed [338.10(B)(4)]	Not within 6 ft (1.8 m) of hole [338.10(B)(4)]	Not within 6 ft (1.8 m) of hole [340.10]
Basement—across bottom or on bottom of joists			2-6 AWG or larger cable [334.15(C)]	Interior wiring not allowed [338.10(B)(4)]	2-6 AWG or larger conductors [338.10(B)(4)]	2-6 AWG or larger conductors [340.10]

Table 8-3 Installation requirements for selected cable assemblies.

"WIREMAN'S GUIDE"
ARMORED CABLE

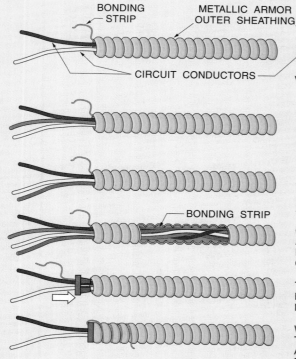

ARMORED CABLE (TYPE AC CABLE) IS CIRCULAR AND HAS A SHEATH OF CONTINUOUS METAL INTERLOCKING TAPE ENCLOSING THE CIRCUIT CONDUCTORS. ALSO INCLUDED IS AN INTERNAL METALLIC BONDING STRIP THAT IS IN CONTACT WITH THE ARMOR SHEATHING FOR ITS ENTIRE LENGTH. THE CONDUCTORS MUST BE 90°C RATED WITH AMPACITIES FROM THE "60°C" COLUMN OF *TABLE 310.16*.

THE OUTER SHEATHING AND THE TERMINAL CONNECTORS OF TYPE AC CABLE ARE LISTED AS AN ACCEPTABLE EQUIPMENT GROUNDING PATH. TYPE AC CABLE IS ALSO AVAILABLE WITH A SEPARATE EQUIPMENT GROUNDING CONDUCTOR. THE BONDING STRIP IS NOT AN EQUIPMENT GROUNDING CONDUCTOR AND CAN BE CUT OFF AT THE ENDS OF THE CABLE SHEATHING.

TYPE AC CABLE IS AVAILABLE WITH THREE INSULATED CURRENT-CARRYING CONDUCTORS, EITHER WITH OR WITHOUT AN EQUIPMENT GROUNDING CONDUCTOR.

CUT-AWAY VIEW: TYPE AC CABLE CONDUCTORS ARE TWISTED AROUND EACH OTHER TO FORM A CIRCULAR CABLE. THE EXTRA SPACE IN THE CABLE ASSEMBLY IS FILLED WITH PAPER OR A LIGHTWEIGHT CARDBOARD.

TYPE AC CABLE CONDUCTORS ARE REQUIRED TO HAVE A BUSHING INSTALLED TO SEPARATE THE CABLES FROM THE METAL EDGE OF THE OUTER SHEATHING.

WITH THE BUSHING IN PLACE, RATHER THAN CUT OFF ALL OF THE BONDING STRIP, MANY ELECTRICIANS WRAP THE STRIP AROUND THE BODY OF THE CABLE. THIS STEP SECURES THE BUSHING TO HELP KEEP IT IN PLACE.

A SPECIAL HAND TOOL IS REQUIRED TO CUT AC CABLE. THIS TOOL CAN CUT A NOTCH IN THE AC CABLE OUTER METALLIC SHEATH, AS SHOWN AT RIGHT, BUT WILL NOT CUT SO DEEP AS TO NICK OR CUT THE CONDUCTORS. BECAUSE THIS NOTCH SPANS ONE COMPLETE TURN OF THE OUTER SHEATHING, THE SHEATHING CAN BE SEPARATED SIMPLY BY PULLING IN OPPOSITE DIRECTIONS.

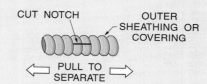

THERE ARE SEVERAL TYPES OF AC CABLE CLAMPS:

TWO-SCREW AC CABLE CLAMPS ARE VERY MUCH LIKE NM CABLE CLAMPS (SEE FIGURE 8-3), EXCEPT THAT THEY HAVE A SEAT FOR SUPPORT OF THE AC CABLE BUSHING; THE NM CABLE CLAMP HAS NO SEAT.

SINGLE-SCREW AC CABLE CLAMPS HAVE NO SEAT. THE AC CABLE IS SUPPORTED BY A SMALL FLANGE INSIDE THE CLAMP.

SOME TYPES OF AC CABLE CLAMPS ALLOW FOR THE INSTALLATION OF TWO CABLES, USING ONE CLAMP AND OCCUPYING ONLY ONE KNOCKOUT IN THE BOX OR CABINET.

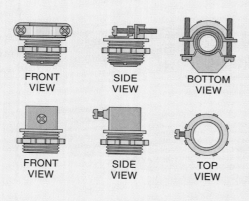

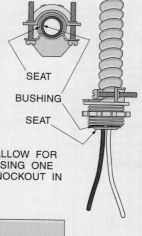

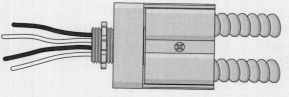

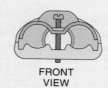

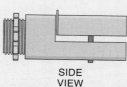

Figure 8-1

"WIREMAN'S GUIDE"
METAL-CLAD CABLE

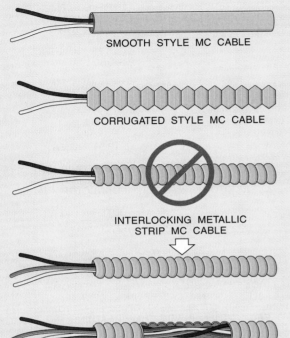

SMOOTH STYLE MC CABLE

CORRUGATED STYLE MC CABLE

INTERLOCKING METALLIC
STRIP MC CABLE

METAL-CLAD (MC) CABLE IS AVAILABLE IN THREE TYPES:
(1) WITH A SMOOTH METAL OUTER JACKET OR SHEATH,
(2) WITH A CORRUGATED OUTER SHEATH, AND
(3) WITH AN INTERLOCKING METAL STRIP.

THE SMOOTH AND CORRUGATED OUTER SHEATH TYPES OF MC CABLE ARE LISTED AS EQUIPMENT GROUNDING CONDUCTORS.

THE INTERLOCKING METAL STRIP OUTER SHEATH TYPE OF MC CABLE IS NOT LISTED AS AN EQUIPMENT GROUNDING CONDUCTOR. THE NECESSARY EQUIPMENT GROUNDING CONDUCTOR MUST BE INCLUDED AS ONE OF THE CONDUCTORS HOUSED BY THE CABLE.

THE MOST POPULAR STYLE OF MC CABLE IS THE INTERLOCKING METAL STRIP.

MC CABLE IS MORE ADAPTABLE TO SPECIAL NEEDS THAN AC CABLE BECAUSE IT CAN BE MANUFACTURED IN THOUSANDS OF COMBINATIONS OF INSULATION TYPES, CONDUCTOR SIZE, AND NUMBER OF CONDUCTORS.

THE CONDUCTORS OF MC CABLE ARE WRAPPED IN A MYLAR SHEET THAT IS USED AS FILLER INSTEAD OF THE PAPER FILLER IN AC CABLE.

NEC® 330 DOES NOT MENTION A REQUIREMENT TO INSTALL BUSHINGS AS WITH AC CABLE. IN MANY CASES, THE AHJ REQUIRES THAT BUSHINGS BE INSTALLED, HOWEVER.

FRONT
VIEW

SIDE
VIEW

BOTTOM
VIEW

TWO-SCREW
MC AND AC CABLE
CONNECTORS

MANY OF THE FITTINGS THAT ARE APPROVED FOR AC CABLE CAN ALSO BE USED WITH THE INTERLOCKING METAL STRIP TYPE OF MC CABLE.

FRONT
VIEW

SIDE
VIEW

TOP
VIEW

SINGLE-SCREW
MC AND AC
CABLE CONNECTORS

Figure 8-2

"WIREMAN'S GUIDE"
NONMETALLIC-SHEATHED (NM) CABLE

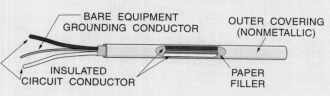

BARE EQUIPMENT GROUNDING CONDUCTOR

OUTER COVERING (NONMETALLIC)

INSULATED CIRCUIT CONDUCTOR

PAPER FILLER

TOP VIEW OF TWO-WIRE NM CABLE. PART OF TOP SHEATHING HAS BEEN REMOVED TO EXPOSE WIRES INSIDE.

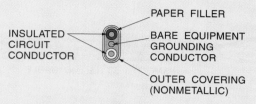

PAPER FILLER

INSULATED CIRCUIT CONDUCTOR

BARE EQUIPMENT GROUNDING CONDUCTOR

OUTER COVERING (NONMETALLIC)

END VIEW OF TWO-WIRE NM CABLE

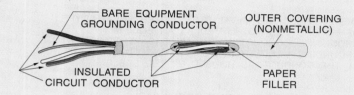

BARE EQUIPMENT GROUNDING CONDUCTOR

OUTER COVERING (NONMETALLIC)

INSULATED CIRCUIT CONDUCTOR

PAPER FILLER

TOP VIEW OF THREE-WIRE NM CABLE. PART OF THE TOP SHEATHING HAS BEEN REMOVED TO EXPOSE WIRES INSIDE.

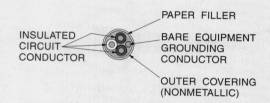

PAPER FILLER

INSULATED CIRCUIT CONDUCTOR

BARE EQUIPMENT GROUNDING CONDUCTOR

OUTER COVERING (NONMETALLIC)

END VIEW OF THREE-WIRE NM CABLE

TYPE NM CABLE (ALSO KNOWN IN THE ELECTRICAL INDUSTRY AS **ROMEX**®—A BRAND NAME OF THE ROME CABLE COMPANY) IS AVAILABLE AS TWO-WIRE AND THREE-WIRE CABLES. ALL NM CABLES INCLUDE AN EQUIPMENT GROUNDING CONDUCTOR, AND THE CURRENT-CARRYING CONDUCTORS MUST BE 90°C RATED. TWO-WIRE CABLES HAVE THE CONDUCTORS ARRANGED NEXT TO EACH OTHER INSIDE THE SHEATHING, WHICH MAKES FOR A FLAT CABLE. IN THREE-WIRE CABLES THE CONDUCTORS ARE TWISTED AROUND EACH OTHER TO MAKE A ROUND CABLE.

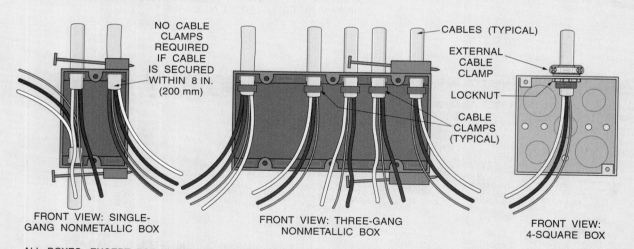

NO CABLE CLAMPS REQUIRED IF CABLE IS SECURED WITHIN 8 IN. (200 mm)

CABLES (TYPICAL)

EXTERNAL CABLE CLAMP

LOCKNUT

CABLE CLAMPS (TYPICAL)

FRONT VIEW: SINGLE-GANG NONMETALLIC BOX

FRONT VIEW: THREE-GANG NONMETALLIC BOX

FRONT VIEW: 4-SQUARE BOX

ALL BOXES, EXCEPT FOR SINGLE-GANG NONMETALLIC BOXES, MUST HAVE PROVISIONS FOR FASTENING THE CABLES TO THE BOX. WITH NONMETALLIC BOXES, CABLE CLAMPS ARE USUALLY INCLUDED AS PART OF THE BOX. WITH METAL BOXES, CABLE CLAMPS MAY BE INCLUDED WITH THE BOX, OR EXTERNAL CABLE CONNECTORS CAN BE UTILIZED.

FRONT VIEW OF A TWO-SCREW CABLE CONNECTOR

SIDE VIEW OF A TWO-SCREW CABLE CONNECTOR

TOP VIEW OF A TWO-SCREW CABLE CONNECTOR

TWO-SCREW CABLE CONNECTORS ARE AVAILABLE FOR ALL SIZES OF NM CABLE. OTHER VARIETIES OF METALLIC AND NONMETALLIC CABLE CONNECTORS ARE AVAILABLE.

Figure 8-3

Figure 8-4 NM cable runs in a dwelling under construction.

NEC® 338—Service-entrance cable (USE): *NEC® 338* covers cables used for service conductors. These include Types underground service-entrance (USE) and service-entrance (SE) cables. Because they are quite different in their makeup and usage, they are considered as separate wiring methods in this book. Type USE conductors are approved as service-entrance conductors in underground systems. They are available as single conductors, as cable assemblies without an outer covering, and as cable assemblies with an outer covering. Type USE cables are not allowed for interior wiring except where the cable may penetrate the structure to supply a distribution system.

NEC® 338—Service-entrance cable (SE): Type SE cable is approved for service-entrance conductors, but in some cases it can also be used for branch circuits and feeders. SE cable can be installed

indoors or outdoors and is available in two popular types, SEU and SER. More information about SE cables is shown in Figure 8-5.

NEC® 340—Underground feeder and branch-circuit cable (UF): Although Type UF is intended for underground wiring, the cable has many of the same advantages as Type NM cable. UF cable can be used outdoors and may be used in place of NM cable indoors if desired, except in attics and other places where the temperature could exceed 60°C. However, Type UF has the conductors molded into the nonmetallic covering itself, rather than separated by a layer of paper, as with NM. This makes UF cable relatively difficult to strip (remove the outer covering), and to expose the conductors. More information about UF cable is presented in Figure 8-6.

8.4.2: Nonmetallic Conduit and Raceway

The majority of residential electrical work involves installing NM cable for lighting, general-purpose outlets, and small-appliance circuits. None of these tasks requires special wiring methods or materials. However, the residential electrician must also provide circuits for air conditioners, furnaces, sump pumps, and a long list of other equipment that may be found in certain dwellings. A working knowledge of wiring methods that can be used for these types of installations is important to the residential electricians.

Table 8-4 (page 208) summarizes those wiring methods utilizing nonmetallic conduits or raceways that may be used with dwelling units (see Table 8-2). There are four methods presented in Table 8-4: two involving rigid conduit and two involving flexible conduit.

NEC® 352—Rigid nonmetallic conduit (RNC): Type RNC is commonly referred to as PVC conduit. RNC is employed for underground feeders and service conductors but can also be used for underground branch circuits. RNC can be bent using a special heating pad or a torch. RNC can also be used above ground, where it must withstand impact, heating from direct sunlight, and other abuses. Figure 8-7 (page 209) shows RNC and some of its uses and restrictions. Type RNC shrinks and swells with changes in temperature, resulting in significant periods of lengthening and shortening. This requires that

"WIREMAN'S GUIDE"
TYPES SEU AND SER CABLES

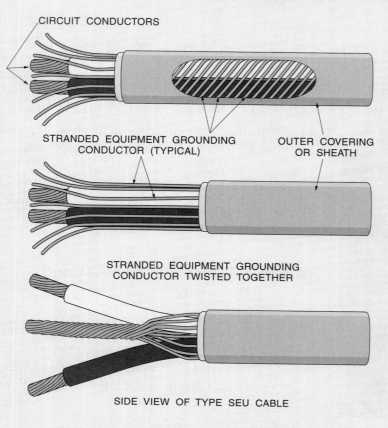

CIRCUIT CONDUCTORS

STRANDED EQUIPMENT GROUNDING CONDUCTOR (TYPICAL)

OUTER COVERING OR SHEATH

STRANDED EQUIPMENT GROUNDING CONDUCTOR TWISTED TOGETHER

SIDE VIEW OF TYPE SEU CABLE

THE CIRCUIT CONDUCTORS OF SEU TYPE CABLE ARE WRAPPED BY A SEPARATELY STRANDED EQUIPMENT GROUNDING CONDUCTOR. THE STRANDS ARE WRAPPED (TWISTED) AROUND EACH OTHER AT THE TERMINATION.

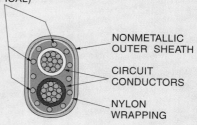

STRANDED EQUIPMENT GROUNDING CONDUCTOR (TYPICAL)

NONMETALLIC OUTER SHEATH

CIRCUIT CONDUCTORS

NYLON WRAPPING

END VIEW OF TYPE SEU CABLE

THE CIRCUIT CONDUCTORS ARE LAID SIDE BY SIDE AND ARE WRAPPED BY THE INDIVIDUAL STRANDS OF THE EQUIPMENT GROUNDING CONDUCTOR. THESE STRANDS ARE TWISTED TOGETHER AT THE TERMINATION, BUT BECAUSE THE OUTER SHEATH DOES NOT CONSTITUTE INSULATION, THE EQUIPMENT GROUNDING CONDUCTOR IS BARE.

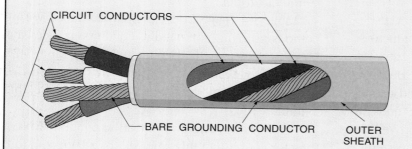

CIRCUIT CONDUCTORS

BARE GROUNDING CONDUCTOR

OUTER SHEATH

SIDE VIEW OF TYPE SER CABLE

TYPE SER CABLE HAS AN UNINSULATED GROUNDING CONDUCTOR AS WELL AS CIRCUIT CONDUCTORS. BECAUSE OF THE GROUNDING SYSTEM USED AT THE SERVICE, THE GROUNDED CIRCUIT CONDUCTOR OF SERVICE CONDUCTORS IS NOT USUALLY REQUIRED TO BE INSULATED. IF THE CABLE IS USED FOR BRANCH CIRCUITS OR FEEDERS, THE BARE CONDUCTOR CANNOT BE USED AS THE GROUNDED CIRCUIT CONDUCTOR AND MUST BE USED AS AN EQUIPMENT GROUNDING CONDUCTOR.

NYLON OR SIMILAR WRAPPING

BARE GROUNDING CONDUCTOR

CIRCUIT CONDUCTORS

OUTER SHEATH

END VIEW OF TYPE SER CABLE

TYPE SER CABLE IS ROUGHLY CIRCULAR IN CROSS SECTION. THE CONDUCTORS ARE WRAPPED WITH A NYLON OR SIMILAR COVERING. THE OUTER SHEATH DOES NOT INSULATE THE BARE CONDUCTOR.

Figure 8-5

"WIREMAN'S GUIDE"
UNDERGROUND FEEDER AND
BRANCH-CIRCUIT CABLE

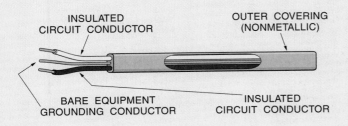

INSULATED
CIRCUIT CONDUCTOR

OUTER COVERING
(NONMETALLIC)

BARE EQUIPMENT
GROUNDING CONDUCTOR

INSULATED
CIRCUIT CONDUCTOR

TOP VIEW OF UNDERGROUND FEEDER (UF)
AND BRANCH-CIRCUIT CABLE

THE OUTER COVERING IS A NONMETALLIC
PLASTIC-LIKE MATERIAL THAT FLEXES EASILY
BUT DOES NOT CRACK OR SPLIT. THE OUTER
COVERING IS TOUGH AND CAN WITHSTAND
THE UNDERGROUND ENVIRONMENT FOR A
LONG TIME.

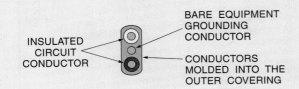

INSULATED
CIRCUIT
CONDUCTOR

BARE EQUIPMENT
GROUNDING
CONDUCTOR

CONDUCTORS
MOLDED INTO THE
OUTER COVERING

END VIEW OF UF AND BRANCH-CIRCUIT CABLE

LIKE NM CABLE, UF CABLE INCLUDES AN
EQUIPMENT GROUNDING CONDUCTOR. THIS
CONDUCTOR CAN BE COVERED, BARE, OR
INSULATED. THE OUTER COVERING IS MOLDED
AROUND THE CIRCUIT CONDUCTORS. THIS
FEATURE MAKES THE CABLE DIFFICULT TO STRIP.

TOP VIEW

FRONT VIEW

SIDE VIEW

TOP VIEW
OF A LOCKNUT

TWO-SCREW CABLE CLAMP

MOST OF THE CABLE CLAMPS APPROVED FOR USE WITH TYPE NM CABLE ARE
ALSO APPROVED FOR USE WITH TYPE UF CABLE. OTHER TYPES OF CABLE CLAMPS
ARE AVAILABLE. IF THE CABLE IS USED OUTDOORS AND ENTERS THE TOP OF A
BOX, A WATERPROOF CLAMP MAY BE REQUIRED.

IF THE UNDERGROUND BRANCH CIRCUIT IS PROTECTED BY A
GFCI SYSTEM, AND IF THE CIRCUIT CURRENT FLOW IS 20 AMPERES
OR LESS, THE UF CABLE CANNOT BE INSTALLED LESS THAN
12 IN. (300 mm) BELOW GRADE LEVEL.

IF THE CONDITIONS LISTED ABOVE
ARE NOT SATISFIED, THE UF CABLE
MUST BE INSTALLED NO LESS THAN
18 IN. (450 mm) BELOW GRADE. IF THE
UF CABLE IS INSTALLED UNDER A
ONE- OR TWO-FAMILY DWELLING UNIT'S
DRIVEWAY, IT CANNOT BE LESS THAN
18 IN. (450 mm) BELOW GRADE.

DRIVEWAY

18 IN. (450 mm)

12 IN. (300 mm)

Figure 8-6

REQUIREMENTS

FEATURE	RNC 352	LFNC 356	ENT 362	NSR 388
Sizes allowed	½ in. (13 mm) thru 6 in. (150 mm) trade size [352.20]	½ in. (13 mm) thru 4 in. (100 mm) trade size [356.20]	½ in. (13 mm) thru 2 in. (50 mm) trade size [362.20]	Per the design Size marked on carton [388.22]
Bends—number of bends allowed in run	360° (4 x 90°) [352.26]	360° (4 x 90°) [356.26]	360°C (4 x 90°) [362.26]	Per listing [388.6]
Bends—minimum radius of bends in conduit or tubing	Per Table 344.24 [352.24]	Per Table 344.24 [356.26]	Per Table 344.24 [356.24]	Per listing—manufactured fittings [388.6]
Supports—at boxes—feet from box	Within 3 ft (900 mm) [352.30(A)]	Not exceeding 1 ft (300 mm) [356.30]	Within 3 ft (900 mm) [362.30]	Per listing [388.6]
Supports—along conduit or tubing run—feet from previous support	Per 352.30(B)	3 ft (900 mm) [356.30]	3 ft (900 mm) [362.30]	Per listing [388.6]
Bushings required at termination	Yes [352.46]	Yes [300.4(F)]	Yes [362.46]	Per listing [388.6]
Available as factory-prewired assembly	No	Yes [356.10]	Yes ½–1 in. (13–25 mm) trade size [362.100]	No
Maximum temperature rating	Not above 50°C ambient [352.12(D)]	Listed temperature rating [356.12]	Not above 50°C [362.12]	Per listing [388.6]
Thermal expansion joints required	Yes; in accordance with Table 352.44 [352.44]	No	No	No
Number of conductors in the conduit or tubing	Per Table 1, Chapter 9 [352.22]	Per Table 1, Chapter 9 [356.22]	Per Table 1, Chapter 9 [362.22]	Per design and listing [388.22]
Joints in sections of conduit or tubing	Approved method [352.48]	Listed terminal fittings No angle fittings if concealed [356.42]	Approved method [362.48]	Per listing [388.6]
Trimming end cuts required	Yes [352.28]	Yes [356.28]	Yes [362.28]	Per listing [388.6]

Table 8-4 Installation requirements for selected nonmetallic conduits and raceways.

"WIREMAN'S GUIDE"
RIGID NONMETALLIC CONDUIT

RIGID NONMETALLIC CONDUIT (RNC) IS AVAILABLE IN 10-FT (3.048-m) LENGTHS (EACH LENGTH IS COMMONLY CALLED A "STICK") WITH A COUPLING FACTORY INSTALLED ON ONE END.

END VIEW
RNC SCHEDULE 40

END VIEW
RNC SCHEDULE 80

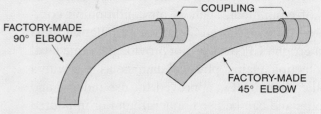

FACTORY-MADE
90° ELBOW

COUPLING

FACTORY-MADE
45° ELBOW

RNC IS AVAILABLE WITH SEVERAL WALL THICKNESSES, CALLED SCHEDULES. SCHEDULE 80 CONDUIT HAS A WALL TWICE AS THICK AS THAT OF SCHEDULE 40 RNC. SCHEDULE 40 CAN BE USED UNDERGROUND AND ALSO ABOVE GROUND IF IT WILL NOT BE SUBJECT TO PHYSICAL DAMAGE. IF THE CONDUIT IS SUBJECT TO PHYSICAL DAMAGE, SCHEDULE 80 MUST BE USED ABOVE GROUND.

TYPE RNC CAN BE BENT BY HEATING A STICK WITH A HEATING BLANKET OR TORCH UNTIL IT IS PLIABLE. IT IS THEN BENT TO THE DESIRED ANGLE AND ALLOWED TO COOL. FACTORY-MADE STANDARD ANGLES, 90° AND 45°, ARE AVAILABLE FOR MOST SIZES.

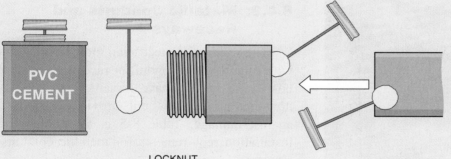

PVC CEMENT

WITH RNC, PVC CEMENT IS USED TO JOIN PIECES TOGETHER. THE CEMENT IS APPLIED TO THE INSIDE OF THE FITTING AND TO THE OUTSIDE OF THE CONDUIT. WHEN THE CEMENT IS APPLIED, THE PIECES ARE PUSHED TOGETHER UNTIL THEY SEAT. THE CEMENT IS ALLOWED TO DRY.

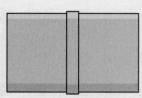

SIDE VIEW: RNC COUPLING

COUPLINGS ARE USED TO JOIN TWO STICKS OF RNC TOGETHER. EACH 10-FT (3.048-m) LENGTH HAS A COUPLING INSTALLED FROM THE FACTORY

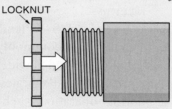

LOCKNUT

SIDE VIEW: RNC MALE ADAPTER (MA)

AN MA IS USED WITH A LOCKNUT TO CONNECT THE RNC TO A BOX, PANEL, OR CABINET.

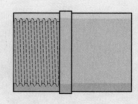

SIDE VIEW: RNC FEMALE ADAPTER (FA)

AN FA IS USED TO CONNECT THE RNC WITH A THREADED NIPPLE OR AN EMT CONNECTOR OR SIMILAR FITTING. THE FA HAS THREADS ON THE INSIDE WALL SURFACE OF THE FITTING.

SEAT

END VIEW OF ANY OF THESE FITTINGS

EACH TYPE OF FITTING HAS A SEAT MOLDED INTO THE INSIDE OF THE FITTING. FOR OPTIMAL ASSEMBLY, THE CONDUIT MUST SEAT COMPLETELY BEFORE THE CEMENT DRIES.

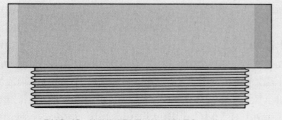

RNC IS AVAILABLE IN UP TO 6 IN. (155 mm) TRADE SIZE DIAMETER.

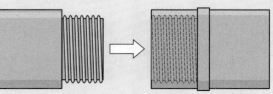

AN MA AND AN FA CAN BE THREADED TOGETHER IF DESIRED TO MAKE A COUPLING.

Figure 8-7

expansion fittings be placed at regular intervals to allow for this movement. Figure 8-8 shows RNC installed as a riser for an underground service.

NEC® 356—Liquidtight flexible nonmetallic conduit (LFNC): Type LFNC is used for the connection to equipment. It is used when the wiring method must be flexible because of the vibration or movement of equipment, such as air-conditioning compressors. It is also used when the installation must be watertight. LFNC does not provide the same level of protection against physical damage as that with a metallic conduit. It is approved for use indoors and outdoors and can be used underground or in poured concrete if listed for that use.

Figure 8-8 An RNC riser installed as part of a dwelling service.

NEC® 362—Electrical nonmetallic tubing (ENT): Type ENT is a nonmetallic tubing that is popular in high-rise multi-family structures. It is recognized by its distinctive blue color and must be used with fittings approved for the purpose. It can be embedded in poured concrete and is approved for installation in wet locations.

NEC® 388—Nonmetallic surface raceway (NSR): NSR is used primarily in wiring for remodeling. NSR installs directly to the surface of finished walls and ceilings and must be used with boxes and fittings manufactured for the purpose. Many types of NSR are set and held in place by glue strips on the back, or on the base of the raceway. The raceway has a high resistance to impact. See Figure 8-9 for more about surface raceways.

8.4.3: Metallic Conduits and Raceways

The residential electrician must also be familiar with four metallic conduit or raceway wiring methods. These methods have the same purpose as for the nonmetallic methods but provide extra strength against damage. Table 8-5 is a summary of the installation requirements for metallic conduits and raceways.

NEC® 348—Flexible metal conduit (FMC): Type FMC is commonly called "Flex" in the electrical industry and is widely used for the connection of equipment that vibrates or otherwise moves during normal operation, such as a forced air gas furnace. There are severe restrictions on the use of FMC as an equipment grounding conductor, and an equipment grounding conductor should be installed in every run of FMC to ensure that the installation complies with the Code. Figure 8-10 provides additional information about FMC. Figure 8-11 (page 214) shows FMC installed on a basement wall protecting Type NM cable as it runs down the wall.

NEC® 349—Liquidtight flexible metal conduit (LFMC): Type LFMC is used to connect equipment that moves or vibrates during normal operation and is located outdoors or in a wet location. Some LFMC is listed for direct burial. Type LFMC can be viewed as FMC with a liquidtight outer covering and special connectors and fittings.

NEC® 358—Electrical metallic tubing (EMT): Type EMT is sometimes called thin-wall in the

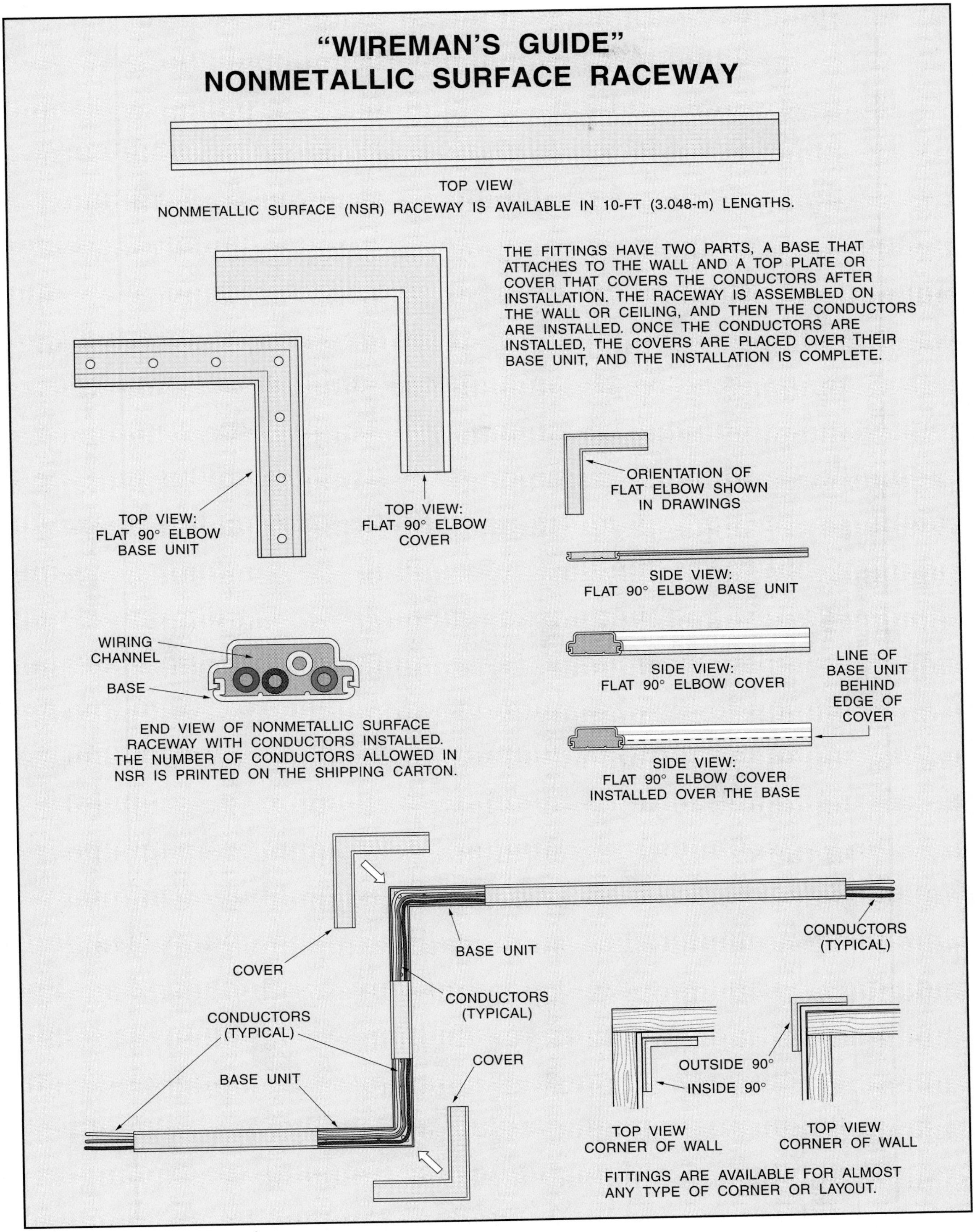

"WIREMAN'S GUIDE"
NONMETALLIC SURFACE RACEWAY

TOP VIEW
NONMETALLIC SURFACE (NSR) RACEWAY IS AVAILABLE IN 10-FT (3.048-m) LENGTHS.

THE FITTINGS HAVE TWO PARTS, A BASE THAT ATTACHES TO THE WALL AND A TOP PLATE OR COVER THAT COVERS THE CONDUCTORS AFTER INSTALLATION. THE RACEWAY IS ASSEMBLED ON THE WALL OR CEILING, AND THEN THE CONDUCTORS ARE INSTALLED. ONCE THE CONDUCTORS ARE INSTALLED, THE COVERS ARE PLACED OVER THEIR BASE UNIT, AND THE INSTALLATION IS COMPLETE.

TOP VIEW:
FLAT 90° ELBOW
BASE UNIT

TOP VIEW:
FLAT 90° ELBOW
COVER

ORIENTATION OF
FLAT ELBOW SHOWN
IN DRAWINGS

SIDE VIEW:
FLAT 90° ELBOW BASE UNIT

WIRING
CHANNEL

BASE

END VIEW OF NONMETALLIC SURFACE
RACEWAY WITH CONDUCTORS INSTALLED.
THE NUMBER OF CONDUCTORS ALLOWED IN
NSR IS PRINTED ON THE SHIPPING CARTON.

SIDE VIEW:
FLAT 90° ELBOW COVER

LINE OF
BASE UNIT
BEHIND
EDGE OF
COVER

SIDE VIEW:
FLAT 90° ELBOW COVER
INSTALLED OVER THE BASE

COVER

BASE UNIT

CONDUCTORS
(TYPICAL)

CONDUCTORS
(TYPICAL)

CONDUCTORS
(TYPICAL)

COVER

BASE UNIT

BASE UNIT

OUTSIDE 90°

INSIDE 90°

TOP VIEW
CORNER OF WALL

TOP VIEW
CORNER OF WALL

FITTINGS ARE AVAILABLE FOR ALMOST
ANY TYPE OF CORNER OR LAYOUT.

Figure 8-9

REQUIREMENTS

FEATURE	FMC 348	LFMC 349	EMT 358	SR 386
Sizes allowed	½ in. (13 mm) thru 4 in. (100 mm) trade size; 3/8 in. (12 mm) trade size restricted [348.20]	½ in. (13 mm) thru 4 in. (100 mm) trade size; 3/8 in. (12 mm) trade size restricted [349.20]	½ in. (13 mm) thru 4 in. (100 mm) trade size [358.20]	As per design size marked on carton [386.22]
Bends—number of bends allowed in run	360° (4 x 90°) [348.26]	360° (4 x 90°) [349.26]	360° (4 x 90°) [358.26]	Per listing [386.6]
Bends—minimum radius of bends in conduit or tubing	Per Table 344.24 [348.24]	Per Table 344.24 [349.24]	Per Table 344.24 [358.24]	Per listing. Manufactured fittings [386.6]
Supports—at boxes—feet from box	Within 1 ft (300 mm) [348.30]	Within 1 ft (300 mm) [351.8]	Within 3 ft (900 mm) [358.30]	Per listing [386.6]
Supports—along cable run—feet from previous support	4½ ft (1.4 m) [348.30]	4½ ft (1.4 m) [349.30]	10 ft (3 m) [358.30]	Per listing [386.6]
Bushings required at termination	Yes; 4 AWG and larger conductors [300.4(F)]	Yes; 4 AWG and larger conductors [300.4(F)]	Yes; 4 AWG and larger conductors [300.4(F)]	Per listing [386.6]
Threadless couplings	Yes	Yes	Yes [358.28]	Yes
Treaded couplings	No	No	No [358.28]	No
Number of conductors in conduit or tubing	Per Table 1, Chapter 9 or Table 348.22 [348.22]	Per Table 1, Chapter 9 or Table 348.22 [349.22]	Per Table 1, Chapter 9 [358.22]	Per design and listing [386.22]
Trimming or reaming of cut ends required	Yes [348.28]	Yes [349.28]	Yes [358.28]	Per listing [386.6]

Table 8-5 Installation requirements for selected metallic conduits and raceways.

"WIREMAN'S GUIDE"
FLEXIBLE METAL CONDUIT

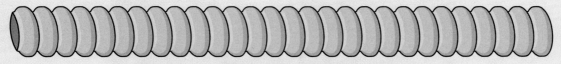

SIDE VIEW: FLEXIBLE METAL CONDUIT (FMC)
FMC IS CONSTRUCTED FROM INTERLOCKING METAL TAPE

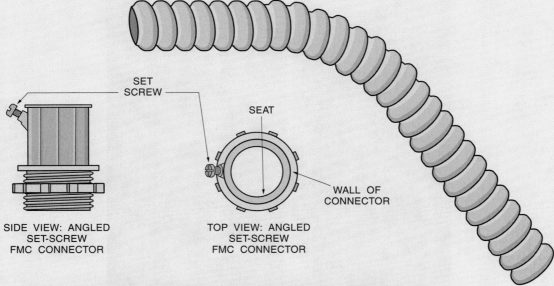

SET
SCREW

SEAT

WALL OF
CONNECTOR

SIDE VIEW: ANGLED
SET-SCREW
FMC CONNECTOR

TOP VIEW: ANGLED
SET-SCREW
FMC CONNECTOR

FMC CONNECTORS ARE AVAILABLE IN A STRAIGHT STYLE, PICTURED ABOVE, AND A 90° STYLE, NOT SHOWN. THE CONNECTORS HAVE A SEAT AT THE BOTTOM OF THE THROAT OF THE FITTING TO SECURE THE END OF THE FMC.

ANGLED SET-SCREW CONNECTORS ARE NOT LISTED FOR GROUNDING EXCEPT IN 3/8 IN. (12 mm) THROUGH 3/4 IN. (19 mm) SIZES IF USED WITH CONDUCTORS PROTECTED BY NO MORE THAN A 20-AMPERE OVERCURRENT-PROTECTIVE DEVICE.

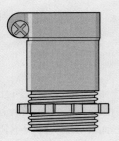

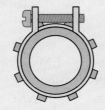

CIRCUMFERENCE-TYPE FMC CONNECTORS ARE *NEC®* LISTED FOR GROUNDING IF USED WITH CONDUCTOR PROTECTED BY A MAXIMUM 20-AMPERE OVERCURRENT-PROTECTIVE DEVICE.

SIDE VIEW:
CIRCUMFERENCE-TYPE
FMC CONNECTOR

TOP VIEW:
CIRCUMFERENCE-TYPE
FMC CONNECTOR

WHEN FMC IS USED TO ALLOW FOR FLEXIBILITY TO SUPPLY EQUIPMENT, THE INSTALLATION MUST INCLUDE AN EQUIPMENT GROUNDING CONDUCTOR (SEE *348.60*). FMC THAT IS NOT LISTED FOR GROUNDING CAN BE USED AS AN EQUIPMENT GROUNDING CONDUCTOR IF: (1) THE CONNECTORS ARE LISTED FOR GROUNDING, (2) IF NO CONDUCTOR IS PROTECTED BY AN OVERCURRENT-PROTECTIVE DEVICE RATED ABOVE 20 AMPERES, (3) IF THE GROUND PATH ON THE FLEXIBLE METAL CONDUIT DOES NOT EXCEED 6 FT (1.8 m), AND (4) IF THE CONDUIT WAS NOT INSTALLED FOR FLEXIBILITY (SEE *250.18*).

Figure 8-10

Figure 8-11 The electricians used FMC to protect the cable run down the wall to the sump pump.

electrical industry, because it is a steel conduit system that has a thinner wall than that of RMC or IMC. EMT cannot be threaded and has fittings designed for use indoors and outdoors exposed to the weather. It is the most popular type of metallic conduit system and can be used in dwellings during remodeling, as protection for conductors that would otherwise be exposed, or for exposed circuiting outdoors. Figure 8-12 has more information concerning EMT and its uses and restrictions. Figure 8-13 shows EMT installed on the side of a warehouse.

NEC® 386—Surface raceway: Type SR is metal raceway that is designed to be installed on the surface of walls and ceilings during additions to the electrical system or during remodeling. It provides the same service as NSR but with the added protection provided by a metal raceway and fittings. See Figure 8-9 for more about surface raceways.

"WIREMAN'S GUIDE"
ELECTRICAL METALLIC TUBING

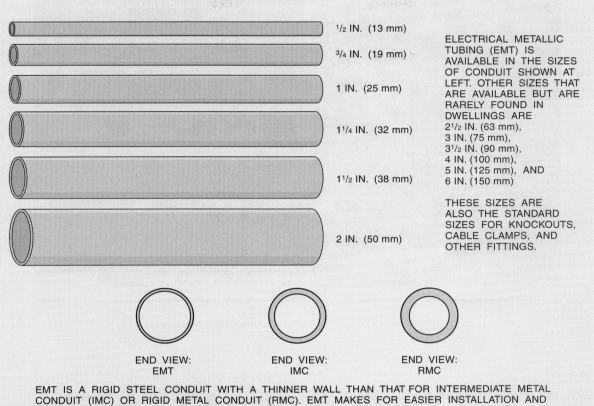

1/2 IN. (13 mm)

3/4 IN. (19 mm)

1 IN. (25 mm)

1 1/4 IN. (32 mm)

1 1/2 IN. (38 mm)

2 IN. (50 mm)

ELECTRICAL METALLIC TUBING (EMT) IS AVAILABLE IN THE SIZES OF CONDUIT SHOWN AT LEFT. OTHER SIZES THAT ARE AVAILABLE BUT ARE RARELY FOUND IN DWELLINGS ARE 2 1/2 IN. (63 mm), 3 IN. (75 mm), 3 1/2 IN. (90 mm), 4 IN. (100 mm), 5 IN. (125 mm), AND 6 IN. (150 mm)

THESE SIZES ARE ALSO THE STANDARD SIZES FOR KNOCKOUTS, CABLE CLAMPS, AND OTHER FITTINGS.

END VIEW: EMT

END VIEW: IMC

END VIEW: RMC

EMT IS A RIGID STEEL CONDUIT WITH A THINNER WALL THAN THAT FOR INTERMEDIATE METAL CONDUIT (IMC) OR RIGID METAL CONDUIT (RMC). EMT MAKES FOR EASIER INSTALLATION AND LOWER COST BUT IS NOT AS STRONG AGAINST PHYSICAL DAMAGE.

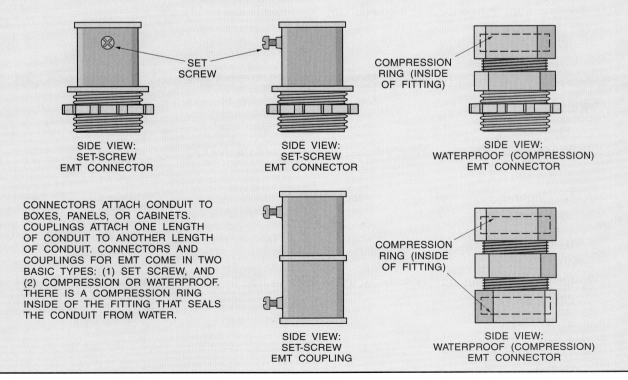

SET SCREW

COMPRESSION RING (INSIDE OF FITTING)

SIDE VIEW: SET-SCREW EMT CONNECTOR

SIDE VIEW: SET-SCREW EMT CONNECTOR

SIDE VIEW: WATERPROOF (COMPRESSION) EMT CONNECTOR

CONNECTORS ATTACH CONDUIT TO BOXES, PANELS, OR CABINETS. COUPLINGS ATTACH ONE LENGTH OF CONDUIT TO ANOTHER LENGTH OF CONDUIT. CONNECTORS AND COUPLINGS FOR EMT COME IN TWO BASIC TYPES: (1) SET SCREW, AND (2) COMPRESSION OR WATERPROOF. THERE IS A COMPRESSION RING INSIDE OF THE FITTING THAT SEALS THE CONDUIT FROM WATER.

COMPRESSION RING (INSIDE OF FITTING)

SIDE VIEW: SET-SCREW EMT COUPLING

SIDE VIEW: WATERPROOF (COMPRESSION) EMT CONNECTOR

Figure 8-12

Figure 8-13   EMT installed on a commercial building.

SUMMARY

Many different wiring methods are detailed in the Code. Of those wiring methods, 21 may be used in dwelling construction, but only 14 methods are commonly used in dwellings. Of these 14, six use cable assemblies, four use metallic conduits (both flexible and rigid), and four use NM conduit (both flexible and rigid).

REVIEW

1. Wiring methods are covered in _____ of the *NEC®*.

   a. *Chapter 2*

   b. *Chapter 3*

   c. *Chapter 4*

   d. *Chapter 5*

2. Of the 14 wiring methods likely to be used in dwellings, how many of them utilize cable assemblies?

   a. 4

   b. 5

   c. 6

   d. 7

3. Which two "possible" or "probable" wiring methods use metallic conduit that can be threaded?

   a. EMT and IMC

   b. IMC and RMC

   c. RMC and FMC

   d. FMC and EMT

4. Which two "possible" or "probable" wiring methods use cables that can be buried directly in a trench without conduit protection?

   a. UF and EMT

   b. EMT and NSR

   c. NSR and USE

   d. USE and UF

5. Installations concealed in walls are prohibited by the *NEC®* for which of the following wiring methods?

   a. AC

   b. NSR

   c. MC

   d. EMT

6. *True or False:* Type UF cable can be used for feeders or branch circuits.

7. *True or False:* Type NM cable must be supported within 12 in. (300 mm) of a box and no more than every 4½ ft (1.4 m) between supports.

8. *True or False:* Type UF cable has 90°C-rated conductor installed in the cable.

9. *True or False:* Type NM cable has 60°C-rated conductors installed in the cable.

10. *True or False:* Type AC and Type MC cables have the same support requirements.

11. *True or False:* RNC is available with factory-installed conductors.

12. *True or False:* EMT is rated for indoor use only.

13. *True or False:* Type NM cable is available with or without an equipment grounding conductor.

14. *True or False:* Type NM cable can be buried directly into a trench without conduit protection.

15. *True or False:* Type NM cable can be used for service conductors.

"WIREMAN'S GUIDE" REVIEW

1. What happens to the bonding strip that is installed in AC cable when the cable is stripped? Can it be cut off completely? Must all of the bonding strip be wrapped around the cable sheath?

2. With AC cable, a bushing must be installed in the end to protect the conductors against chafing or other damage over time. The Code does not specify requirement for MC cable. Can you think of any reasons why?

3. Type SE cable comes in two styles, SER and SEU. Describe the differences between the two styles.

4. How is the maximum number of conductors allowed in surface raceway determined?

5. Describe the conditions specified by the Code for use of FMC and an equipment grounding conductor.

CHAPTER 9

Cable Installation

On completion of this chapter, the student will be able to:

☑ Apply the *NEC*® and UBC requirements for notching and drilling studs, joists, and rafters.

☑ State the requirements for installing conductors in environmental air spaces as applied to residential wiring.

☑ Apply the rules for bundling cables through holes or notches.

☑ Apply the rules for running cables parallel to framing members.

☑ Describe the proper way to splice and pigtail conductors with wire nuts or splice caps.

☑ Identify conductors for makeup using a uniform system of markings.

☑ Explain the importance of conductor organization in junction or device boxes.

INTRODUCTION

This chapter covers the requirements for drilling or notching of wall studs and ceiling joists or rafters for the installation of electrical wiring in dwelling units. The *NEC*® has rules for keeping conductors away from the edges of studs, joists, and rafters to help protect them from physical damage. This code is not the only regulation that places limits on the notching or drilling of studs, joists, and rafters, however. The Uniform Building Code (UBC) also has requirements for drilling and notching framing members. The UBC document is recognized throughout the United States as the accepted guide for the construction of dwellings and other structures.

KEY TERMS

Splice   The act of connecting two individual conductors together to form one continuous conductor, or the location of that connection. Splices usually occur in boxes and are accomplished using proper methods and materials.

Pigtail   (as referring to splices) An extra conductor that is added to a splice for connection to a device.

Box makeup   The act of preparing the conductors contained in a device box for the installation of a device. Makeup is accomplished with the intent of making subsequent installation of the device as easy as possible.

This chapter also examines the connection of conductors using **splices** and **pigtails** and the proper procedures to use when such connections are made. In addition, the organization of the conductors in the various boxes is covered. **Box makeup** is one of the most important functions of the electrician during rough-in, and proper organization and marking of the various conductors are central to the makeup process. An understanding of the procedures for installing the cables and for makeup of the boxes is necessary before proceeding to the actual cable routings covered in Chapters 10 and 11.

9.1: *NEC®* REQUIREMENTS FOR DRILLING OR NOTCHING STUDS, RAFTERS, AND JOISTS

Drilled holes Holes that are drilled in framing members for the installation of electrical conductors or cables. Holes that are drilled in studs or joists must meet the requirements of the *NEC®* and the UBC.

Cut notches Sections along the edge of a joist or stud for the installation of cables or raceways. Notches must be cut with a saw, and usually the opening must be covered with a plate to protect the cable or raceway. Notches must meet the requirements of the *NEC®* and the UBC.

Stud A framing member that makes up a part of a wall. Framed walls also usually have a top plate and a bottom plate.

Rafter A framing member that makes up part of a roof support system.

Joist A framing member that makes up part of a flooring system.

Notch plate A metal plate that is installed over a notch or over a hole that is closer than 1¼ in. (32 mm) from the edge of a framing member in order to protect the cables or raceway.

The *NEC®* requirements for **drilled holes** or **cut notches** in framing members for dwellings are intended to protect electric cable from physical damage. Damage may occur during the construction process but may also occur during the life of the structure as the result of penetration by nails or screws during remodeling, the installation of wall-mounted shelving, the hanging of pictures, or any number of other events. According to *300.4(A)*, any cable must be kept at least 1¼ in. (32 mm) from the edge of a **stud**, **rafter**, or **joist**. This restriction provides an area in the center of the framing member that satisfies the requirements of the *NEC®*. This area is called the *drilling zone* and is shown in Figure 9-1. Figure 9-2 shows an NM cable running through a drilled hole. If the hole must be drilled closer than 1¼ in. (32 mm) to the edge of a stud, rafter, or joist, or if a notch must be cut, the cable must be protected by a steel plate at least ¹⁄₁₆ in. (1.6 mm) thick. Such a plate is called a **notch plate** and is illustrated in Figures 9-3 and 9-4 (page 223 and page 224).

9.2: UBC REQUIREMENTS FOR DRILLING OR CUTTING STUDS, JOISTS, AND RAFTERS

The UBC has an entirely different interest in the notching and drilling of framing members. The UBC is concerned with a building's structural integrity—whether the building can withstand the stresses likely to be encountered in use of the structure. Any notch or hole in a stud, rafter, or joist potentially weakens the structure; therefore, the UBC requirements for cutting notches and drilling holes are not intended to protect the wiring but are intended to protect the building. The residential electrician must be aware of these UBC requirements and must follow them when drilling in preparation for cable installation.

Not all dwellings are constructed using wooden studs and rafters. Several construction methods employ manufactured systems such as floor trusses and manufactured wooden beams, among others. The drilling requirements for these systems should be obtained from the general contractor, the UBC, or the manufacturer of the structural members. Some of these structural members cannot be drilled or notched at all. Some dwellings may use metal studs instead of wooden studs. When NM cable is installed in holes in metal studs, bushings need to be installed in the hole to protect the cable. Bushings for standard-size holes are readily available at most electrical supply houses.

The UBC separates the functions of drilling and notching. Drilling involves making round holes, with wood on all sides, whereas notching involves removing some of the framing member's edge. Generally, notches weaken the structure of the dwelling more than holes do; therefore, they are more restricted in size and location. Structural members of a dwelling are divided by the UBC into two distinct groups: (1) wall supports (studs) and (2) floor supports or ceiling supports (joist or rafter).

9.2.1: Notching Wall Studs

The UBC says that a notch can be cut into any wall stud to a depth of 25% of the stud's width and anywhere up the length of the stud, except back to back in the same spot on the stud. The cutting is to be done with a saw. If the stud is not in a load-bearing wall, the notch can take up to 40% of the

"WIREMAN'S GUIDE"
NEC® DRILLING REQUIREMENTS

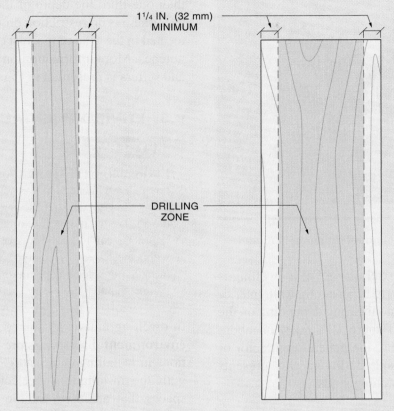

2 X 4 WALL STUD:
THE ACTUAL MEASUREMENTS OF
A 2 X 4 STUD ARE APPROXIMATELY
1¹⁄₂ IN. (38 mm) X 3¹⁄₂ IN. (90 mm)

2 X 6 WALL STUD:
THE ACTUAL MEASUREMENTS
OF A 2 X 6 STUD ARE APPROXIMATELY
1¹⁄₂ IN. (38 mm) X 5¹⁄₂ IN. (140 mm)

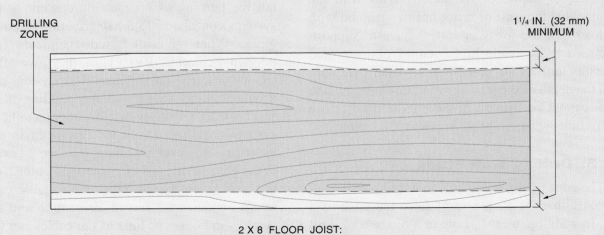

2 X 8 FLOOR JOIST:
THE ACTUAL MEASUREMENTS OF A 2 X 8 JOIST IS
APPROXIMATELY 1¹⁄₂ IN. (38 mm) X 7¹⁄₂ IN. (191 mm)

Figure 9-1

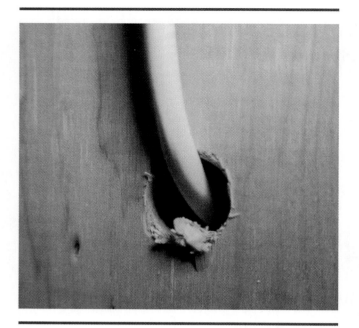

Figure 9-2   A drilled hole containing an NM cable.

depth of the stud. Any exterior wall of a dwelling is automatically considered to be a bearing wall. Interior walls can be bearing or nonbearing depending on the design of the house. If there is a question about whether a wall is bearing, the general contractor or builder should be consulted. Figure 9-5 presents more information on notching studs.

9.2.2: Notching Joists and Rafters

The requirements for notching floor joists, ceiling joists, and rafters are discussed in *UBC 2320.12.4*. The size, or depth, of the notch varies with the distance that the joist or rafter has to span and with the load that the joists or rafters have to support. Notches can be made in the edges of joists and rafters as long as the notch is not deeper than one-sixth the depth of the joist or rafter. Furthermore, the notch cannot be located in the middle third of the joist or rafter span, as shown in Figure 9-6 (page 226).

9.2.3: Drilling Wall Studs

The edge of any bored hole in a stud wall cannot be closer than ⅝ in. (16 mm) from the edge of a stud. In addition, a bored hole in a nonbearing wall cannot take up more than 60% of the stud depth. If the wall is a bearing wall, the hole cannot take more than 40% of the stud depth. Figure 9-7 (page 227) presents more information on drilling wall studs.

9.2.4: Drilling Joists and Rafters

The requirements for drilling floor joists and rafters are discussed in *UBC 2320.12.4*. Holes cannot be made within 2 in. (50 mm) of the edges of joists and rafters. Furthermore, the hole cannot be larger than one-third the depth of the joist or rafter. However, there is no restriction on where the holes can be located in the span of the joist or rafter as there is with notches. More information on drilling holes in joists and rafters is presented in Figure 9-8 (page 228).

9.3: ENVIRONMENTAL AIRSPACES

> **KEY TERMS**
>
> **Environmental airspaces**   Airspaces within a structure that are intended as part of the structure's heating and cooling systems. These areas are used to circulate the air through the furnace or air conditioner and then return the air to the rooms of the building. Wiring within these areas is restricted because of the possibility of the rapid spreading of fire through the air-handling system should a fire occur.

In dwelling units with forced air heat, a number of **environmental airspaces** present special considerations in installing cable. *NEC® 300.22* is concerned with the installation of electrical conductors into spaces that are used for the delivery or return of heated or cooled air. The Code is very careful not to allow nonmetallic wiring methods in airspaces where the air will circulate through the building. Not only can this circulating air increase the spread of a fire, but the burning of nonmetallic conduit or cabling systems contributes to the distribution of poisonous gases. When chloride-based chemicals such as polyvinyl chloride (PVC) burn, they give off a poisonous gas. The environmental air system can distribute this gas throughout the building in a very short period. Therefore, usually only metallic wiring systems are allowed to be installed in environmental airspaces. However, dwelling units are an exception to the general rule. The *Exception* to *300.22(C)* says that it is acceptable to install nonmetallic cables in return air ducts for forced air heating and cooling systems in houses as long as the cables run through the environmental airspaces using the shortest route. Figure 9-9 (page 229) shows the method usually employed to convert a joist space into a return airspace for a forced air heating and cooling system.

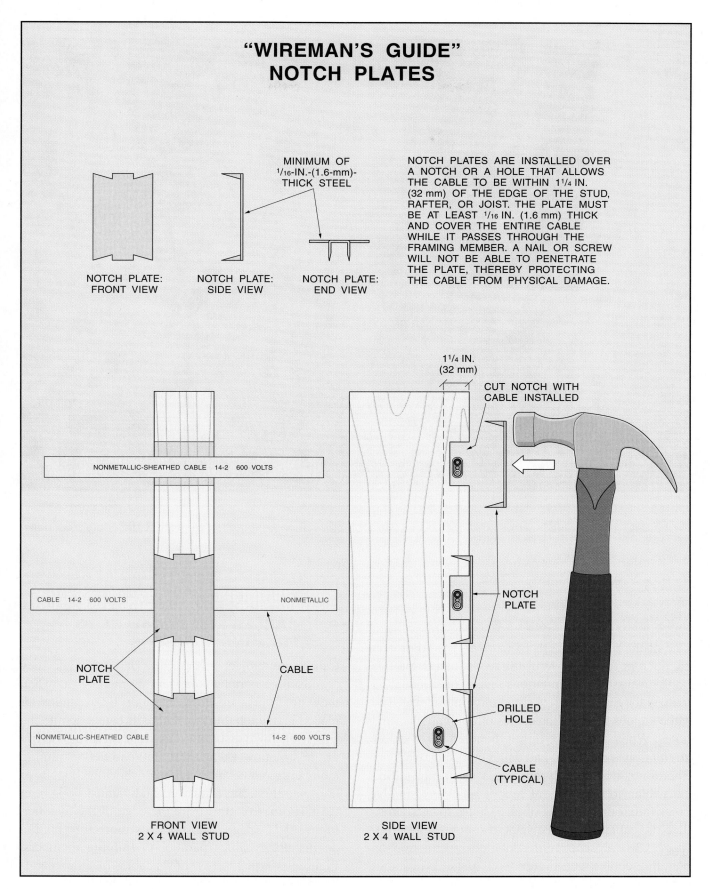

"WIREMAN'S GUIDE" NOTCH PLATES

MINIMUM OF 1/16-IN.-(1.6-mm)-THICK STEEL

NOTCH PLATE: FRONT VIEW

NOTCH PLATE: SIDE VIEW

NOTCH PLATE: END VIEW

NOTCH PLATES ARE INSTALLED OVER A NOTCH OR A HOLE THAT ALLOWS THE CABLE TO BE WITHIN 1¼ IN. (32 mm) OF THE EDGE OF THE STUD, RAFTER, OR JOIST. THE PLATE MUST BE AT LEAST 1/16 IN. (1.6 mm) THICK AND COVER THE ENTIRE CABLE WHILE IT PASSES THROUGH THE FRAMING MEMBER. A NAIL OR SCREW WILL NOT BE ABLE TO PENETRATE THE PLATE, THEREBY PROTECTING THE CABLE FROM PHYSICAL DAMAGE.

NONMETALLIC-SHEATHED CABLE 14-2 600 VOLTS

CABLE 14-2 600 VOLTS   NONMETALLIC

NOTCH PLATE

CABLE

NONMETALLIC-SHEATHED CABLE   14-2 600 VOLTS

FRONT VIEW 2 X 4 WALL STUD

1¼ IN. (32 mm)

CUT NOTCH WITH CABLE INSTALLED

NOTCH PLATE

DRILLED HOLE

CABLE (TYPICAL)

SIDE VIEW 2 X 4 WALL STUD

Figure 9-3

Figure 9-4   If the cable is within 1¼ in. (32 mm), a notch plate must be installed to protect the cable from nails.

9.4: BUNDLING CABLES

One of the principal aims of the *NEC*® is to protect conductors against overheating. Chapter 5 of this book discusses the limitations placed on conductors to keep the heating below some maximum level. The same applies to installing cables through notches and holes. When cables are tightly bundled together, air circulation is restricted and conductors can overheat. Therefore, it is usually advisable for the installing electrician to drill several sets of relatively smaller holes rather than one set of large holes to accommodate the cables using the same routing. Figure 9-10 (page 230) presents more detailed information on bundling cables, and Figure 9-11 (page 231) shows cables in a dwelling that are closely bundled.

9.5: CABLES RUN PARALLEL TO FRAMING MEMBERS

Cables are subject to the same damage when they are installed parallel to a framing member as when they are when installed through bored holes or placed in a notch. Screws and nails can penetrate the cables during construction as well as during the life of the structure. Cables installed parallel to framing members—cables that run up or down wall studs or horizontally along a joist or stud—must be kept at least 1¼ in. (32 mm) from the edge of the member. Figure 9-12 (page 232) presents more information on installing cables parallel to framing members. Figure 9-13 (page 233) shows cables installed parallel to framing members in a typical dwelling.

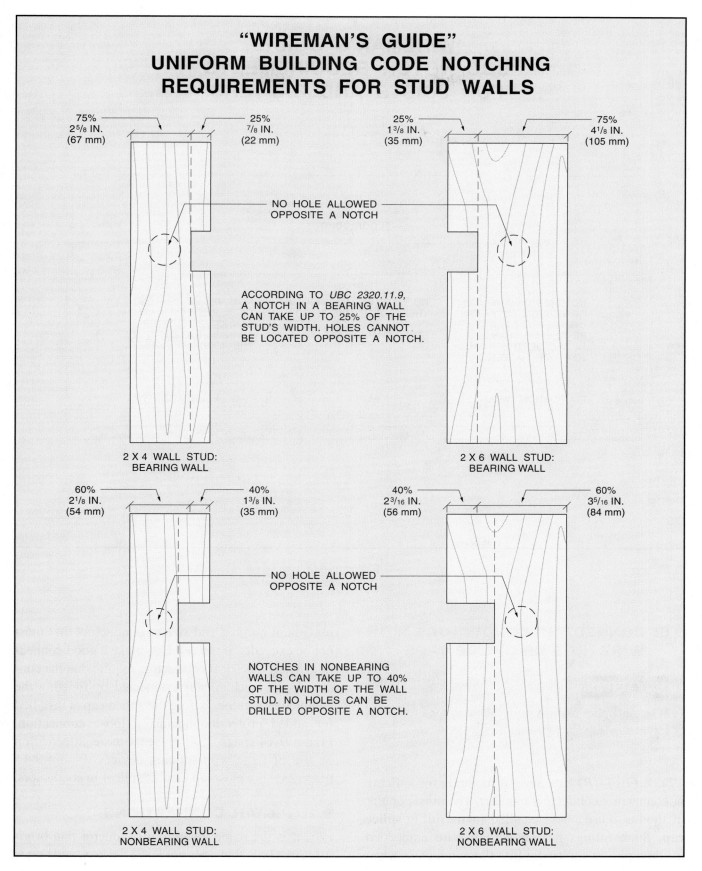

"WIREMAN'S GUIDE"
UNIFORM BUILDING CODE NOTCHING REQUIREMENTS FOR STUD WALLS

75%
2⅝ IN.
(67 mm)

25%
⅞ IN.
(22 mm)

25%
1⅜ IN.
(35 mm)

75%
4⅛ IN.
(105 mm)

NO HOLE ALLOWED
OPPOSITE A NOTCH

ACCORDING TO *UBC 2320.11.9,*
A NOTCH IN A BEARING WALL
CAN TAKE UP TO 25% OF THE
STUD'S WIDTH. HOLES CANNOT
BE LOCATED OPPOSITE A NOTCH.

2 X 4 WALL STUD:
BEARING WALL

2 X 6 WALL STUD:
BEARING WALL

60%
2⅛ IN.
(54 mm)

40%
1⅜ IN.
(35 mm)

40%
2³⁄₁₆ IN.
(56 mm)

60%
3⁵⁄₁₆ IN.
(84 mm)

NO HOLE ALLOWED
OPPOSITE A NOTCH

NOTCHES IN NONBEARING
WALLS CAN TAKE UP TO 40%
OF THE WIDTH OF THE WALL
STUD. NO HOLES CAN BE
DRILLED OPPOSITE A NOTCH.

2 X 4 WALL STUD:
NONBEARING WALL

2 X 6 WALL STUD:
NONBEARING WALL

Figure 9-5

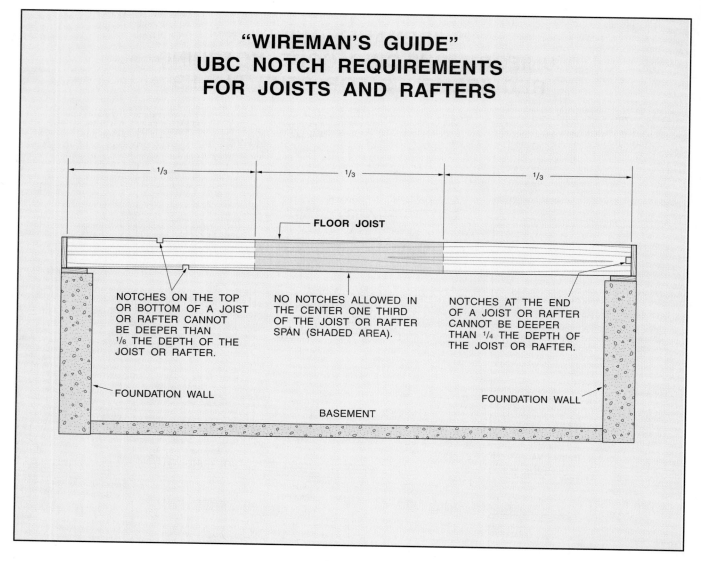

Figure 9-6

9.6: CONNECTING CONDUCTORS WITH WIRE NUTS OR SPLICE CAPS

> **KEY TERMS**
>
> **Wire nut/splice cap**   A fitting allowing for a solder-less method of splicing conductors.

NEC® 110.14(B) lists several methods for splicing or connecting conductors together. The most popular method is to use a device called a **wire nut** or **splice cap**. Such fittings have a coil of wire embedded inside a plastic cap to dig into the conductors when they are joined, thereby holding them tightly together. The different sizes of splice caps are listed for a maximum number and size of conductors that must not be exceeded if the cap is to make a good connection. Many electrical contractors require that the conductors be mechanically connected by twisting the conductors together before the splice cap is installed for added protection against a loose connection. Figure 9-14 (page 234) presents more information on use of these wire-splicing devices. Figure 9-15 (page 235) is a close-up view of splices in conductors.

9.7: PIGTAIL CONNECTIONS

Pigtail is the name given to a conductor that originates in a box, that does not leave the box, and that is included with other conductors in a splice. Pigtails provide an easy and effective means for tapping into

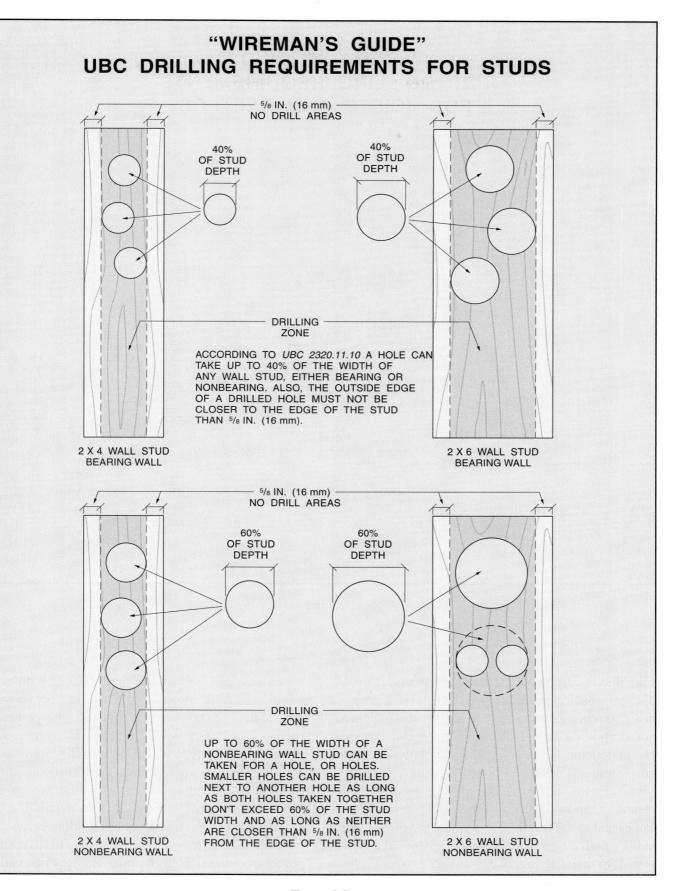

"WIREMAN'S GUIDE"
UBC DRILLING REQUIREMENTS FOR STUDS

⁵/₈ IN. (16 mm)
NO DRILL AREAS

40%
OF STUD
DEPTH

40%
OF STUD
DEPTH

DRILLING
ZONE

ACCORDING TO *UBC 2320.11.10* A HOLE CAN
TAKE UP TO 40% OF THE WIDTH OF
ANY WALL STUD, EITHER BEARING OR
NONBEARING. ALSO, THE OUTSIDE EDGE
OF A DRILLED HOLE MUST NOT BE
CLOSER TO THE EDGE OF THE STUD
THAN ⁵/₈ IN. (16 mm).

2 X 4 WALL STUD
BEARING WALL

2 X 6 WALL STUD
BEARING WALL

⁵/₈ IN. (16 mm)
NO DRILL AREAS

60%
OF STUD
DEPTH

60%
OF STUD
DEPTH

DRILLING
ZONE

UP TO 60% OF THE WIDTH OF A
NONBEARING WALL STUD CAN BE
TAKEN FOR A HOLE, OR HOLES.
SMALLER HOLES CAN BE DRILLED
NEXT TO ANOTHER HOLE AS LONG
AS BOTH HOLES TAKEN TOGETHER
DON'T EXCEED 60% OF THE STUD
WIDTH AND AS LONG AS NEITHER
ARE CLOSER THAN ⁵/₈ IN. (16 mm)
FROM THE EDGE OF THE STUD.

2 X 4 WALL STUD
NONBEARING WALL

2 X 6 WALL STUD
NONBEARING WALL

Figure 9-7

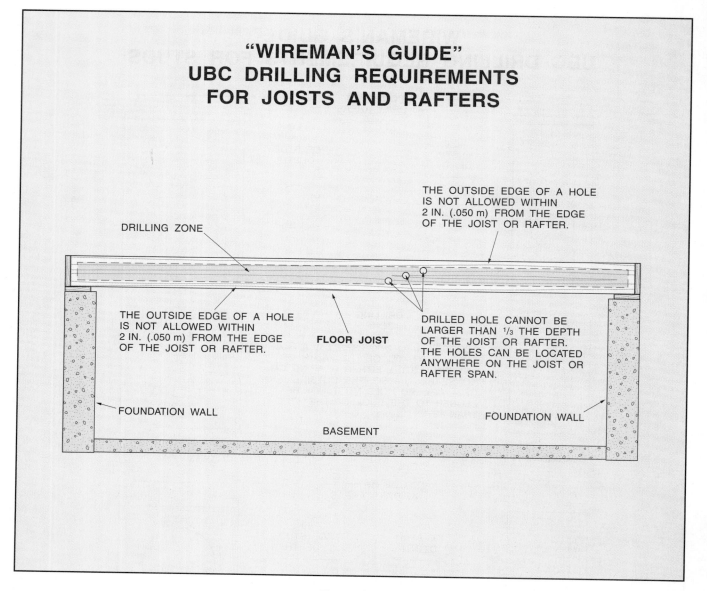

**"WIREMAN'S GUIDE"
UBC DRILLING REQUIREMENTS
FOR JOISTS AND RAFTERS**

THE OUTSIDE EDGE OF A HOLE IS NOT ALLOWED WITHIN 2 IN. (.050 m) FROM THE EDGE OF THE JOIST OR RAFTER.

DRILLING ZONE

THE OUTSIDE EDGE OF A HOLE IS NOT ALLOWED WITHIN 2 IN. (.050 m) FROM THE EDGE OF THE JOIST OR RAFTER.

FLOOR JOIST

DRILLED HOLE CANNOT BE LARGER THAN 1/3 THE DEPTH OF THE JOIST OR RAFTER. THE HOLES CAN BE LOCATED ANYWHERE ON THE JOIST OR RAFTER SPAN.

FOUNDATION WALL

FOUNDATION WALL

BASEMENT

Figure 9-8

a conductor system (such as the power conductors to connect a device or luminaire [fixture]. For example, in a box containing a cable that provides line side power (phase conductor) and neutral (grounded circuit conductor), another cable containing a load side power and neutral, and a cable to supply a luminaire (fixture) with switched power, there needs to be a means to tap into the power available in the power cables and to use that tap to operate the switch. A pigtail is an effective means of accomplishing the tap.

More than one pigtail can be taken from a splice bundle. In multigang switch boxes, several switches may all be powered by the same circuit. With such an arrangement, multiple pigtails are necessary in order for the power cable to provide the power to each of the switches. With other arrangements, a pigtail may be needed from each of the circuit conductors, power conductors, neutral conductors, equipment grounding conductors, and switch leg conductors. Figure 9-16 (page 236) details how to make a pigtail. The equipment grounding conductor is used as an example of the pigtail assembly process, but the same techniques are used in creating pigtails of other types of conductors as well.

"WIREMAN'S GUIDE"
JOIST SPACE USED AS ENVIRONMENT AIRSPACE

SIDE VIEW OF COLD AIR RETURN METAL DUCT

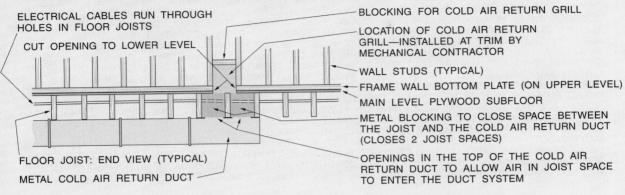

ELECTRICAL CABLES RUN THROUGH HOLES IN FLOOR JOISTS

CUT OPENING TO LOWER LEVEL

FLOOR JOIST: END VIEW (TYPICAL)

METAL COLD AIR RETURN DUCT

BLOCKING FOR COLD AIR RETURN GRILL

LOCATION OF COLD AIR RETURN GRILL—INSTALLED AT TRIM BY MECHANICAL CONTRACTOR

WALL STUDS (TYPICAL)

FRAME WALL BOTTOM PLATE (ON UPPER LEVEL)

MAIN LEVEL PLYWOOD SUBFLOOR

METAL BLOCKING TO CLOSE SPACE BETWEEN THE JOIST AND THE COLD AIR RETURN DUCT (CLOSES 2 JOIST SPACES)

OPENINGS IN THE TOP OF THE COLD AIR RETURN DUCT TO ALLOW AIR IN JOIST SPACE TO ENTER THE DUCT SYSTEM

LOOKING DOWN FROM THE MAIN FLOOR WITH SUBFLOORING REMOVED

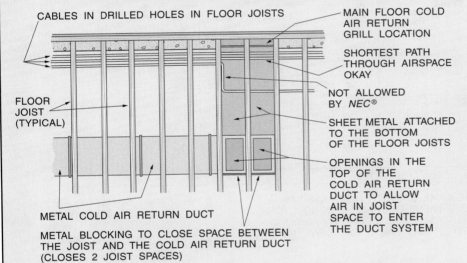

CABLES IN DRILLED HOLES IN FLOOR JOISTS

FLOOR JOIST (TYPICAL)

METAL COLD AIR RETURN DUCT

METAL BLOCKING TO CLOSE SPACE BETWEEN THE JOIST AND THE COLD AIR RETURN DUCT (CLOSES 2 JOIST SPACES)

MAIN FLOOR COLD AIR RETURN GRILL LOCATION

SHORTEST PATH THROUGH AIRSPACE OKAY

NOT ALLOWED BY *NEC*®

SHEET METAL ATTACHED TO THE BOTTOM OF THE FLOOR JOISTS

OPENINGS IN THE TOP OF THE COLD AIR RETURN DUCT TO ALLOW AIR IN JOIST SPACE TO ENTER THE DUCT SYSTEM

MECHANICAL OR HEATING AND COOLING CONTRACTORS SOMETIMES USE THE SPACE BETWEEN FLOOR JOISTS AS PART OF THE COLD AIR RETURN SYSTEM WHEN HEATING AND COOLING WITH FORCED AIR. THESE JOIST SPACES BECOME WHAT IS TERMED ENVIRONMENT AIRSPACES, AND ANY NONMETALLIC WIRING SYSTEM IN THAT SPACE MUST TAKE THE SHORTEST ROUTE THROUGH THE SPACE, ACCORDING TO 300.22(C), EXCEPTION.

AN OPENING IS CUT INTO THE SUBFLOOR TO ALLOW ACCESS TO THE COLD AIR RETURN BETWEEN THE JOISTS IN THE LOWER LEVEL. THIS OPENING IS USUALLY INSIDE AN UPPER-LEVEL WALL STUD SPACE. THE WALL STUD SPACE IS BLOCKED OFF ABOVE THE GRILL LOCATION TO CLOSE THE COLD AIR RETURN SYSTEM.

THE LOCAL AHJ SHOULD BE CONSULTED TO ENSURE THAT THERE ARE NO LOCAL RESTRICTIONS REGARDING INSTALLING CABLES IN ENVIRONMENTAL AIRSPACES.

END VIEW OF COLD AIR DUCT AND JOIST SPACE

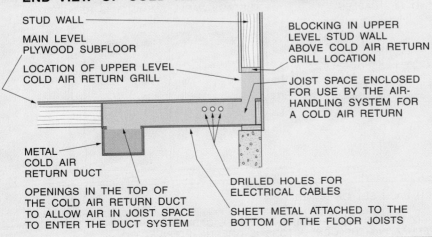

STUD WALL

MAIN LEVEL PLYWOOD SUBFLOOR

LOCATION OF UPPER LEVEL COLD AIR RETURN GRILL

METAL COLD AIR RETURN DUCT

OPENINGS IN THE TOP OF THE COLD AIR RETURN DUCT TO ALLOW AIR IN JOIST SPACE TO ENTER THE DUCT SYSTEM

BLOCKING IN UPPER LEVEL STUD WALL ABOVE COLD AIR RETURN GRILL LOCATION

JOIST SPACE ENCLOSED FOR USE BY THE AIR-HANDLING SYSTEM FOR A COLD AIR RETURN

DRILLED HOLES FOR ELECTRICAL CABLES

SHEET METAL ATTACHED TO THE BOTTOM OF THE FLOOR JOISTS

Figure 9-9

"WIREMAN'S GUIDE"
BUNDLING CABLES

NEC® 336.26 SAYS THAT TYPE NM CABLE SHALL BE CONSTRUCTED FROM CONDUCTORS WITH 90°C INSULATION BUT THAT THE AMPACITY MUST COME FROM THE 60°C COLUMN OF THE AMPACITY TABLES. ON THE 90°C COLUMN OF *TABLE 310.16,* THE MAXIMUM ALLOWABLE AMPACITY OF 14 AWG 90°C COPPER CONDUCTORS IS 25 AMPERES, AND FOR 12 AWG IS 30 AMPERES. *TABLE 310.15(B)(2)(a)* SAYS THAT UP TO NINE CONDUCTORS CAN OCCUPY THE SAME RACEWAY WITH A 70% REDUCTION IN ALLOWABLE AMPACITY. THAT WOULD MEAN THAT 14 AWG CONDUCTORS ARE ALLOWED TO CARRY 17.5 AMPERES (25 A X .70 = 17.5 A) AND 12 AWG CONDUCTORS CAN CARRY UP TO 21.0 AMPERES (30 A X .70 = 21.0). SINCE THERE IS A 15-AMPERE LIMIT ON 14 AWG CIRCUIT AMPACITY AND A 20-AMPERE LIMIT ON 12 AWG CIRCUIT AMPACITY REQUIRED BY *240.4(D),* IT WOULD SEEM THAT CABLES WITH A TOTAL OF NINE CURRENT-CARRYING CONDUCTORS WOULD BE ALLOWED TO OCCUPY THE SAME SET OF HOLES. THIS COULD BE FOUR 12-2 NM CABLES, OR THREE 14-2 CABLES AND ONE 14-3 CABLE, OR ANY OTHER POSSIBLE COMBINATION THAT TOTAL NINE OR FEWER CURRENT-CARRYING CONDUCTORS. THIS ANALYSIS MAY NOT SATISFY THE LOCAL INSPECTOR. IN MANY LOCALES, THE AHJ MAY CONSIDER THE SHEATHING ON THE CABLES AS A HINDRANCE TO COOLING THE CONDUCTORS AND WILL NOT ALLOW EVEN THIS MANY CABLES BUNDLED TOGETHER. *NEC® 310.15(B)(2) EXCEPTION 5* ALLOWS UP TO 20 CURRENT-CARRYING CONDUCTORS IN AC AND MC CABLES TO BE BUNDLED TOGETHER WITHOUT HAVING TO ADJUST THE ALLOWABLE AMPACITY. IF THERE IS A QUESTION CONCERNING HOW THE LOCAL ELECTRICAL INSPECTOR WILL ENFORCE THIS PROVISION OF THE CODE, THE AHJ SHOULD BE CONSULTED FOR MORE DETAILS.

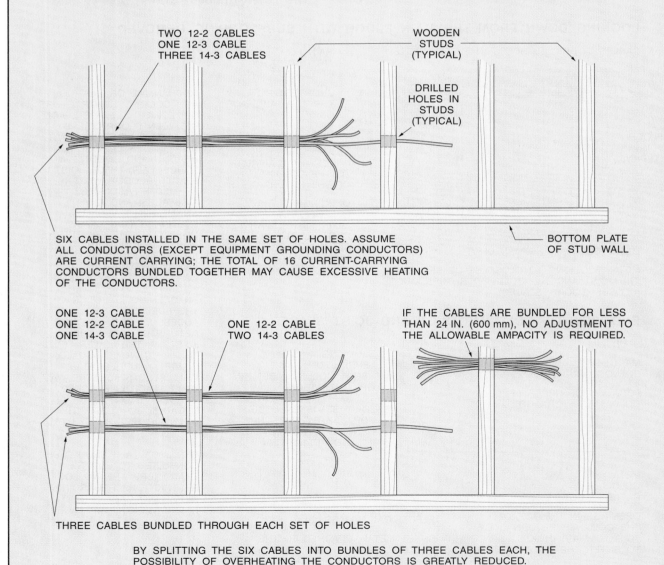

TWO 12-2 CABLES
ONE 12-3 CABLE
THREE 14-3 CABLES

WOODEN STUDS (TYPICAL)

DRILLED HOLES IN STUDS (TYPICAL)

SIX CABLES INSTALLED IN THE SAME SET OF HOLES. ASSUME ALL CONDUCTORS (EXCEPT EQUIPMENT GROUNDING CONDUCTORS) ARE CURRENT CARRYING; THE TOTAL OF 16 CURRENT-CARRYING CONDUCTORS BUNDLED TOGETHER MAY CAUSE EXCESSIVE HEATING OF THE CONDUCTORS.

BOTTOM PLATE OF STUD WALL

ONE 12-3 CABLE
ONE 12-2 CABLE
ONE 14-3 CABLE

ONE 12-2 CABLE
TWO 14-3 CABLES

IF THE CABLES ARE BUNDLED FOR LESS THAN 24 IN. (600 mm), NO ADJUSTMENT TO THE ALLOWABLE AMPACITY IS REQUIRED.

THREE CABLES BUNDLED THROUGH EACH SET OF HOLES

BY SPLITTING THE SIX CABLES INTO BUNDLES OF THREE CABLES EACH, THE POSSIBILITY OF OVERHEATING THE CONDUCTORS IS GREATLY REDUCED.

Figure 9-10

Figure 9-11   Bundled cables installed in TGI-type floor joists.

9.8: IDENTIFICATION OF CONDUCTORS

There are several different uses for cable in any device or switch point box. The cable may contain a phase conductor (power) and a grounded circuit conductor (neutral), or a switch leg and a neutral cable, or a switch leg and power for the switch, or a cable that contains travelers with a neutral, a switch leg, or a power feed to the switch. Each of these conductors must be positively identified at each box during cable installation in order to ensure proper box makeup.

Box makeup is the act of preparing the box to receive the switch or receptacle to be installed at trim. The electrician who installs the devices (several weeks or months after rough-in) must know

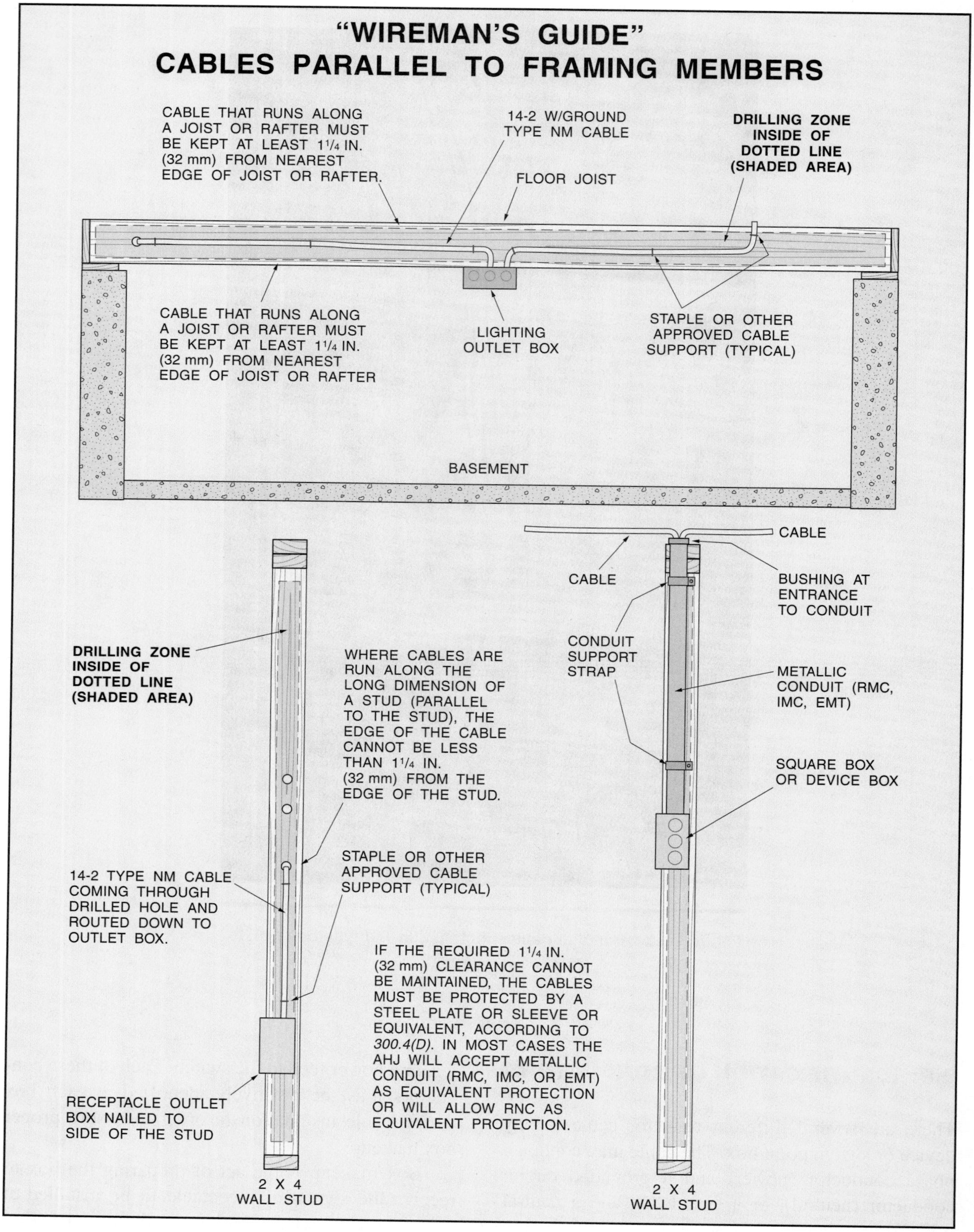

"WIREMAN'S GUIDE"
CABLES PARALLEL TO FRAMING MEMBERS

CABLE THAT RUNS ALONG A JOIST OR RAFTER MUST BE KEPT AT LEAST 1¼ IN. (32 mm) FROM NEAREST EDGE OF JOIST OR RAFTER.

14-2 W/GROUND TYPE NM CABLE

DRILLING ZONE INSIDE OF DOTTED LINE (SHADED AREA)

FLOOR JOIST

CABLE THAT RUNS ALONG A JOIST OR RAFTER MUST BE KEPT AT LEAST 1¼ IN. (32 mm) FROM NEAREST EDGE OF JOIST OR RAFTER

LIGHTING OUTLET BOX

STAPLE OR OTHER APPROVED CABLE SUPPORT (TYPICAL)

BASEMENT

CABLE

CABLE

BUSHING AT ENTRANCE TO CONDUIT

DRILLING ZONE INSIDE OF DOTTED LINE (SHADED AREA)

WHERE CABLES ARE RUN ALONG THE LONG DIMENSION OF A STUD (PARALLEL TO THE STUD), THE EDGE OF THE CABLE CANNOT BE LESS THAN 1¼ IN. (32 mm) FROM THE EDGE OF THE STUD.

CONDUIT SUPPORT STRAP

METALLIC CONDUIT (RMC, IMC, EMT)

SQUARE BOX OR DEVICE BOX

14-2 TYPE NM CABLE COMING THROUGH DRILLED HOLE AND ROUTED DOWN TO OUTLET BOX.

STAPLE OR OTHER APPROVED CABLE SUPPORT (TYPICAL)

IF THE REQUIRED 1¼ IN. (32 mm) CLEARANCE CANNOT BE MAINTAINED, THE CABLES MUST BE PROTECTED BY A STEEL PLATE OR SLEEVE OR EQUIVALENT, ACCORDING TO *300.4(D)*. IN MOST CASES THE AHJ WILL ACCEPT METALLIC CONDUIT (RMC, IMC, OR EMT) AS EQUIVALENT PROTECTION OR WILL ALLOW RNC AS EQUIVALENT PROTECTION.

RECEPTACLE OUTLET BOX NAILED TO SIDE OF THE STUD

2 X 4 WALL STUD

2 X 4 WALL STUD

Figure 9-12

Figure 9-13   Cables run parallel with framing members.

exactly what each conductor is being used for, and what conductors to connect to which terminals on the switch or receptacle. A white wire may be a traveler, or it may be a power (phase) conductor in some circuits. A red conductor may be a switch leg, a power, or a traveler. A black conductor may be a switch leg or a power conductor. A system to positively identify each conductor in a box is needed in order to eliminate confusion and error-plagued installations. The particular system used is not of major importance, but a systematic approach to conductor identification is essential. It is also important that everyone working on the project use the same coding system, at both rough-in and trim-out. The system presented in this chapter is simple and is the standard in many parts of the country and with many different contractors, although it is not the only system employed. This system can be used with wiring methods using Type NM cable as well as AC and MC cable, or other wiring methods. The system specifies five rules to be followed in making up a box at rough-in. The terms used in these five rules are the same terms that are employed in the field to identify each conductor. Brief definitions of these terms as used in this identification system follow:

- *Power:* The phase conductor—*not* the phase conductor plus the neutral. Neutrals are considered separately.

- *Neutral:* The grounded circuit conductor.

- *Switch leg:* The conductor that carries the power to the equipment or appliance using the power (the load) if the load is controlled by a switch of any kind. Again, the neutral required by the load is considered separately.

- *Traveler:* Two of the three conductors needed to complete a 3-way or 4-way switching system. A traveler is not to be connected to the common terminal of a 3-way switch.

- *In:* The line side of the box—the conductors entering the box. For example, power in means the line side phase conductor.

"WIREMAN'S GUIDE"
SPLICING WITH WIRE NUTS OR SPLICE CAPS

WIRE CONNECTIONS MAY BE MADE EITHER BY SOLDERING OR BY THE USE OF SOLDERLESS CONNECTORS. THE CODE ALLOWS SOLDERED CONNECTIONS BUT REQUIRES THAT THE CONNECTION BE MECHANICALLY SOUND BEFORE THE SOLDER IS APPLIED—THE SPLICE CANNOT DEPEND SOLELY ON THE SOLDER FOR THE CONNECTION. SOLDERLESS CONNECTIONS MAY BE MADE BY EXOTHERMIC WELDING OR WITH THE USE OF CONNECTORS SUCH AS LUGS OR CRIMP-ON SLEEVES. THE MOST POPULAR CONNECTORS ARE SCREW-ON CONNECTORS, COMMONLY REFERRED TO AS WIRE NUTS OR SPLICE CAPS.

WIRE NUTS OR SPLICE CAPS COME IN A NUMBER OF SIZES THAT CAN ACCOMMODATE ALMOST ANY SPLICING REQUIREMENT FOR THE CONDUCTORS COMMONLY FOUND IN DWELLING UNITS, EXCEPT FOR THE SERVICE AND DISTRIBUTION SYSTEM. YELLOW IS THE SMALLEST PRACTICAL WIRE NUT FOR EVERYDAY WIRING. THERE IS A SMALLER SERIES OF WIRE NUTS INCLUDING A BLUE WIRE NUT FOR SPLICING CONDUCTORS SMALLER THAN 14 AWG. THERE IS ALSO A BLUE WIRE NUT THAT IS LARGER THAN THE GRAY SIZE. YELLOW AND RED ARE THE MOST COMMONLY USED SIZES FOR 12 AWG AND 14 AWG CONDUCTORS. THERE IS EVEN A GREEN ONE WITH A HOLE IN THE SMALL END TO ASSIST IN THE MAKEUP OF THE EQUIPMENT GROUNDING CONDUCTORS.

A CUT-AWAY SIDE VIEW OF A TYPICAL WIRE NUT IS SHOWN AT LEFT. IT CONSISTS OF A METAL COIL, OR CONE, THAT IS TIGHTLY WRAPPED AND WITH SHARP EDGES EXPOSED TO THE INTERIOR OF THE CONE. PLASTIC OR OTHER INSULATING MATERIAL IS MOLDED OVER THE METAL CONE SO THAT THE CONE IS HELD FAST BY THE PLASTIC. THE LARGE END OF THE CONE IS LEFT EXPOSED TO RECEIVE THE WIRES.

WIRE NUTS ARE SAFE, EASY TO USE, AND RELATIVELY INEXPENSIVE. THE WIRES TO BE SPLICED ARE FORMED INTO A NEAT BUNDLE WITH THE INSULATION REMOVED FROM ALL OF THE CONDUCTORS FOR APPROXIMATELY ½ IN. (.013 m) TO 1 IN. (.0254 m) AND THE TOPS OF ALL THE CONDUCTORS AT THE SAME LEVEL. THE PROPERLY SIZED WIRE NUT IS THEN TWISTED ONTO THE EXPOSED CONDUCTORS WITH A CLOCKWISE TURN AND DOWNWARD PRESSURE APPLIED UNTIL THE WIRE NUT IS TIGHT. MAKE SURE TO FOLLOW ANY AND ALL INSTALLATION INSTRUCTIONS THAT MAY ACCOMPANY THE WIRE NUTS AT PURCHASE. ALSO, IT IS VERY IMPORTANT THAT THE WIRE NUT BE PROPERLY SIZED.

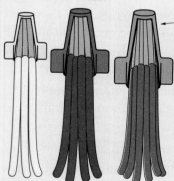

PROPERLY SIZED WIRE NUTS FIT SNUGLY WITHOUT BEING FORCED ONTO THE WIRES. THE LARGER THE CONDUCTOR SIZE, OR THE MORE CONDUCTORS IN THE SPLICE, THE LARGER THE WIRE NUT HAS TO BE. PROPERLY SIZED AND INSTALLED WIRE NUTS WILL PROVIDE A SAFE AND SECURE SPLICE.

IN MANY LOCALES, THE AHJ MAY REQUIRE THAT THE CONDUCTORS BE TWISTED TOGETHER BEFORE THE WIRE NUT IS INSTALLED. THE SPLICE MUST BE MECHANICALLY SECURE BEFORE THE WIRE NUT IS INSTALLED. THIS HAS THE EFFECT OF MAKING THE WIRE NUT NOTHING MORE THAN AN INSULATING FITTING AND A BACKUP CONNECTION METHOD TO THE TWISTED SPLICE.

SOME COMMON PROBLEMS EXPERIENCED WITH WIRE NUT INSTALLATION:
- OVERSIZING THE WIRE NUT. TOO MUCH SPACE INSIDE THE WIRE NUT CAN YIELD A LOOSE CONNECTION.
- OVERSTRIPPING THE CONDUCTORS. THE BARE CONDUCTOR IS THEN TOO LONG TO BE COVERED BY THE WIRE NUT, THUS LEAVING BARE CONDUCTORS EXPOSED.
- UNDERSIZING THE WIRE NUT. USE OF TOO MANY CONDUCTORS OR CONDUCTORS THAT ARE TOO LARGE CAN YIELD A LOOSE CONNECTION.

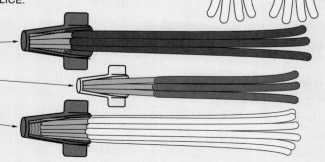

Figure 9-14

Figure 9-15 Conductor makeup in a box. These conductors are spliced using wire nuts or splice caps.

- *Out:* The load side of the box. For example, the switch leg out is the conductor that powers the load when the switch is closed. In most circumstances, a power in is accompanied by a neutral in and a power out is accompanied by a neutral out, but not always.

These six terms describe all of the possible combinations of conductors that can exist in a box. The following five rules apply to roughing-in cables, either nonmetallic-sheathed or metallic-sheathed:

- *Rule 1:* If the conductor is a power in, a power out, a neutral in, a neutral out, or an equipment grounding conductor, no further identification is necessary. The neutrals are identified by their white or gray coloring, the power conductors are identified by their black or red coloring, and

the equipment grounding conductor is identified by the green coloring or by the absence of insulation (bare conductor).

- *Rule 2:* If the conductor is a switch leg or if the conductor is common in a 3-way or 4-way switching system, strip about ½ in. (13 mm) of insulation from the end of the conductor.

- *Rule 3:* If the conductor is one of a pair of travelers, locate the other traveler of the pair and lightly twist (using approximately one twist for each inch [26 mm] of conductor) the two travelers together.

- *Rule 4:* If the conductor is an out (load side) of a ground-fault circuit interrupter device or the out (load side) of an arc-fault circuit-interrupter device, strip about ½ in. (13 mm) of the insulation off both the load side power and the load side neutral conductors. Associate the power in and the neutral out by lightly twisting (using one twist for each inch [26 mm]) the two neutral and the two power conductors together.

- *Rule 5:* If the conductor is white or gray in color but is being used as something other than a neutral in a single-pole or two-pole, 3-way or 4-way switching system, permanently re-identify the conductor with tape, paint, or other effective means at each box where the conductor is visible, and at each termination, to indicate its use.

It is obvious from the rules that the electrician doing the rough-in must be aware of the use assigned to each conductor as the cables are being pulled. The rough-in electricians must have a wiring plan, and that plan must be followed or errors will undoubtedly occur. For example, if a switch leg for a luminaire (fixture) is not properly identified when it is installed into the box, it is possible for the switch leg to be mistaken for a power conductor; then the luminaire (fixture) that it supplies will not be switched and will burn all the time. More information about the conductor identification system is presented in Figure 9-17.

9.9: ORGANIZING THE BOX

The system for box organization presented in this book is certainly not the only system, but it is reasonably efficient and provides satisfactory results. Regardless of the system employed, good box

"WIREMAN'S GUIDE"
PIGTAILS AND GROUND SCREW CONNECTIONS

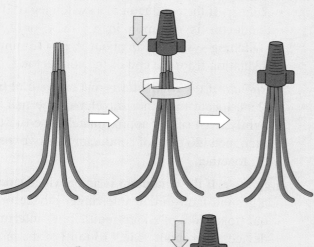

MAKING PIGTAILS IS FUNDAMENTAL TO BOX MAKEUP. THE FOLLOWING DESCRIPTION OF HOW TO MAKE A PIGTAIL USES THE EQUIPMENT GROUNDING CONDUCTOR AS AN EXAMPLE, BUT THE SAME TECHNIQUES APPLY TO PIGTAILS FOR POWER CONDUCTORS, NEUTRAL CONDUCTORS, OR OTHER CONDUCTOR TYPES THAT MAY REQUIRE PIGTAILS.

ACCORDING TO THE *NEC*® ALL EQUIPMENT GROUNDING CONDUCTORS THAT ENTER A BOX, REGARDLESS OF THE SIZE OR THE CIRCUITING OF THE CONDUCTORS, MUST BE CONNECTED TOGETHER TO FORM ONE EQUIPMENT GROUNDING SYSTEM. THERE IS ONE EXCEPTION TO THIS CONCERNING ISOLATED GROUNDING RECEPTACLES.

IN A NORMAL SPLICE, ALL OF THE EQUIPMENT GROUNDING CONDUCTORS HAVE THE INSULATION, IF ANY, REMOVED FROM THE ENDS OF THE WIRE. THE WIRES ARE THEN INCLUDED IN THE SPLICE BUNDLED AND CONNECTED USING A WIRE NUT.

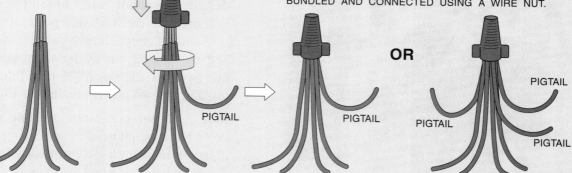

ONE OR MANY PIGTAILS CAN BE CONNECTED TO THE SPLICE BUNDLE. THIS WIRE IS THEN CONNECTED TO THE EQUIPMENT GROUNDING SYSTEM, AND THE END OF THE PIGTAIL CAN BE CONNECTED TO A BOX OR TO A DEVICE. IF MORE THAN ONE PIGTAIL IS INCLUDED IN THE BUNDLE, THERE COULD BE A SEPARATE PIGTAIL FOR EACH OF THE ITEMS THAT NEED CONNECTION TO THE GROUNDING SYSTEM.

THE *NEC*® ALSO REQUIRES THAT ALL METALLIC BOXES BE CONNECTED TO THE EQUIPMENT GROUNDING CONDUCTOR SYSTEM. THIS IS ACCOMPLISHED BY INSTALLING AN EQUIPMENT GROUNDING SCREW INTO A THREADED HOLE IN THE BACK OF THE METAL BOX. THERE ARE TWO POSSIBILITIES OR A COMBINATION OF POSSIBILITIES MAKING THESE CONNECTIONS:
1. A SEPARATE WIRE FOR THE PIGTAIL IS INCLUDED IN THE SPLICE WITH THE OTHER CONDUCTORS. THE PIGTAIL WILL THEN BE CONNECTED TO THE BOX USING THE GROUNDING SCREW.
2. STRIP ABOUT 1 IN. (.0254 m) OF INSULATION OFF THE CENTER OF ONE OF THE EQUIPMENT GROUNDING CONDUCTORS, AND INSTEAD OF GOING DIRECTLY TO THE SPLICING BUNDLE, LOOP THE CONDUCTOR AROUND THE GROUNDING SCREW FIRST. WHEN THE SCREW IS TIGHTENED IT WILL SECURE THE BOX TO THE EQUIPMENT GROUNDING SYSTEM.

THE CODE ALSO REQUIRES THAT AN EQUIPMENT GROUNDING SYSTEM CONNECTION BE MADE TO ALL DEVICES INSTALLED IN THE BOX. THIS IS BEST ACCOMPLISHED USING A PIGTAIL. IF THE DEVICE IS REMOVED FOR MAINTENANCE OR INSPECTION THE EQUIPMENT GROUNDING CONNECTION WILL REMAIN INTACT.

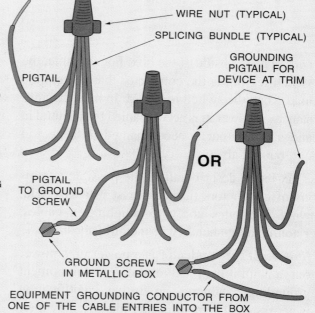

Figure 9-16

"WIREMAN'S GUIDE"
CONDUCTOR IDENTIFICATION AND MAKEUP

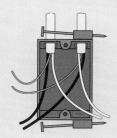

RULE 1: IF THE CONDUCTOR IS A POWER IN, A POWER OUT, A NEUTRAL IN, A NEUTRAL OUT, OR AN EQUIPMENT GROUNDING CONDUCTOR, NO FURTHER IDENTIFICATION IS NECESSARY.

THE BOX AT LEFT CONTAINS TWO 14-2 CABLES. IT CAN BE IDENTIFIED AS A RECEPTACLE OUTLET BOX BECAUSE ALL OF THE POWERS AND NEUTRALS ARE UNMARKED.

TO COMPLETE THE MAKEUP OF THE BOX: CONNECT AND PIGTAIL THE EQUIPMENT GROUNDING CONDUCTORS. THE TWO POWER CONDUCTORS, THE TWO NEUTRAL CONDUCTORS, AND THE EQUIPMENT GROUNDING CONDUCTOR PIGTAIL WILL TERMINATE ON THE RECEPTACLE.

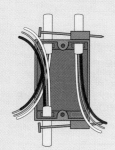

RULE 2: IF THE CONDUCTOR IS A SWITCH LEG, OR IF THE CONDUCTOR IS A COMMON IN A 3-WAY OR 4-WAY SWITCHING SYSTEM, REMOVE APPROXIMATELY 1/2 IN. (13 mm) OF INSULATION FROM THE END OF THE CONDUCTOR.

THE BOX AT LEFT CONTAINS THREE 14-2 CABLES. IT CAN BE IDENTIFIED AS A SINGLE-POLE SWITCH POINT BOX BECAUSE ONE OF THE BLACK CONDUCTORS HAS HAD THE INSULATION REMOVED TO IDENTIFY IT AS A SWITCH LEG.

TO COMPLETE THE MAKEUP OF THE BOX: CONNECT AND PIGTAIL ALL OF THE EQUIPMENT GROUNDING CONDUCTORS. CONNECT ALL OF THE NEUTRAL CONDUCTORS. CONNECT AND PIGTAIL THE TWO UNIDENTIFIED POWER CONDUCTORS. THE POWER PIGTAIL, THE SWITCH LEG, AND THE EQUIPMENT GROUNDING CONDUCTOR PIGTAIL ALL TERMINATE ON THE SINGLE-POLE SWITCH.

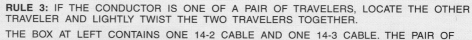
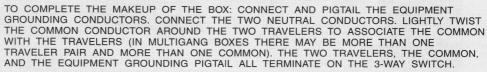

RULE 3: IF THE CONDUCTOR IS ONE OF A PAIR OF TRAVELERS, LOCATE THE OTHER TRAVELER AND LIGHTLY TWIST THE TWO TRAVELERS TOGETHER.

THE BOX AT LEFT CONTAINS ONE 14-2 CABLE AND ONE 14-3 CABLE. THE PAIR OF TRAVELERS ARE TWISTED TOGETHER TO IDENTIFY THEM AS A TRAVELER PAIR. NOTICE THAT THE POWER CONDUCTOR FROM THE 2-WIRE CABLE HAS THE INSULATION REMOVED FROM THE END, IDENTIFYING IT AS THE COMMON CONDUCTOR TO THE TRAVELERS.

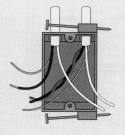

TO COMPLETE THE MAKEUP OF THE BOX: CONNECT AND PIGTAIL THE EQUIPMENT GROUNDING CONDUCTORS. CONNECT THE TWO NEUTRAL CONDUCTORS. LIGHTLY TWIST THE COMMON CONDUCTOR AROUND THE TWO TRAVELERS TO ASSOCIATE THE COMMON WITH THE TRAVELERS (IN MULTIGANG BOXES THERE MAY BE MORE THAN ONE TRAVELER PAIR AND MORE THAN ONE COMMON). THE TWO TRAVELERS, THE COMMON, AND THE EQUIPMENT GROUNDING PIGTAIL ALL TERMINATE ON THE 3-WAY SWITCH.

RULE 4: IF THE CONDUCTOR IS AN OUT (LOAD SIDE) CONDUCTOR OF A GFCI OR AN AFCI DEVICE, REMOVE ABOUT 1/2 IN. (13 mm) OF THE INSULATION FROM THE END OF BOTH THE POWER CONDUCTOR AND THE NEUTRAL CONDUCTOR. ASSOCIATE THE POWER OUT AND THE NEUTRAL OUT CONDUCTORS BY LIGHTLY TWISTING THEM TOGETHER. IF THE CONDUCTOR IS A LINE SIDE FOR A GFCI OR AN AFCI DEVICE THAT HAS NO LOAD CONDUCTORS (ONLY ONE CABLE IN THE BOX), REMOVE APPROXIMATELY 1/2 IN. (13 mm) THE INSULATION ON BOTH THE POWER AND THE NEUTRAL AS NOTIFICATION THAT THIS IS A SPECIAL RECEPTACLE.

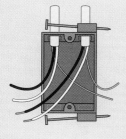

THE BOX AT LEFT CONTAINS TWO 14-2 CABLES. THE POWER AND THE NEUTRAL FROM ONE OF THE CABLES HAVE HAD THE INSULATION REMOVED FROM THE END, IDENTIFYING THEM AS LOAD SIDE CONDUCTORS FROM A GFCI OR AN AFCI RECEPTACLE. THE DETERMINATION OF WHICH DEVICE CAN BE MADE BY THE LOCATION OF THE BOX.

TO COMPLETE THE MAKEUP OF THE BOX: CONNECT AND PIGTAIL THE EQUIPMENT GROUNDING CONDUCTORS. THE TWO POWER CONDUCTORS, THE TWO NEUTRAL CONDUCTORS, AND THE EQUIPMENT GROUNDING PIGTAIL WILL ALL TERMINATE ON THE DEVICE.

RULE 5: IF THE CONDUCTOR IS WHITE OR GRAY IN COLOR BUT IS BEING USED AS SOMETHING OTHER THAN A GROUNDED CIRCUIT CONDUCTOR, PERMANENTLY RE-IDENTIFY THE CONDUCTOR WITH TAPE, PAINT, OR OTHER EFFECTIVE MEANS.

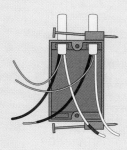

THE BOX TO THE LEFT CONTAINS TWO 14-2 CABLES. THE NEUTRAL CONDUCTOR FROM ONE OF THE CABLES IS MARKED WITH BLACK TAPE AND THE POWER CONDUCTOR FROM THE SAME CABLE IS MARKED AS A SWITCH LEG. THE WHITE CONDUCTOR IS BEING USED AS A POWER TO SUPPLY A SINGLE-POLE SWITCH (IN ANOTHER BOX) AND THE RETURN BLACK CABLE IS BEING USED AS THE SWITCH LEG. THIS INDICATES THAT THE RECEPTACLE TO BE INSTALLED IN THIS DEVICE BOX IS SWITCH CONTROLLED.

TO COMPLETE THE MAKEUP OF THE BOX: CONNECT AND PIGTAIL THE EQUIPMENT GROUNDING CONDUCTORS AND CONNECT THE RE-IDENTIFIED WHITE CONDUCTOR WITH THE POWER IN CONDUCTOR FROM THE OTHER CABLE. THE REMAINING NEUTRAL CONDUCTOR, THE SWITCH LEG, AND THE EQUIPMENT GROUNDING PIGTAIL ARE TO BE TERMINATED ON THE DEVICE.

Figure 9-17

"WIREMAN'S GUIDE"
DEVICE AND WIRE MANAGEMENT

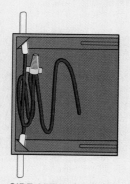

SIDE VIEW OF BOX

PIGTAIL STORED ACCORDION STYLE FOR CONNECTION TO THE DEVICE AT TRIM

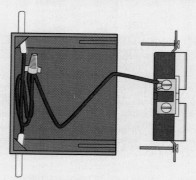

SIDE VIEW OF BOX

PULL THE CONDUCTORS OUT OF THE BOX AND ATTACH THE DEVICE—IN THIS CASE, A RECEPTACLE. THE CONDUCTORS PULL EASILY FROM THE BOX AND ARE READY TO CONNECT. (ONLY THE BLACK CONDUCTOR IS SHOWN FOR CLARITY.)

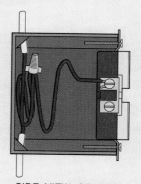

SIDE VIEW OF BOX

WHEN THE RECEPTACLE IS INSTALLED INTO THE BOX, THE WIRE PUSHES IN EASILY AND RE-FOLDS LIKE AN ACCORDION.

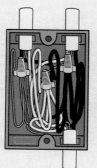

SPLICE BUNDLE AND CONDUCTORS STORED ACCORDION STYLE AND IN BACK OF THE BOX. THERE ARE PIGTAILS ON ALL THREE SPLICE BUNDLES FOR CONNECTION TO A RECEPTACLE.

FRONT VIEW OF BOX

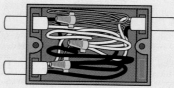

IF THE RECEPTACLE IS TO BE INSTALLED ON THE SIDE WITH THE EQUIPMENT GROUNDING CONDUCTOR TO THE LEFT, MAKE UP THE BOX WITH THE EQUIPMENT GROUNDING CONDUCTORS ON THE TOP, THE GROUNDED CIRCUIT CONDUCTOR IN THE MIDDLE, AND THE PHASE CONDUCTOR ON THE BOTTOM. THIS IS THE RECOMMENDED INSTALLATION POSITION FOR HORIZONTALLY MOUNTED RECEPTACLES.

IF THE RECEPTACLE IS TO BE INSTALLED WITH THE EQUIPMENT GROUNDING CONNECTION IN THE DOWN POSITION, MAKE UP THE BOX WITH THE EQUIPMENT GROUNDING CONDUCTORS ON THE LEFT, THE GROUNDED CIRCUIT CONDUCTOR IN THE MIDDLE, AND THE PHASE CONDUCTOR ON THE RIGHT. THIS INSTALLATION POSITION IS NOT RECOMMENDED.

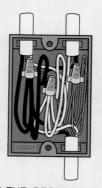 

IF THE RECEPTACLE IS TO BE INSTALLED WITH THE EQUIPMENT GROUNDING CONNECTION IN THE UP POSITION, MAKE UP THE BOX WITH THE EQUIPMENT GROUNDING CONDUCTORS ON THE RIGHT, THE GROUNDED CIRCUIT CONDUCTOR IN THE MIDDLE, AND THE PHASE CONDUCTOR ON THE LEFT. THIS IS THE RECOMMENDED INSTALLATION POSITION FOR VERTICALLY MOUNTED RECEPTACLES.

Figure 9-18

organization is critical to device installation and an efficient trim. Conductors occupy a considerable amount of space within the box, and many devices such as GFCI receptacles and dimmer switches take up a lot of space. The conductors should not have to be excessively compressed to make room in the box for the device, and the device should not have to be forced into the box because of the location of the conductors. Additionally, the trim electrician should be able to pull out of the box just those conductors that are needed to terminate to the device, and the other conductors should remain in the box where they were placed at makeup during rough-in. When the device is connected, it and the conductors it is connected to should install into the box with a minimum of effort.

An efficient way of storing for the conductors attached to the device to be stored in the box is in accordion fashion. This arrangement allows the conductors some flexibility so that the device can easily be pulled from the box and easily returned to the box. Using this method allows the device to be easily positioned both horizontally and vertically for proper alignment with the mounting screw holes of the box. This method also allows the trim electrician to remove only those conductors necessary for connection to the device, and does not require expenditure of time on unrolling or reshaping the conductors. More details on box management are presented in Figure 9-18.

SUMMARY

There are requirements for the drilling and notching of framing members to protect the cables from physical damage and to ensure the structural integrity of the building. There are different rules for drilled holes and cut notches. Cable should not be bundled when installed in holes or notches. A system needs to be used for identifying conductors for box makeup and trim. The system must positively identify each conductor, power, switch leg, neutral, and equipment grounding conductor at every outlet or switch point box. Cables are usually spliced using wire nuts or splice caps. Circuit taps can be accomplished using pigtails from splices. Proper box makeup and wire management will ensure a trouble-free and efficient trim.

REVIEW

1. According to the *NEC*®, if the edge of a drilled hole is closer than _____ to the edge of a wall stud, a notch plate must be installed to protect the cable.

   a. ¾ in. (19 mm)

   b. 1 in. (25 mm)

   c. 1¼ in. (32 mm)

   d. 1½ in. (38 mm)

2. According to the UBC, a notch can remove up to _____ percent of the depth of a wall stud if it is a bearing wall.

   a. 25

   b. 33

   c. 50

   d. 67

3. According to the UBC, a notch cannot be located within the middle _____ of a joist or rafter.

    a. half

    b. third

    c. quarter

    d. none of the above—a notch can be made anywhere in a joist or rafter

4. According to the UBC, a hole cannot be located within the middle _____ of a joist or rafter.

    a. half

    b. third

    c. quarter

    d. none of the above—a hole can be drilled anywhere in a joist or rafter, assuming it has the proper clearances from the edges

5. According to the *NEC®*, a cable that is run parallel to a stud must be located at least _____ from the edge of the stud.

    a. ¾ in. (19 mm)

    b. 1 in. (25 mm)

    c. 1¼ in. (32 mm)

    d. 1½ in. (38 mm)

6. Nonmetallic cables can _____ in environmental airspaces.

    a. never be installed

    b. always be installed

    c. be installed with protection against physical damage

    d. be installed if run using the shortest possible route

7. A white conductor _____ a grounded circuit conductor.

    a. is always

    b. is never

    c. can be re-identified for uses other than as

    d. is an equipment grounding conductor or

8. A wire nut or splice cap is intended to _____ .

    a. connect conductors together in a splice

    b. be used as insulation only

    c. supply a method to splice conductors without the use of a box

    d. connect the box to the phase conductor

9. A method of tapping a splice is called a _____ .

   a. pig ear

   b. pigtail

   c. dog leg

   d. dog ear

10. According to the conductor identification system used in this book, a power (phase) conductor is _____ .

    a. always identified by a stripped end

    b. twisted together with the neutral (grounded circuit) conductor

    c. left unidentified by any marking

    d. always identified by a tight curl at the end

11. *True or False:* According to the *NEC®*, no hole can be drilled less than ⅝ in. (16 mm) from the edge of a wall stud.

12. *True or False:* Conductors must be twisted together before splicing with a wire nut or splice cap.

13. *True or False:* No hole larger than 2 in. (50 mm) can be drilled in a floor joist or ceiling rafter.

14. *True or False:* According to the UBC, no hole can be drilled so that its edge is closer than 2 in. (50 mm) from the edge of a rafter or joist.

15. *True or False:* The length of the rafter determines the maximum size of a drilled hole.

"WIREMAN'S GUIDE" REVIEW

1. Draw the makeup of a single-pole switch box with a 2-wire power and neutral cable in, a 2-wire power and neutral cable out, and a switch leg and neutral cable out. All cables include an equipment grounding conductor. Make sure to identify all of the conductors that must be connected to the switch.

2. Draw the makeup of a receptacle outlet with a 2-wire power and neutral cable in and a power and a neutral cable out. All cables include an equipment grounding conductor. Make sure to identify all of the conductors that must be connected to the receptacle.

3. Draw the makeup of a half-switched receptacle with a 3-wire cable in containing a power, a switch leg, and a neutral, with a 2-wire power and neutral cable out. All cables include an equipment grounding conductor. Make sure to identify all of the conductors that must be connected to the receptacle.

4. Draw the makeup of a 3-way switch box with a 2-wire cable with power and neutral in, a 2-wire cable with power and neutral out, and a 3-wire cable with travelers and a neutral out. All cables include an equipment grounding conductor. Make sure to identify all of the conductors that must be connected to the switch.

5. Draw the makeup of a two-gang switch box with a 2-wire power and neutral cable in, two 2-wire power and neutral cables out, and one 2-wire switch leg and neutral cable out. All cables include an equipment grounding conductor. Make sure to identify all of the conductors that must be connected to the switches.

6. Draw the makeup for a switch box with a 3-way switch and a single-pole switch with a 2-wire power and neutral cable in, a 2-wire power and neutral cable out, a 2-wire switch leg and neutral cable out, and a 3-wire cable with the travelers and a power conductor out. All cables include an equipment grounding conductor. Make sure to identify all of the conductors that must be connected to the switch.

CHAPTER 10

Branch Circuits and Required Outlets

OBJECTIVES

On completion of this chapter, the student will be able to:

☑ Explain the *NEC*® and UBC requirements regarding branch circuits that are required to be installed in a dwelling.

☑ Differentiate among connected loads, calculated loads, and demand loads.

☑ Calculate the area of a dwelling to determine the minimum number of required general lighting and general-use receptacle branch circuits.

☑ Determine the general lighting and general-use receptacle feeder loads for the dwelling.

☑ Draw the final box makeup for the general lighting and general-use receptacle outlets, lighting outlets, smoke detection outlets, small-appliance receptacle outlets, and switch points in the sample house.

INTRODUCTION

This chapter is concerned with the installation of the required circuits for the sample house. The layouts of the circuits for the sample house are shown in Figure 1-8, p. 31, for the basement and in Figure 1-9, p. 32, for the main floor. Every electrician will lay out and wire a specific dwelling differently, and there is possibly no "best" way. However, some layouts are certainly better than others, and the one illustrated here is reasonably efficient from both a material and a labor perspective. The wiring in this chapter concerns only the required circuits.

In order to determine the circuits that are required to be installed, knowledge of branch-circuit loads and branch-circuit load calculations is necessary. In this chapter, following the determination of the number and type of required loads, each circuit is covered in some detail. Included in this chapter are drawings of each box, receptacle outlet, lighting outlet, and switch point, with the necessary conductor identification coding as presented in Chapter 9.

10.1: THE REQUIRED CIRCUITS

The circuits that are required in a dwelling are stipulated by *NEC*® *210.11*. Most of the outlets installed on these required circuits are also required outlets, but some outlets will be installed according to the wishes or preferences of the owner, including receptacles installed for added convenience and luminaires (fixtures) installed for improved lighting. The smoke detection and alarm circuit is required to be installed by the UBC, as shown in Figure 10-1.

The following circuits are considered required circuits by the *NEC*® for the sample house, plus the smoke detection and alarm circuit required by the UBC.

1. Circuits required by the *NEC*®:
   - General lighting and general-use receptacle outlet branch circuits
   - Small-appliance branch circuits
   - Laundry branch circuit
   - Bathroom receptacle outlet circuit

"WIREMAN'S GUIDE"
UBC SMOKE DETECTOR AND ALARM CIRCUIT REQUIREMENTS FOR DWELLINGS

1997 UNIFORM BUILDING CODE, SECTION 310.9.1: SMOKE DETECTOR LOCATIONS WITHIN DWELLING UNITS. IN DWELLING UNITS, A DETECTOR SHALL BE INSTALLED IN EACH SLEEPING ROOM AND AT A POINT CENTRALLY LOCATED IN THE CORRIDOR OR AREA GIVING ACCESS TO EACH SEPARATE SLEEPING AREA. WHEN THE DWELLING UNIT HAS MORE THAN ONE STORY AND IN DWELLINGS WITH BASEMENTS, A DETECTOR SHALL BE INSTALLED ON EACH STORY AND IN THE BASEMENT. IN DWELLING UNITS WHERE A STORY OR BASEMENT IS SPLIT INTO TWO OR MORE LEVELS, THE SMOKE DETECTOR SHALL BE INSTALLED ON THE UPPER LEVEL EXCEPT THAT, WHEN THE LOWER LEVEL CONTAINS A SLEEPING AREA, A DETECTOR SHALL BE INSTALLED ON EACH LEVEL. WHEN SLEEPING ROOMS ARE ON AN UPPER LEVEL, THE DETECTOR SHALL BE PLACED AT THE CEILING OF THE UPPER LEVEL IN CLOSE PROXIMITY TO THE STAIRWAY. IN DWELLING UNITS WHERE THE CEILING HEIGHT OF A ROOM OPEN TO THE HALLWAY SERVING THE BEDROOMS EXCEEDS THAT OF THE HALLWAY BY 24 IN. (600 mm) OR MORE, SMOKE DETECTORS SHALL BE INSTALLED IN THE HALLWAY AND IN THE ADJACENT ROOM. DETECTORS SHALL SOUND AN ALARM AUDIBLE IN ALL SLEEPING AREAS OF THE DWELLING UNIT IN WHICH THEY ARE LOCATED. IN NEW CONSTRUCTION, REQUIRED SMOKE DETECTORS SHALL RECEIVE THEIR PRIMARY POWER FROM A COMMERCIAL SOURCE AND HAVE A BATTERY BACKUP.

Figure 10-1

2. Circuits required by the UBC:
   - Smoke detection and alarm circuit

Each of these types of circuits will be considered later in this chapter.

10.2: LOADS

NEC® 210.11 requires that branch circuits be provided for lighting and appliance circuits according to their loads calculated by the rules stipulated in *220.3*. The term *load* is used to define the power drawn from the supply circuit by a particular piece of equipment, appliance, or luminaire (fixture). A load can best be depicted as voltage drop. Wherever there is voltage drop in an electrical circuit, there is a load being served.

10.2.1: Loads as Current Flow

The amount of current flowing in a circuit is of primary importance to the electrician because the current flow is the phenomenon that causes heating of the conductors. Heating the conductors beyond their maximum allowable temperature causes thermal breakdown of the insulation and eventually system failure or fire. Conductors have to be sized to handle the expected current flow in order to limit conductor heating. For any circuit, the larger the loads, the higher the current flow will be at a given voltage. Because all general-use receptacle outlets and lighting outlets in a dwelling are supplied by a 120-volt circuit (see *210.6*), the power used in those circuits is best described by the level of current flow. Any piece of electrical equipment that consumes power is required to be marked with enough technical electrical information to allow the electrician or engineer to determine the expected current flow at the rated voltage.

Two methods are used by the *NEC®* in determining current requirements for the various loads found in a dwelling unit: connected load and calculated load. Both types of loads are commonly found in dwelling units.

Connected Loads

KEY TERMS

Connected load The load, measured in amperes or watts (volt-amperes), that is actually required by a load to operate properly. Connect load is the nameplate load rating of an appliance or other type of electrical utilization equipment.

To find **connected load**, the information from the nameplate of the equipment, luminaire (fixture), or appliance itself is used. The current levels can be taken directly from the nameplate of the appliance, luminaire (fixture), or equipment. If the rated current flow is not specifically included on the nameplate, it can easily be calculated from the information given there. This current value is called connected load because it is defined as the amount of load actually connected to the circuit. For example, a 60-watt light bulb connected to a 120-volt circuit will draw ½ ampere from the voltage source. This is the actual level of current flow obtained by dividing the 60 watts by the 120 volts (60 watts/120 volts = .5 ampere) as required by the power formula. There is a load of 60 watts, or ½ ampere, on this circuit due to this one load. Many and various types of loads may be connected to the same circuit, of course. The total load for the circuit is determined by adding together all of the connected loads. The loads of most appliances and luminaires (fixtures) installed in dwellings are calculated as connected loads.

Calculated Loads

KEY TERMS

Calculated load The load of utilization equipment or outlets as determined by procedures allowed or required to be employed by the *NEC®*. For example, the general lighting load is calculated from the total floor area of a building, or a motor load is determined to be 125% of the motor's FLA rating.

The second method of load determination is to use a **calculated load**. Calculated loads arise from the *NEC®* rules, and there are many different methods employed by the *NEC®* for determining calculated loads. General lighting and general-use receptacle loads are examples of calculated loads. The Code provides a value to be used for the load based on the total floor area of the structure. Some load calculations are very complex, but most calculations dealing with dwelling unit loads are relatively elementary and straightforward.

Demand Loads

KEY TERMS

Demand load The load remaining after the factors allowed by the *NEC®* for a number of like or similar loads have been employed. For example, a demand load reduction is allowed for the total load for the ranges of an apartment building on the theory that not all ranges will be in use at the same time.

A special type of calculated load found in the Code is a **demand load**. Demand loads apply when a load that is available to many areas of the structure will not be fully loaded at the same time. For example, the general lighting loads for dwellings must be calculated under a given set of rules that ensure installation of an adequate number of branch circuits in each dwelling. However, it is recognized by the NFPA that not all of the outlets on all of the general-use circuits will be in use at the same time. Therefore, for dealing with feeders, the Code allows for de-rating of those loads to some lower level. With the exception of the branch circuits for household electric ranges, demand loads apply to feeders and service conductors only. The demand calculation can be used in sizing a branch circuit for a range as well as in sizing the feeder load for a single range.

10.2.2: Loads as Circuits

The *NEC®* classifies some loads as circuits. These loads do not have any particular loading value where branch circuits are concerned. This type of loading occurs when there is no practical way to determine the actual load on a circuit. The general-use receptacle outlets that are part of the general lighting load are distributed throughout the dwelling to allow convenient access. Not all of the receptacle outlets will actually have loads connected, however, and a load that is connected can be replaced or eliminated by the occupants at any time simply by unplugging the load. This fact makes the accurate determination of a specific branch-circuit load impossible to obtain. The receptacles located above the kitchen countertops present the same problem. These outlets can be extensively used during cooking, but the level of the current flow to these outlets can vary widely from dwelling to dwelling. Therefore, the Code sometimes simply states that a circuit of a certain size must be installed to satisfy a given load. There is no specific branch-circuit load calculation because the branch-circuit load is satisfied by requiring a specific circuit size. All of the required circuits in the foregoing list are of this type. There is no specific branch-circuit load attached to any of the circuits, but a minimum number of circuits of a given size must be installed to satisfy the actual loads. There may be, however, specific feeder load requirements from such circuits.

10.3: GENERAL LIGHTING AND GENERAL-USE RECEPTACLE BRANCH CIRCUITS

KEY TERMS

General lighting outlets Outlets intended for general use as lighting outlets for fixed-in-place luminaires (fixtures). Specialized task lighting is not included in this category.

General lighting outlets include the lighting outlets required by *210.70* (see Chapter 4 of this book). These lighting outlets do not include any special lighting for workshops, photography labs, or studios that may be located in the dwelling; they are only those used for lighting for the normal use of the occupants. Although the locations for the minimum number of luminaires (fixtures) is determined by the requirements of the *NEC®*, it is likely that more luminaires (fixtures) will be actually installed than the minimum required.

KEY TERMS

General-use receptacle outlets Receptacle outlets installed for general use as outlets for portable lighting (table lamps) as well as general-use appliances, such as televisions or stereo equipment. Kitchen equipment and other specialized utilization equipment are not included in this category.

General-use receptacle outlets are those outlets required by *210.52(A)* (see Chapter 3 of this book) to be installed on branch circuits required by *210.11*. These are the receptacle outlets used by the occupants to power lamps, stereo equipment, televisions, and the other trappings of modern life. They do not include receptacles in the kitchen, dining room, bathroom, or the receptacle required for the clothes washer.

Because there is no way of determining the load presented by these outlets, the Code provides a method for calculating the minimum loads as volt-amperes per ft² (m²) for the entire dwelling. This power load can then be converted to amperes by dividing out the voltage. Once the load in amperes has been determined, it is possible to determine the minimum number of general lighting and general-use receptacle circuits required.

10.3.1: Determining the Number of General Lighting and General-Use Receptacle Circuits

According to *220.3*, the general lighting and general-use receptacle loads for all structures, including dwellings, are calculated using the total ft² (m²) of the structure and the unit values for each type of occupancy listed in *Table 220.3(A)*.

Determining the Total Floor Area

The first step in determining the minimum number of general-use branch circuits, both for lighting and for receptacles, is to determine the area of the dwelling in ft² (m²). Figure 10-2 provides measurements for the basement and crawl space foundations in both English and SI units. These are the same measurements as those provided in Construction Drawings A-2 and A-6SI, only simplified. The simplified measurements for the main floor dimensions are provided in Figure 10-3.

In determining the area of the dwelling, it is often most practical to divide the structure into separate areas. There are many different ways to divide the floor area in order to calculate the total floor area; the method used here is presented as an example. There is no right or wrong way to divide the floor for doing the calculation as long as all of the area is included and no part of the area is counted twice. Dividing the floor plan into rectangles is recommended for easy calculation. The number of divisions should be kept as few as is practical, and the dimensions shown should be used, if possible, rather than trying to scale the distance off the plan with a ruler or measuring tape.

The area of the garage and the area of the open porch are not included in the total floor area. The basement, although unfinished, is adaptable to future use and is included in the total floor area. Both the exclusion of the garage and porch and the inclusion of the basement are according to *220.3(A)*.

Figure 10-2 shows that the area of the basement is 512 ft² (47.5 m²). According to Figure 10-3, the area of the main floor is 1304.5 ft² (121.2 m²). Therefore, the total area of the dwelling is:

Main floor area:	1304.5 ft²	(121.2 m²)
Basement area:	512.0 ft²	(47.5 m²)
Total Area:	1816.5 ft²	(168.7 m²)

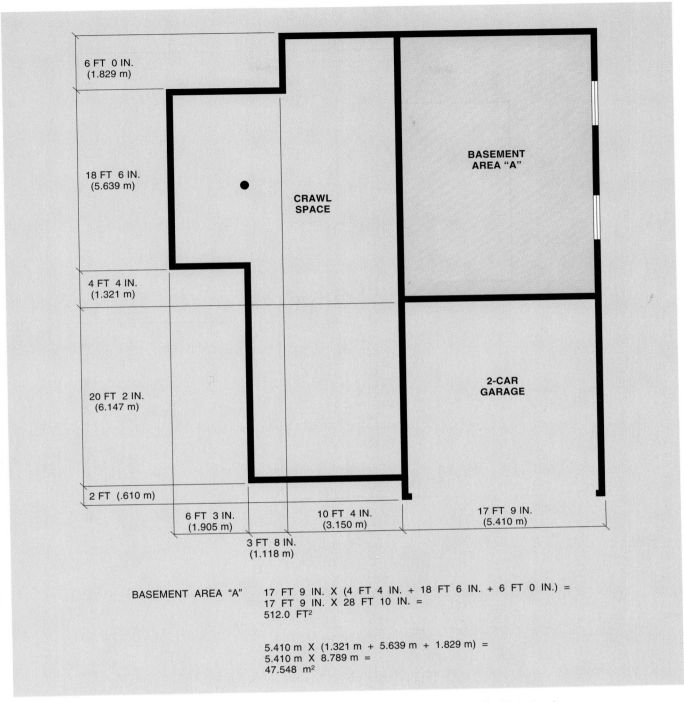

Figure 10-2   Basement floor area for calculating general lighting load.

This area measurement will now be used, along with the unit load per ft² (m²), to determine the general-use receptacle and lighting outlet load.

Determining the General Lighting Load

Calculating the general lighting and general-use receptacle circuit loads uses the total area of the dwelling as required by *220.3*. The unit load per ft² (m²) is obtained from *Table 220.3(A)*, which lists the general lighting load for virtually every commercial and residential structure. The unit load constitutes the minimum calculated load. It is usually necessary to actually install more than the minimum number of circuits to make the dwelling functional.

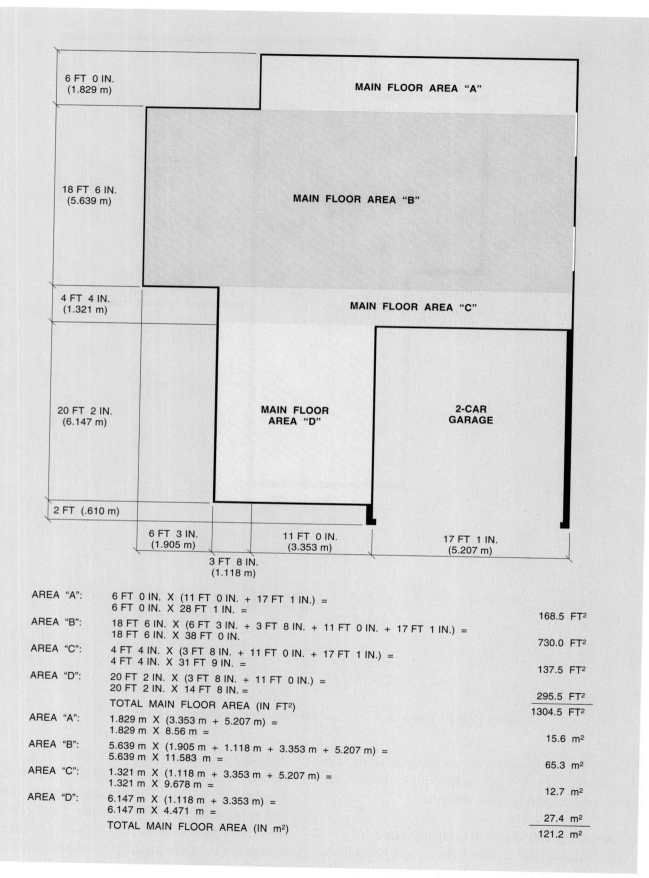

AREA "A":	6 FT 0 IN. X (11 FT 0 IN. + 17 FT 1 IN.) = 6 FT 0 IN. X 28 FT 1 IN. =	
		168.5 FT²
AREA "B":	18 FT 6 IN. X (6 FT 3 IN. + 3 FT 8 IN. + 11 FT 0 IN. + 17 FT 1 IN.) = 18 FT 6 IN. X 38 FT 0 IN.	
		730.0 FT²
AREA "C":	4 FT 4 IN. X (3 FT 8 IN. + 11 FT 0 IN. + 17 FT 1 IN.) = 4 FT 4 IN. X 31 FT 9 IN. =	
		137.5 FT²
AREA "D":	20 FT 2 IN. X (3 FT 8 IN. + 11 FT 0 IN.) = 20 FT 2 IN. X 14 FT 8 IN. =	
		295.5 FT²
	TOTAL MAIN FLOOR AREA (IN FT²)	1304.5 FT²
AREA "A":	1.829 m X (3.353 m + 5.207 m) = 1.829 m X 8.56 m =	
		15.6 m²
AREA "B":	5.639 m X (1.905 m + 1.118 m + 3.353 m + 5.207 m) = 5.639 m X 11.583 m =	
		65.3 m²
AREA "C":	1.321 m X (1.118 m + 3.353 m + 5.207 m) = 1.321 m X 9.678 m =	
		12.7 m²
AREA "D":	6.147 m X (1.118 m + 3.353 m) = 6.147 m X 4.471 m =	
		27.4 m²
	TOTAL MAIN FLOOR AREA (IN m²)	121.2 m²

Figure 10-3   Main floor area for calculating general lighting load.

The unit load listed in *Table 220.3(A)* shows 3 volt-amperes per ft² (33 volt-amperes per m²) for dwelling units. For the purposes of calculating the loads on dwelling-unit circuits, it is assumed that the volt-ampere load is equal to the watt load. Note "a," following the dwelling unit occupancy, refers to the bottom of the table, which in turn references *220.3(B)(10)*, NEC® *220.3(B)(10)* says that the general-use receptacle loads and the lighting loads of dwellings are both included in the designated 3 watts per ft² (33 watts per m²). Therefore, the general lighting and general-use receptacle outlet load for the sample house is calculated as follows:

$$1816.5 \text{ ft}^2 \times 3 \text{ watts per ft}^2 = 5449.5 \text{ watts}$$
$$(168.7 \text{ m}^2 \times 33 \text{ watts per m}^2 = 5567.1 \text{ watts})$$

The difference between the two calculated values is due to rounding used by the *NEC®* to determine the load as 33 watts per m². The actual load is closer to 32.28 watts per m². The ft² watt rating is used in the remainder of the calculations.

Once the unit load per ft² has been converted into total general lighting load for the dwelling, the minimum number of general lighting circuits can be calculated.

Determining the Minimum Number of General Lighting Branch Circuits

The total general lighting load for the sample house is 5449.5 watts. In order to convert watts into the number of circuits, it is necessary to eliminate both volts and amperes from the units of measure, leaving only the number of circuits.

- 5449.5 watts divided by 120 volts = 45.415 amperes. This is the calculated load for the general-use receptacles and general lighting for the dwelling measured in amperes.

- 45.415 amperes divided by the general lighting branch-circuit size gives the minimum number of branch circuits. The *NEC®* allows either 15-ampere or 20-ampere branch circuits for general lighting and general-use receptacle circuits. Therefore,

  - For 15-ampere circuits: 45.415 amperes divided by 15 amperes = 3.03 circuits.

  - For 20-ampere circuits: 45.415 amperes divided by 20 amperes = 2.27 circuits.

Because it is not possible to have a partially powered electrical circuit that operates properly, and because these are minimum units, the total number of branch circuits with use of 15-ampere circuits is four circuits (3.03 circuits rounded up to 4 circuits) and three circuits with use of 20-ampere branch circuits (2.27 circuits rounded up to 3 circuits).

10.3.2 The General Lighting and General-Use Receptacle Outlet Circuits as Installed in the Sample House

The general lighting and general-use receptacle outlet circuits for the sample house are to be wired using 14 AWG Type NM cable. Although it has been determined that four circuits will satisfy the general-use outlet load requirements, there are actually five general lighting and general-use receptacle circuits being installed in the sample house. This determination is based on the fact that the house is relatively small for a three-bedroom house; therefore, the outlets will be used more intensively than those of a larger house occupied by the same-size family. Additionally, the basement is unfinished but is likely to be finished at some future date. A circuit has been included for the basement to provide the necessary power if and when the basement is finished.

For the study of the sample house, each circuit has its own drawing, a wiring diagram, to illustrate the extent of the circuit in isolation without other cables shown. This circuiting is the same as that shown in Figures 1-8 and 1-9. The colored lines on the wiring diagrams represent NM cable runs according to the following color codes:

- Green: a 15-ampere (14 AWG) or a 20-ampere (12 AWG), 2-wire NM cable of one black conductor for power and one white conductor for the neutral. Also included in the cable, but not included in the conductor count in specifying a particular cable, is one green or bare equipment grounding conductor.

- Red: a 15-ampere (14 AWG) or a 20-ampere (12 AWG), 2-wire cable used as a switch leg. This is a cable with one black switch leg conductor and a white conductor usually used as a neutral, although if re-identified it may function as another type of conductor. This cable also includes an equipment grounding conductor.

- Blue: a 15-ampere (14 AWG) or a 20-ampere (12 AWG), 3-wire cable, with an equipment grounding conductor. The 3-wire cable may include travelers with a switch leg, neutral, or phase conductor, depending on the makeup of the system.

- Green and red double lines: a 15-ampere (14 AWG) or a 20-ampere (12 AWG), 3-wire cable with a power, a switch leg, and a neutral conductor. The cable contains a black phase conductor, a red switch leg conductor, and a white neutral (grounded) conductor as well as an equipment grounding conductor.

KEY TERMS

Opening A term used to describe a receptacle outlet or a switch point.

Each individual outlet or switch point is identified in the wiring diagram by the circuit number followed by the count of that particular opening on that particular circuit. **Opening** is a term used to include the number of switch points with the count of the receptacle and lighting outlets. For example, opening 6.2 represents the second opening in circuit 6. Also included in the shaded box in the diagrams is a letter code identification for the outlet or switch point. The letter R designates the box for a receptacle that appears in Figure 10-4, an S designates a box for a switch that appears in Figure 10-5, and an L designates a box for a luminaire (fixture) that appears in Figure 10-6. The number that follows designates a specific drawing within the respective figure. For example, S3 refers to drawing 3 in Figure 10-5.

Each outlet or switch point box must be closely examined to determine several bits of information, including the minimum size of the box in in.^{3} (cm^3) required, the wire coding to be used in installing the cables to allow easy makeup, and the final makeup of the box to facilitate ease of trim. The guidelines for cable identification and final makeup are addressed in Chapters 9 and 16 of this book. Most of the boxes shown are nonmetallic boxes. If the cable runs are to be in AC or MC cable and metal boxes are required to terminate the cable, or if a metal box is used with NM cable, it should be understood that the boxes must be connected to the equipment grounding conductor system. The boxes shown in the gray area on the diagram are boxes served by 20-ampere circuits, using 12 AWG copper conductors.

Circuit 18—Subpanel A: Basement

The general lighting and general-use outlet circuit for the sample house basement is circuit 18 from subpanel A. A detailed illustration of circuit 18 is presented in Figure 10-7 (page 254). There are ten total openings on the circuit: six luminaires (fixtures), three switches, and one receptacle. The basement general lighting circuit homerun (the run of cable from the distribution panel and the first load or switch point) terminates in the two-gang switch box labeled 18.1 at the bottom of the basement stairs. The height above the floor for receptacle outlets and the switch point is as stated in the specifications on Construction Drawing E-1.

Circuit 7—Subpanel A: Master Bedroom and Master Bath

The master bedroom and the master bath are supplied from a 15-ampere circuit. This circuit is designated circuit 7 from subpanel A, and a detailed illustration of the circuit is shown in Figure 10-8 (page 255). The homerun is to a duplex receptacle box in the master bedroom, and the circuit ends in the switch box for the light over the toilet area. There are five receptacles, five switches, and four luminaires (fixtures) powered by circuit 7. One of the luminaires (fixtures) is a recessed can, installed in the closet. Figure 10-9 (page 256) shows an example of a recessed can rough-in. The Code requires that the circuits providing 120-volt power to outlets in a bedroom be AFCI protected. Because of the circuit routing, the AFCI protector must be installed in Subpanel A as part of an AFCI in combination with a circuit breaker to protect the entire circuit. If an AFCI device receptacle-type protector is used, the line side of the homerun termination in box 7.1 will not be properly protected. The outlets on circuit 7 are to be installed at 13 in. (.331 m) to bottom above the finished floor (AFF), and the switch points are installed at 44 in. (1.1 m) to bottom AFF, according to the specifications on Construction Drawing E-1.

Circuit 3—Subpanel A: Kitchen, Dining Room, Living Room, and Entry

Circuit 3 from subpanel A provides lighting for the kitchen, the dining room, and the outdoor

"WIREMAN'S GUIDE"
RECEPTACLE BOX MAKEUP

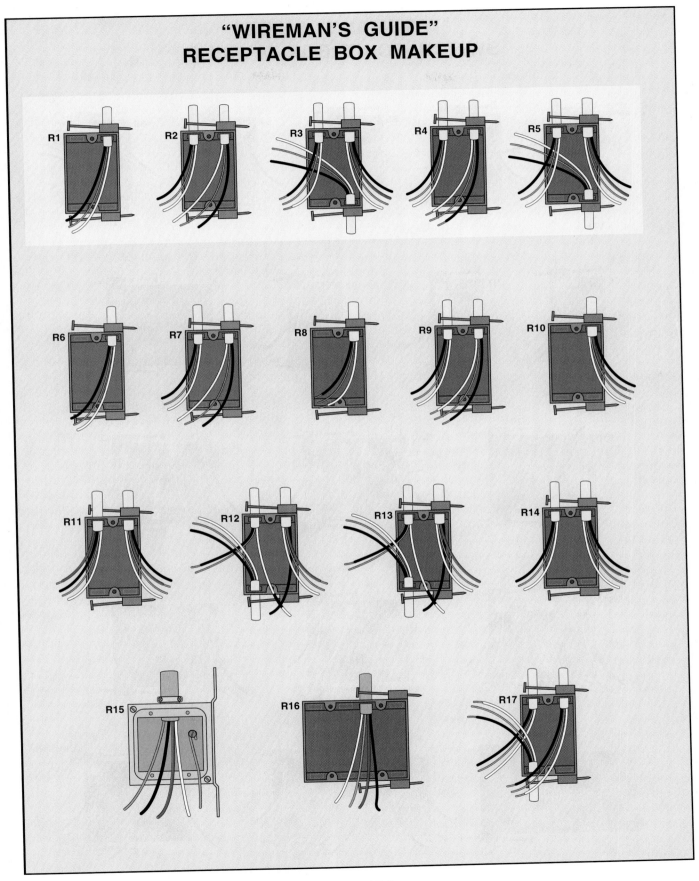

Figure 10-4

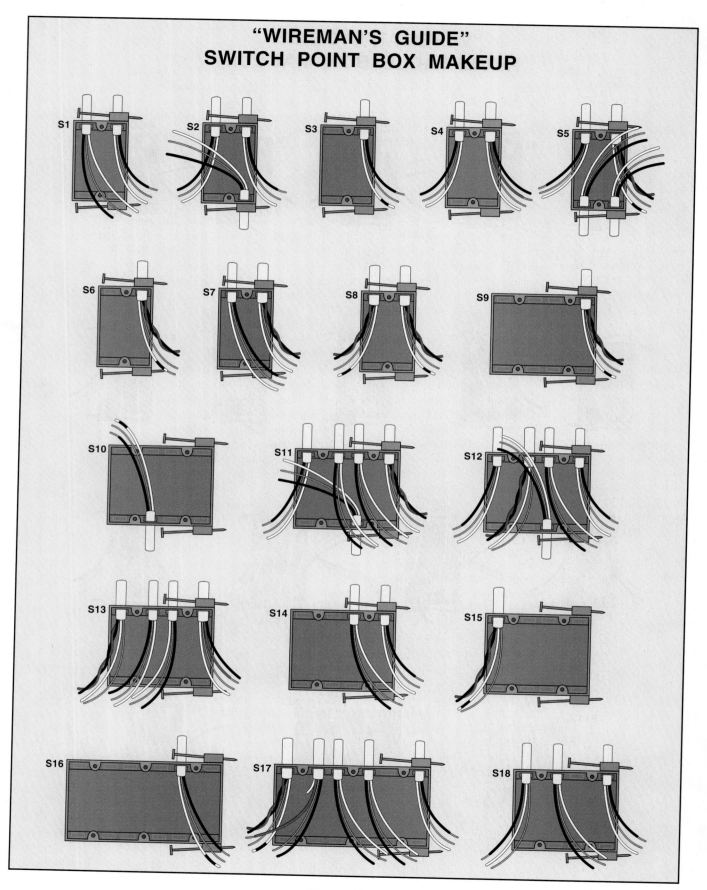

Figure 10-5

"WIREMAN'S GUIDE"
LUMINAIRE (FIXTURE) BOX CONDUCTOR IDENTIFICATION

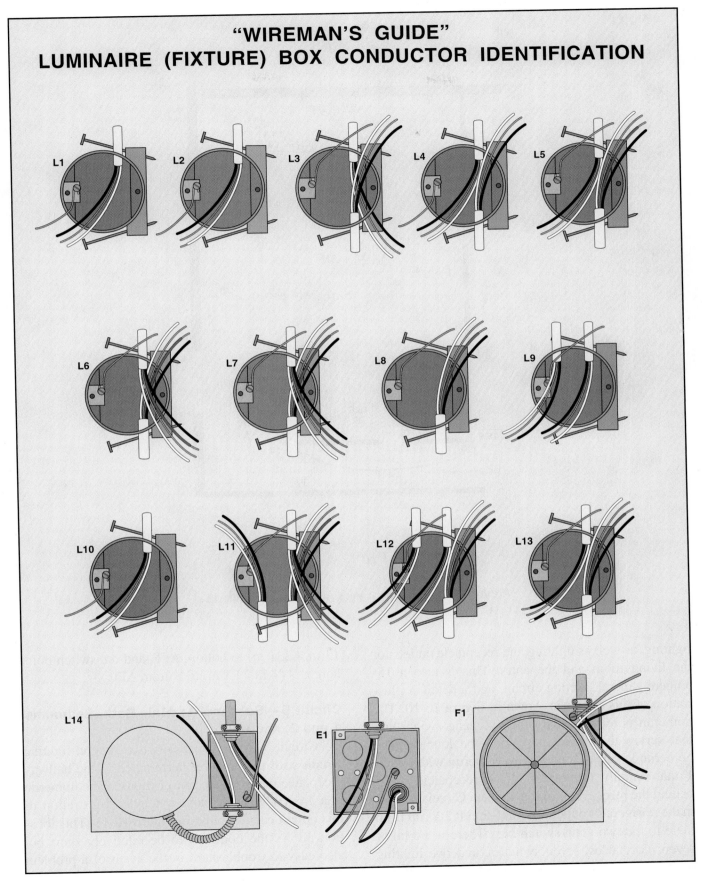

Figure 10-6

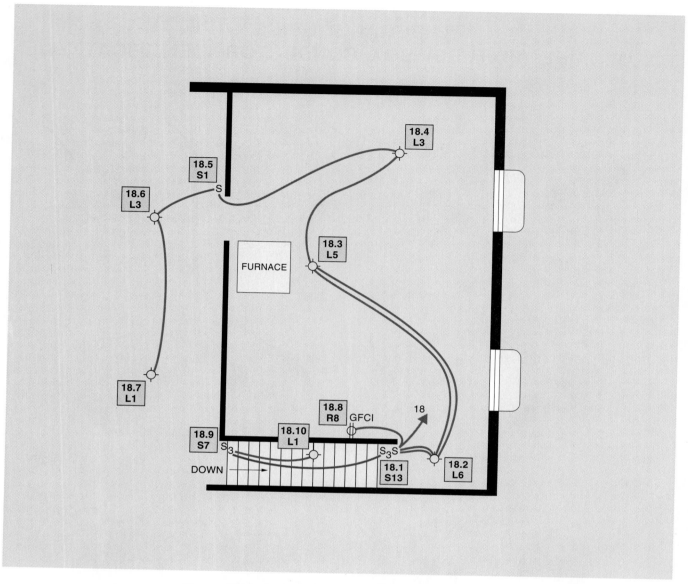

Figure 10-7   Basement general lighting: circuit 18.

lighting, as well as lighting and receptacle outlets for the living room and the entry. This is also a 15-ampere general lighting circuit, and a detailed illustration of this circuit is shown in Figure 10-10. The homerun is terminated in the two-gang switch box that serves the dining room and the kitchen. The homerun is part of a multiwire homerun with circuit 1 that initially terminates in the receptacle outlet behind the refrigerator in the kitchen. Circuit 1 is left in the refrigerator outlet box, and circuit 3 is run from there to the two-gang switch box. There is a total of seven receptacles, seven switches, and five lighting outlets on circuit 3. Outlet height for receptacles is

13 in. (.331 m) to bottom AFF, and the switch point height is 44 in. (1.1 m) to bottom AFF.

Circuit 5—Subpanel A: Main Bath, Bedrooms 1 and 2, and Hall

Circuit 5 from subpanel A is a general lighting circuit with the homerun terminated in the hallway receptacle outlet of the sample house. The homerun is a 14 AWG 2-wire, and a detailed illustration of this circuit can be found in Figure 10-11. This location allows the homerun to be relatively short but also easy to troubleshoot in the event of a problem at trim.

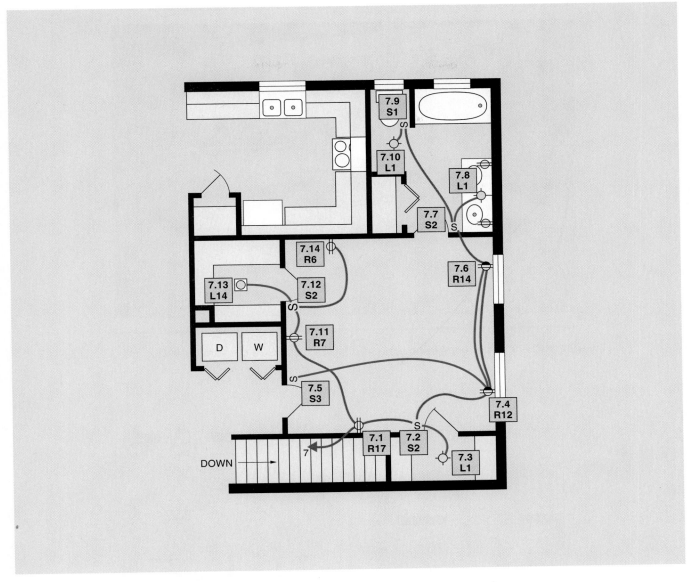

Figure 10-8 Main floor general lighting: circuit 7.

Bedroom outlets, both lighting and receptacle, must be AFCI protected according to *210.12*. The wiring for the outlets in bedroom 1 and bedroom 2 are routed so that only one AFCI receptacle needs to be installed in box 5.1 in the circuit in order to protect all of the receptacle and lighting outlets. The circuit routing allows the use of an AFCI receptacle rather than a circuit breaker because the protected receptacle is installed in the hallway, which does not require AFCI protection. Therefore, all of the bedroom outlets in the circuit are protected on the line and load sides. This installation technique has the advantage that if the AFCI device opens, it is readily

accessible in the hallway. There is a total of nine receptacles, two switches, and two lighting outlets on circuit 5. The heights for the switch points and receptacle outlets are according to the specifications in Construction Drawing E-1.

Circuit 14—Subpanel A: Garage, Main Bath and Hallway Lighting, and Front Outdoor Receptacle and Luminaires (Fixtures)

The homerun for circuit 14 from subpanel A terminates in one of the lighting outlets in the ceiling of the garage. There are two fluorescent luminaires (fixtures) in the garage. The garage general-use recepta-

Figure 10-9 The rough-in housing of a typical recessed can luminaire (fixture).

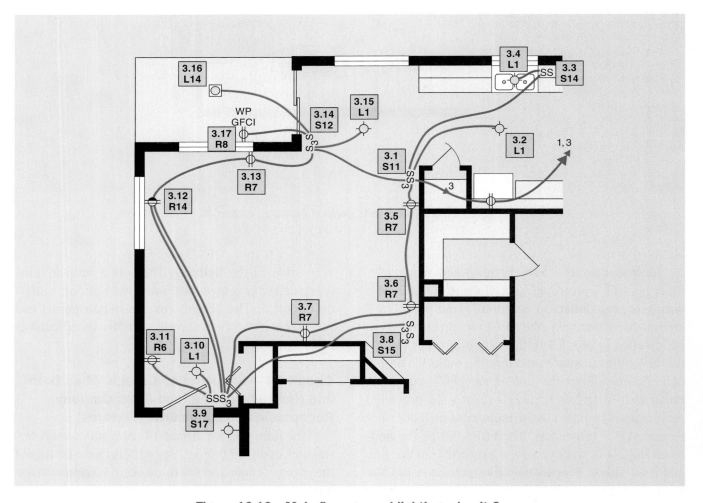

Figure 10-10 Main floor general lighting: circuit 3.

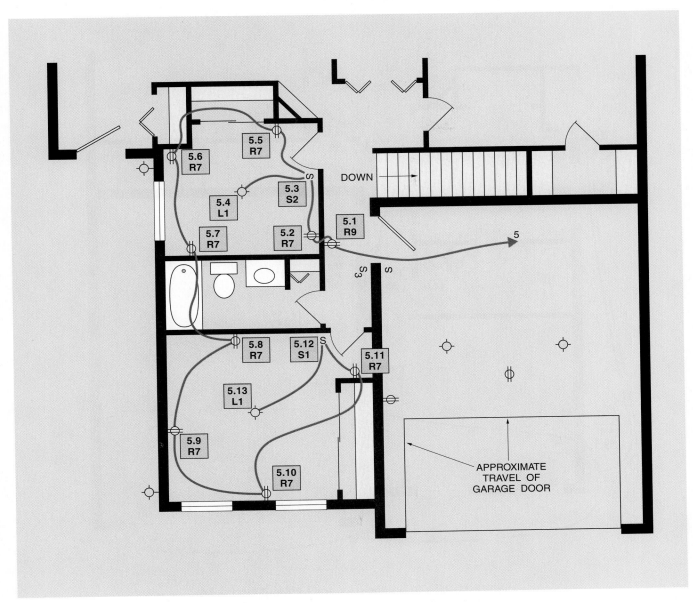

Figure 10-11   Main floor general lighting: circuit 5.

cle outlet is on the wall by the garage entry door, and the outdoor receptacle is located along the walk on the west side of the house. This circuit also powers the outdoor luminaires (fixtures) that are switched from the three-gang switch box at the main entry door, as shown in Figure 10-12. The garage general-use receptacle and the outdoor receptacle must be GFCI protected. Installation of a GFCI receptacle in the garage that also protects the front outdoor receptacle and the front walkway lighting is permissible. Examination of the circuit routing shows that it is also possible to install a GFCI receptacle both in

the garage and at the outdoor receptacle and not have GFCI protection for the luminaires (fixtures) along the main entry walkway. There is a total of two receptacle outlets, seven luminaires (fixtures), and seven switches on circuit 14. The height AFF for the switch point is 44 in. (1.1 m) according to the specifications. The height for the receptacle outlets is also 44 in. (1.1 m) AFF for the garage, also according to the specifications. The height for the outdoor receptacle is 13 in. (.331 m). The mounting height for the outdoor luminaires (fixtures) is to be even with the top of the garage door.

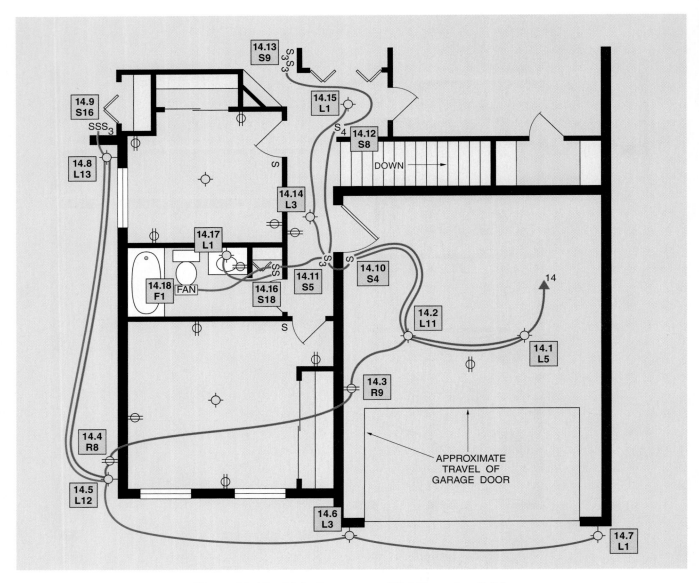

Figure 10-12 Main floor general lighting: circuit 14.

10.4: SMALL-APPLIANCE BRANCH CIRCUITS

Three small-appliance circuits service the kitchen and dining area in the sample house, all supplied from subpanel A. The small-appliance branch circuits are circuits 2 and 4. As with the general lighting and general-use receptacle outlets, there is no specific branch-circuit load for the small-appliance circuits. According to *210.11(C)(1)*, there must be a minimum of two 20-ampere circuits installed in the kitchen, and according to *210.52(B)(3)*, the kitchen countertop receptacles must be serviced by no fewer than two circuits, although those circuits can serve other small-appliance loads such as the dining room. The result of these requirements is that there never

needs to be more than two small-appliance circuits installed in a dwelling unit, according to the *NEC®*. In many locales, the AHJ places limits on the number of receptacle outlets that can be placed on a small-appliance circuit, however, so this item needs to be discussed with the AHJ before installation begins.

The specifications require the above-countertop receptacles to be 4 in. (.100 m) above the finished countertops. The counter height plus the countertop depth comes to 38 in. (.965 m) AFF, according to the detail on Construction Drawings A-4 and A-5, so the bottom of the box should be 42 in. (1.1 m) AFF. Notice that the backsplash on countertops can vary in height from one house to another. Care must be taken to ensure that the boxes are installed sufficiently high to clear the countertop backsplashes, even if it means increasing the height from the specification height requirement.

10.4.1: Circuit 2—Subpanel A: Small Appliance

Circuit 2 from subpanel A supplies power to the small-appliance receptacle outlets on part of the west half of the north wall and the entire west wall of the kitchen, as detailed in Figure 10-13. All of these outlets, except for receptacle 2.5, are located above the kitchen countertops. According to *210.8(A)(6)*, all above-counter receptacle outlets in the kitchen must have GFCI protection. The only receptacle outlet that is installed on a small-appliance circuit but is not above a kitchen countertop is the receptacle in the dining room; this outlet is located at 13 in. (.331 m) AFF according to the specifications. The cable routing is such that a GFCI receptacle can be installed in box 2.1. The feed-through GFCI protection will protect the remainder of the circuit. The dining room receptacle outlet does not have to be GFCI protected, but because of the routing of the cable for circuit 2, it will be protected along with the rest of the circuit.

10.4.2: Circuit 4—Subpanel A: Small Appliance

Circuit 4 from subpanel A supplies the east half of the north wall and the entire east wall of the kitchen, as shown on Figure 10-14. All of the receptacle outlets on circuit 4 are above countertops and therefore must be GFCI protected. The routing of the cable enables the protection of the circuit against ground faults with the installation of a receptacle GFCI device in box 4.1.

10.4.3: Circuit 1—Subpanel A: The Refrigerator Circuit

The receptacle behind the refrigerator is not a countertop outlet and thus can be installed at any height up to 5½ ft (1.7 m) AFF. It will be installed at 44 in. (1.1 m) AFF with the rest of the outlets on this circuit for ease of installation and for access to the refrigerator plug. The refrigerator receptacle outlet does not have to be GFCI protected because it is not above the countertops, and an exception to *210.52(B)(1)* allows a 15-ampere circuit to feed the refrigerator if it is an individual branch circuit, meaning that it has no other outlets. The refrigerator receptacle is supplied from one circuit of a multiwire branch circuit with circuit 3 from subpanel A, as allowed by *210.4(A)*. Details of circuit 1 are shown in Figure 10-15.

10.5: CIRCUIT 16—SUBPANEL A: LAUNDRY (CLOTHES WASHER) CIRCUIT

NEC® 210.11(C)(2) requires that a laundry circuit be installed in all dwelling units. In the sample house,

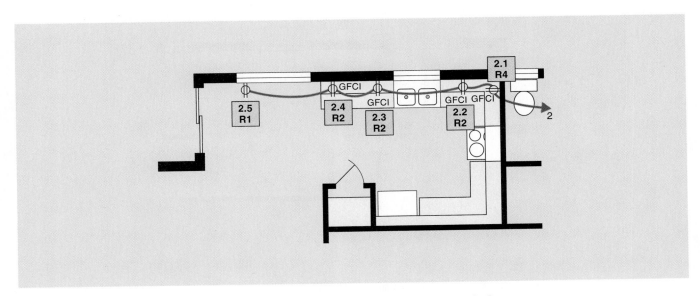

Figure 10-13   Small-appliance circuit: circuit 2.

the laundry receptacle is located in a closet-like area that opens onto the hallway just outside the master bedroom door, as shown in Figure 10-16. There is no specific branch-circuit load required for the laundry circuit because the Code simply requires that a 20-ampere circuit be installed. It goes on to say that the laundry circuit is to have no other outlets other than the laundry. There are several ways in which the AHJ can interpret this requirement. The details of the different interpretations are shown in Figure 10-17. It is important that the residential electrician know how the local AHJ will enforce the laundry receptacle installation before beginning the installation.

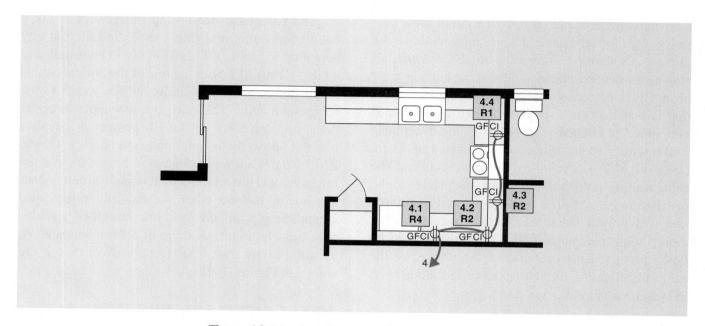

Figure 10-14   Small-appliance circuit: circuit 4.

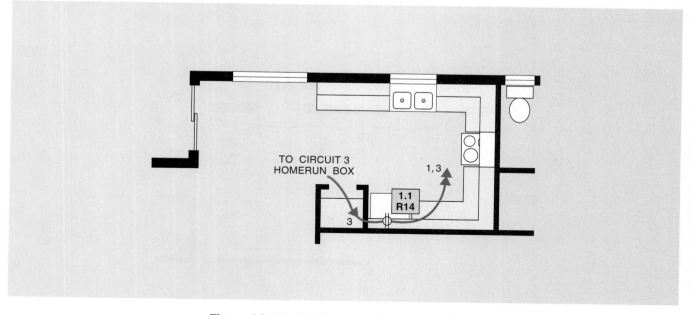

Figure 10-15   Refrigerator circuit: circuit 1.

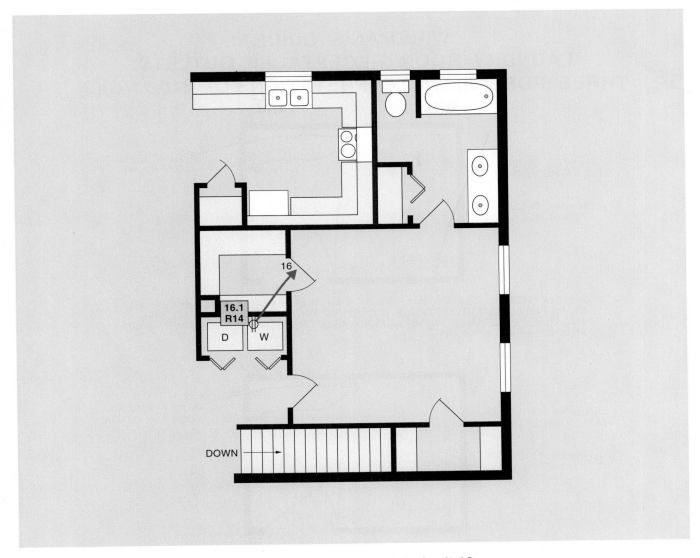

Figure 10-16 Laundry outlet: circuit 16.

10.6: CIRCUIT 12—SUBPANEL A: BATHROOM RECEPTACLE CIRCUIT

NEC® 210.11(C)(3) requires that a 20-ampere branch circuit be installed to service the general-use receptacle outlets in the bathrooms. There is no branch-circuit load for this circuit; the load is included in the general lighting and general-use receptacle loads calculated earlier. Circuit 12 from subpanel A is used for this circuit, and the details of the circuit are shown in Figure 10-18. This circuit includes only the bathroom receptacles and has no other openings.

There is another possibility for wiring the bathroom receptacles, as detailed in Figure 10-19 (page 264) and the exception to *210.11(C)(3)*. This excep-

tion allows the power for the bathroom receptacle outlets to be supplied by a circuit that supplies no other loads except that bathroom. The bathroom lighting and receptacles, as well as any exhaust fan, may be powered by the same homerun, but that homerun cannot power any other bathroom or any other part of the dwelling.

10.7: CIRCUIT 19—SUBPANEL A: SMOKE DETECTION AND ALARM CIRCUIT

The circuit that is required by the UBC for smoke and fire detection is powered by circuit 19 from subpanel A. The homerun for this circuit terminates in

"WIREMAN'S GUIDE"
LAUNDRY ROOM RECEPTACLE OUTLETS:
THREE POSSIBILE INTERPRETATIONS OF *210.11(C)(2)*

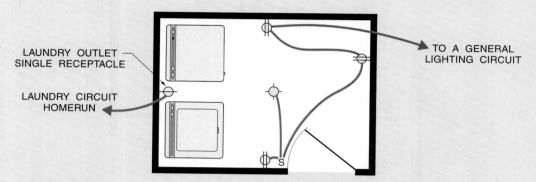

INTERPRETATION 1: THE AHJ MAY REQUIRE THAT THE LAUNDRY CIRCUIT BE SERVED BY AN INDIVIDUAL OUTLET (SINGLE OUTLET RATHER THAN A DUPLEX), AND THAT ANY OTHER RECEPTACLE IN THE LAUNDRY AREA OR ROOM MUST BE SERVED FROM ANOTHER BRANCH CIRCUIT.

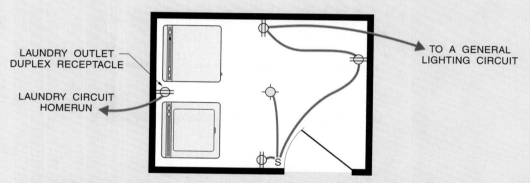

INTERPRETATION 2: THE AHJ MAY ALLOW THE LAUNDRY RECEPTACLE TO BE A DUPLEX RECEPTACLE, BUT ANY OTHER RECEPTACLES IN THE LAUNDRY AREA OR ROOM MUST BE SUPPLIED FROM ANOTHER BRANCH CIRCUIT.

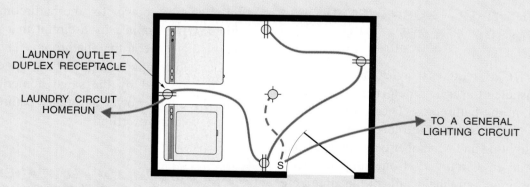

INTERPRETATION 3: THE AHJ MAY ALLOW ANY RECEPTACLE OUTLET IN THE LAUNDRY AREA OR ROOM TO BE SERVED BY THE LAUNDRY CIRCUIT. NOTICE THAT THE LAUNDRY ROOM LIGHTING IS NOT INCLUDED WITH THE RECEPTACLES ON THE LAUNDRY OUTLET CIRCUIT.

Figure 10-17

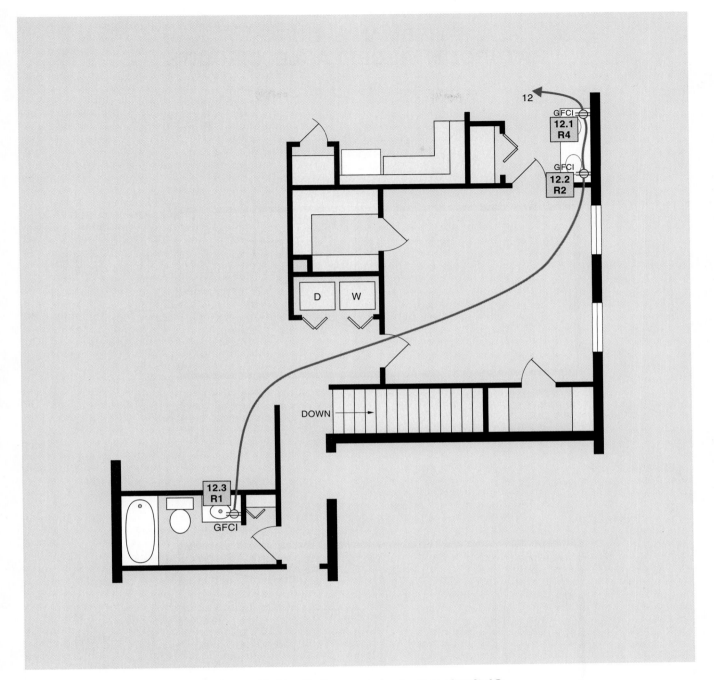

Figure 10-18 Bathroom receptacles: circuit 12.

the smoke detector in the master bedroom and continues to the other three detectors on the main floor and the detector in the basement, as seen in Figure 10-20. The smoke detectors have a terminal for a switch leg as well as for power. The switch leg will cause all smoke detectors to sound the alarm if any detect fire or smoke. All smoke detectors in the dwelling must be wired together with 3-wire cable to have all of the alarms sound even if only one detector senses smoke.

"WIREMAN'S GUIDE"
BATHROOM RECEPTACLE CIRCUITS

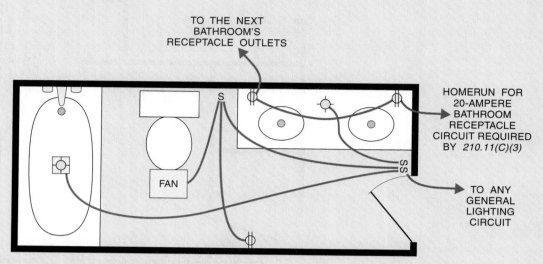

THE *NEC®* REQUIRES A 20-AMPERE CIRCUIT TO SUPPLY POWER TO THE BATHROOM RECEPTACLE OUTLETS. THERE CAN BE NO OTHER OUTLETS ON THIS CIRCUIT. THE OTHER OUTLETS IN A BATHROOM CAN BE SERVED FROM A GENERAL LIGHTING BRANCH CIRCUIT.

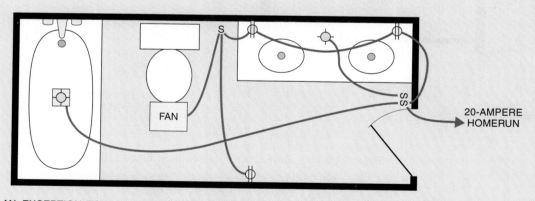

AN EXCEPTION TO *210.11(C)(3)* ALLOWS THAT A DEDICATED CIRCUIT MAY SERVE ALL OF THE OUTLETS, INCLUDING RECEPTACLE OUTLETS, IN A BATHROOM AS LONG AS THEY SERVE NO OTHER LOADS.

Figure 10-19

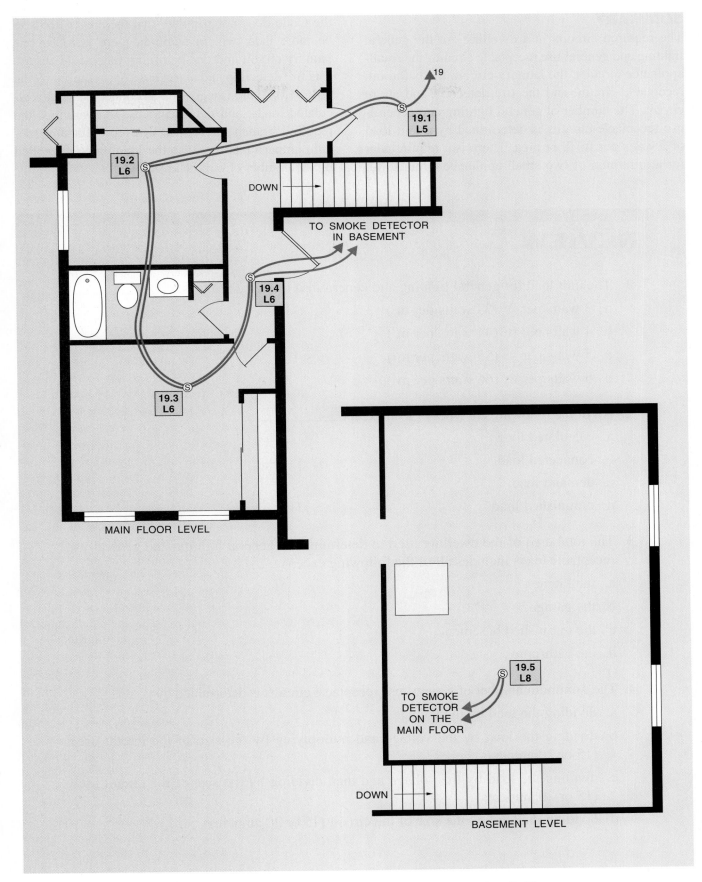

Figure 10-20 Smoke detection and alarm circuit: circuit 19.

SUMMARY

The required circuits in a dwelling are the general lighting and general-use receptacle circuits, the small-appliance circuits, the laundry circuit, the bathroom receptacle circuit, and the fire detection and alarm circuit. The number of general lighting and general-use receptacle circuits is determined by a unit load of 3 watts per ft² floor area. There is a requirement for a minimum of two small-appliance circuits, but the Code does not stipulate that there ever needs to be more than two. In addition, there is to be one laundry circuit and one bathroom receptacle circuit. The fire detection and alarm circuit is required by the UBC. Loads are categorized as connected loads, calculated loads, and demand loads. In dwellings, the required branch circuits do not represent any specific ampere load because the loading is controlled by the number of circuits installed.

REVIEW

1. The unit load for general lighting and general-use branch circuits is _____.
   a. 3 watts per ft² (33 watts per m²)
   b. 4 watts per ft² (44 watts per m²)
   c. 5 watts per ft² (55 watts per m²)
   d. 6 watts per ft² (66 watts per m²)

2. All of the following are types of load except _____ .
   a. calculated load
   b. connected load
   c. demand load
   d. diminished load

3. The total area of the dwelling used to determine the general lighting and general-use receptacle loads includes all of the following except _____.
   a. the kitchen
   b. the garage
   c. the unfinished basement
   d. the bathrooms

4. The minimum number of general-use receptacle circuits is determined by _____.
   a. dividing the load by the voltage
   b. dividing the load by the voltage and multiplying by the size of the circuit used (15 or 20 amperes)
   c. dividing the load by the voltage and then dividing by the size of the circuit used (15 or 20 amperes)
   d. dividing the load by the size of the circuit (15 or 20 amperes)

5. If the calculation does not yield a whole number, the remainder must be _____ .

   a. rounded up

   b. rounded down

   c. dropped from the number

   d. rounded up if .5 or larger

6. All of the following circuits are required in a dwelling unit except _____ .

   a. the laundry branch circuit

   b. the small-appliance circuits

   c. the general lighting and general-use branch circuit

   d. the furnace circuit

7. The fire detection and alarm circuit required by the UBC applies to _____ .

   a. dwellings larger than 1000 ft^2 (93 m^2) only

   b. dwellings with more than two levels only

   c. dwellings with electric heating or air conditioning only

   d. all dwellings

8. The connected load can be obtained from _____ .

   a. the general contractor or builder

   b. the nameplate rating of the appliance or equipment

   c. testing the appliance or equipment and measuring the current flow with an ammeter

   d. the AHJ

9. Small-appliance circuits have a branch-circuit load rating of _____ each.

   a. 1000 watts (volt-amperes)

   b. 1500 watts (volt-amperes)

   c. 2400 watts (volt-amperes)

   d. none of the above—there is no specific rating

10. Each receptacle outlet on the general lighting and general-use receptacle circuits has a rating of _____ .

    a. 120 watts (volt-amperes)

    b. 1800 watts (volt-amperes) or 1.5 amperes

    c. 20 amperes

    d. none of the above—there is no specific rating

"WIREMAN'S GUIDE" REVIEW

1. A dwelling has 2500 ft² of habitable floor area, a 600 ft² garage, and a 2500 ft² unfinished basement that could someday be finished. What is the general lighting and general-use receptacle load?

2. Draw the final box makeup for box 18.5 (fifth box on circuit 18). The rough-in makeup identification is found in drawing S1 in Figure 10-5. Be sure to include the equipment grounding conductors.

3. Draw the final box makeup for box 14.1. Be sure to include equipment grounding conductors.

4. Draw the final box makeup for box 14.8. Be careful to re-identify any white conductors that may be used for other than a neutral. Be sure to include equipment grounding conductors.

5. Box 3.1 and box 3.14 are the two ends of a three-way switching system. The rough makeup coding for box 3.1 is S11 and for 3.14, S12. From these boxes, determine if the power for the system is taken from box 3.1 or from 3.14. How do you know?

6. Box 3.9 is a three-gang box, but only two of the three switches are powered from circuit 3. Which circuit powers the third switch? Do the two circuits need to be terminated on a 2-pole breaker in subpanel A so that both circuits are opened at the same time to protect against electrocution should someone need to work on the box?

7. Draw the final makeup of box 3.9. Be sure to include the equipment grounding conductors. How will the power and the switch leg for the outdoor luminaires (fixtures) be separated from the switches powered by circuit 3?

CHAPTER 11

Branch Circuits and Branch-Circuit Loads for Nonrequired Outlets

OBJECTIVES

On completion of this chapter, the student will be able to:

☑ Identify loads and ampere ratings for specific appliances.

☑ State the requirements for connecting appliances with an attachment plug-and-cord.

☑ State the requirements for disconnecting appliances and other utilization equipment other than luminaires (fixtures).

☑ Calculate the load for and install a branch circuit for electric clothes dryers, ranges, motors, and other appliances.

☑ Utilize elevation drawings of kitchens and bathrooms in wiring applications.

☑ Define a split-wired receptacle, and describe the dangers presented by two circuits connected to the same device.

INTRODUCTION

This chapter is concerned with those outlets that are not required by the *NEC®* or the UBC. Many outlets are installed in dwelling units that are not required outlets. Kitchen appliances such as dishwashers and ranges, space-heating equipment such as gas-fired furnaces, equipment such as pumps for groundwater control or irrigation systems, and fans for moving air are routinely installed in dwellings. These loads need to be supplied with electrical power. Furthermore, all appliances require some sort of disconnecting means to isolate the equipment from the electrical supply during maintenance or equipment failure. Some appliances are allowed to be cord-and-plug connected, but many typically are connected by a permanent wiring method, and all appliances must be installed according to the manufacturer's installation instructions. The wiring of the branch circuits for the sample house is complete with the end of this chapter. Attention is then directed at feeders and service conductors in Chapter 12.

11.1: BRANCH-CIRCUIT LOADING AND DISCONNECTING MEANS NEEDED FOR NONREQUIRED ELECTRICAL EQUIPMENT AND APPLIANCE CIRCUITS

KEY TERMS

Other Loads—All Occupancies The loads for all utilization equipment other than general lighting and general-use receptacle outlet loads.

NEC® 210.11 requires that branch circuits be installed to supply any additional nonrequired loads. The circuit size shall be based on the load calculation detailed in *220.3*. There are 11 types of branch-circuit loads other than general lighting loads—other loads—listed in *220.3*. They are referred to as **Other Loads—All Occupancies**, but only five apply in any way to dwelling-unit branch-circuit calculations.

11.1.1: Specific Appliances or Loads

The first *220.3(B)*-listed category of other loads is *specific appliances or loads*. This category includes any appliances not specified in one of the other 10 (out of the 11 total) items listed. Kitchen appliances, except for electric ranges, counter-mounted cooktops, and wall-mounted ovens, are not listed as requiring a specific branch-circuit load in *220.3(B)*. Therefore, specific kitchen appliances, such as the garbage disposal and the dishwasher, have their branch-circuit loads calculated under this load type.

Ampacity Ratings of Specific Appliances

The nameplate rating of the appliance, or connected load, is used to determine the minimum branch-circuit ampacity and the maximum rating for the overcurrent-protective device. *NEC® 422.60* requires that each appliance nameplate include the appliance rating, either in volts or amperes or in volts and watts, as shown in Figure 11-1.

During rough-in, it is not always possible to determine the rated ampacity of the appliance that will be installed at trim. Fortunately, many appliances that are fastened in place in dwellings are engineered by the manufacturer so that the appliance can be powered by no larger than a 20-ampere cir-

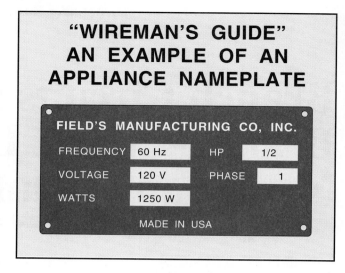

"WIREMAN'S GUIDE" AN EXAMPLE OF AN APPLIANCE NAMEPLATE

FIELD'S MANUFACTURING CO, INC.

FREQUENCY	60 Hz	HP	1/2
VOLTAGE	120 V	PHASE	1
WATTS	1250 W		

MADE IN USA

Figure 11-1

cuit, and in a majority of cases, these appliances can be powered by a 15-ampere circuit. The manufacturers realize that a 15-ampere circuit is the minimum size circuit installed in dwellings and therefore size their appliances to operate on a 15-ampere circuit so that the purchaser of the appliance will not have to have additional circuits installed. If there is some doubt about the circuit required for a given appliance, installing a 20-ampere circuit will virtually ensure enough power to supply the residential rated appliance, except for a range, a counter-mounted cooktop, a wall-mounted oven, or an electric clothes dryer. If there are concerns that an upgraded appliance—one not specifically marketed for installation in dwellings—is to be installed, the general contractor or the builder should be contacted to obtain a copy of the manufacturer's installation instructions. Figure 11-2 presents additional information about appliance circuits.

Cord-and-Attachment Plug Connected Specific Appliances

KEY TERMS

Attachment plug-and-cord (See *NEC® Article 100*): An assembly of a male cord cap (plug) and the necessary cable intended to connect utilization equipment to a receptacle outlet.

NEC® 422.16(B) details the requirements for connection of certain specific appliances with an

"WIREMAN'S GUIDE"
CORD-CONNECTED AND PERMANENTLY
CONNECTED APPLIANCE CIRCUITS

IT IS GENERALLY NECESSARY TO DETERMINE THE RATED AMPACITY OF THE APPLIANCE TO DETERMINE THE MINIMUM CIRCUIT AMPACITY AND THE MAXIMUM OVERCURRENT DEVICE RATINGS. IF IT IS NOT LISTED ON THE APPLIANCE NAMEPLATE, IT CAN BE CALCULATED FROM THE OTHER RATINGS SHOWN ON THE NAMEPLATE.

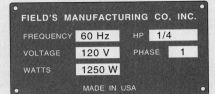

FIELD'S MANUFACTURING CO, INC.

FREQUENCY	60 Hz	HP	1/4
VOLTAGE	120 V	PHASE	1
WATTS	1250 W		

MADE IN USA

1250 WATTS/120 VOLTS = 10.42 AMPERES

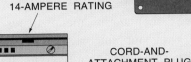

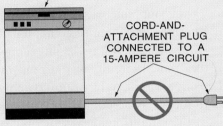

APPLIANCE WITH A 14-AMPERE RATING

CORD-AND-ATTACHMENT PLUG CONNECTED TO A 15-AMPERE CIRCUIT

IT IS IMPORTANT TO IDENTIFY IF THE APPLIANCE IS CONNECTED WITH A CORD-AND-ATTACHMENT PLUG TO A RECEPTACLE. IF THE APPLIANCE IS CORD-AND-ATTACHMENT PLUG CONNECTED, THE CIRCUIT CAN BE LOADED ONLY TO 80% OF THE TOTAL AMPACITY ACCORDING TO *210.23(A)(1)*. FOR EXAMPLE, IF A CORD-AND-ATTACHMENT PLUG CONNECTED APPLIANCE HAS A NAMEPLATE CURRENT RATING OF 14 AMPERES, IT CANNOT BE CONNECTED TO A 15-AMPERE BRANCH CIRCUIT BECAUSE 80% OF THE 15-AMPERE CIRCUIT EQUALS 12 AMPERES. THIS IS THE MAXIMUM FOR ONE CORD-AND-ATTACHMENT PLUG CONNECT LOAD.

THIS SAME APPLIANCE CAN BE CONNECTED TO A 15-AMPERE CIRCUIT IF IT IS CONNECTED DIRECTLY TO THE CIRCUIT USING A PERMANENT WIRING METHOD, BECAUSE IT IS PERMISSIBLE TO LOAD A CIRCUIT TO ITS FULL AMPERE RATING UNLESS THE LOAD IS CONTINUOUS DUTY RATED. IT IS WIDESPREAD THROUGHOUT THE COUNTRY THAT THERE ARE NO CIRCUITS IN A DWELLING THAT ARE ASSUMED TO BE CONTINUOUS DUTY RATED BECAUSE THERE IS NO INDICATION FROM THE EXAMPLES IN *APPENDIX D* OF THE *NEC®* THAT ANY RESIDENTIAL LOAD IS TO BE CONSIDERED A CONTINUOUS DUTY LOAD. THIS IS NOT A UNIVERSAL INTERPRETATION, HOWEVER, AND IF THERE IS A QUESTION ABOUT THE RULES ENFORCED IN ANY PARTICULAR AREA, THE AHJ SHOULD BE CONSULTED.

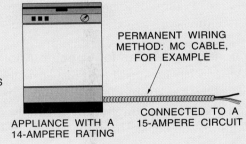

PERMANENT WIRING METHOD: MC CABLE, FOR EXAMPLE

CONNECTED TO A 15-AMPERE CIRCUIT

APPLIANCE WITH A 14-AMPERE RATING

ALSO, IF THE APPLIANCE IS FASTENED IN PLACE AND THE CIRCUIT TO WHICH THE APPLIANCE IS CONNECTED ALSO FEEDS LIGHTING, RECEPTACLE, OTHER CORD-CONNECTED UTILIZATION EQUIPMENT NOT FASTENED IN PLACE, THE LOAD OF THE FASTENED-IN-PLACE APPLIANCE CANNOT BE MORE THAN 50% OF THE TOTAL CIRCUIT AMPACITY RATING. ACCORDING TO *422.60(A)*, IF THE APPLIANCE IS MOTOR OPERATED AND IF THE APPLIANCE REQUIRES SEPARATE OVERCURRENT PROTECTION AGAINST OVERLOADS FOR THE MOTOR, IT SHALL BE IDENTIFIED ON THE APPLIANCE NAMEPLATE.

IN THIS EXAMPLE, THE TOTAL LOAD OF THE FASTENED-IN-PLACE APPLIANCES CANNOT BE MORE THAN 7.5 AMPERES (15 AMPERES X 50%).

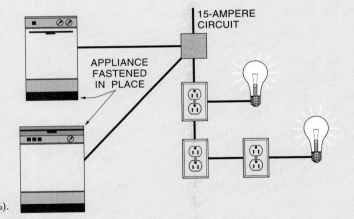

15-AMPERE CIRCUIT

APPLIANCE FASTENED IN PLACE

THE CODE ALSO REQUIRES THAT THE NAMEPLATE INCLUDE THE MINIMUM SUPPLY CONDUCTOR AMPACITY AND THE MAXIMUM RATING OF THE BRANCH-CIRCUIT OVERCURRENT-PROTECTIVE DEVICE IF THE APPLIANCE DOES NOT HAVE A FACTORY-INSTALLED ATTACHMENT PLUG, OR IF EITHER THE MINIMUM CONDUCTOR AMPACITY OR THE MAXIMUM OVERCURRENT-PROTECTIVE DEVICE RATING (OR BOTH) IS LARGER THAN 15 AMPERES. IF THE APPLIANCE HAS AN ATTACHMENT PLUG THAT WAS FACTORY INSTALLED, IT CAN BE ASSUMED THAT THE APPLIANCE CAN BE USED ON ANY CIRCUIT THAT WILL SUPPORT THE SIZE AND CONFIGURATION OF THE ATTACHMENT PLUG.

Figure 11-2

attachment plug-and-cord. The kitchen appliances in the sample house are intended to be cord-and-plug connected to receptacle outlets where allowed. The proper device boxes, as well as the necessary cable, must be installed at rough-in to provide a receptacle outlet to supply the appliances. Boxes may also be necessary to provide control switching for the appliance. The following appliances, other than ranges and dryers, may be cord-and-plug connected:

- Electrically operated kitchen waste disposals: *NEC® 422.16(B)(1)* details four requirements that must be satisfied if the appliance is to be cord-and-plug connected. All of these four conditions are satisfied for the sample house installation. Figure 11-3 presents details of the four conditions.

- Built-in dishwashers and trash compactors: *NEC® 422.16(B)(2)* details five requirements that must be satisfied if the appliance is to be cord-and-plug connected. All five of these conditions are satisfied for the sample house installation. Figure 11-3 presents details of the five conditions.

Disconnecting Means for Specific Appliances

All appliances must be furnished with a means of disconnection from all ungrounded phase conductors. A disconnecting means is simply a way of isolating the appliance from the circuit so that during maintenance the electrical worker can be assured that the appliance will not suddenly start and that the power will not present an electrocution hazard.

Disconnecting Cord-and-Attachment Plug Connected Appliances. For a cord-and-plug connected appliance, the disconnecting means can be the attachment plug, as illustrated in Figure 11-3. Unplugging the appliance effectively disconnects it from the circuit. The only requirement for use of the plug and receptacle as a disconnecting means is that the appliance plug and the receptacle must be accessible and must be rated for the ampere rating of the appliance.

Disconnecting Permanently Connected Appliances. Permanently connected appliances—appliances connected with other than a cord and plug using an approved wiring method—fall in one of two categories: (1) 300 volt-amperes, or ⅛ horsepower or less, and (2) over 300 volt-amperes, or ⅛ horsepower.

- For permanently connected appliances rated at not over 300 volt-amperes or ⅛ horsepower, the branch-circuit overcurrent-protective device (a circuit breaker, for example) is permitted to serve as the disconnecting means. No other disconnects are required for these smaller appliances.

- For permanently connected appliances rated over 300 volt-amperes or ⅛ horsepower, the disconnecting means can be the branch-circuit overcurrent-protective device (a circuit breaker, for example) if the overcurrent-protective device is within sight from the appliance. If the disconnect is within sight from the appliance, no other disconnecting means needs to be provided for this appliance.

For permanently connected appliances rated over 300 volt-amperes or ⅛ horsepower and when the branch-circuit overcurrent-protective device is not within sight from the appliance, it can still be considered as the disconnecting means if the circuit breaker is capable of being locked in the open (off) position. Many electrical distribution panels can be locked in the open position using either a built-in lock or an external padlock. When the circuit breaker for the appliance is opened and the panel is closed and locked, the requirements for the disconnection means for these appliances are satisfied.

If the circuit breaker cannot be locked in the open position, an additional disconnecting means must be provided for the appliance. This additional disconnect can be a switch that opens all ungrounded conductors and that is within sight from the appliance or a switch that is not within sight from the appliance but that can be locked in the open position. An on-off switch that is part of the appliance, and that disconnects all of the ungrounded conductors from the branch circuit, can also be used to satisfy this requirement.

11.1.2: Electric Clothes Dryers and Household Electric Cooking Appliances

The second type of other load listed in *220.3(B)* includes electric clothes dryers and household electric cooking appliances. For each of these loads, the branch-circuit load is determined from the feeder

"WIREMAN'S GUIDE"
REQUIREMENTS TO CONNECT
SPECIFIC APPLIANCES WITH A
CORD-AND-ATTACHMENT PLUG

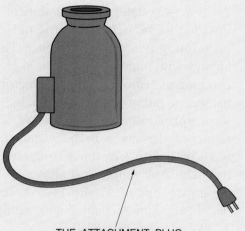

THE ATTACHMENT PLUG
CAN SERVE AS THE
REQUIRED DISCONNECTING
MEANS FOR THE APPLIANCE.

FOUR CONDITIONS THAT MUST BE SATISFIED
TO POWER A GARBAGE DISPOSAL USING A
CORD-AND-ATTACHMENT PLUG:
1. THE CORD MUST BE LONGER
   THAN 18 IN. (450 mm) BUT NOT
   LONGER THAN 36 IN. (900 mm).
2. THE ATTACHMENT PLUG MUST BE
   OF THE GROUNDING TYPE.
3. THE OUTLET MUST BE LOCATED
   TO AVOID DAMAGE TO THE CORD
   OR ATTACHMENT PLUG.
4. THE RECEPTACLE SUPPLYING THE
   DISPOSAL MUST BE ACCESSIBLE.

FIVE CONDITIONS THAT MUST BE SATISFIED
TO POWER A DISHWASHER USING A CORD-
AND-ATTACHMENT PLUG:
1. THE LENGTH OF THE CORD MUST BE
   BETWEEN 3 AND 4 FT (.9 AND 1.2 m).
2. THE ATTACHMENT PLUG MUST BE OF
   THE GROUNDING TYPE.
3. THE OUTLET MUST BE LOCATED
   TO AVOID DAMAGE TO THE CORD
   OR ATTACHMENT PLUG.
4. THE RECEPTACLE SUPPLYING THE
   DISHWASHER MUST BE ACCESSIBLE.
5. THE RECEPTACLE MUST BE LOCATED
   IN THE SPACE THAT WILL BE
   OCCUPIED BY THE DISHWASHER, OR
   IN AN IMMEDIATELY ADJACENT SPACE.

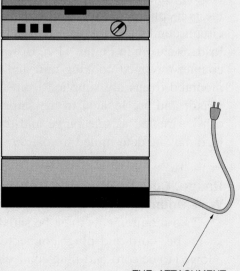

THE ATTACHMENT
PLUG CAN SERVE
AS THE REQUIRED
DISCONNECTING
MEANS FOR
THE APPLIANCE.

THE SAME REQUIREMENTS THAT APPLY
TO THE DISHWASHER ALSO APPLY TO A
TRASH COMPACTOR.

Figure 11-3

load procedures. The branch-circuit load for dryers can be found in *220.18*, and that for ranges can be found in *220.19*. Each of these loads is considered separately in this chapter.

Electric Clothes Dryer Loads

The branch-circuit load for one dryer is to be 5000 volt-amperes (watts) or the nameplate rating of the dryer, whichever is larger. The branch-circuit load is based on a calculated load (5000 volt-amperes or watts) if the nameplate rating is less than 5000 volt-amperes and on a connected load if the dryer's ampacity rating is larger than 5000 volt-amperes. At 240 volts the 5000-volt-ampere dryer load is 20.83 amperes (5000 VA / 240 V = 20.83 A), so the minimum branch-circuit rating for the dryer must be at least 25 amperes. Conductors of size 12 AWG, although rated for 25 amperes in the ampacity tables, cannot supply more than 20 amperes to a dryer load [see *240.4(D)*]. Therefore, the use of 10-3 NM, AC, or MC cable with 30-ampere overcurrent protection has become relatively standard for dryer branch circuits. The cable must also include an equipment grounding conductor.

Electric Clothes Dryer Disconnecting Means

Electric clothes dryers can also be cord-and-plug connected. The attachment plug serves as the required disconnecting means. If the dryer is cord-and-attachment plug connected, the plug must be a 4-wire device with a separate terminal for the grounded circuit conductor and the equipment ground conductor. If the dryer is connected using permanent wiring methods, the disconnecting requirements detailed earlier in this chapter for appliances fastened in place and with a rating of greater than 300 volt-amperes apply.

Household Electric Cooking Appliances

Household cooking appliances are of two types: (1) ranges, which include a cooktop and an oven in the same appliance, and (2) those with a separate counter-mounted cooktop and wall-mounted oven. Although the load calculations for these different types of appliances are similar, they need to be considered individually.

Range Loads. The branch-circuit load for an electric range is calculated from *Table 220.19* using the nameplate kilowatt (kW) rating of the range. Residential electric range loads are always calculated loads. The branch-circuit load is determined by applying the nameplate load rating to the values given in *Table 220.19*. Determining the rated load from *Table 220.19* is potentially a complicated process, and Figure 11-4 shows a logic diagram for determining range loads.

However, for range circuits in dwellings, *210.19(A)(3)* states that a 40-ampere circuit must be installed for household electric ranges if the nameplate rating is 8¾ kW or higher. This requirement applies with almost all household electric ranges in common use (most household electric ranges have a nameplate rating of 11.5 kW), including ranges with a nameplate rating of up to 16 kW. If the nameplate rating of the range to be installed falls between 8¾ kW and 16 kW, the 40-ampere circuit satisfies the load. Any range load above 16 kW requires a 50-ampere circuit. Figure 11-4 presents calculation details concerning ranges with ratings higher than 12 kW or lower than 8¾ kW.

Separate Counter-Mounted Cooktops and Wall-Mounted Oven Loads. Note 4 to *Table 220.19* says that if the appliance is served by an individual branch circuit, the rating for one counter-mounted cooking unit or for one wall-mounted oven will be the nameplate rating of the appliance. The branch-circuit conductors are determined by the connected load, as shown in Figure 11-5. However, if that same counter-mounted cooking unit and up to two wall-mounted ovens are supplied from the same branch circuit and are located in the same room, the load can be treated as one range and the branch-circuit load may be determined from *Table 220.19*.

Range, Counter-Mounted Cooktop, and Wall-Mounted Oven Disconnecting Means

Two requirements must be satisfied if the range is to be cord-and-plug connected. Ranges are allowed to be cord-and-plug connected by the *NEC®* if the receptacle is accessible from the front of the range by the removal of a drawer. The plug on this attachment assembly can be used as the disconnecting means for the range. The range receptacle outlet can be flush mounted or surface mounted, and the cord-and-attachment plug must be rated for the ampere load of the range. This ampere rating is the

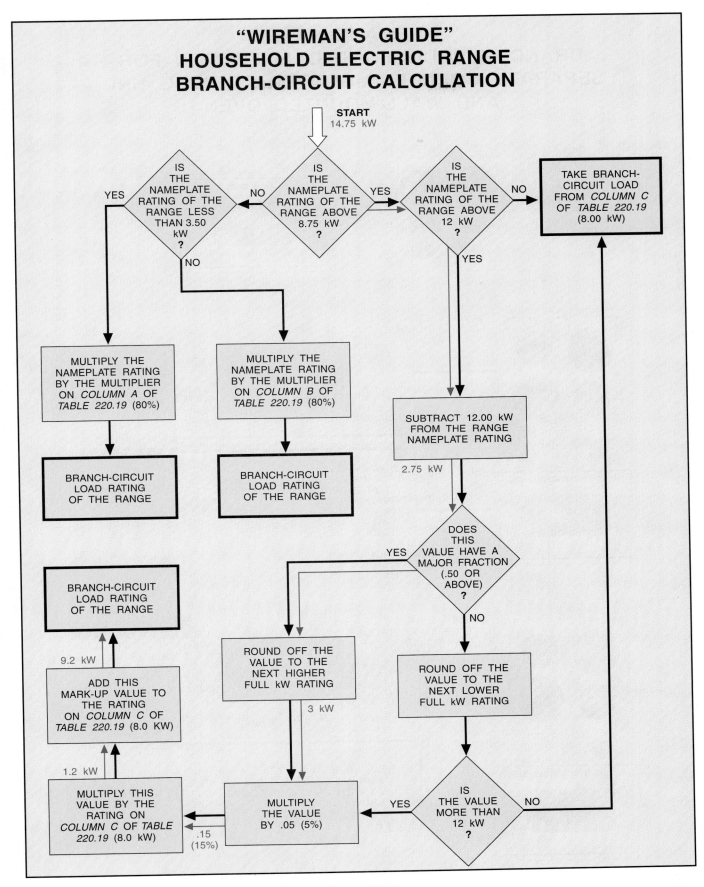

"WIREMAN'S GUIDE"
HOUSEHOLD ELECTRIC RANGE
BRANCH-CIRCUIT CALCULATION

START
14.75 kW

IS THE NAMEPLATE RATING OF THE RANGE LESS THAN 3.50 kW ?

YES

NO

IS THE NAMEPLATE RATING OF THE RANGE ABOVE 8.75 kW ?

YES

IS THE NAMEPLATE RATING OF THE RANGE ABOVE 12 kW ?

NO

YES

TAKE BRANCH-CIRCUIT LOAD FROM *COLUMN C* OF *TABLE 220.19* (8.00 kW)

MULTIPLY THE NAMEPLATE RATING BY THE MULTIPLIER ON *COLUMN A* OF *TABLE 220.19* (80%)

MULTIPLY THE NAMEPLATE RATING BY THE MULTIPLIER ON *COLUMN B* OF *TABLE 220.19* (80%)

SUBTRACT 12.00 kW FROM THE RANGE NAMEPLATE RATING

2.75 kW

BRANCH-CIRCUIT LOAD RATING OF THE RANGE

BRANCH-CIRCUIT LOAD RATING OF THE RANGE

DOES THIS VALUE HAVE A MAJOR FRACTION (.50 OR ABOVE) ?

YES

NO

BRANCH-CIRCUIT LOAD RATING OF THE RANGE

9.2 kW

ADD THIS MARK-UP VALUE TO THE RATING ON *COLUMN C* OF *TABLE 220.19* (8.0 KW)

ROUND OFF THE VALUE TO THE NEXT HIGHER FULL kW RATING

3 kW

ROUND OFF THE VALUE TO THE NEXT LOWER FULL kW RATING

1.2 kW

MULTIPLY THIS VALUE BY THE RATING ON *COLUMN C OF TABLE 220.19* (8.0 kW)

MULTIPLY THE VALUE BY .05 (5%)

.15 (15%)

YES

IS THE VALUE MORE THAN 12 kW ?

NO

Figure 11-4

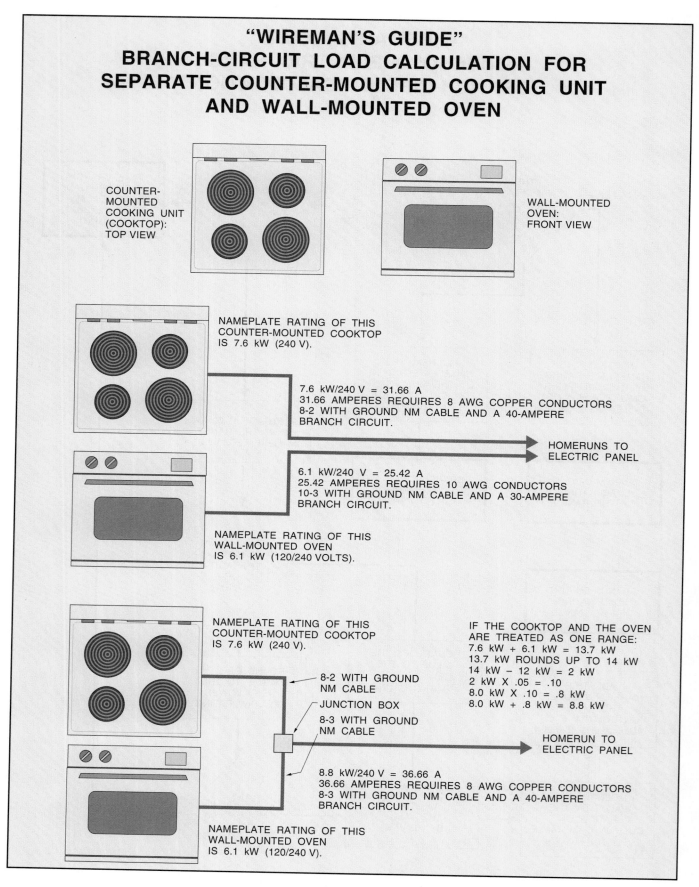

"WIREMAN'S GUIDE"
BRANCH-CIRCUIT LOAD CALCULATION FOR SEPARATE COUNTER-MOUNTED COOKING UNIT AND WALL-MOUNTED OVEN

COUNTER-MOUNTED COOKING UNIT (COOKTOP): TOP VIEW

WALL-MOUNTED OVEN: FRONT VIEW

NAMEPLATE RATING OF THIS COUNTER-MOUNTED COOKTOP IS 7.6 kW (240 V).

7.6 kW/240 V = 31.66 A
31.66 AMPERES REQUIRES 8 AWG COPPER CONDUCTORS
8-2 WITH GROUND NM CABLE AND A 40-AMPERE
BRANCH CIRCUIT.

HOMERUNS TO ELECTRIC PANEL

6.1 kW/240 V = 25.42 A
25.42 AMPERES REQUIRES 10 AWG CONDUCTORS
10-3 WITH GROUND NM CABLE AND A 30-AMPERE
BRANCH CIRCUIT.

NAMEPLATE RATING OF THIS WALL-MOUNTED OVEN IS 6.1 kW (120/240 VOLTS).

NAMEPLATE RATING OF THIS COUNTER-MOUNTED COOKTOP IS 7.6 kW (240 V).

IF THE COOKTOP AND THE OVEN ARE TREATED AS ONE RANGE:
7.6 kW + 6.1 kW = 13.7 kW
13.7 kW ROUNDS UP TO 14 kW
14 kW – 12 kW = 2 kW
2 kW X .05 = .10
8.0 kW X .10 = .8 kW
8.0 kW + .8 kW = 8.8 kW

8-2 WITH GROUND NM CABLE

JUNCTION BOX

8-3 WITH GROUND NM CABLE

HOMERUN TO ELECTRIC PANEL

8.8 kW/240 V = 36.66 A
36.66 AMPERES REQUIRES 8 AWG COPPER CONDUCTORS
8-3 WITH GROUND NM CABLE AND A 40-AMPERE
BRANCH CIRCUIT.

NAMEPLATE RATING OF THIS WALL-MOUNTED OVEN IS 6.1 kW (120/240 V).

Figure 11-5

calculated load for the range rather than for the connected load. There is a range in the sample house.

Separate counter-mounted cooktops and wall-mounted ovens can be cord-and-plug connected if the cord-and-attachment plug are rated for the temperature of the space the appliance occupies and if the cord-and-attachment plug are rated for the ampere load of the appliance. This ampere load is based on the calculated load, not the connected load, when the counter-mounted cooktop and the wall-mounted oven are powered from the same branch circuit.

11.1.3: Motor Circuits

The third type of other loads listed in *220.3(B)* is *motor loads*. For the most part, the equipment and appliances that are installed in dwellings are listed appliances. When the motor or motors are part of a listed appliance, and when the nameplate of the appliance shows both the horsepower (hp) rating and the full-load amperes (FLA) of the appliance, the branch circuit is sized according to the appliance's nameplate FLA, as stipulated by *430.6 Exception 3*. As detailed earlier in this chapter for appliances, this appliance ampere rating is used to determine the conductor ampacity, short-circuit and ground-fault overcurrent-protective devices, and disconnect and control switch ratings for the listed appliance unless these ratings are marked otherwise on the appliance nameplate.

If the motor is not part of a listed appliance, *430* of the Code details the requirements for sizing motor circuits, switches, and overcurrent protection. For general motor applications, the branch-circuit load represented by the motor must be determined from the motor horsepower rating from the motor's nameplate. Also on the nameplate is an ampacity rating, usually marked FLA. The nameplate ampere rating must not be used for determining the branch-circuit load of a motor under general motor applications; instead, the motor nameplate horsepower rating is used as the base ampacity for motor calculations, as shown in Figure 11-6. The horsepower rating is then converted into amperes using *Table 430.148*. Figure 11-7 shows the procedures for sizing circuit conductors, switch rating, and overcurrent protection for general-use motor loads.

There are also different types of motor loads other than general motor applications. Torque motor

branch circuits are rated according to the locked-rotor current of the motor. Alternating-current adjustable voltage motors have branch circuits sized according to the motor nameplate FLA. Multispeed motors have branch circuits sized according to *430.22(B)* and *430.52*. Motors that are used for short-time, intermittent, periodic, or variable duty have branch circuits and are labeled with their duty rating on the nameplate and sized according to *430.22(E)*.

The *NEC*® requires that circuits that supply power to more than one motor, or to motors and other nonmotor loads, be sized to 125% of the largest motor load plus 100% of the other motor and nonmotor loads. This load is called the *largest motor load* and must be calculated in determining feeder or service-entrance conductor sizes for the dwelling and also for branch circuits servicing more than one motor load. If the largest motor load is a listed appliance, with the hp and FLA rating of the motor included on the nameplate, the FLA on the nameplate is used to determine the largest motor load. If the motor is not part of the listed appliance, the FLA shown in *Table 430.148* is used.

11.1.4: Dwelling Occupancies

Load type 10 of *220.3(B)* is concerned with dwelling occupancy loads. It establishes that the receptacle outlets installed on general-use receptacle and lighting circuits are included in the stipulated 3 watts per ft^2 (33 watts per m^2) general lighting load calculations. It also establishes that the required luminaire (fixture) outlets, the required bathroom receptacle circuit, the garage general-use receptacle outlets, the outdoor general-use receptacle outlets, and the basement general-use receptacle outlet are included with the general lighting load and that no other branch-circuit load calculation needs to be undertaken.

11.1.5: Other Outlets

Load type 11 in *220.3(B)* encompasses *other outlets*. The Code makes clear that if an outlet being installed does not supply a load as defined for the specific load types 1 through 10 in *220.3(B)*, 180 volt-amperes is to be used as the minimum outlet load. This grouping is a catch-all for loads not specifically referred to elsewhere in *220.3(B)*.

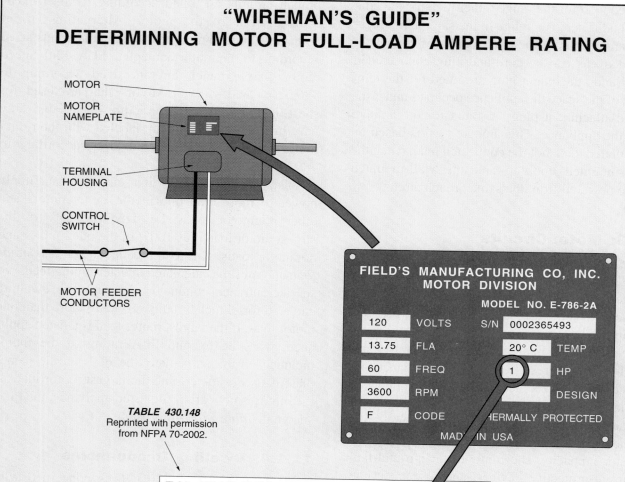

"WIREMAN'S GUIDE"
DETERMINING MOTOR FULL-LOAD AMPERE RATING

MOTOR

MOTOR NAMEPLATE

TERMINAL HOUSING

CONTROL SWITCH

MOTOR FEEDER CONDUCTORS

FIELD'S MANUFACTURING CO, INC.
MOTOR DIVISION

MODEL NO. E-786-2A

120	VOLTS	S/N	0002365493
13.75	FLA	20° C	TEMP
60	FREQ	1	HP
3600	RPM		DESIGN
F	CODE		THERMALLY PROTECTED

MADE IN USA

TABLE 430.148
Reprinted with permission
from NFPA 70-2002.

Table 430.148 Full-Load Currents in Amperes, Single-Phase Alternating-Current Motors

The following values of full-load currents are for motors running at usual speeds and motors with normal torque characteristics. Motors built for especially low speeds or high torques may have higher full-load currents, and multispeed motors will have full-load current varying with speed, in which case the nameplate current ratings shall be used.

The voltages listed are rated motor voltages. The currents listed shall be permitted for system voltage ranges of 110 to 120 and 220 to 240 volts.

Horsepower	115 Volts	200 Volts	208 Volts	230 Volts
⅙	4.4	2.5	2.4	2.2
¼	5.8	3.3	3.2	2.9
⅓	7.2	4.1	4.0	3.6
½	9.8	5.6	5.4	4.9
¾	13.8	7.9	7.6	6.9
1	16	9.2	8.8	8.0
1½	20	11.5	11.0	10
2	24	13.8	13.2	12
3	34	19.6	18.7	17
5	56	32.2	30.8	28
7½	80	46.0	44.0	40
10	100	57.5	55.0	50

THE FULL-LOAD CURRENT RATING TO USE FOR THIS MOTOR FOR SIZING CONDUCTORS, SHORT-CIRCUIT AND GROUND-FAULT PROTECTION, AND SWITCHES IS THE FLA OBTAINED FROM *TABLE 430.148*. THE MOTOR NAMEPLATE RATING FLA IS USED IN SIZING THE OVERLOAD OVERCURRENT-PROTECTIVE PROTECTOR.

Figure 11-6

"WIREMAN'S GUIDE"
SIZING SWITCHES, SHORT-CIRCUIT AND GROUND-FAULT OVERCURRENT PROTECTION AND CONDUCTOR SIZE FOR GENERAL MOTOR APPLICATIONS

Table 430.148 Full-Load Currents in Amperes, Single-Phase Alternating-Current Motors

The following values of full-load currents are for motors running at usual speeds and motors with normal torque characteristics. Motors built for especially low speeds or high torques may have higher full-load currents, and multispeed motors will have full-load current varying with speed, in which case the nameplate current ratings shall be used.

The voltages listed are rated motor voltages. The currents listed shall be permitted for system voltage ranges of 110 to 120 and 220 to 240 volts.

Horsepower	115 Volts	200 Volts	208 Volts	230 Volts
1/6	4.4	2.5	2.4	.2
1/4	5.8	3.3	3.2	2.9
1/3	7.2	4.1	.0	3.6
1/2	9.8	5.6	5.4	4.9
3/4	13.8	.9	7.6	6.9
1	16	9.2	8.8	8.0
1½	20	11.5	11.0	10
2	24	13.8	13.2	12
3	34	19.6	18.7	17
5	56	32.2	30.8	28
7½	80	46.0	44.0	40
10	100	57.5	55.0	50

Reprinted with permission from NFPA 70-2002.

FOR GENERAL MOTOR APPLICATIONS, THE FLA VALUE GIVEN IN *TABLE 430.148* IS USED INSTEAD OF THE ACTUAL FLA SHOWN ON THE MOTOR NAMEPLATE. THE VALUE FROM THE TABLE IS USED TO DETERMINE THE RATING OF SWITCHES, THE RATING OF THE SHORT-CIRCUIT AND GROUND-FAULT OVERCURRENT-PROTECTIVE DEVICES, AND THE SIZE OF THE CIRCUIT CONDUCTORS.

TO DETERMINE THE MINIMUM MOTOR CONDUCTOR SIZE:

MULTIPLY THE MOTOR TABLE FLA BY 125% (1.25) TO DETERMINE THE MINIMUM CIRCUIT AMPACITY. FOR EXAMPLE: A 1-HP MOTOR HAS A TABLE FLA OF 16 AMPERES.

16 AMPERES X 1.25 = 20 AMPERES

20 AMPERES IS THE RATING OF THE SMALLEST CONDUCTOR THAT CAN SUPPLY THE MOTOR.

TO DETERMINE THE RATING OF THE MOTOR SHORT-CIRCUIT AND GROUND-FAULT OVERCURRENT-PROTECTIVE DEVICE:

MULTIPLY THE TABLE FLA BY THE VALUE GIVEN IN *TABLE 430.52* FOR THE TYPE OF OVERCURRENT-PROTECTIVE DEVICE USED AND THE TYPE OF MOTOR BEING PROTECTED. THE MOST COMMON FORM OF SHORT-CIRCUIT AND GROUND-FAULT OVERCURRENT PROTECTION IS THE INVERSE TIME CIRCUIT BREAKER. FOR THE 1-HP MOTOR USED IN THE EXAMPLES ABOVE THE MULTIPLIER WILL BE 250% OF THE FLA AS DETERMINED FROM *TABLE 430.148*.

16 AMPERES X 250% = 40 AMPERES

THE MAXIMUM RATING OF THE CIRCUIT BREAKER WILL BE 40 AMPERES.

TO DETERMINE THE MINIMUM SWITCH SIZE (DISCONNECT SWITCH):

MULTIPLY THE TABLE FLA BY 115% (1.15) OF THE FLA OF THE MOTOR.

16 AMPERES X 115% = 18.4 AMPERES

THE MINIMUM RATING OF THE DISCONNECT SWITCH FOR GENERAL MOTOR APPLICATIONS IS 18.4 AMPERES.

NOTE: MOTORS AND MOTOR CIRCUITS VARY TREMENDOUSLY, AND DIFFERENT REQUIREMENTS APPLY TO DIFFERENT MOTOR TYPES AND APPLICATIONS. THE RULES STATED ABOVE APPLY ONLY TO THE SIMPLEST APPLICATIONS. THE *NEC*® SHOULD BE CONSULTED FOR COMPLETE REQUIREMENTS BEFORE THE INSTALLATION OF ANY MOTOR LOAD.

Figure 11-7

11.2: NONREQUIRED OUTLETS AS USED IN THE SAMPLE HOUSE

The sample house includes several circuits that are not required. These circuits must be installed using the guidelines provided in 11.1 of this chapter. According to the sample houses's electrical specifications, all fastened-in-place appliances in the kitchen must be supplied by an individual branch circuit. Multiwire branch circuits can meet these requirements because each phase conductor does not have to serve more than one load. For the purposes of defining an individual branch circuit, it is not material that the two circuits share the grounded circuit conductor.

11.2.1: Circuits 13 and 15— Subpanel A: Electric Clothes Dryer Circuits

KEY TERMS

Multiwire branch circuit (See *NEC® Article 100*): (Branch circuit—multiwire) More than one branch circuit sharing a grounded circuit conductor. This is usually referred to as a "3-wire homerun" in residential work or a "full boat" in commercial and industrial work. Extreme care must be employed in installing multiwire branch circuits to avoid overloading the neutral conductor.

The electric clothes dryer is installed on circuits 13 and 15 from subpanel A of the sample house's electrical system. The branch circuit is a **multiwire branch circuit** when there are two ungrounded circuit conductors sharing the neutral (grounded circuit conductor). This shared neutral is necessary because some of the components of the dryer are rated for 240 volts, mainly the heating elements that hasten the drying process, and some are rated for 120 volts. The motor that turns the drum inside the dryer and many of the controls for temperature and duration are usually 120 volts rated. The cable will terminate in a two-gang box with a minimum capacity of 17.5 in.^{3} (287 cm^3) if the cable clamps are on the inside of the box sized for the 10-3 AWG cable. If the plastic box is not listed for that large a cable, then a square metal box with a plaster ring can be employed. A square box with cable clamps already installed inside the box

can be used, or a separate cable clamp can be installed on the exterior of the box. If the square box without clamps is used and an external clamp is installed, then the minimum volume capacity of the box needs to be only 15.0 in.^{3} (246 cm^3). The dryer circuit and the box makeup are shown in Figure 11-8.

The dryer is an appliance and as such must have some kind of approved disconnecting means. The dryer for the sample house is to be cord-and-plug connected to a flush-mounted receptacle installed on the wall behind the dryer. The 4-wire cord-and-attachment plug function serves as the required disconnecting means.

The *NEC®* requires that the load for a clothes dryer be considered to be 5000 volt-amperes or the nameplate current, whichever is larger. The nameplate rating of the dryer in the sample house is 120/240 volts, 5500 volt-amperes. The dryer also has 120-volt components; therefore, a 3-wire cable must be installed. Cables with 10 AWG conductors are allowed to supply up to 30 amperes, so cables with 10-3 AWG conductors (plus an equipment grounding conductor) are used.

11.2.2: Circuits 9 and 11— Subpanel A: Household Electric Range Circuits

Many of the procedures described for installation of the dryer are duplicated when the cable for the range, wall-mounted oven, or counter-mounted cooking appliance is installed. For the sample house, the range is supplied from circuits 9 and 11 from subpanel A, as shown in Figure 11-9.

The *NEC®* requires that the load of the range be 8000 volt-amperes if the rating of the range is higher than 8¾ kW or equal to or lower than 12 kW. The nameplate rating of the range in the sample house is 11.5 kW, 120/240 volts; therefore, it is included in the feeder and service load calculation as 8000 volt-amperes (8 kW). The sample house electrical panel schedule (Construction Drawing E-6) shows a 40-ampere, 240-volt circuit. The range also has 120-volt components; therefore, a 3-wire cable must be installed. Cables with 8 AWG conductors are allowed to supply up to 40 amperes, so cables with 8-3 AWG conductors (plus a 10 AWG bare or green equipment grounding conductor) are used.

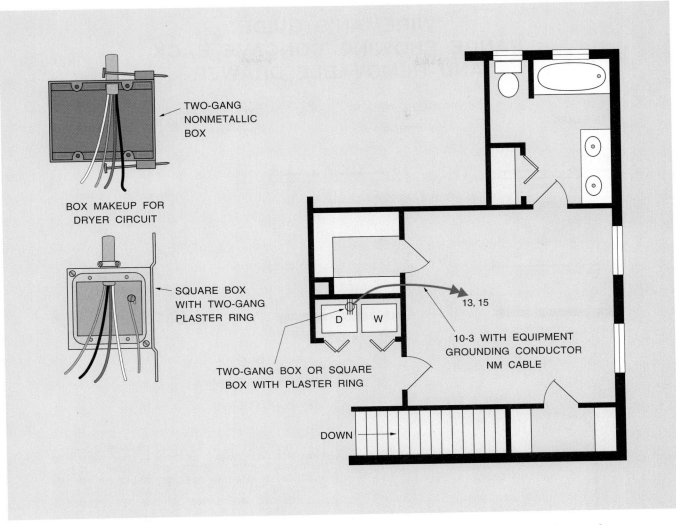

Figure 11-8   Electric clothes dryer branch circuit: circuits 13 and 15 and homerun box makeup.

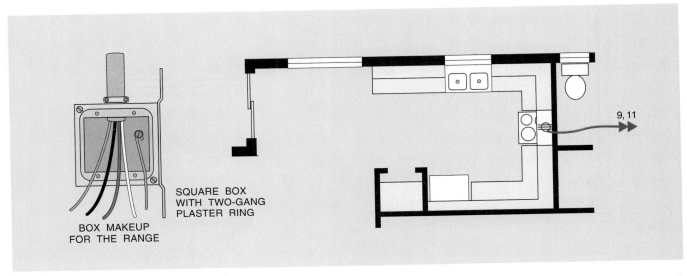

Figure 11-9   Electric range branch circuit: circuits 9 and 11 and homerun box makeup.

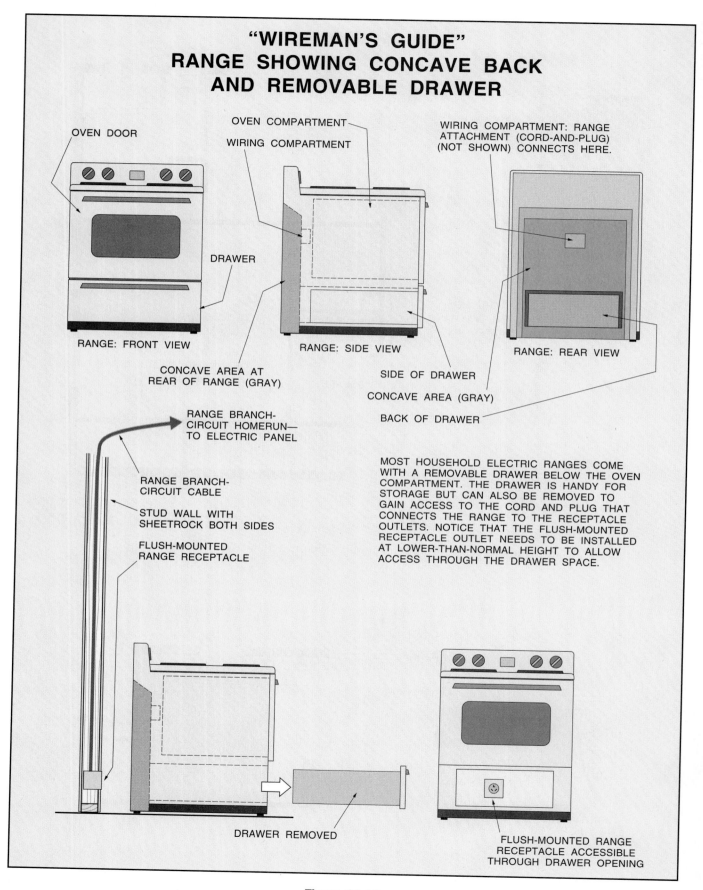

"WIREMAN'S GUIDE"
RANGE SHOWING CONCAVE BACK
AND REMOVABLE DRAWER

OVEN DOOR

OVEN COMPARTMENT

WIRING COMPARTMENT

WIRING COMPARTMENT: RANGE ATTACHMENT (CORD-AND-PLUG) (NOT SHOWN) CONNECTS HERE.

DRAWER

RANGE: FRONT VIEW

RANGE: SIDE VIEW

RANGE: REAR VIEW

CONCAVE AREA AT REAR OF RANGE (GRAY)

SIDE OF DRAWER

CONCAVE AREA (GRAY)

BACK OF DRAWER

RANGE BRANCH-CIRCUIT HOMERUN— TO ELECTRIC PANEL

RANGE BRANCH-CIRCUIT CABLE

STUD WALL WITH SHEETROCK BOTH SIDES

FLUSH-MOUNTED RANGE RECEPTACLE

MOST HOUSEHOLD ELECTRIC RANGES COME WITH A REMOVABLE DRAWER BELOW THE OVEN COMPARTMENT. THE DRAWER IS HANDY FOR STORAGE BUT CAN ALSO BE REMOVED TO GAIN ACCESS TO THE CORD AND PLUG THAT CONNECTS THE RANGE TO THE RECEPTACLE OUTLETS. NOTICE THAT THE FLUSH-MOUNTED RECEPTACLE OUTLET NEEDS TO BE INSTALLED AT LOWER-THAN-NORMAL HEIGHT TO ALLOW ACCESS THROUGH THE DRAWER SPACE.

DRAWER REMOVED

FLUSH-MOUNTED RANGE RECEPTACLE ACCESSIBLE THROUGH DRAWER OPENING

Figure 11-10

For a nonmetallic box with internal cable clamps, the minimum volume capacity of the box is 20.5 in.^{3} (336 cm^3). The box must also be listed for that particular size of cable. These same requirements also hold for a metal box with internal cable clamps. If the cable clamp is external to the box, the minimum capacity is 17.5 in.^{3} (286 cm^3).

The range, like any appliance, needs to be supplied with a disconnecting means. The disconnect for the range of the sample house is the cord-and-attachment plug. The access to the receptacle is obtained through the bottom section of the range by the removal of a drawer. Most ranges have this type of drawer, and ranges are constructed with a concave back to allow room for the receptacle and the cord-and-attachment plug, as shown in Figure 11-10. The receptacle must be located so that the plug is accessible through the drawer opening and therefore must be installed as low as possible on the wall behind the range.

With another routinely used installation technique, a surface-mounting range receptacle attached to the floor is used to supply the range instead of the flush mount in the wall, as shown in Figure 11-11. If this type of installation is planned, it is only necessary at rough-in to stub the range cable out of the wall in the center of the space where the range is to be located. The receptacle is installed at trim after the finished floor has been installed.

11.2.3: Circuit 6—Subpanel A: Microwave Circuit

The microwave oven is installed on circuit 6 from subpanel A, according to the panel schedule. The location of the microwave and the box makeup are shown in Figure 11-12. According to the mechanical schedule, Construction Drawing M-1, the nameplate rating of the microwave oven is 120 volts, 1750 volt-amperes, or 14.6 amperes. Construction Drawings E-3 and A-4 show the microwave installed above the range. The electrical panel schedule in Construction Drawing E-6 shows a 20-ampere circuit breaker protecting the circuit. The microwave oven will also serve as the range hood exhaust fan. The microwave operates on 120 volts, so a 2-wire, 12 AWG copper cable, plus an equipment grounding conductor, is all that is required.

Detailed installation instructions for the microwave may not be available at electrical rough-in. If the microwave is to be installed above the range, as in the sample house, the homerun cable for the microwave should be stubbed out of the wall above the range in a location that allows the cable to enter the back of the upper cabinet supporting the microwave, as shown in Figure 11-13 (page 286). This location allows the most flexibility for access to the wiring compartment of the appliance at trim. Most residential microwave ovens are supplied by the factory with an attachment cord-and-plug for use in a standard 20-ampere, 120-volt receptacle outlet. Microwave ovens that are intended to be mounted above the range usually have the cord leaving the wiring compartment through the top of the appliance so that it can be routed into the cabinet above the microwave. A receptacle outlet is then installed inside the cabinet. Those microwaves that are intended for mounting at other locations may have the cord and wiring compartment in the back or side of the appliance, but detailed installation instructions must be consulted before an accurate determination can be made.

11.2.4: Circuits 8 and 10— Subpanel A: Dishwasher and Garbage Disposal Split-Wire Receptacle Circuits

The dishwasher and the garbage disposal are wired from a multiwire branch circuit. The nameplate rating for the dishwasher is 120 volts, 1620 volt-amperes, or 13.5 amperes, and the nameplate rating for the garbage disposal is 120 volts, 3/4 horsepower, 1440 volt-amperes, or 12.0 amperes. Circuit 8 from subpanel A is used to power the dishwasher, and circuit 10 is used to power the garbage disposal, as shown in the electrical panel schedule for Construction Drawing E-6. The homerun terminates in a duplex receptacle outlet box beneath the kitchen sink. Both the dishwasher and the garbage disposal get their power from the same duplex receptacle, so a 12-3 cable will be installed for the homerun, as shown in Figure 11-14 (page 287). This outlet box is for a split-wired duplex receptacle, the top half providing switched power to the garbage disposal and the bottom half providing unswitched power to the dishwasher. At trim, the bus that connects the two

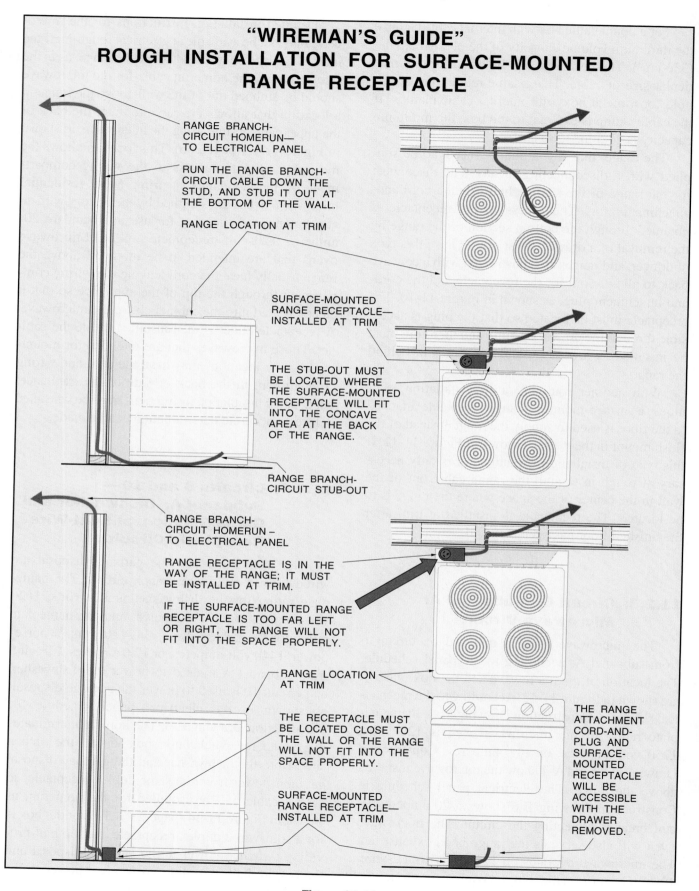

"WIREMAN'S GUIDE"
ROUGH INSTALLATION FOR SURFACE-MOUNTED RANGE RECEPTACLE

RANGE BRANCH-CIRCUIT HOMERUN—TO ELECTRICAL PANEL

RUN THE RANGE BRANCH-CIRCUIT CABLE DOWN THE STUD, AND STUB IT OUT AT THE BOTTOM OF THE WALL.

RANGE LOCATION AT TRIM

SURFACE-MOUNTED RANGE RECEPTACLE—INSTALLED AT TRIM

THE STUB-OUT MUST BE LOCATED WHERE THE SURFACE-MOUNTED RECEPTACLE WILL FIT INTO THE CONCAVE AREA AT THE BACK OF THE RANGE.

RANGE BRANCH-CIRCUIT STUB-OUT

RANGE BRANCH-CIRCUIT HOMERUN – TO ELECTRICAL PANEL

RANGE RECEPTACLE IS IN THE WAY OF THE RANGE; IT MUST BE INSTALLED AT TRIM.

IF THE SURFACE-MOUNTED RANGE RECEPTACLE IS TOO FAR LEFT OR RIGHT, THE RANGE WILL NOT FIT INTO THE SPACE PROPERLY.

RANGE LOCATION AT TRIM

THE RECEPTACLE MUST BE LOCATED CLOSE TO THE WALL OR THE RANGE WILL NOT FIT INTO THE SPACE PROPERLY.

SURFACE-MOUNTED RANGE RECEPTACLE—INSTALLED AT TRIM

THE RANGE ATTACHMENT CORD-AND-PLUG AND SURFACE-MOUNTED RECEPTACLE WILL BE ACCESSIBLE WITH THE DRAWER REMOVED.

Figure 11-11

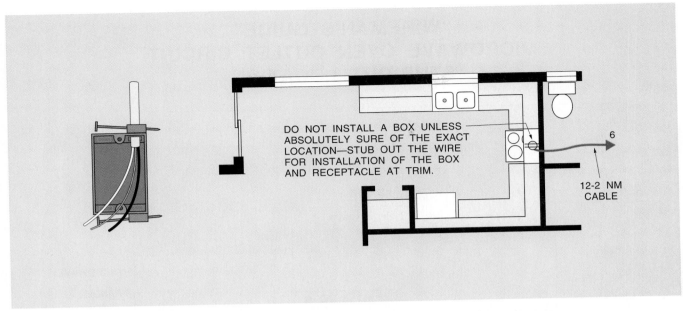

Figure 11-12   Microwave oven circuit: circuit 6 and homerun box makeup.

halves of the phase conductor termination side of the duplex receptacle will be removed. This effectively separates the duplex receptacle into two single receptacles, allowing them to function from their own circuits. The bus on the neutral side of the duplex receptacle will remain intact because both circuits are sharing that conductor. Both appliances are cord-and-plug connected; the plug assembly also serves as the disconnecting means for each appliance.

Other receptacle outlets may appear to be split wired in the sample house, but in fact they are not truly split wired. These are half-switched receptacles rather than split wired, and the difference is important to note. The half-switched outlets in the master bedroom and in the living room appear to be split wired because the top and bottom halves of the receptacles operate independently of each other, but the two halves of these receptacles are on the same circuit. With a split-wired receptacle, two separate circuits are involved, and two circuit breakers must be opened before all of the power to the receptacle is cut off. The Code states that the two circuit breakers that supply any split-wired receptacle in dwellings must be arranged so that both circuits are opened simultaneously. When one of these circuits is opened, the other circuit must also open automatically. Figure 11-15 (page 288) presents additional information about split-wired receptacles.

11.2.5: Circuit 17—Subpanel A: Furnace Branch Circuit

The furnace in the sample house is a natural gas–fueled (fired), forced hot air furnace that is located in the basement area. The furnace utilizes sheet metal ductwork to distribute hot air throughout the dwelling. This ductwork system is also employed by an air conditioner to distribute cold air. The sheet metal ducts are (or should have been) installed before the electrical rough-in begins.

NEC® *422.12* requires that an individual homerun be provided for the furnace or any other central space-heating equipment. The homerun for the furnace, circuit 17 from subpanel A, runs across the basement ceiling to the furnace location, as shown in Figure 11-16 (page 289). The nameplate rating on the furnace according to the mechanical schedule on Construction Drawing M-1 is 120 volts, ¾ horsepower, and 9.2 amperes. Construction Drawing E-6, the electrical panel schedule, shows a 20-ampere circuit. Therefore, a 12-2 AWG with ground NM cable is to be used for the furnace homerun. Aside from control and thermostat circuits, the furnace has no other load except for the blower motor load.

The furnace is usually a stand-alone unit; therefore, the branch-circuit conductors either will drop from the ceiling or, if the furnace is to be located close to a wall, can be installed on or in the wall. The latter is the case with the sample house. There is a concrete

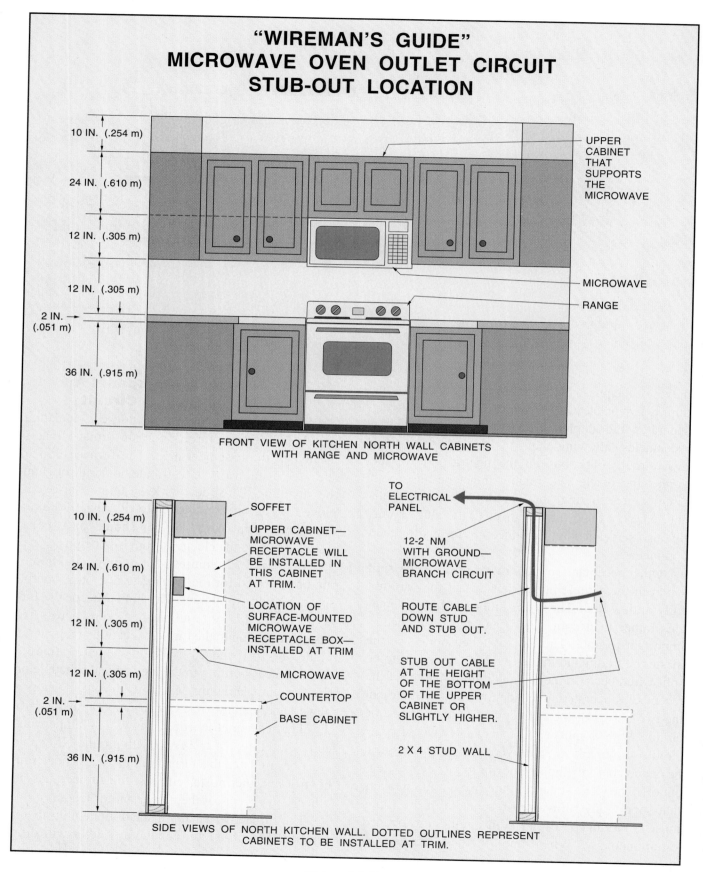

"WIREMAN'S GUIDE"
MICROWAVE OVEN OUTLET CIRCUIT
STUB-OUT LOCATION

10 IN. (.254 m)

24 IN. (.610 m)

12 IN. (.305 m)

12 IN. (.305 m)

2 IN. (.051 m)

36 IN. (.915 m)

UPPER CABINET THAT SUPPORTS THE MICROWAVE

MICROWAVE

RANGE

FRONT VIEW OF KITCHEN NORTH WALL CABINETS WITH RANGE AND MICROWAVE

10 IN. (.254 m)

24 IN. (.610 m)

12 IN. (.305 m)

12 IN. (.305 m)

2 IN. (.051 m)

36 IN. (.915 m)

SOFFET

UPPER CABINET— MICROWAVE RECEPTACLE WILL BE INSTALLED IN THIS CABINET AT TRIM.

LOCATION OF SURFACE-MOUNTED MICROWAVE RECEPTACLE BOX— INSTALLED AT TRIM

MICROWAVE

COUNTERTOP

BASE CABINET

TO ELECTRICAL PANEL

12-2 NM WITH GROUND— MICROWAVE BRANCH CIRCUIT

ROUTE CABLE DOWN STUD AND STUB OUT.

STUB OUT CABLE AT THE HEIGHT OF THE BOTTOM OF THE UPPER CABINET OR SLIGHTLY HIGHER.

2 X 4 STUD WALL

SIDE VIEWS OF NORTH KITCHEN WALL. DOTTED OUTLINES REPRESENT CABINETS TO BE INSTALLED AT TRIM.

Figure 11-13

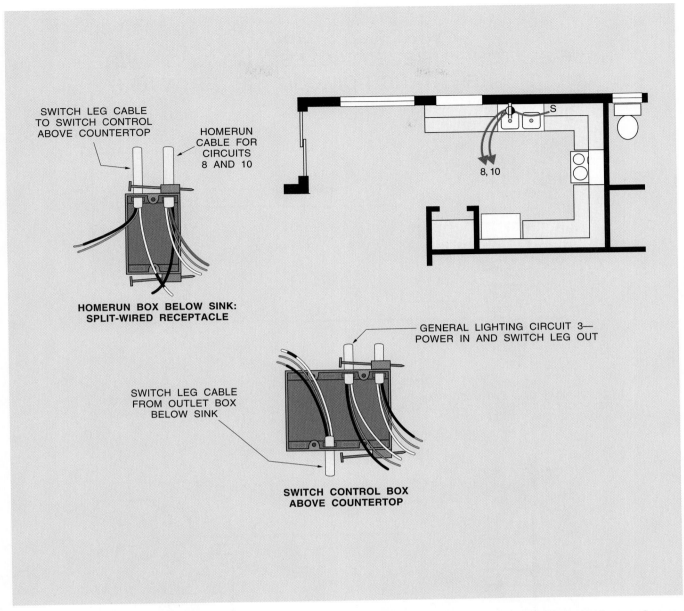

Figure 11-14   Kitchen dishwasher and garbage disposal circuits: circuits 8 and 10 and homerun box makeup.

wall just to the south of the furnace, and the branch-circuit conductors for the furnace will be run down the surface of the cement wall into a square box used as the furnace outlet box. The conductors must be protected against damage where they run down the wall. This is accomplished by using a length of electrical metallic tubing (EMT), as shown in Figure 11-17 (page 290). The conduit is used simply to provide protection for the cable, and no conduit fill calculations need to be performed. However, *300.15(C)* requires that whenever conductors or cables enter or exit from a conduit, a substantial bushing be installed to protect the conductors or cable from abrasion.

The furnace must also be provided with a disconnect. Some furnace installation instructions require a fuse to protect the furnace against overcurrent, in addition to the protection provided by the branch-circuit overcurrent-protective device (the circuit breaker) and the motor overload protection built into the motor. A switch or a combination switch and fuse device is installed at trim in the square box to satisfy the disconnect requirements.

"WIREMAN'S GUIDE"
HALF-SWITCHED RECEPTACLES AND SPLIT-WIRED MULTIWIRE BRANCH-CIRCUIT RECEPTACLES

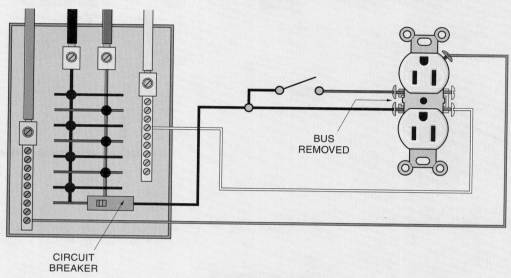

CIRCUIT BREAKER

BUS REMOVED

HALF-SWITCHED RECEPTACLE

THE TYPICAL HALF-SWITCHED RECEPTACLES, LIKE THE ONES IN THE SAMPLE HOUSE MASTER BEDROOM AND THE LIVING ROOM, USE THE SAME CIRCUIT FOR BOTH THE UPPER AND LOWER HALVES OF THE RECEPTACLE.

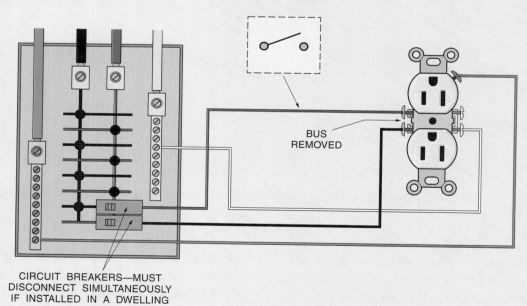

BUS REMOVED

CIRCUIT BREAKERS—MUST DISCONNECT SIMULTANEOUSLY IF INSTALLED IN A DWELLING

SPLIT-WIRED MULTIWIRE BRANCH-CIRCUIT RECEPTACLE

THE TYPICAL SPLIT-WIRED MULTIWIRE BRANCH-CIRCUIT RECEPTACLE IS SUPPLIED FROM TWO SEPARATE CIRCUITS, ONE FOR THE UPPER AND ONE FOR THE LOWER HALVES OF THE RECEPTACLE. THEY SHARE THE GROUNDED CIRCUIT CONDUCTOR. EVEN IF A SWITCH IS PLACED ON EITHER OR BOTH CIRCUITS, THE RECEPTACLE IS STILL CLASSIFIED AS A SPLIT-WIRED RECEPTACLE INSTEAD OF A HALF-SWITCHED RECEPTACLE. FOR DWELLINGS, ACCORDING TO *210.4(B)*, IF THE TWO CIRCUITS SUPPLY LOADS ON THE SAME YOKE OR STRAP, AS IN THIS EXAMPLE, BOTH CIRCUITS MUST BE DISCONNECTED SIMULTANEOUSLY AT THE PANEL WHERE THE CIRCUIT ORIGINATED.

Figure 11-15

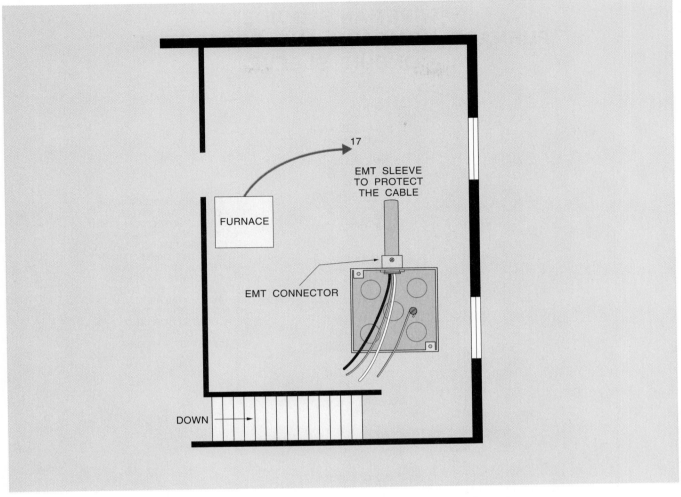

Figure 11-16 Furnace circuit: circuit 17 and homerun box makeup.

11.2.6: Circuit 21—Subpanel A: Garage Door Opener Circuit

The garage door opener circuit is terminated in a device box in the ceiling of the garage, as shown in Figure 11-18 from subpanel A. The nameplate rating of the garage door opener from Construction Drawing M-1 is 120 volts, ¾ horsepower, 1476 volt-amperes, and 12.3 amperes. The garage door opener is installed on a separate circuit because the opener load can exceed 50% of the circuit ampacity; this consideration eliminates the possibility of installing additional lighting or receptacle outlets on that circuit. A receptacle outlet is to be installed into the device box at trim, and the garage door opener will be connected to the receptacle with a cord-and-attachment plug. This connection also provides for the required disconnecting means for the opener.

No branch-circuit load is provided for the opener in the plans, but the electrical panel schedule in Construction Drawing E-6 shows the circuit to be a 15-ampere circuit.

11.2.7: Circuit 24—Subpanel A: Sump Pump Circuit

A sump pump is installed in the basement of the sample house to control excess groundwater. The sump pump is a listed assembly and is supplied by a 20-ampere individual circuit using a 12-2 copper NM cable, as shown in Figure 11-19 (page 292). As with the furnace, the cable requires protection from physical damage where it runs down the foundation wall. The same technique shown in Figure 11-17 for the furnace can be employed with the sump pump

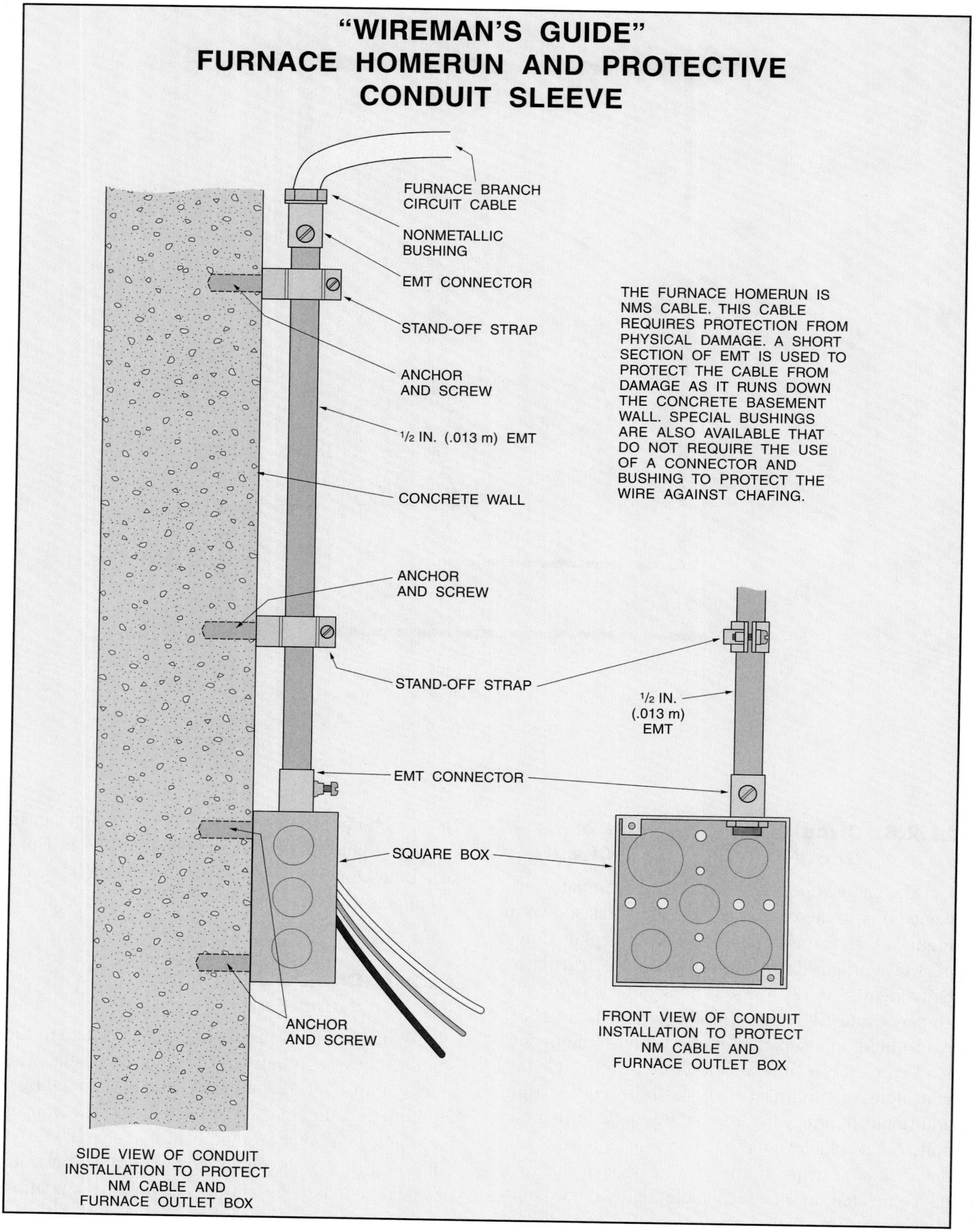

"WIREMAN'S GUIDE"
FURNACE HOMERUN AND PROTECTIVE CONDUIT SLEEVE

FURNACE BRANCH CIRCUIT CABLE

NONMETALLIC BUSHING

EMT CONNECTOR

STAND-OFF STRAP

ANCHOR AND SCREW

½ IN. (.013 m) EMT

CONCRETE WALL

ANCHOR AND SCREW

STAND-OFF STRAP

EMT CONNECTOR

SQUARE BOX

ANCHOR AND SCREW

THE FURNACE HOMERUN IS NMS CABLE. THIS CABLE REQUIRES PROTECTION FROM PHYSICAL DAMAGE. A SHORT SECTION OF EMT IS USED TO PROTECT THE CABLE FROM DAMAGE AS IT RUNS DOWN THE CONCRETE BASEMENT WALL. SPECIAL BUSHINGS ARE ALSO AVAILABLE THAT DO NOT REQUIRE THE USE OF A CONNECTOR AND BUSHING TO PROTECT THE WIRE AGAINST CHAFING.

½ IN. (.013 m) EMT

FRONT VIEW OF CONDUIT INSTALLATION TO PROTECT NM CABLE AND FURNACE OUTLET BOX

SIDE VIEW OF CONDUIT INSTALLATION TO PROTECT NM CABLE AND FURNACE OUTLET BOX

Figure 11-17

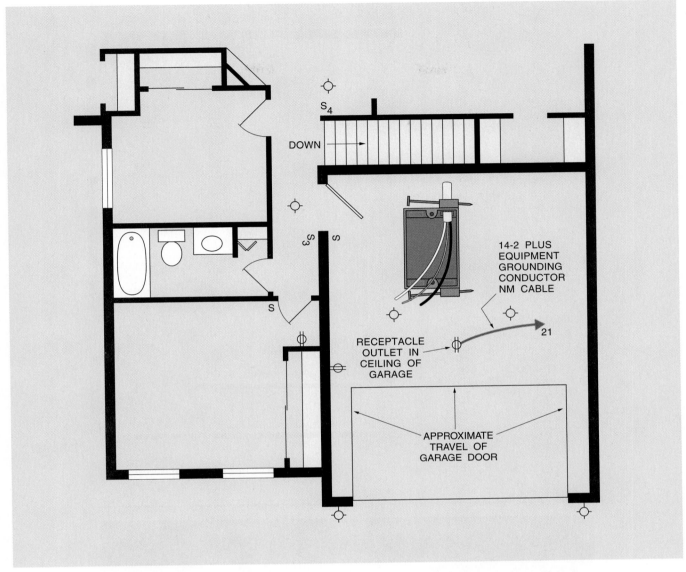

Figure 11-18 Garage door opener circuit: circuit 21 and homerun box makeup.

cable. Figure 11-20 shows a typical sump pump installation below a basement floor.

The sample house mechanical schedule and the pump nameplate show the sump pump to be a 120-volt, ¾ horsepower load. According to ampacity *Table 430.148*, a rating of ¾ horsepower is equivalent to 13.8 amperes or 1476 volt-amperes. The electrical panel schedule for the sample house, (Construction Drawing E-6) shows that a 20-ampere, 120-volt circuit is required. A 12-2 NM cable will be run from the panel to the sump pump location. The sump pump will connect at trim to this circuit either by means of a receptacle and cord-and-attachment plug or by an approved permanent wiring method. If the motor does not have internal thermal protection against overload overcurrents, it must be provided at trim. If the sump pump is not connected with a cord-and-attachment plug, another type of approved disconnect must be provided.

11.2.8: Circuits 20 and 22— Subpanel A: Electric Baseboard Heater

There is an electric baseboard heating unit in the garage along the wall backing up to the stairway, as shown in Figure 11-21. This is rated 2000 watts at 240 volts on the unit nameplate and in Construction

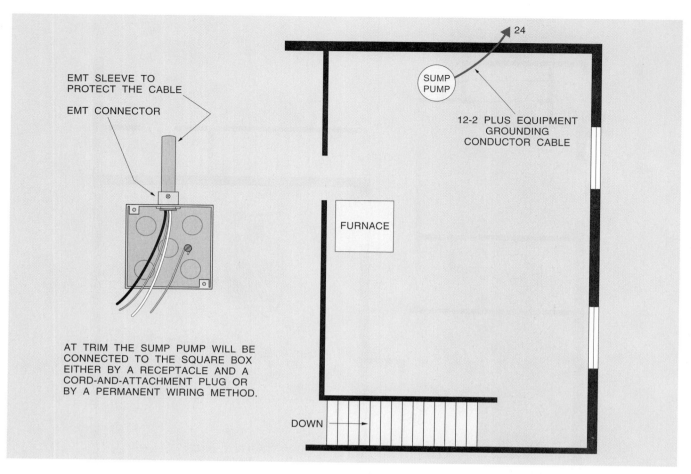

EMT SLEEVE TO PROTECT THE CABLE

EMT CONNECTOR

24

SUMP PUMP

12-2 PLUS EQUIPMENT GROUNDING CONDUCTOR CABLE

FURNACE

AT TRIM THE SUMP PUMP WILL BE CONNECTED TO THE SQUARE BOX EITHER BY A RECEPTACLE AND A CORD-AND-ATTACHMENT PLUG OR BY A PERMANENT WIRING METHOD.

DOWN

Figure 11-19   Sump pump circuit: circuit 24 and homerun box makeup.

Figure 11-20   A typical sump pump pit with electric power supply and controls.

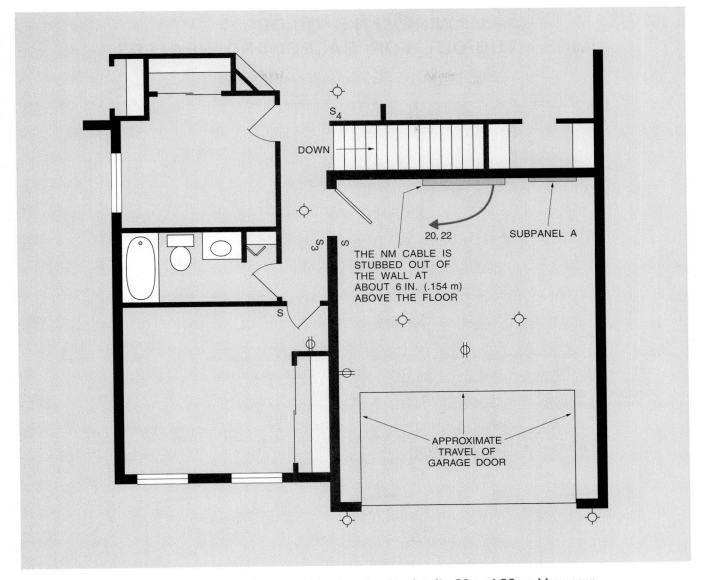

Figure 11-21   Electric baseboard heater circuit: circuits 20 and 22 and homerun.

Drawing M-1. According to the electrical panel schedule, the baseboard heater is installed on a 14-2 AWG circuit with a 2-pole, 15-ampere circuit breaker. As stipulated in *424.3(B)*, the branch-circuit conductors must be sized for at least 125% of the heater current rating. A 2000 volt-ampere heater at 240 volts draws 8.3 amperes, so the branch-circuit rating must be at least 10.4 amperes (8.3 A × 125%). The heater is controlled by a thermostat that is built into one of the end plates of the heater. The off position on the in-line thermostat, in combination with the branch-circuit overcurrent-protective device in subpanel A, will satisfy the disconnecting requirements.

The cable is to be stubbed out of the wall at the location of one of the ends of the intended heater location. The heater is supplied with junction boxes at either end of the unit for connection to the power supply. At trim, the cable will be installed into one of the junction boxes and the heater screwed to the wall surface or framing. Figure 11-22 presents additional information about electric heating.

11.2.9: Circuits 1 and 3— Main Panel: Air Conditioner Circuit

The compressor unit for the air conditioner is located on the side of the house about 15 ft (4.6 m) to the east of the main electrical distribution panel and

"WIREMAN'S GUIDE"
WIRE STUB-OUT FOR BASEBOARD HEATERS

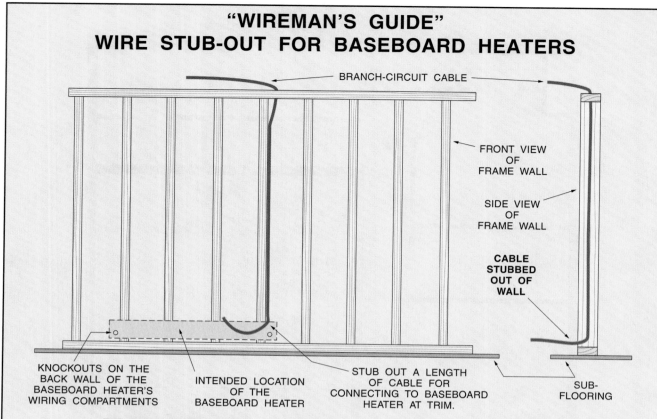

BRANCH-CIRCUIT CABLE

FRONT VIEW OF FRAME WALL

SIDE VIEW OF FRAME WALL

CABLE STUBBED OUT OF WALL

KNOCKOUTS ON THE BACK WALL OF THE BASEBOARD HEATER'S WIRING COMPARTMENTS

INTENDED LOCATION OF THE BASEBOARD HEATER

STUB OUT A LENGTH OF CABLE FOR CONNECTING TO BASEBOARD HEATER AT TRIM.

SUB-FLOORING

THE BRANCH-CIRCUIT CABLE IS STUBBED OUT OF THE FRAME WALL AT THE POSITION FOR ONE OF THE ENDS OF THE BASEBOARD HEATER. THE BASEBOARD HEATERS HAVE A WIRING COMPARTMENT AT EACH END TO CONNECT TO THE BRANCH-CIRCUIT CONDUCTORS AND TO PROVIDE A BOX FOR THE IN-LINE THERMOSTAT IF ONE IS INSTALLED.

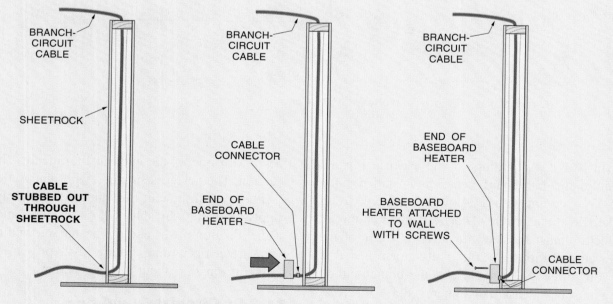

BRANCH-CIRCUIT CABLE

SHEETROCK

CABLE STUBBED OUT THROUGH SHEETROCK

BRANCH-CIRCUIT CABLE

CABLE CONNECTOR

END OF BASEBOARD HEATER

BRANCH-CIRCUIT CABLE

END OF BASEBOARD HEATER

BASEBOARD HEATER ATTACHED TO WALL WITH SCREWS

CABLE CONNECTOR

SIDE VIEW OF FRAME WALL WITH SHEETROCK INSTALLED ON BOTH SIDES. THE BRANCH CIRCUIT IS STUBBED OUT THROUGH THE SHEETROCK OR WALL BOARD BY THE WALL FINISH CONTRACTOR. AT TRIM, A CABLE CONNECTOR IS PLACED ON THE CABLE AND THE BASEBOARD HEATER IS INSTALLED TO THE WALL WITH THE CABLE COMING THROUGH THE KNOCKOUT IN THE REAR WALL OF THE HEATER. THE CABLE CLAMP IS THEN SECURED, AND THE BASEBOARD HEATER IS LEVELED AND SCREWED TO THE WALLBOARD.

Figure 11-22

electrical service, as shown in Figure 11-23. The air conditioner compressor is on a 30-ampere, 240-volt supply circuit, as shown in Construction Drawing E-6. Air conditioners are installed according to *440*. The air conditioner nameplate is marked with the maximum allowable short-circuit and ground-fault overcurrent-protective device rating and the minimum ampacity of the supply conductors. According to the air conditioner nameplate, the minimum circuit ampacity is 28.8 amperes, with an overcurrent-

protective device provided in the form of a 35-ampere fuse. The air conditioner does not need a neutral (grounded circuit conductor) because all components operate at 240 volts. Therefore, a 10-2 NM cable with equipment grounding conductor is installed from the panel through the master bathroom and bedroom wall to the air conditioner compressor's location. The cable is to be stubbed out of the wall and connected to a junction box or a disconnect switch enclosure at trim. See Figure 11-24.

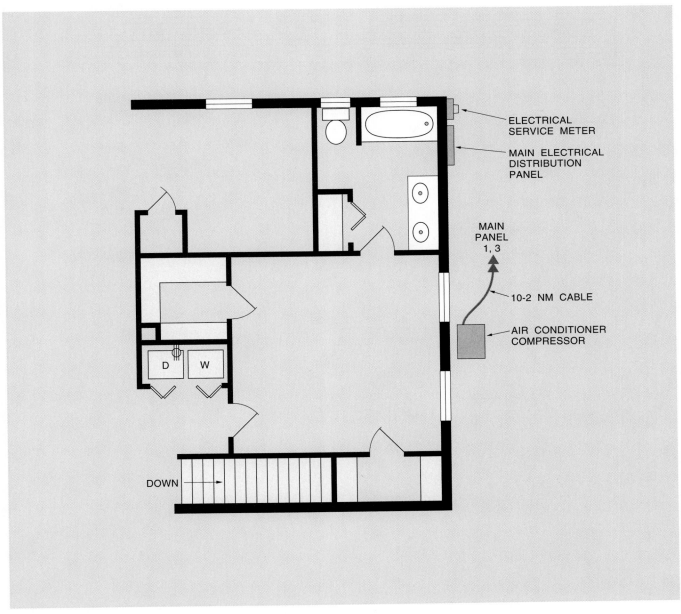

Figure 11-23   Air conditioner circuit: main electrical panel circuits 1 and 3.

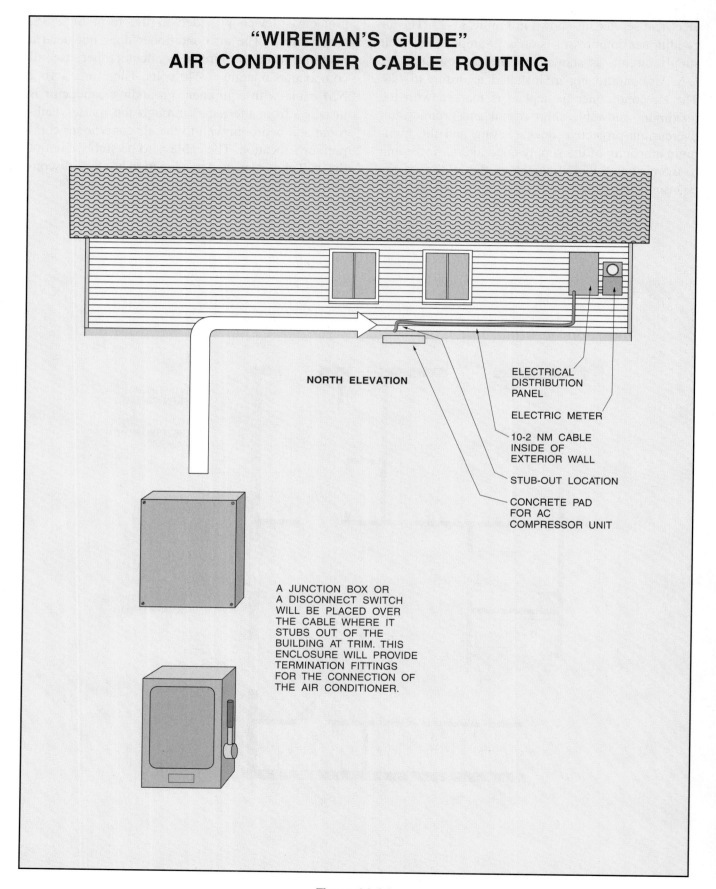

"WIREMAN'S GUIDE"
AIR CONDITIONER CABLE ROUTING

NORTH ELEVATION

ELECTRICAL
DISTRIBUTION
PANEL

ELECTRIC METER

10-2 NM CABLE
INSIDE OF
EXTERIOR WALL

STUB-OUT LOCATION

CONCRETE PAD
FOR AC
COMPRESSOR UNIT

A JUNCTION BOX OR
A DISCONNECT SWITCH
WILL BE PLACED OVER
THE CABLE WHERE IT
STUBS OUT OF THE
BUILDING AT TRIM. THIS
ENCLOSURE WILL PROVIDE
TERMINATION FITTINGS
FOR THE CONNECTION OF
THE AIR CONDITIONER.

Figure 11-24

SUMMARY

The branch-circuit loads for nonrequired appliances are, with the exception of household electric ranges, counter-mounted cooktops, wall-mounted ovens, and electric clothes dryers, determined from the nameplate rating of the appliance. A demand factor is applied to electric ranges, and to counter-mounted cooktops and wall-mounted ovens if they are supplied from the same branch circuit. The load for a dryer is 5000 watts (volt-amperes) or the nameplate rating, whichever is larger. Many household appliances that are fastened in place can be cord-and-attachment plug connected, including a range and a dryer. An attachment plug-and-cord used for connecting the appliance also satisfies the requirements for the disconnecting means. If the appliance is permanently connected, the disconnecting means can be a combination of the following: the branch-circuit overcurrent-protective device if it is within sight from the appliance or if it can be locked in an open position; an on-off switch on the appliance; a control switch for the appliance that is within sight from the appliance; and possibly the building's main disconnecting means. The Code details the differences between a motor load and a motor-driven appliance load. The load for a motor is determined by the horsepower rating of the motor. The appliance load rating is taken from the appliance ampere rating on the nameplate.

The sample house has a dishwasher and garbage disposal that are cord-and-attachment plug connected on a split-wired duplex receptacle. The microwave oven is installed above the range; the furnace and the sump pump are located in the basement, and installation of each requires some conduit work to protect the branch-circuit cables from physical damage. A garage door opener outlet is installed in the garage ceiling, and an electric baseboard heater is installed on the garage wall close to subpanel A.

REVIEW

1. For ranges rated over 8¾ kW, the minimum size branch circuit allowed by the Code is
   _____ .

   a. 30 amperes

   b. 40 amperes

   c. 50 amperes

   d. the nameplate rating

2. Dryer loads are determined by a calculated load unless the nameplate rating is greater than _____ when the load becomes the nameplate rating.

   a. 20 amperes

   b. 5000 watts (volt-amperes)

   c. 30 amperes

   d. 8000 watts (volt-amperes)

3. Household appliance loads are usually determined from the appliance _____ .

   a. nameplate

   b. fix load table from the *NEC*®

   c. voltage rating

   d. installation instructions

4. A dwelling-unit appliance does not require a separate disconnecting means if it is _____.
   a. installed in the kitchen
   b. a listed and labeled appliance
   c. under 300 volt-ampere or ⅛ horsepower rated
   d. cord-and-attachment plug connected

5. The dryer in the sample house is powered by a/an _____ cable.
   a. 12-3 NM
   b. 10-2 NM
   c. 10-3 NM
   d. 8-3 NM

6. *True or False:* The range requires a 4-wire circuit.

7. *True or False:* The overcurrent-protective devices for both of the circuits for the split-wired receptacle that serves the disposal and dishwasher branch circuits must be opened at the same time.

8. *True or False:* The two circuits that supply the split-wired receptacle for the dishwasher and the disposal must be on the same phase (or leg).

9. *True or False:* The range can be cord-and-attachment plug connected only if the receptacle is flush mounted in the wall.

10. *True or False:* The sump pump in the sample house is considered to be a motor load rather than an appliance load.

11. *True or False:* The *NEC®* requires the microwave oven circuit to have a minimum rating of 20 amperes.

12. *True or False:* The furnace can be installed on the same branch circuit as the sump pump.

13. *True or False:* The *NEC®* requires that the furnace circuit have a minimum rating of 20 amperes.

14. *True or False:* Electric clothes dryers can be supplied with a 2-wire NM cable because they have no components that operate on 120 volts.

15. *True or False:* The baseboard heater in the garage does not need to have a separate disconnecting means because the branch-circuit overcurrent-protective device is within sight from the heater.

"WIREMAN'S GUIDE" REVIEW

1. Explain how a range can be cord-and-attachment plug connected to meet Code requirements. What if the receptacle is surface mounted?

2. What is the branch-circuit load for a range with a nameplate rating of 16.5 kW? What is the minimum branch-circuit size if NM cable is used to supply the range?

3. Explain why the garage door opener is installed on an individual branch circuit.

4. Explain why the split-wired duplex receptacle for the disposal and dishwasher circuits must be a 20-ampere-rated receptacle. Must the supply circuits also be rated at 20 amperes?

5. Explain how the minimum feeder rating for conductors supplying an air conditioner are determined. What section of the *NEC*® details that requirement?

CHAPTER 12

Feeder and Service Conductor Loads

OBJECTIVES

On completion of this chapter, the student will be able to:

☑ Explain how feeders and service conductors are sized, and state how they differ from branch circuits.

☑ Identify the basis for the de-rating (demand load) allowed for a dwelling's general lighting and general-use receptacle loads.

☑ Explain the demand factors that can sometimes be applied to appliances in dwelling units.

☑ Define noncoincident loads and describe how they apply to feeder and service conductor load calculations.

☑ Calculate a feeder or a service-entrance conductor load for a dwelling.

☑ Explain the differences to be considered in sizing feeder and service phase (power) conductors and grounded circuit or service (neutral) conductors.

INTRODUCTION

This chapter is concerned with feeder and service loads. Branch circuits are not the only classification of conductors that provide power to the loads. Feeders and service conductors provide power to the branch circuits. Because all branch circuits are supplied by feeders or service conductors, they must also be limited in their current rating. Feeders and service conductors follow the same calculation procedures prescribed by the *NEC®*, so they can be considered together. However, it is important to remember that service conductors have overcurrent protection on the load side only, and that feeders have overcurrent protection on both the line side and the load side. The sample house has a feeder circuit from the main electrical panel to subpanel A in the garage.

12.1: FEEDER AND SERVICE LOAD PHASE CONDUCTOR CALCULATION

KEY TERMS

Demand factor (See *NEC® Article 100*): Any of the multipliers used in calculating demand loads.

NEC® 220.10 says that the load on the feeders or service conductors is equal to the sum of the branch-circuit loads (the loads addressed in Chapters 10 and 11 of this book). However, some modifications to the total branch-circuit loads are allowed. The modifiers take into account certain realities that will lower the total phase conductor load. The reductions in the calculated load are called **demand factors**.

12.1.1: General Lighting and General-Use Receptacle Outlet Feeder and Service Demand Factors

Demand factors for general lighting and general use receptacle loads apply because it is very unlikely that all of the appliances and luminaires (fixtures) installed in the dwelling will be used at the same time. The demand factors allowed by the Code for general lighting are detailed in *Table 220.11*. The first 3000 volt-amperes of general lighting and general-use receptacle loads are included in the feeder and service load calculations at 100%. The next 117,000 volt-amperes are included at 35% of their normal load, and any load general lighting and general-use receptacle loads over 120,000 volt-amperes is included in the calculation at 25% of normal load. It is very unlikely that a single-family dwelling service load will approach 120,000 volt-amperes. This reduced general lighting load cannot be used in determining the minimum number of general lighting and general-use receptacle outlet circuits. Small-appliance and laundry loads are also included in the demand factors that are allowed by *220.11*.

12.1.2: Small-Appliance Circuits and Laundry Feeder and Service Conductor Loads

No specific branch-circuit load is established for small-appliance circuits or for the laundry circuit. The branch-circuit loads that these circuits represent are included in the 3 volt-amperes per ft^2 (33 volt-amperes per m^2) used to calculate the general lighting and general-use receptacle branch-circuit loads. These branch circuits are required to be 20-ampere circuits, and *210.11(C)* requires a minimum number of these circuits in each dwelling: two small-appliance circuits and one laundry circuit at a minimum. According to *220.16*, 1500 volt-amperes must be included in the feeder and service load calculation for each small-appliance circuit and for each laundry circuit installed in the residence. A minimum of two small-appliance circuits is required, and a minimum of one laundry circuit is required, giving a total of 4500 volt-amperes (1500 VA for each circuit × 3 circuits = 4500 VA) minimum. If additional circuits are actually installed, they must also be included in the feeder and service load calculation

at 1500 volt-amperes each. One exception applies to small-appliance circuits. *NEC® 210.52(B)(1), Exception No. 2*, allows for the refrigerator to be placed on an individual branch circuit rated at 15 amperes or greater rather than included on a regular small-appliance circuit. When this option is exercised, the load from this individual circuit does not need to be included in the feeder load calculations for the dwelling, in accordance with *220.16, Exception*. In the sample house, the refrigerator is on an individual branch circuit.

Figure 12-1 shows the de-rating of the general lighting and general-use receptacle circuit loads, including the small-appliance and laundry loads, for the sample dwelling. The sample dwelling has two small-appliance circuits that serve the countertops and the dining room.

12.1.3: Motor Feeder and Service Conductor Loads

The FLA rating for motors is determined in Chapter 11, and each motor load is included in the feeder and service load calculation at 100% of the FLA or the current rating if it is a motor load. Listed appliances are included in the load calculations at 100% of the appliance ampere rating. Appliances are subject to a feeder and service demand factor, however, and motor loads are not.

12.1.4: Fixed Electric Space-Heating Feeder and Service Conductor Loads

The loads for fixed electric space-heating loads are included in the feeder and service conductor load calculation at 100% of the total connected load. The sample house has a baseboard heater in the garage.

12.1.5: Appliance Feeder and Service Conductor Load Demand Factor

The fastened-in-place appliance load for the feeder and service load calculation is taken at 100% of the connected appliance branch-circuit load for a total of three appliances. Starting with the fourth appliance, the total appliance load can be reduced to only 75% of the connected load. This demand factor is included because it is unlikely that all of the appliances will be functioning at the same time. The

"WIREMAN'S GUIDE"
CALCULATING FEEDER/SERVICE CONDUCTOR LOAD FOR GENERAL LIGHTING LOAD IN SAMPLE HOUSE

CALCULATION FOR DETERMINING THE GENERAL LIGHTING DEMAND LOAD PER *220.11*

TOTAL AREA OF THE DWELLING SUBJECT TO GENERAL LIGHTING LOADS:	1304.5 FT²
LOAD PER SQUARE FOOT ACCORDING TO *220.3*	3 VOLT-AMPERES
TOTAL GENERAL LIGHTING AND GENERAL-USE RECEPTACLE LOAD:	3913.5 VOLT-AMPERES
TOTAL SMALL APPLIANCE LOAD (TWO SMALL-APPLIANCE CIRCUITS AT 1500 VOLT-AMPERES EACH)	3000.0 VOLT-AMPERES
TOTAL LAUNDRY LOAD (ONE LAUNDRY CIRCUIT AT 1500 VOLT-AMPERES)	1500.0 VOLT-AMPERES
TOTAL LOAD SUBJECT TO THE DEMAND FACTORS ALLOWED IN *220.11*	8413.5 VOLT-AMPERES

APPLICATION OF THE DEMAND FACTOR ALLOWED IN *220.11*:

FIRST 3000 VOLT-AMPERES AT 100%	3000 VOLT-AMPERES
FROM 3001 VOLT-AMPERES TO 120,000 VOLT-AMPERES AT 35% (8413.5 − 3000 = 5413.5 × .35)	1895 VOLT-AMPERES
FROM 120,000 VOLT-AMPERES AND UP	0 VOLT-AMPERE
FEEDER/SERVICE CONDUCTOR GENERAL LIGHTING LOAD:	4895 VOLT-AMPERES

CALCULATION FOR DETERMINING THE GENERAL LIGHTING DEMAND LOAD PER *220.11*

TOTAL AREA OF THE DWELLING SUBJECT TO GENERAL LIGHTING LOADS:	121.2 m²
LOAD PER SQUARE METER ACCORDING TO *220.3*	33 VOLT-AMPERES
TOTAL GENERAL LIGHTING AND GENERAL-USE RECEPTACLE LOAD:	3999.6 VOLT-AMPERES
TOTAL SMALL-APPLIANCE LOAD (TWO SMAL-APPLIANCE CIRCUITS AT 1500 VOLT-AMPERES EACH)	3000.0 VOLT-AMPERES
TOTAL LAUNDRY LOAD (ONE LAUNDRY CIRCUIT AT 1500 VOLT-AMPERES)	1500.0 VOLT-AMPERES
TOTAL LOAD SUBJECT TO THE DEMAND FACTORS ALLOWED IN *220.11*	8499.6 VOLT-AMPERES

APPLICATION OF THE DEMAND FACTOR ALLOWED IN *220.11*:

FIRST 3000 VOLT-AMPERES AT 100%	3000 VOLT-AMPERES
FROM 3001 VOLT-AMPERES TO 120,000 VOLT-AMPERES AT 35% (8499.6 − 3000 = 5499.6 × .35)	1925 VOLT-AMPERES
FROM 120,000 VOLT-AMPERES AND UP	0 VOLT-AMPERE
FEEDER/SERVICE CONDUCTOR GENERAL LIGHTING LOAD:	4925 VOLT-AMPERES

THE DIFFERENCE BETWEEN THE TWO CALCULATIONS (4925 VA − 4895 VA = 30 VA) IS DUE TO THE "SOFT CONVERSION" OF 3 VOLT-AMPERES PER FT² TO 33 VOLT-AMPERES PER m²

Figure 12-1

appliances that are subject to this demand factor do not include electric ranges, counter-mounted cook-tops, wall-mounted ovens, electric clothes dryers, space-heating equipment (both electric space-heating equipment and any gas- or oil-fired furnace blowers or controls), and air-conditioning equipment. It also does not include appliances that are not fastened in place such as portable dishwashers. Figure 12-2 presents more information on the appliance demand factor.

12.1.6: Household Electric Range and Electric Clothes Dryer Feeder and Service Conductor Loads

The feeder and service conductor loads for a household electric range, counter-mounted cooking unit, wall-mounted oven, and electric clothes dryer are the same as the branch-circuit loads. For multi-family buildings, the feeder and service loads are calculated using *Table 220.18* for clothes dryers and *Table 220.19* for ranges and other cooking appliances rated over 1750 volt-amperes.

12.1.7: Noncoincident Loads

If two or more loads will not be used at the same time, the smaller of the loads can be eliminated from the load calculation. The usual examples of these types of loads are the furnace load and the air-conditioning load. If the furnace is heating the dwelling, the air conditioner will not also be working to cool the dwelling. However, no two loads are noncoincident in the sample house. There are an air conditioner and a furnace in the sample house, but they represent coincident loads because the air-conditioning system uses the blower from the furnace to distribute the cold air throughout the dwelling in summer as well as the hot air from the furnace in winter. While the air conditioner is working, the furnace fan (the vast majority of the load represented by the gas furnace) is also working.

12.2: LARGEST MOTOR FEEDER AND SERVICE CONDUCTOR LOAD

All of the appliance loads and the motor loads are included in the feeder and service conductor load calculation at 100% of the nameplate or FLA rating. However, *430.24* requires that the largest motor load be included in the load calculation at 125% of the appliance nameplate current or the motor FLA. It is necessary to determine the largest motor load and then add 25% of that value to the feeder and service conductor load. Only 25% is added to the load calculation because the 100% that represents the name-plate rating or the FLA of the motor is already included in the load calculation. Figure 12-3 presents more details on this process and on the largest motor calculation for the sample house.

12.3: TOTAL PHASE CONDUCTOR FEEDER AND SERVICE LOAD CALCULATION FOR THE SAMPLE HOUSE

There are actually two load calculations related to the feeders and service conductors in the sample house. The first calculation is for the feeder conductors between the electrical panel and subpanel A. Subpanel A includes all the circuits in the dwelling except for the circuit that supplies the air conditioner. The main panel feeds only the air conditioner and subpanel A. Therefore, a feeder or service load calculation must be accomplished for subpanel A and for the main panel, the service equipment. The extra circuit-breaker spaces included in the main panel are for future circuits that may be needed to finish the basement, to add a swimming pool or hot tub, or for other uses. There is no *NEC®* requirement to have extra spaces in the service panel for future expansion, however, but in some locales, the AHJ may require one or two additional spaces.

12.3.1: Feeder Load Calculation for Subpanel A

The feeder conductor load calculations for the sample house are shown in Figure 12-4 (page 306). The general lighting and general-use receptacle loads are calculated in Chapter 10. The loading of each appliance is included in Construction Drawing M-1 and in Chapter 11. However, this information may not be available at rough-in, and initial load calculations can be completed by using the maximum load that a circuit of a given size (12 AWG, for example) may possibly provide (20 amperes).

The total calculated load for subpanel A in the sample house is 2573 volt-amperes. With 240-volt

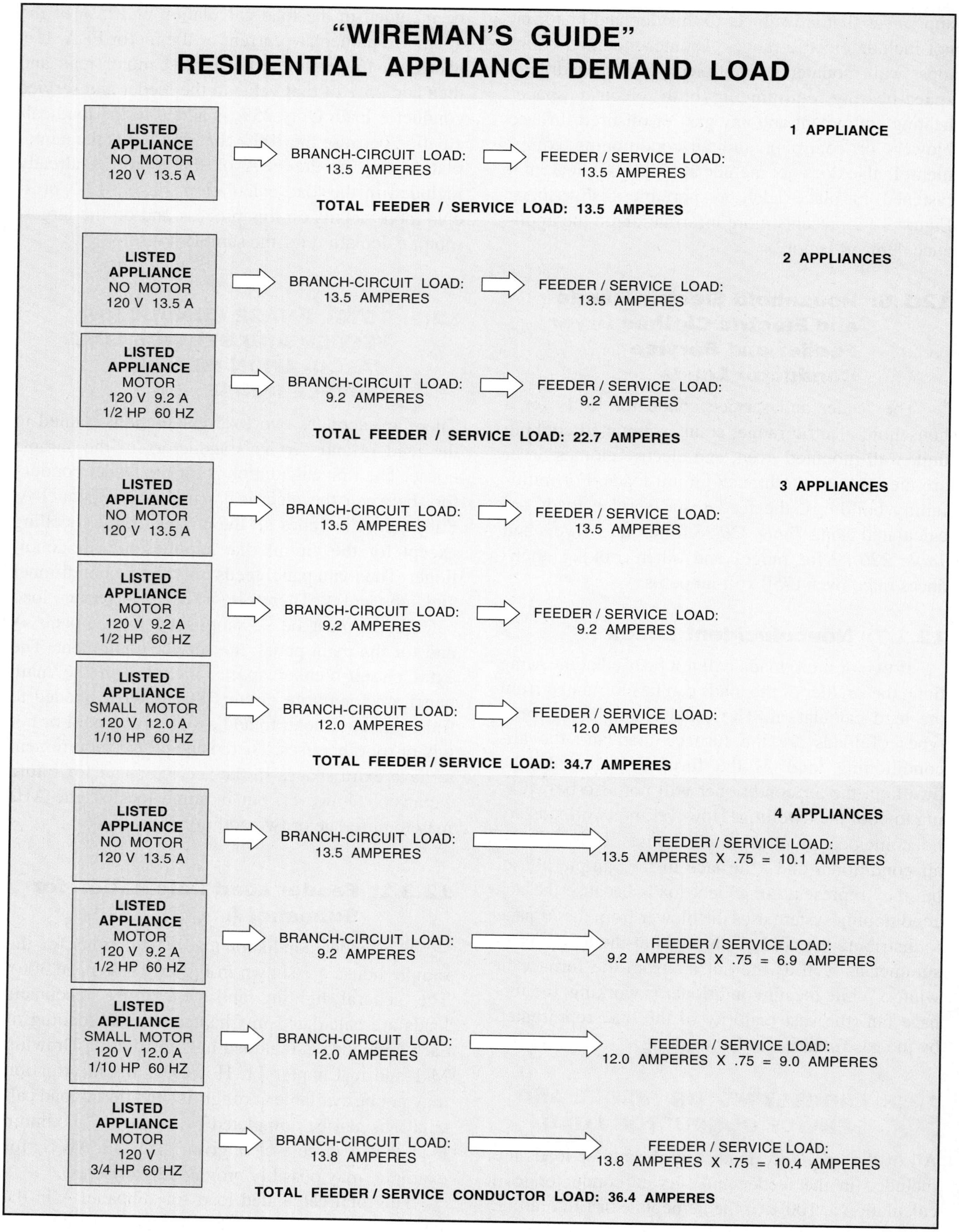

"WIREMAN'S GUIDE"
RESIDENTIAL APPLIANCE DEMAND LOAD

1 APPLIANCE

| LISTED APPLIANCE NO MOTOR 120 V 13.5 A | ⟹ | BRANCH-CIRCUIT LOAD: 13.5 AMPERES | ⟹ | FEEDER / SERVICE LOAD: 13.5 AMPERES |

TOTAL FEEDER / SERVICE LOAD: 13.5 AMPERES

2 APPLIANCES

| LISTED APPLIANCE NO MOTOR 120 V 13.5 A | ⟹ | BRANCH-CIRCUIT LOAD: 13.5 AMPERES | ⟹ | FEEDER / SERVICE LOAD: 13.5 AMPERES |

| LISTED APPLIANCE MOTOR 120 V 9.2 A 1/2 HP 60 HZ | ⟹ | BRANCH-CIRCUIT LOAD: 9.2 AMPERES | ⟹ | FEEDER / SERVICE LOAD: 9.2 AMPERES |

TOTAL FEEDER / SERVICE LOAD: 22.7 AMPERES

3 APPLIANCES

| LISTED APPLIANCE NO MOTOR 120 V 13.5 A | ⟹ | BRANCH-CIRCUIT LOAD: 13.5 AMPERES | ⟹ | FEEDER / SERVICE LOAD: 13.5 AMPERES |

| LISTED APPLIANCE MOTOR 120 V 9.2 A 1/2 HP 60 HZ | ⟹ | BRANCH-CIRCUIT LOAD: 9.2 AMPERES | ⟹ | FEEDER / SERVICE LOAD: 9.2 AMPERES |

| LISTED APPLIANCE SMALL MOTOR 120 V 12.0 A 1/10 HP 60 HZ | ⟹ | BRANCH-CIRCUIT LOAD: 12.0 AMPERES | ⟹ | FEEDER / SERVICE LOAD: 12.0 AMPERES |

TOTAL FEEDER / SERVICE LOAD: 34.7 AMPERES

4 APPLIANCES

| LISTED APPLIANCE NO MOTOR 120 V 13.5 A | ⟹ | BRANCH-CIRCUIT LOAD: 13.5 AMPERES | ⟹ | FEEDER / SERVICE LOAD: 13.5 AMPERES X .75 = 10.1 AMPERES |

| LISTED APPLIANCE MOTOR 120 V 9.2 A 1/2 HP 60 HZ | ⟹ | BRANCH-CIRCUIT LOAD: 9.2 AMPERES | ⟹ | FEEDER / SERVICE LOAD: 9.2 AMPERES X .75 = 6.9 AMPERES |

| LISTED APPLIANCE SMALL MOTOR 120 V 12.0 A 1/10 HP 60 HZ | ⟹ | BRANCH-CIRCUIT LOAD: 12.0 AMPERES | ⟹ | FEEDER / SERVICE LOAD: 12.0 AMPERES X .75 = 9.0 AMPERES |

| LISTED APPLIANCE MOTOR 120 V 3/4 HP 60 HZ | ⟹ | BRANCH-CIRCUIT LOAD: 13.8 AMPERES | ⟹ | FEEDER / SERVICE LOAD: 13.8 AMPERES X .75 = 10.4 AMPERES |

TOTAL FEEDER / SERVICE CONDUCTOR LOAD: 36.4 AMPERES

Figure 12-2

"WIREMAN'S GUIDE" LARGEST MOTOR LOAD

FOR GENERAL MOTOR APPLICATIONS, THE MOTOR BRANCH-CIRCUIT LOAD IS EQUAL TO THE MOTOR FLA FROM *TABLE 430.148* FOR THE PARTICULAR MOTOR HORSEPOWER AND VOLTAGE RATINGS FROM THE NAMEPLATE TIMES 125%.

BRANCH-CIRCUIT LOAD = 8.6 AMPERES FLA X 1.25 = 10.8 AMPERES

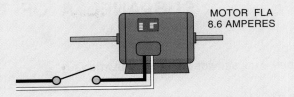

MOTOR FLA 8.6 AMPERES

FOR GENERAL MOTOR APPLICATIONS, THE LOADS FOR FEEDERS OR SERVICE CONDUCTORS THAT PROVIDE POWER TO MORE THAN ONE MOTOR ARE INCLUDED IN THE CALCULATION AT 100% OF THE MOTOR FLA FROM *TABLE 148* PLUS 25% OF THE FLA OF THE LARGEST MOTOR.

FEEDER/SERVICE LOAD = 8.6 A + 9.0 A + 5.5 A + (9.0 A X .25)

FEEDER/SERVICE LOAD = 8.6 A + 9.0 A + 5.5 A + 2.25 A

FEEDER/SERVICE LOAD = 23.1 A + 2.25 A

= 25.35 AMPERES

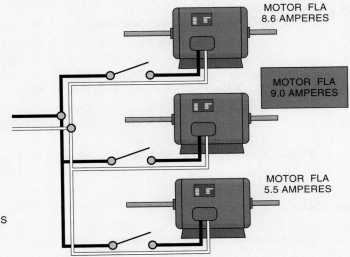

MOTOR FLA 8.6 AMPERES

MOTOR FLA 9.0 AMPERES

MOTOR FLA 5.5 AMPERES

ANY FEEDER OR SERVICE CONDUCTOR THAT PROVIDES POWER TO A MOTOR, SEVERAL MOTORS, A MOTOR AND ANOTHER LOAD, OR MOTORS AND OTHER LOADS IS SIZED ACCORDING TO THE CURRENT RATINGS OF ALL OF THE LOADS (INCLUDING THE MOTOR OR MOTORS) PLUS 25% OF THE LARGEST MOTOR LOAD.

IF THE MOTOR IS PART OF A LISTED MOTOR-DRIVEN APPLIANCE (FOR EXAMPLE, A GARBAGE DISPOSAL), THE FLA CURRENT RATING IS OBTAINED FROM THE APPLIANCE NAMEPLATE USING THE FLA OF THE APPLIANCE, *NOT* THE HORSEPOWER RATING OF THE MOTOR. THE REMAINDER OF THE CALCULATIONS ARE THE SAME.

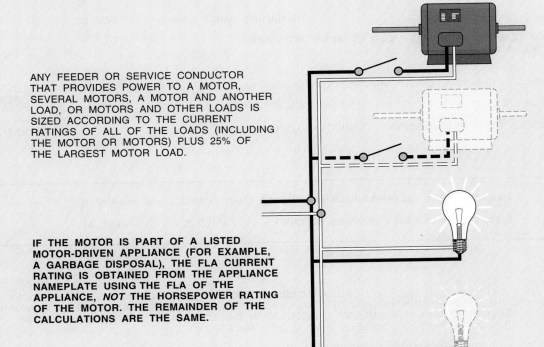

Figure 12-3

"WIREMAN'S GUIDE"
FEEDER LOAD CALCULATIONS FOR
SUBPANEL A OF THE SAMPLE HOUSE

FEEDER AND SERVICE LOAD CALCULATIONS:

GENERAL LIGHTING AND GENERAL-USE RECEPTACLE LOAD:
   1304 FT² AT 3 VOLT-AMPERES PER FT²   3913.5 VA
SMALL APPLIANCE LOAD:
   TWO SMALL APPLIANCE CIRCUITS AT 1500 VOLT-AMPERES EACH   3000.0 VA
LAUNDRY CIRCUIT LOAD:
   ONE LAUNDRY CIRCUIT AT 1500 VOLT-AMPERES EACH   1500.0 VA
                                                                         _____
                                          TOTAL LOAD: 8413.5 VA

APPLICATION OF GENERAL LIGHTING DEMAND FACTOR:
   8413.5 TOTAL LOAD: FIRST 3000 VOLT-AMPERES AT 100%   3000 VA
   5413.5 VOLT-AMPERES AT 35%   1895 VA
                                                                         _____
                                          TOTAL LOAD: 4895 VA

TOTAL GENERAL LIGHTING CALCULATED LOADS: .... **4895 VA**

OTHER LOADS:

RANGE LOAD: (40-AMPERE, 120/240-VOLT CIRCUIT) .... 8000 VA
CLOTHES DRYER: (30-AMPERE 120/240-VOLT CIRCUIT) .... 5500 VA
APPLIANCES: DISHWASHER (20-AMPERE CIRCUIT AT 120 VOLTS)   1620 VA
            GARBAGE DISPOSAL (20-AMPERE CIRCUIT AT 120 VOLTS)   1440 VA
            MICROWAVE OVEN (20-AMPERE CIRCUIT AT 120 VOLTS)   1750 VA
            SUMP PUMP (20-AMPERE CIRCUIT AT 120 VOLTS)   1650 VA
                                                                             _____
                                   TOTAL APPLIANCE LOAD: 6220 VA

APPLICATION OF DEMAND FACTOR FOR FOUR OR MORE APPLIANCES:
   6220 VOLT-AMPERES X .75 .... 4665 VA

BASEBOARD HEATER (15-AMPERE CIRCUIT AT 240 VOLTS) .... 2000 VA
FURNACE (20-AMPERE CIRCUIT AT 120 VOLTS) .... 1100 VA
LARGEST MOTOR LOAD (SUMP PUMP AT 1650 VOLT-AMPERES X 25%) .... 413 VA
                                                                                              _____
MAXIMUM POSSIBLE TOTAL FEEDER/SERVICE CONDUCTOR LOAD: .... **26,573 VA**

MAXIMUM TOTAL FEEDER/SERVICE CONDUCTOR LOAD COMPUTED IN AMPERES:

26,573 VA (LOAD) / 240 V (SERVICE VOLTAGE) = 110.7 A (LOAD ON SUBPANEL A)

THERE IS A SLIGHT DIFFERENCE IN THE LOAD CALCULATIONS IF 33 VOLT-AMPERES PER m² IS USED FOR THE LOAD CALCULATION. THE GENERAL LIGHTING AND GENERAL-USE RECEPTACLE LOADS USING m² ARE AN ADDITIONAL 30 VOLT-AMPERES (SEE FIGURE 12-1). THIS AMOUNT IS INCLUDED IN THE LOAD CALCULATION AT 35% FOR AN ADDITIONAL LOAD OF 10.5 VOLT-AMPERES OR LESS THAN ¹⁄₁₀ OF 1 AMPERE.

Figure 12-4

incoming service from the power-providing utility, the total ampere load for the dwelling is 110.7 amperes.

12.3.2: Service Conductor Load Calculation for the Sample House

The sample house has a total load calculation for the service conductors of 33,485 volt-amperes, as shown in Figure 12-5. The calculation includes the 113 amperes that are distributed by subpanel A and the additional load of the air conditioner that is supplied from the main panel. At 240 volts, the service conductor ampere load is 140 amperes.

12.4: FEEDER AND SERVICE GROUNDED CIRCUIT CONDUCTOR LOAD FOR THE SAMPLE HOUSE

In any multiwire system, the grounded circuit conductor (neutral) exists solely to carry the unbalanced current from the phase conductors. This is also the case with feeders and service conductors that are multiwire circuits. Just as no neutral conductor is required for a load that is balanced between phases or legs (for example, the baseboard heater in the garage of the sample house), no allowance for neutral current needs to be made for the feeder or service conductors supplying only 240-volt branch-circuit loads. *NEC®* 220.22 details the rules for de-rating the neutral conductor on feeder or service conductors.

In the sample dwelling, the air conditioner load is a balanced, 240-volt load calculated for a maximum load of 6720 volt-amperes (28.8 A × 240 V). Because the neutral leg will carry no current (and a neutral conductor is not even supplied to the air conditioner), the 6912 volt-amperes of the phase conductor load can be removed from the service neutral load. This represents a reduction in the neutral conductor of 28.8 amperes.

For dwellings, with the multiwire branch circuit that serves the range, counter-mounted cooking unit, wall-mounted oven, or electric clothes dryer, the load for the branch-circuit neutral conductor can be reduced to 70% of the load on the ungrounded conductors. The sample house has a range with a calculated load of 8000 volt-amperes. This allows a reduction in the neutral feeder or service conductor of 2400 volt-amperes (8000 × .30), or 10 amperes.

The dwelling also has an electric clothes dryer with a nameplate rating of 5500 volt-amperes. The neutral load can be reduced by 30%, to 70% of the phase conductor load. This represents a reduction in the neutral conductor load of 1650 volt-amperes, or 6.9 amperes.

The baseboard heater in the garage is a 240-volt load. No neutral is required because there is no unbalanced current. This represents a reduction in the neutral load of 2000 volt-amperes, or 8.3 amperes.

No other load in the sample house is subject to these reductions because they are all 120-volt loads. These loads use the neutral as a circuit conductor and therefore are subject to unbalancing the feeders or service conductor loads.

"WIREMAN'S GUIDE"
SERVICE LOAD CALCULATIONS FOR MAIN PANEL OF THE SAMPLE HOUSE

FEEDER AND SERVICE LOAD CALCULATIONS:

GENERAL LIGHTING AND GENERAL-USE RECEPTACLE LOAD:

1304 FT² AT 3 VOLT-AMPERES PER FT²	3913.5 VA
SMALL APPLIANCE LOAD:	
TWO SMALL APPLIANCE CIRCUITS AT 1500 VOLT-AMPERES EACH	3000.0 VA
LAUNDRY CIRCUIT LOAD:	
ONE LAUNDRY CIRCUIT AT 1500 VOLT-AMPERES EACH	1500.0 VA
TOTAL LOAD:	8413.5 VA

APPLICATION OF GENERAL LIGHTING DEMAND FACTOR:

8413.5 TOTAL LOAD: FIRST 3000 VOLT-AMPERES AT 100%	3000 VA
5413.5 VOLT-AMPERES AT 35%	1895 VA
TOTAL LOAD:	4895 VA

TOTAL GENERAL LIGHTING CALCULATED LOADS: .... **4895 VA**

OTHER LOADS:

RANGE LOAD: (40-AMPERE, 240-VOLT CIRCUIT EQUALS 9600 VOLT-AMPERES)	8000 VA
CLOTHES DRYER: (30-AMPERE 240-VOLT CIRCUIT EQUALS 7200 VOLT-AMPERES)	5500 VA
APPLIANCES: DISHWASHER (20-AMPERE CIRCUIT AT 120-VOLTS)	1620 VA
GARBAGE DISPOSAL (20-AMPERE CIRCUIT AT 120 VOLTS)	1440 VA
MICROWAVE OVEN (20-AMPERE CIRCUIT AT 120 VOLTS)	1750 VA
SUMP PUMP (20-AMPERE CIRCUIT AT 120 VOLTS)	1650 VA
TOTAL APPLIANCE LOAD:	6220 VA

APPLICATION OF DEMAND FACTOR FOR FOUR OR MORE APPLIANCES:

6220 VOLT-AMPERES X .75 .... 4665 VA

BASEBOARD HEATER (15-AMPERE CIRCUIT AT 240 VOLTS)	2000 VA
FURNACE (20-AMPERE CIRCUIT AT 120 VOLTS)	1100 VA
LARGEST MOTOR LOAD (SUMP PUMP AT 1650 VOLT-AMPERES X 25%)	413 VA
AIR CONDITIONER LOAD (30-AMPERE CIRCUIT AT 240 VOLTS)	6912 VA

MAXIMUM POSSIBLE TOTAL FEEDER/SERVICE CONDUCTOR LOAD: .... **33,485 VA**

MAXIMUM TOTAL FEEDER/SERVICE CONDUCTOR LOAD COMPUTED IN AMPERES:

33485 VA (LOAD) / 240 V (SERVICE VOLTAGE) = 140 A (LOAD ON THE MAIN PANEL)

THERE IS A SLIGHT DIFFERENCE IN THE LOAD CALCULATIONS IF 33 VOLT-AMPERES PER M² IS USED FOR THE LOAD CALCULATION. THE GENERAL LIGHTING AND GENERAL-USE RECEPTACLE LOAD USING M² IS AN ADDITIONAL 30 VOLT-AMPERES (SEE FIGURE 12-1). THIS AMOUNT IS INCLUDED IN THE LOAD CALCULATION AT 35% FOR AN ADDITIONAL LOAD OF 10.5 VOLT-AMPERES, OR LESS THAN ¹⁄₁₀ OF 1 AMPERE.

Figure 12-5

SUMMARY

The feeder and service loads are the sum of the branch-circuit loads after the calculations allowed for demand factors have been completed. Demand factors are allowed for general lighting and general-use receptacle loads that also include small-appliance circuit and laundry circuit loads of 1500 volt-amperes each. A demand factor can also be applied to the total load with fixed-in-place appliances, other than ranges, dryers, space-heating equipment, and air-conditioning equipment, for a reduction to 75% of the nameplate rating. A 25% increase in the ampere rating of the largest motor load must be included in the feeder and service load calculations. A further decrease in the size of the grounded service conductor or grounded feeder conductor is allowed for all 240-volt loads that do not have any 120-volt components. The grounded conductor load can also be reduced for ranges and dryers to 70% for the range and dryer branch-circuit loads.

REVIEW

1. *True or False:* Feeder loads are calculated using the same procedures as for service-entrance conductor loads.

2. *True or False:* Service loads are calculated using the same set of procedures as for branch circuits.

3. *True or False:* Branch-circuit loads are calculated under the same set of rules as for feeder-conductor loads.

4. *True or False:* The feeder load and the branch-circuit load are the same for a range installed in a dwelling unit.

5. *True or False:* The feeder load for an electric clothes dryer can never be less than 5,000 volt-amperes.

6. *True or False:* Luminaires (fixtures) and general-use receptacles can be installed on the same branch circuit in a dwelling.

7. Calculate the feeder load for the general lighting if the general lighting and general-use receptacle load for a dwelling of 13,500 volt-amperes (4500 ft^2 × 3 watts [volt-amperes] per ft^2 [33 watts per m^2]). Be sure to include all circuits that are allowed to be reduced in accordance with *220.11*.
   a. 4725 volt-amperes
   b. 5250 volt-amperes
   c. 8250 volt-amperes
   d. 13,500 volt-amperes

8. The feeder and service load for a small-appliance circuit is _____ if it is to power a garbage disposal rated at 900 volt-amperes, a dishwasher rated at 1500 volt-amperes, and a trash compactor rated at 1300 volt-amperes, all at 120 volts.
   a. 3700 volt-amperes
   b. 2590 volt-amperes
   c. 5000 volt-amperes
   a. none of the above—the load cannot be determined from the information given

9. If the dwelling has a detached garage with electrical power, the unit load for general lighting and general-use receptacle circuit is _____.

    a. 0 watt per ft^2 (0 watt per m^2)

    b. 1 watt per ft^2 (11 watts per m^2)

    c. 2 watts per ft^2 (22 watts per m^2)

    d. 3 watts per ft^2 (33 watts per m^2)

10. The conductors that serve subpanel A for the sample house are _____.

    a. branch circuits

    b. feeders

    c. service conductors

    d. laterals

11. The conductors that service the main electrical panel from the meter enclosure for the sample house are _____.

    a. branch circuits

    b. feeders

    c. service conductors

    d. laterals

12. The conductors that supply the meter in the sample house are _____.

    a. branch circuits

    b. feeders

    c. service conductors

    d. laterals

"WIREMAN'S GUIDE" REVIEW

1. Explain how the general lighting load is calculated for a dwelling.

2. Explain why the furnace and the air conditioner are not considered noncoincident loads for the sample house.

3. Explain the need to increase the dwelling load by 25% of the largest motor load. What if two motors are tied for the largest? What if the largest motor load is a garbage disposal with a load reduced to 70% of the normal load because of the demand factors stipulated in *220.17*?

4. Explain how *220.3(B)* does not restrict the number of general-use outlets that can be installed on a circuit for dwellings. Theoretically, how many receptacles can be installed on a general lighting and general-use receptacle circuit?

CHAPTER 13

The Electrical Service

OBJECTIVES

On completion of this chapter, the student will be able to:

☑ **Define service drops, service laterals, and service-entrance conductors.**

☑ **Explain the difference between service equipment and equipment.**

☑ **State the rationale for placement of the service disconnecting means and overcurrent-protective devices.**

☑ **State the rationale for placement of electrical metering equipment.**

☑ **Properly install residential service equipment.**

☑ **Design and install a simple residential service.**

INTRODUCTION

This chapter is concerned with the dwelling-unit service. A service can be defined as that wiring and equipment necessary to connect the premises' wiring system to the power utility provider's wiring system. Some of the requirements for services are covered in Chapter 2 of this book, on installing the temporary construction power. Chapter 7 of this text explores some aspects of services and the inherent problem of overcurrent protection for service conductors. Installation of permanent services for dwellings follows many of the same rules and must address the same basic problem: service conductors and service equipment are unprotected against faults. If there is a fault condition in the service equipment or on the line side of the service equipment, there is no overcurrent-protective device to open in order to eliminate the fault. Therefore, the service is a special location in the wiring system.

13.1: LOCAL POWER-PROVIDING UTILITY

The *NEC®* and the AHJ have obvious control over the way in which electrical systems are installed, not only in dwellings but also in all other types of structures. The local power provider—the utility that generates and distributes the electrical service—also has rules and regulations that must be followed in connecting the dwelling to the permanent power. Different utility companies have quite different rules, and the electrician must be aware of the local

rules when planning and installing the service. Figure 13-1 shows some of the different ways in which a utility can arrange a service to a dwelling; there are many other possibilities. Figure 13-2 shows a typical overhead service drop to a mast. Figure 13-3 shows a typical underground service lateral to a combination meter housing/distribution panel.

The exact location of the service—where the service equipment is installed in the dwelling—and the method of service, either overhead or underground, are largely determined by the power utility.

"WIREMAN'S GUIDE"
DWELLING-UNIT OVERHEAD AND
UNDERGROUND SERVICES

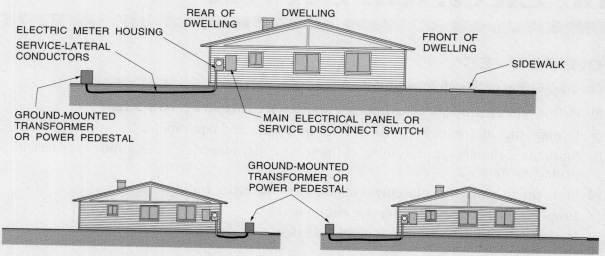

ELECTRIC METER HOUSING

SERVICE-LATERAL
CONDUCTORS

REAR OF DWELLING

DWELLING

FRONT OF DWELLING

SIDEWALK

GROUND-MOUNTED
TRANSFORMER
OR POWER PEDESTAL

MAIN ELECTRICAL PANEL OR
SERVICE DISCONNECT SWITCH

GROUND-MOUNTED
TRANSFORMER OR
POWER PEDESTAL

THE LOCATION OF THE ELECTRICAL SERVICE IS DETERMINED LARGELY BY THE REQUIREMENTS OF THE POWER-PROVIDING UTILITY. WITH SOME DEVELOPMENTS, THE ELECTRICAL UTILITIES ARE IN THE FRONT OF THE DWELLING, AND WITH OTHERS, THE DISTRIBUTION SYSTEM IS IN THE BACK OF THE DWELLINGS. THE POWER PROVIDER MAY WANT THE SERVICE EITHER ON THE REAR OF THE HOUSE, OR ON THE SIDE WALL OF THE HOUSE. THE UTILITY IS CONCERNED WITH THE METER LOCATION SO THAT THE METER READER CAN EASILY AND SAFELY ACCESS THE METER WITHOUT HAVING TO ENTER PRIVATE YARDS OR OPEN GATES IN FENCES.

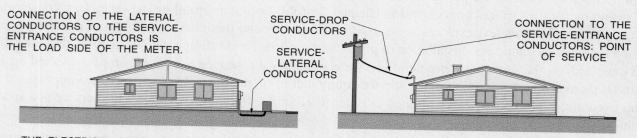

CONNECTION OF THE LATERAL CONDUCTORS TO THE SERVICE-ENTRANCE CONDUCTORS IS THE LOAD SIDE OF THE METER.

SERVICE-DROP CONDUCTORS

SERVICE-LATERAL CONDUCTORS

CONNECTION TO THE SERVICE-ENTRANCE CONDUCTORS: POINT OF SERVICE

THE ELECTRICAL SERVICE MAY BE EITHER OF TWO TYPES: (1) OVERHEAD, OR (2) UNDERGROUND. THE CONDUCTORS THAT EXTEND FROM THE OVERHEAD DISTRIBUTION SYSTEM TO THE DWELLING ARE CALLED DROP CONDUCTORS OR SERVICE-DROP CONDUCTORS. THE UNDERGROUND CONDUCTORS FROM THE DISTRIBUTION SYSTEM TO THE DWELLING SERVICE ARE CALLED LATERAL CONDUCTORS.

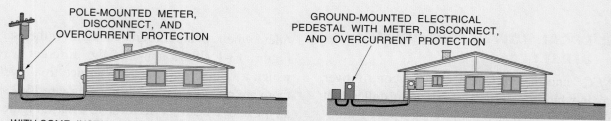

POLE-MOUNTED METER,
DISCONNECT, AND
OVERCURRENT PROTECTION

GROUND-MOUNTED ELECTRICAL
PEDESTAL WITH METER, DISCONNECT,
AND OVERCURRENT PROTECTION

WITH SOME INSTALLATIONS, THE POWER UTILITY MAY REQUIRE THAT THE METER, THE OVERCURRENT DEVICE, AND THE SERVICE DISCONNECTING MEANS BE INSTALLED WHERE THE CONDUCTORS THAT FEED THE DWELLING ARE SUPPLIED. THIS CHANGES THE DEFINITION OF THE CONDUCTORS THAT CARRY THE CURRENT TO THE DWELLING FROM LATERALS AND DROPS TO SERVICE-ENTRANCE CONDUCTORS OR FEEDERS. IN SOME CASES, THE AHJ MAY REQUIRE A MAIN OVERCURRENT-PROTECTIVE DEVICE ON THE DWELLING AS WELL AS AT THE SOURCE OR MAY REQUIRE ONLY A MAIN DISCONNECTING MEANS AT THE DWELLING. THE AHJ MAY TREAT THE CONDUCTORS EITHER AS SERVICE-ENTRANCE CONDUCTORS OR AS FEEDERS.

Figure 13-1

Figure 13-2   An overhead service drop. Note the drip loops. There are no guy wire supports on this mast.

Figure 13-3   An underground service (lateral) to a combination meter and electrical panel. Note the main breaker in the panel.

Part of the development process, the taking of undeveloped land and installing the infrastructure necessary for connection of the individual dwellings to water, sewer, telephone, and cable television, includes the installation of an electrical power distribution system. This system may be installed as an underground distribution system with ground-mounted transformers and power pedestals. Such underground systems can be designed to supply the individual dwelling from the front or from the back. Alternatively, the development may provide overhead distribution conductors. Overhead systems are installed predominantly to the back of the houses, but many distribution systems supply to the front of the dwelling instead. Some utilities require that the electrical contractor provide and install the lateral and drop conductors; others will provide and install these conductors. Obviously, it is necessary to closely coordinate the service installation with the utility before beginning the installation. Figure 13-1 presents additional information about the location of the service.

The ampacity of the lateral and drop conductors must be large enough for them to carry the load to be served. The sizes of these conductors are controlled not by the *NEC®* but by the rules and requirements under which the utility operates.

13.2: COMMON ELEMENTS OF OVERHEAD AND UNDERGROUND SERVICES

Services are of two types: (1) underground and (2) overhead. Different procedures must be followed in installing each type of service, but many of the requirements are common to both types of services.

13.2.1: Service-Entrance Conductors

Each building served is provided with conductors to connect to the power utility conductors. Part of that wiring—part owned or controlled by the utility—is defined as lateral or drop conductors. The location at which the lateral or drop conductors end is where service-entrance conductors begin. This location in the wiring is called the *point of service* or *service point* by the *NEC®*. For dwellings, with overhead services, the point of service is generally at the weatherhead; with underground services, this point is usually in the meter enclosure.

Service conductors are sized according to the service load calculation. The calculated load for the dwelling is the minimum size for the service conductors. The calculated load for the sample dwelling is 33,485 volt-amperes or 140 amperes, according to the calculations in Chapter 12 of this book. This is the minimum ampacity for the service conductors for the sample house.

Ampacities for service conductors for dwellings are not obtained from *Table 310.16*, however. According to *310.15(B)(6)*, the ampacity for service and main power feeder conductors for dwellings can be sized using *Table 310.15(B)(6)*. This table applies to dwellings only if the service is 120/240-volt, 3-wire, single-phase service; otherwise, the ampacity of the conductors is obtained from *Table 310.16*. *Table 310.15(B)(6)* shows that for a 142-ampere service load, the minimum required conductor size is 1 AWG copper or 2/0 AWG aluminum. The ampere ratings for conductors shown in *Table 310.15(B)(6)* are also considered by the *NEC*® to be standard service sizes. An absolute minimum service conductor size for service conductors is stipulated by the *NEC*®. This minimum size is determined by the size of the disconnecting means for the dwelling. This subject is covered in more detail later in this chapter in the discussion of the required disconnect.

Service conductors can be installed on the outside of the dwelling as long as they are protected from physical damage. The Code does not limit the length of the service-entrance conductors but does stipulate that the service conductors are to be terminated in an overcurrent-protective device immediately on entering the interior of a dwelling. The service equipment may also be on the outside of the house, so that the service conductors never enter the house.

13.2.2: Service Disconnecting Means

The *NEC*® requires some method of disconnecting the dwelling from all of the ungrounded service conductors. The disconnect device is a switch or circuit breaker that is rated as service equipment and that opens all ungrounded conductors simultaneously. The grounded circuit conductor is not intended to be disconnected with the ungrounded conductors (the only equipment allowed to disconnect a service neutral is a multipole circuit breaker that will open all of the ungrounded conductors at the same time that the grounded conductor is opened). However, provi-

sions must be made to allow the disconnecting of the neutral conductor at the service disconnect if necessary. This is usually accomplished with a lug connected to the service disconnect enclosure. Because this is service equipment, the lug can be attached directly to the metal disconnect housing without being insulated from the rest of the grounding system.

The handle of the disconnect switch must be installed so that it is no more than 6 ft 7 in. (2 m) above the ground and is readily accessible. There can be up to six service disconnects on any one building. The use of more than one service disconnect on any single-family dwelling unit is unusual but is certainly possible, particularly with larger custom homes. Each service disconnect device is permanently marked to identify it as a service disconnect switch. If there is more than one service disconnect switch, the switches must be grouped and labeled so that emergency workers can disconnect the entire building without having to search for the switches, and to ensure that all of the switches are opened. Whether the disconnect consists of multipole circuit breakers connected together with a handle tie or is merely a circuit breaker constructed with a master handle, it is considered to be one switch regardless of the number of poles.

The service disconnect switch (or switches) can be installed on the exterior of the building or inside the building at the first readily accessible location closest to where the service-entrance conductors enter the dwelling. Because the conductors are not protected against fault overcurrent, it is logical to allow as little of the service conductors as possible inside a building. No service switch can be installed in a bathroom, however.

According to *230.79(C)*, the ampere rating of the disconnect switch for a single-family dwelling unit service is 100 amperes. This is the minimum size service allowed for a single-family dwelling. Switches rated above 100 amperes must be sized to the load served, according to the load calculation discussed in Chapter 12 of this book.

In a typical residential service installation, only the meter is installed on the line side of the disconnecting switch. Several types of equipment are allowed in the line side, as listed in *230.82*, but none of these is commonly used in dwellings. Figure 13-4 presents additional information about service disconnects.

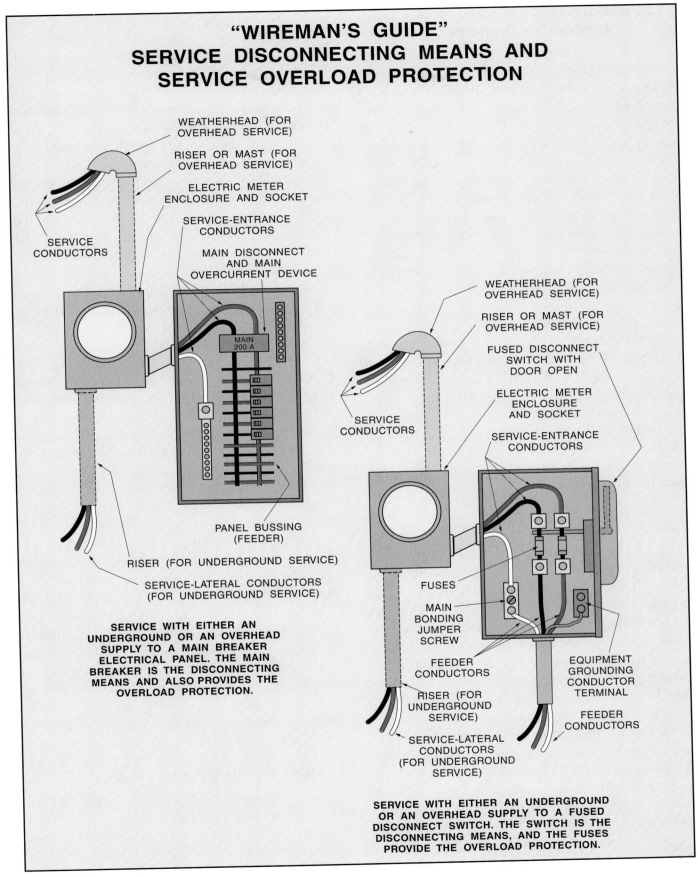

"WIREMAN'S GUIDE"
SERVICE DISCONNECTING MEANS AND
SERVICE OVERLOAD PROTECTION

WEATHERHEAD (FOR OVERHEAD SERVICE)

RISER OR MAST (FOR OVERHEAD SERVICE)

ELECTRIC METER ENCLOSURE AND SOCKET

SERVICE-ENTRANCE CONDUCTORS

MAIN DISCONNECT AND MAIN OVERCURRENT DEVICE

SERVICE CONDUCTORS

MAIN 200 A

PANEL BUSSING (FEEDER)

RISER (FOR UNDERGROUND SERVICE)

SERVICE-LATERAL CONDUCTORS (FOR UNDERGROUND SERVICE)

SERVICE WITH EITHER AN UNDERGROUND OR AN OVERHEAD SUPPLY TO A MAIN BREAKER ELECTRICAL PANEL. THE MAIN BREAKER IS THE DISCONNECTING MEANS AND ALSO PROVIDES THE OVERLOAD PROTECTION.

WEATHERHEAD (FOR OVERHEAD SERVICE)

RISER OR MAST (FOR OVERHEAD SERVICE)

FUSED DISCONNECT SWITCH WITH DOOR OPEN

ELECTRIC METER ENCLOSURE AND SOCKET

SERVICE-ENTRANCE CONDUCTORS

SERVICE CONDUCTORS

FUSES

MAIN BONDING JUMPER SCREW

FEEDER CONDUCTORS

RISER (FOR UNDERGROUND SERVICE)

SERVICE-LATERAL CONDUCTORS (FOR UNDERGROUND SERVICE)

EQUIPMENT GROUNDING CONDUCTOR TERMINAL

FEEDER CONDUCTORS

SERVICE WITH EITHER AN UNDERGROUND OR AN OVERHEAD SUPPLY TO A FUSED DISCONNECT SWITCH. THE SWITCH IS THE DISCONNECTING MEANS, AND THE FUSES PROVIDE THE OVERLOAD PROTECTION.

Figure 13-4

13.2.3: Service Overcurrent-Protective Devices

In addition to the required disconnect switch, the service conductors are also required to be protected against overcurrents due to overloads. The service overcurrent-protective device cannot protect the service against overcurrents due to ground faults or short circuits because it is installed on the load end of the conductors. It does, however, guard against running overcurrents—operating overloads on the service.

This overload overcurrent-protective device must be in series with each ungrounded service conductor and must be an integral part of, or immediately adjacent to, the service disconnect switch. A fused disconnect switch or a main breaker in the electrical panel satisfies this requirement. The maximum rating of the overload-protective device, whether a set of fuses or a circuit breaker, must be no more than the ampere rating of the service conductors. For dwellings, the ampere ratings for the conductors can be taken from *310.15(B)(6)*. If there is more than one disconnecting means, there must also be more than one overcurrent-protective device. The sum of the ratings of the circuit breakers or fuses can be larger than the ampere rating of the service conductors as long as the conductors are sized to supply the calculated load.

13.2.4: Electrical Metering Equipment

The electric meter is installed as part of the service equipment, although individual meter socket enclosures are not considered by the *NEC®* to be service equipment. The meter is located in the circuit before the service disconnect switch to minimize tampering with the conductors by the occupant of the dwelling. The meters are also sealed or locked by the electric utility to further discourage tampering.

The height of the electric meter above the finished grade or finished floor is a requirement of the local power utility. The employees of the utility must have access to the meter in order to obtain a monthly reading for determining the amount of electricity used for billing purposes. Furthermore, the meter should be at eye level to minimize stooping by the meter reader. The normally accepted height is between a maximum of 6 ft 3 in. (1.9 m) and a minimum of 5 ft 0 in. (1.5 m) above the finished surface,

as shown in Figure 13-5. If there is any question about the acceptable mounting height of the meter, the local power provider should be consulted for its particular height requirements before the service is designed.

The electric meter records the power, measured in watts, used in the dwelling over some fixed period. Thus, these meters are often referred to as watt-hour meters. It is easy to convert an ordinary electric bill into a current load by dividing out the relevant units from the number of kilowatt-hours recorded: the hours (there are 720 hours in a 30-day month), the 1000 (kilo), and the service voltage (240 volts nominal is the standard residential voltage). The load obtained using this procedure is an average ampere loading on the service conductors and does not accurately measure the peak or maximum loading.

The meter enclosure is the location at which the service-lateral conductors usually terminate and become service-entrance conductors. The lateral conductors or the service-entrance conductors from the drop conductors terminate on the top set of lugs, and the service-entrance conductors terminate on the lower set of lugs (the load side) in the meter enclosure. The grounded service conductors (neutrals) also have terminals in the enclosure. The grounded service conductor terminals are bonded (electrically connected) to the meter enclosure at this point.

The power-providing utility usually provides and installs the meter into the enclosure, whereas the enclosure itself is usually provided and installed by the electrical contractor. It is usually possible to eliminate the power to a dwelling by disconnecting the meter from the meter enclosure. Most residential meters are flow-through meters, in which the entire load of the service flows through the meter. Figures 13-6 and 13-7 present more detail about meter enclosures and sockets. Figure 13-8 (page 320) shows the terminals for an electric meter that is part of a combination meter enclosure and distribution panel.

13.2.5: Service Grounding

The grounding of the service equipment and enclosures constitutes a very large part of the design and installation of the electrical service. However, there is more to grounding than just the service. Therefore, grounding is covered as a separate area of study in Chapter 14 of this book.

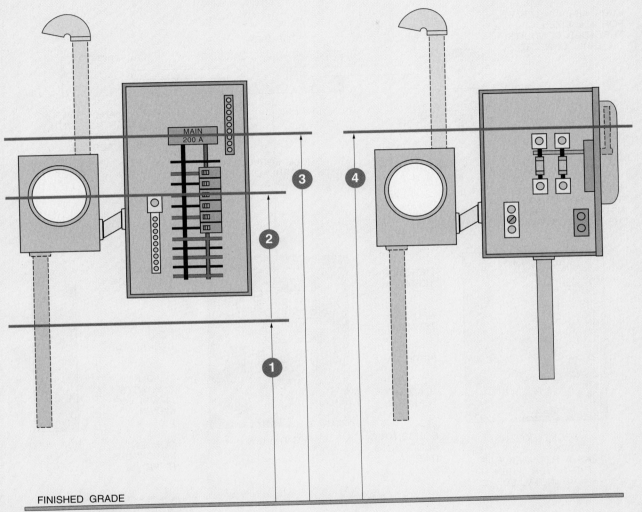

"WIREMAN'S GUIDE"
MOUNTING HEIGHTS FOR METER ENCLOSURES
AND DISCONNECTING MEANS

FINISHED GRADE
OR FLOOR LEVEL

1 THE CENTER OF THE METER SHOULD BE AT EYE LEVEL. EACH UTILITY HAS ITS OWN STANDARDS, BUT NO LOWER THAN 5 FT 0 IN. (1.524 m) IS A COMMON LOWER LIMIT.

2 THE METER SHOULD NOT BE INSTALLED SO THAT THE CENTER IS MORE THAN 6 FT 3 IN. (1.90 m) ABOVE FINISHED GRADE OR FLOOR. THE LOCAL UTILITY HAS STANDARDS FOR METER MAXIMUM HEIGHTS AND SHOULD BE CONSULTED IF THERE IS A QUESTION BEFORE THE SERVICE IS INSTALLED.

3 THE HEIGHT ABOVE FINISHED GRADE OR FINISHED FLOOR CANNOT BE MORE THAN 6 FT 7 IN. (2 m) TO THE CENTER CIRCUIT BREAKER HANDLE. THERE IS NO LOWER LIMIT ESTABLISHED BY THE CODE.

4 THE HEIGHT ABOVE FINISHED GRADE OR FINISHED FLOOR CANNOT BE MORE THAN 6 FT 7 IN. (2 m) TO THE MIDPOINT OF TRAVEL OF THE DISCONNECT HANDLE. THERE IS NO LOWER LIMIT ESTABLISHED BY THE CODE.

Figure 13-5

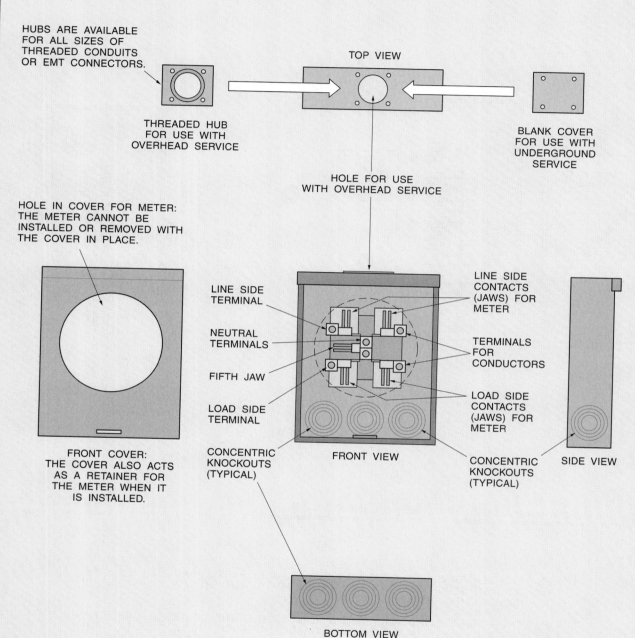

"WIREMAN'S GUIDE"
TYPICAL METER ENCLOSURE FOR OVERHEAD OR UNDERGROUND SERVICES

HUBS ARE AVAILABLE FOR ALL SIZES OF THREADED CONDUITS OR EMT CONNECTORS.

TOP VIEW

THREADED HUB FOR USE WITH OVERHEAD SERVICE

BLANK COVER FOR USE WITH UNDERGROUND SERVICE

HOLE FOR USE WITH OVERHEAD SERVICE

HOLE IN COVER FOR METER: THE METER CANNOT BE INSTALLED OR REMOVED WITH THE COVER IN PLACE.

LINE SIDE TERMINAL

LINE SIDE CONTACTS (JAWS) FOR METER

NEUTRAL TERMINALS

TERMINALS FOR CONDUCTORS

FIFTH JAW

LOAD SIDE TERMINAL

LOAD SIDE CONTACTS (JAWS) FOR METER

FRONT COVER: THE COVER ALSO ACTS AS A RETAINER FOR THE METER WHEN IT IS INSTALLED.

CONCENTRIC KNOCKOUTS (TYPICAL)

FRONT VIEW

CONCENTRIC KNOCKOUTS (TYPICAL)

SIDE VIEW

BOTTOM VIEW

THE FIFTH JAW: TRADITIONALLY, ELECTRIC METERS OPERATED BY MEASURING THE CURRENT FLOW WITH A COIL BETWEEN THE TWO PHASE CONDUCTORS, (240 VOLTS). MANY UTILITIES NOW METER WITH TWO COILS, ONE FROM EACH PHASE CONDUCTOR TO GROUND (120 VOLTS). THE FIFTH JAW PROVIDES THE GROUNDED TERMINAL FOR THE 120-VOLT METERS. FOR THOSE UTILITIES USING 240-VOLT METERING, THE FIFTH JAW IS NOT INCLUDED.

Figure 13-6

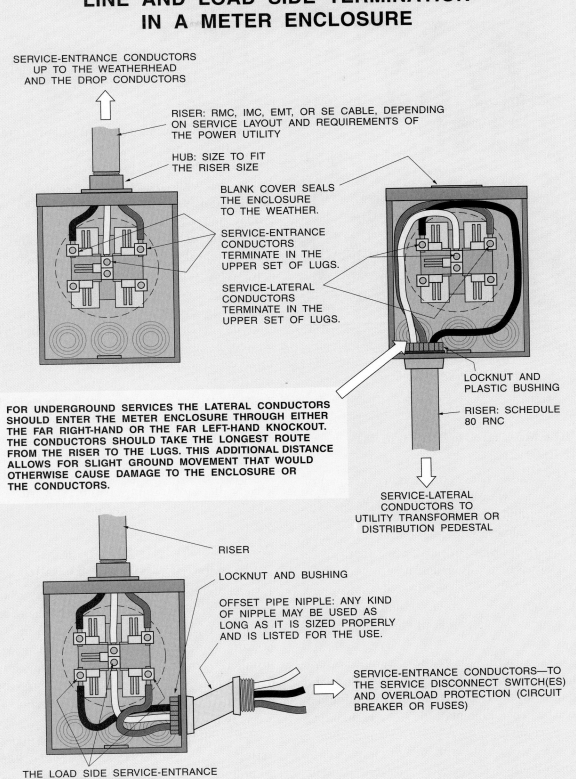

"WIREMAN'S GUIDE"
LINE AND LOAD SIDE TERMINATION
IN A METER ENCLOSURE

SERVICE-ENTRANCE CONDUCTORS
UP TO THE WEATHERHEAD
AND THE DROP CONDUCTORS

RISER: RMC, IMC, EMT, OR SE CABLE, DEPENDING
ON SERVICE LAYOUT AND REQUIREMENTS OF
THE POWER UTILITY

HUB: SIZE TO FIT
THE RISER SIZE

BLANK COVER SEALS
THE ENCLOSURE
TO THE WEATHER.

SERVICE-ENTRANCE
CONDUCTORS
TERMINATE IN THE
UPPER SET OF LUGS.

SERVICE-LATERAL
CONDUCTORS
TERMINATE IN THE
UPPER SET OF LUGS.

FOR UNDERGROUND SERVICES THE LATERAL CONDUCTORS
SHOULD ENTER THE METER ENCLOSURE THROUGH EITHER
THE FAR RIGHT-HAND OR THE FAR LEFT-HAND KNOCKOUT.
THE CONDUCTORS SHOULD TAKE THE LONGEST ROUTE
FROM THE RISER TO THE LUGS. THIS ADDITIONAL DISTANCE
ALLOWS FOR SLIGHT GROUND MOVEMENT THAT WOULD
OTHERWISE CAUSE DAMAGE TO THE ENCLOSURE OR
THE CONDUCTORS.

LOCKNUT AND
PLASTIC BUSHING

RISER: SCHEDULE
80 RNC

SERVICE-LATERAL
CONDUCTORS TO
UTILITY TRANSFORMER OR
DISTRIBUTION PEDESTAL

RISER

LOCKNUT AND BUSHING

OFFSET PIPE NIPPLE: ANY KIND
OF NIPPLE MAY BE USED AS
LONG AS IT IS SIZED PROPERLY
AND IS LISTED FOR THE USE.

SERVICE-ENTRANCE CONDUCTORS—TO
THE SERVICE DISCONNECT SWITCH(ES)
AND OVERLOAD PROTECTION (CIRCUIT
BREAKER OR FUSES)

THE LOAD SIDE SERVICE-ENTRANCE
CONDUCTORS TERMINATE
IN THE LOWER SET OF LUGS.

Figure 13-7

Figure 13-8 The meter enclosure of a typical combination meter and distribution panel. Note the bussing on the load side of the meter.

13.3: ELEMENTS OF OVERHEAD SERVICE

Overhead services are very common in every part of the country. They were used almost exclusively for dwellings up until the late 1960s, when underground installations became more popular. An overhead service consists of a riser or mast, a weatherhead, a meter enclosure (or a gutter and a series of meter enclosures to distribute the service power to more than one meter), service-entrance conductors, and a service disconnect or a series of disconnects. See Figure 13-9.

The *NEC®* specifies very strict clearances from the ground, from the roof, and from windows or other openings where overhead conductors might be reachable. The general intent of the Code regulations concerning clearances for service-drop conductors is to prevent possible contact with the wiring by pedestrians or motor vehicles. The clearances are listed in detail in *230.9, 230.24,* and *230.26* of the Code and include clearances from ground level; from porches, windows, landings, and like areas of structures;

from roofs; and from roadways and parking areas, and for the point of attachment. These clearances are also summarized in Figure 13-10.

Overhead services may take many different configurations, and it is not possible to include all of them in this book. However, the outside overhead service with the riser installed through the overhang of the roof is the most common configuration used today and is examined more closely here. Figure 13-11 illustrates the simplest of these installations. Notice that the installation is configured to prevent any water from entering the riser. The length of conductor allowed for the **drip loop** and the connection to the drop conductors is shown as 36 in. (900 mm) minimum. The local power utility should be consulted about requirements for conductor length before the installation is completed. Extreme care must be taken in cutting the holes in the roof for the riser. Usually two holes are all that is needed, one on the top of the roof and the other in the bottom of the soffit of the overhang. These holes must align exactly; otherwise, the riser will not fit properly. In addition, care must be taken before the holes are drilled to ensure that any framing members or trusses do not block the intended riser location.

13.4: ELEMENTS OF AN UNDERGROUND SERVICE

Instead of a riser that goes up, the underground service has a "riser" that goes down. The riser for underground service conductors is usually 2-in., schedule 80, rigid nonmetallic conduit (PVC). This conduit is installed to protect the conductors against physical damage and is not considered to be a conduit system. The elements of a simple underground service are shown in Figure 13-12 (page 324).

The minimum burial depths for underground wiring methods are detailed in *Table 300.5*. The minimum required depth for direct buried service-lateral conductors is 24 in. (600 mm) minimum. If the fill to be used in filling the trench contains large rocks or rocks with sharp edges, conductors must be

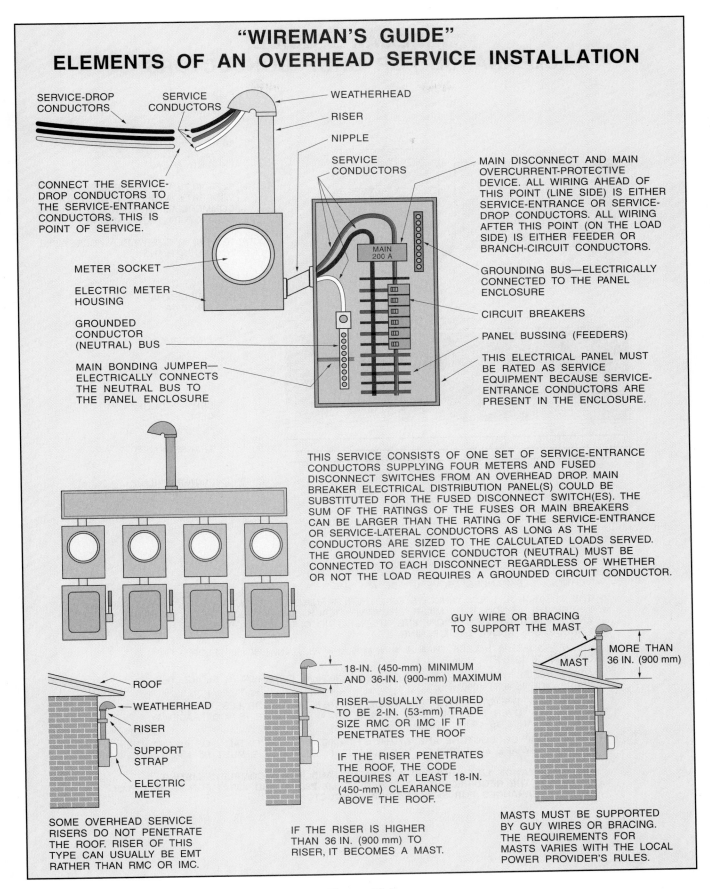

"WIREMAN'S GUIDE"
ELEMENTS OF AN OVERHEAD SERVICE INSTALLATION

SERVICE-DROP CONDUCTORS

SERVICE CONDUCTORS

WEATHERHEAD

RISER

NIPPLE

SERVICE CONDUCTORS

CONNECT THE SERVICE-DROP CONDUCTORS TO THE SERVICE-ENTRANCE CONDUCTORS. THIS IS POINT OF SERVICE.

MAIN DISCONNECT AND MAIN OVERCURRENT-PROTECTIVE DEVICE. ALL WIRING AHEAD OF THIS POINT (LINE SIDE) IS EITHER SERVICE-ENTRANCE OR SERVICE-DROP CONDUCTORS. ALL WIRING AFTER THIS POINT (ON THE LOAD SIDE) IS EITHER FEEDER OR BRANCH-CIRCUIT CONDUCTORS.

METER SOCKET

ELECTRIC METER HOUSING

GROUNDED CONDUCTOR (NEUTRAL) BUS

MAIN BONDING JUMPER—ELECTRICALLY CONNECTS THE NEUTRAL BUS TO THE PANEL ENCLOSURE

MAIN 200 A

GROUNDING BUS—ELECTRICALLY CONNECTED TO THE PANEL ENCLOSURE

CIRCUIT BREAKERS

PANEL BUSSING (FEEDERS)

THIS ELECTRICAL PANEL MUST BE RATED AS SERVICE EQUIPMENT BECAUSE SERVICE-ENTRANCE CONDUCTORS ARE PRESENT IN THE ENCLOSURE.

THIS SERVICE CONSISTS OF ONE SET OF SERVICE-ENTRANCE CONDUCTORS SUPPLYING FOUR METERS AND FUSED DISCONNECT SWITCHES FROM AN OVERHEAD DROP. MAIN BREAKER ELECTRICAL DISTRIBUTION PANEL(S) COULD BE SUBSTITUTED FOR THE FUSED DISCONNECT SWITCH(ES). THE SUM OF THE RATINGS OF THE FUSES OR MAIN BREAKERS CAN BE LARGER THAN THE RATING OF THE SERVICE-ENTRANCE OR SERVICE-LATERAL CONDUCTORS AS LONG AS THE CONDUCTORS ARE SIZED TO THE CALCULATED LOADS SERVED. THE GROUNDED SERVICE CONDUCTOR (NEUTRAL) MUST BE CONNECTED TO EACH DISCONNECT REGARDLESS OF WHETHER OR NOT THE LOAD REQUIRES A GROUNDED CIRCUIT CONDUCTOR.

ROOF

WEATHERHEAD

RISER

SUPPORT STRAP

ELECTRIC METER

SOME OVERHEAD SERVICE RISERS DO NOT PENETRATE THE ROOF. RISER OF THIS TYPE CAN USUALLY BE EMT RATHER THAN RMC OR IMC.

18-IN. (450-mm) MINIMUM AND 36-IN. (900-mm) MAXIMUM

RISER—USUALLY REQUIRED TO BE 2-IN. (53-mm) TRADE SIZE RMC OR IMC IF IT PENETRATES THE ROOF

IF THE RISER PENETRATES THE ROOF, THE CODE REQUIRES AT LEAST 18-IN. (450-mm) CLEARANCE ABOVE THE ROOF.

IF THE RISER IS HIGHER THAN 36 IN. (900 mm) TO RISER, IT BECOMES A MAST.

GUY WIRE OR BRACING TO SUPPORT THE MAST

MAST

MORE THAN 36 IN. (900 mm)

MASTS MUST BE SUPPORTED BY GUY WIRES OR BRACING. THE REQUIREMENTS FOR MASTS VARIES WITH THE LOCAL POWER PROVIDER'S RULES.

Figure 13-9

"WIREMAN'S GUIDE"
CLEARANCE FOR DROP CONDUCTORS

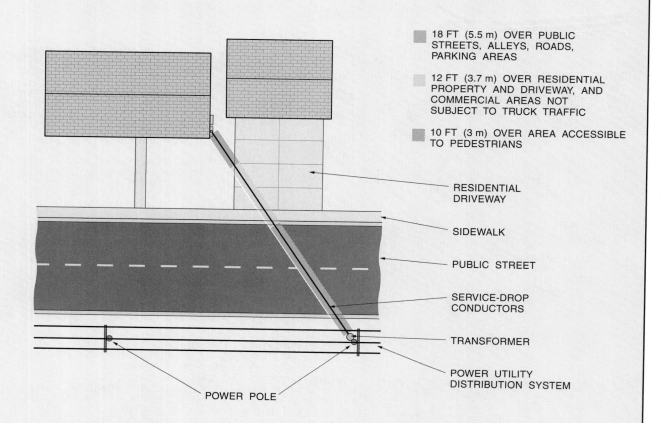

■ 18 FT (5.5 m) OVER PUBLIC STREETS, ALLEYS, ROADS, PARKING AREAS

■ 12 FT (3.7 m) OVER RESIDENTIAL PROPERTY AND DRIVEWAY, AND COMMERCIAL AREAS NOT SUBJECT TO TRUCK TRAFFIC

■ 10 FT (3 m) OVER AREA ACCESSIBLE TO PEDESTRIANS

RESIDENTIAL DRIVEWAY

SIDEWALK

PUBLIC STREET

SERVICE-DROP CONDUCTORS

TRANSFORMER

POWER UTILITY DISTRIBUTION SYSTEM

POWER POLE

- VERTICAL CLEARANCES ABOVE GRADE ARE AS SHOWN.

- THE CLEARANCE FROM WINDOWS OR OTHER OPENINGS, PARTICULARLY WINDOWS, WHERE THE CONDUCTORS MIGHT BE WITHIN REACH IS 3 FT (.900 m) FROM THE SIDES AND THE BOTTOM OF THE OPENING. THERE IS NO CLEARANCE GIVEN FOR CONDUCTORS ABOVE THE OPENING.

- ABOVE ROOFS WITH LESS THAN A 4-IN. (100 mm) TO 12-IN. (300 mm) SLOPE, THE CLEARANCE IS 8 FT (2.5 m).

- ABOVE ROOFS WITH 4-IN. (100 mm) TO 12-IN. (300 mm) OR GREATER SLOPE, THE CLEARANCE IS 3 FT (900 mm).

- ABOVE ONLY THE OVERHANG OF A ROOF, WITH 6 FT (1.8 m) OR LESS OF DROP CONDUCTOR OVER THE ROOF, AND SERVICE BY A THROUGH-THE-ROOF RACEWAY, THE CLEARANCE IS 18 IN. (450 mm).

- IF THE DROP CONDUCTORS ARE WITHIN 3 FT (900 mm) HORIZONTALLY OF ANY PLACE WHERE THEY ARE ACCESSIBLE, THESE VERTICAL SEPARATIONS MUST BE MAINTAINED.

THE INFORMATION LIST ABOVE IS A SUMMARY AND NOT A COMPLETE LISTING OF ALL OF THE REQUIREMENTS. SEE *NEC®* 230.9, 230.10, AND 230.24 FOR MORE DETAIL ABOUT CLEARANCES FOR SERVICE-DROP CONDUCTORS.

Figure 13-10

"WIREMAN'S GUIDE"
SOME REQUIREMENTS FOR INSTALLATION
OF AN OVERHEAD SERVICE

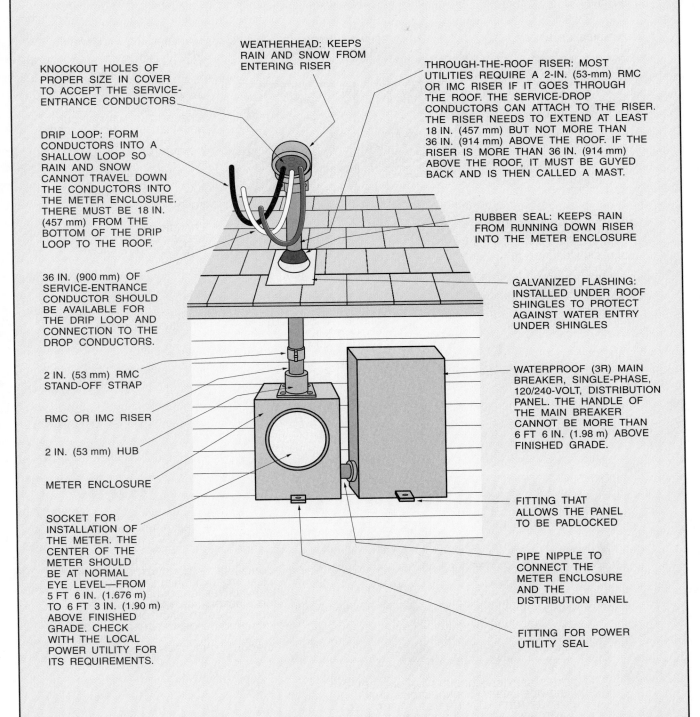

KNOCKOUT HOLES OF PROPER SIZE IN COVER TO ACCEPT THE SERVICE-ENTRANCE CONDUCTORS

DRIP LOOP: FORM CONDUCTORS INTO A SHALLOW LOOP SO RAIN AND SNOW CANNOT TRAVEL DOWN THE CONDUCTORS INTO THE METER ENCLOSURE. THERE MUST BE 18 IN. (457 mm) FROM THE BOTTOM OF THE DRIP LOOP TO THE ROOF.

36 IN. (900 mm) OF SERVICE-ENTRANCE CONDUCTOR SHOULD BE AVAILABLE FOR THE DRIP LOOP AND CONNECTION TO THE DROP CONDUCTORS.

2 IN. (53 mm) RMC STAND-OFF STRAP

RMC OR IMC RISER

2 IN. (53 mm) HUB

METER ENCLOSURE

SOCKET FOR INSTALLATION OF THE METER. THE CENTER OF THE METER SHOULD BE AT NORMAL EYE LEVEL—FROM 5 FT 6 IN. (1.676 m) TO 6 FT 3 IN. (1.90 m) ABOVE FINISHED GRADE. CHECK WITH THE LOCAL POWER UTILITY FOR ITS REQUIREMENTS.

WEATHERHEAD: KEEPS RAIN AND SNOW FROM ENTERING RISER

THROUGH-THE-ROOF RISER: MOST UTILITIES REQUIRE A 2-IN. (53-mm) RMC OR IMC RISER IF IT GOES THROUGH THE ROOF. THE SERVICE-DROP CONDUCTORS CAN ATTACH TO THE RISER. THE RISER NEEDS TO EXTEND AT LEAST 18 IN. (457 mm) BUT NOT MORE THAN 36 IN. (914 mm) ABOVE THE ROOF. IF THE RISER IS MORE THAN 36 IN. (914 mm) ABOVE THE ROOF, IT MUST BE GUYED BACK AND IS THEN CALLED A MAST.

RUBBER SEAL: KEEPS RAIN FROM RUNNING DOWN RISER INTO THE METER ENCLOSURE

GALVANIZED FLASHING: INSTALLED UNDER ROOF SHINGLES TO PROTECT AGAINST WATER ENTRY UNDER SHINGLES

WATERPROOF (3R) MAIN BREAKER, SINGLE-PHASE, 120/240-VOLT, DISTRIBUTION PANEL. THE HANDLE OF THE MAIN BREAKER CANNOT BE MORE THAN 6 FT 6 IN. (1.98 m) ABOVE FINISHED GRADE.

FITTING THAT ALLOWS THE PANEL TO BE PADLOCKED

PIPE NIPPLE TO CONNECT THE METER ENCLOSURE AND THE DISTRIBUTION PANEL

FITTING FOR POWER UTILITY SEAL

Figure 13-11

"WIREMAN'S GUIDE"
ELEMENTS OF AN UNDERGROUND
SERVICE INSTALLATION

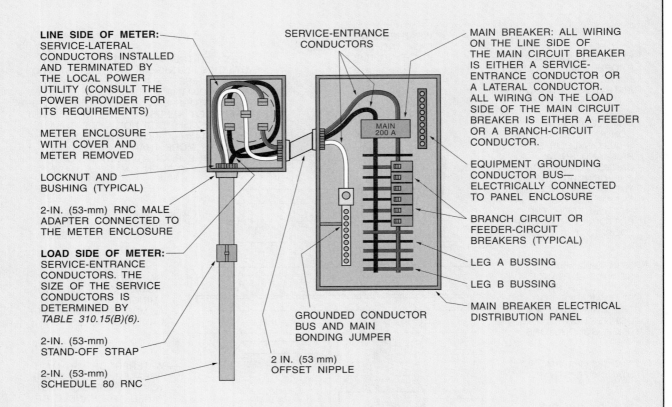

LINE SIDE OF METER: SERVICE-LATERAL CONDUCTORS INSTALLED AND TERMINATED BY THE LOCAL POWER UTILITY (CONSULT THE POWER PROVIDER FOR ITS REQUIREMENTS)

METER ENCLOSURE WITH COVER AND METER REMOVED

LOCKNUT AND BUSHING (TYPICAL)

2-IN. (53-mm) RNC MALE ADAPTER CONNECTED TO THE METER ENCLOSURE

LOAD SIDE OF METER: SERVICE-ENTRANCE CONDUCTORS. THE SIZE OF THE SERVICE CONDUCTORS IS DETERMINED BY *TABLE 310.15(B)(6).*

2-IN. (53-mm) STAND-OFF STRAP

2-IN. (53-mm) SCHEDULE 80 RNC

SERVICE-ENTRANCE CONDUCTORS

GROUNDED CONDUCTOR BUS AND MAIN BONDING JUMPER

2 IN. (53 mm) OFFSET NIPPLE

MAIN BREAKER: ALL WIRING ON THE LINE SIDE OF THE MAIN CIRCUIT BREAKER IS EITHER A SERVICE-ENTRANCE CONDUCTOR OR A LATERAL CONDUCTOR. ALL WIRING ON THE LOAD SIDE OF THE MAIN CIRCUIT BREAKER IS EITHER A FEEDER OR A BRANCH-CIRCUIT CONDUCTOR.

EQUIPMENT GROUNDING CONDUCTOR BUS— ELECTRICALLY CONNECTED TO PANEL ENCLOSURE

BRANCH CIRCUIT OR FEEDER-CIRCUIT BREAKERS (TYPICAL)

LEG A BUSSING

LEG B BUSSING

MAIN BREAKER ELECTRICAL DISTRIBUTION PANEL

MAIN 200 A

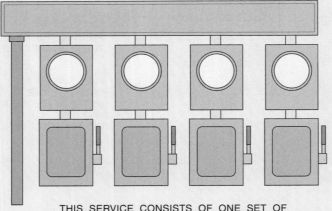

THE SUM OF THE RATINGS OF THE FUSES CAN BE LARGER THAN THE RATING OF THE SERVICE-ENTRANCE OR LATERAL CONDUCTORS AS LONG AS THE CONDUCTORS ARE SIZED TO THE CALCULATED LOAD SERVED. THE GROUNDED SERVICE CONDUCTOR (NEUTRAL) MUST BE CONNECTED TO EACH DISCONNECT REGARDLESS OF WHETHER OR NOT THE LOAD REQUIRES A GROUNDED CIRCUIT CONDUCTOR.

THIS SERVICE CONSISTS OF ONE SET OF SERVICE-LATERAL CONDUCTORS SUPPLYING FOUR METERS AND FUSED DISCONNECT SWITCHES FROM AN UNDERGROUND LATERAL.

Figure 13-12

covered with sand or other fine fill to prevent damage to the conductor insulation. Additionally, if the conductors are not encased in concrete, their location must be marked with warning tape placed in the trench at least 12 in. (300 mm) directly above the conductors. If the lateral is to be installed under a patio or concrete deck, the electrical contractor should provide a conduit sleeve under the patio or deck for the lateral conductors. The standard size conduit for a PVC riser and any necessary sleeve for dwellings is 2 in. (50 mm). See Figure 13-13.

The line side end of the lateral conductor terminates in a ground-mounted transformer or a power pedestal, as shown in Figure 13-1. The power pedestal sometimes includes a circuit breaker to protect the conductors and the equipment belonging to the power provider from damage from faults that could occur at the dwelling.

13.5: COMBINATION METER HOUSINGS AND DISTRIBUTION PANELS

Many manufacturers market combination meter housings and distribution panels for dwellings. This type of equipment is designed for fast installation and reliable service and is very popular in many areas of the country. Figure 13-14 shows some examples of such combination of service equipment, and Figure 13-15 (page 328) shows an example of a typical combination enclosure. The actual design details depend on the particular manufacturer.

13.6: THE SERVICE ON THE SAMPLE HOUSE

The sample house has an underground service that terminates on the north wall at the east end of the dwelling. Figure 13-16 (page 329) shows a layout of the sample house service. A formal plan must be completed before work is begun on installing the service equipment and conductors. A simple service like the one for the sample house does not require a very extensive plan, but large or more complicated services may require a comprehensive plan to prevent mistakes. The meter enclosure is connected to the main panel using a 2-in. (50-mm) offset nipple. The main panel of the sample house has only two loads: (1) the air-conditioning compressor unit, located about 15 ft (4.6 m) to the east of the service, and (2)

subpanel A, located in the garage. Construction Drawing E-6 presents more information about the main panel. The main panel has a 150-ampere main circuit breaker. The 150-ampere load represents the minimum size service conductors, service overcurrent-protective device rating, and disconnect switch rating as required from the load calculations in Chapter 12 of this book for the sample house.

The feeder conductors to the air-conditioning unit are 10 AWG copper in a 2-wire (plus equipment grounding conductor) NMS cable assembly. The conductors are fed from a 2-pole circuit breaker installed on circuits 1 and 3 of the main panel.

Subpanel A is supplied by a 125-ampere circuit breaker connected to circuits 5 and 7 of the main panel. The feeder conductors are sized from the 125-ampere load calculated in Chapter 12 for subpanel A. The conductors are size 0 AWG (1/0 AWG aluminum) in a 4-wire (two phase conductors, one grounded circuit conductor, and one equipment grounding conductor) Type SER cable. The grounded circuit conductor within the SER cable is usually smaller than the phase conductors, as allowed in *220.22*, and the equipment grounding conductor is sized according to *250.122*. The cable is run overhead through the attic to subpanel A in the garage. The unused breaker spaces in the main panel (circuits 2, 4, 6, and 8) are for future expansion should the owner install a hot tub, finish the basement, or install ground and site lighting or other home improvements requiring electricity.

Although subpanel A is not part of the service equipment, it is included in this part of the book because it deals with distribution equipment. Subpanel A is mounted in a stud wall in the garage and is supplied from the main panel. The cables for the subpanel's supply and for the various homeruns into subpanel A enter the stud space using holes drilled in the top and bottom plates. These cables are secured to the panel using cable connectors. The *NEC®* requires that all cables entering the panel, whether it be a service panel or a subpanel, be individually secured to the panel.

The NM cable is stripped of its outer sheathing once inside the panel. The paper wrappings are removed, and the individual conductors are prepared for termination. It is very important that the electrician wiring the dwelling be able to positively identify

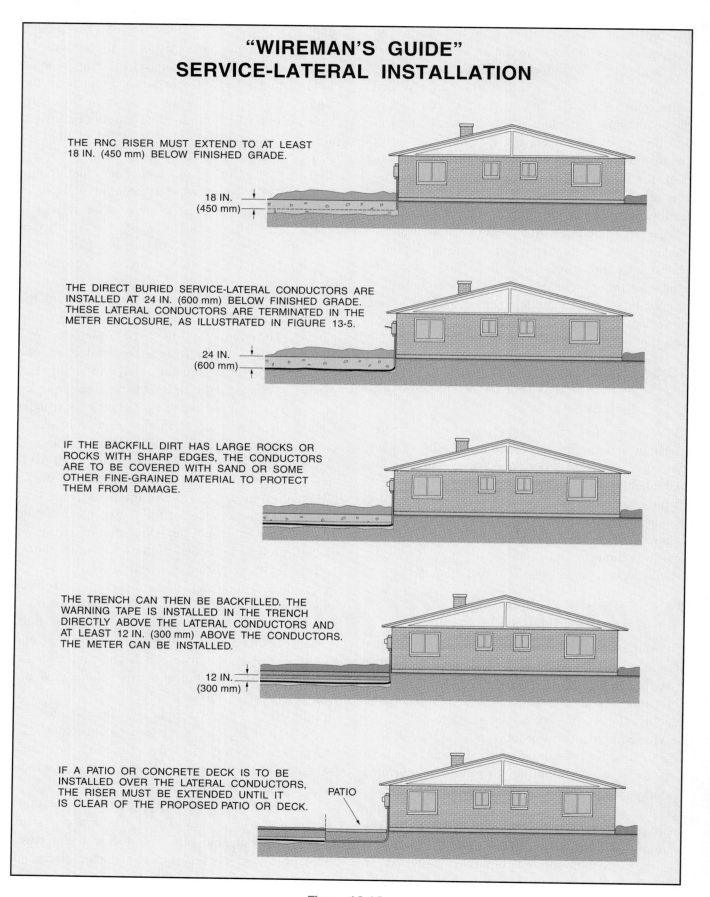

"WIREMAN'S GUIDE"
SERVICE-LATERAL INSTALLATION

THE RNC RISER MUST EXTEND TO AT LEAST 18 IN. (450 mm) BELOW FINISHED GRADE.

18 IN. (450 mm)

THE DIRECT BURIED SERVICE-LATERAL CONDUCTORS ARE INSTALLED AT 24 IN. (600 mm) BELOW FINISHED GRADE. THESE LATERAL CONDUCTORS ARE TERMINATED IN THE METER ENCLOSURE, AS ILLUSTRATED IN FIGURE 13-5.

24 IN. (600 mm)

IF THE BACKFILL DIRT HAS LARGE ROCKS OR ROCKS WITH SHARP EDGES, THE CONDUCTORS ARE TO BE COVERED WITH SAND OR SOME OTHER FINE-GRAINED MATERIAL TO PROTECT THEM FROM DAMAGE.

THE TRENCH CAN THEN BE BACKFILLED. THE WARNING TAPE IS INSTALLED IN THE TRENCH DIRECTLY ABOVE THE LATERAL CONDUCTORS AND AT LEAST 12 IN. (300 mm) ABOVE THE CONDUCTORS. THE METER CAN BE INSTALLED.

12 IN. (300 mm)

IF A PATIO OR CONCRETE DECK IS TO BE INSTALLED OVER THE LATERAL CONDUCTORS, THE RISER MUST BE EXTENDED UNTIL IT IS CLEAR OF THE PROPOSED PATIO OR DECK.

PATIO

Figure 13-13

"WIREMAN'S GUIDE"
COMBINATION METER AND DISTRIBUTION PANEL ENCLOSURES

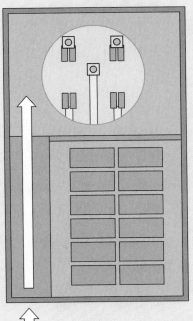

A COMBINATION DISTRIBUTION PANEL AND METER HOUSING COMBINES BOTH FUNCTIONS INTO ONE ENCLOSURE. THE INCOMING SERVICE-LATERAL CONDUCTORS ENTER AT THE BOTTOM OF THE ENCLOSURE AND ARE ROUTED THROUGH A SPECIAL CHANNEL TO THE METER ENCLOSURE SECTION. THIS CHANNEL IS TO SEPARATE THE SERVICE CONDUCTORS (LINE SIDE OF THE METER) FROM THE BUSSING (LOAD SIDE OF THE METER) IN THE DISTRIBUTION PANEL SECTION OF THE ENCLOSURE. THE SERVICE CONDUCTORS TERMINATE INTO LUGS ON THE TOP SIDE OF THE METER. THE LOAD SIDE OF THE METER IS CONNECTED TO BUSSING THAT CARRIES THE CURRENT TO THE DISTRIBUTION SECTION OF THE ENCLOSURE FOR CONNECTION TO CIRCUIT BREAKERS. IN THE COMBINATION ENCLOSURE SHOWN AT RIGHT, ALL OF THE BREAKER SPACES ARE FOR CONNECTION TO SERVICE EQUIPMENT; THEREFORE, NO BRANCH OR FEEDER CIRCUITS CAN ORIGINATE IN THIS ENCLOSURE.

METER ENCLOSURE SECTION

MAIN CIRCUIT BREAKER

ENCLOSURE SECTION

SPACES FOR CIRCUIT BREAKER (TYPICAL)

SERVICE CONDUCTOR CHANNEL

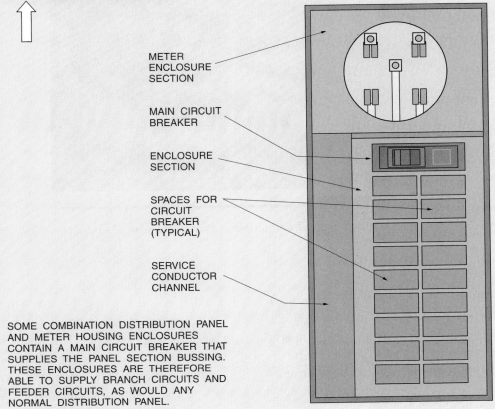

SOME COMBINATION DISTRIBUTION PANEL AND METER HOUSING ENCLOSURES CONTAIN A MAIN CIRCUIT BREAKER THAT SUPPLIES THE PANEL SECTION BUSSING. THESE ENCLOSURES ARE THEREFORE ABLE TO SUPPLY BRANCH CIRCUITS AND FEEDER CIRCUITS, AS WOULD ANY NORMAL DISTRIBUTION PANEL.

Figure 13-14

Figure 13-15 A typical combination enclosure with the electric meter installed.

each homerun. The Code requires that a schedule of circuits be posted in each panel identifying the loads that each circuit serves. For example, numbered markers may be wrapped around each homerun to identify the circuit number (the circuit breaker number) where that conductor terminates. The NM cables must be secured within 12 in. (300 mm) of its entry into the panel. The cables can be stapled to a scrap of wood installed for this purpose, as shown in Figure 13-17.

Also included in the service of the sample house are the system interface boxes for the telephone and cable television systems. These special systems are considered in Chapter 15 of this book, but the location where the electrical service receives its supply is usually the same location where the special systems also connect to their distribution systems. The special systems also require access to the grounding system for the electrical service, as covered in Chapter 14 of this book.

"WIREMAN'S GUIDE"
THE SAMPLE HOUSE SERVICE

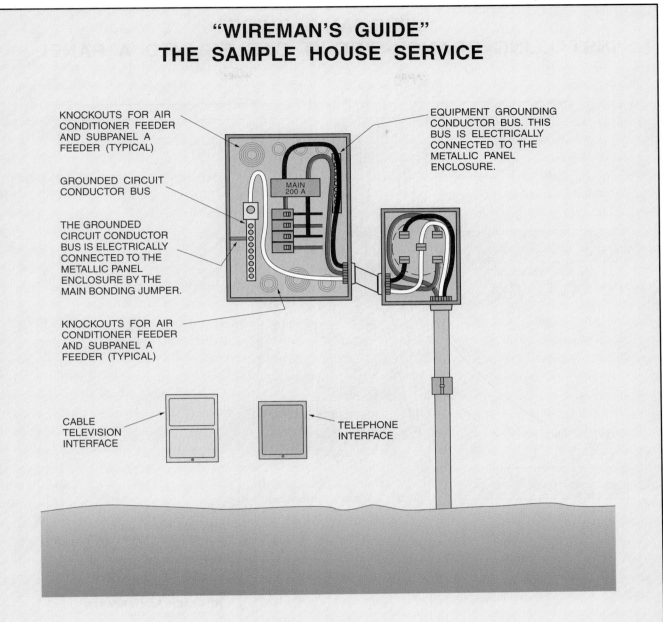

KNOCKOUTS FOR AIR CONDITIONER FEEDER AND SUBPANEL A FEEDER (TYPICAL)

GROUNDED CIRCUIT CONDUCTOR BUS

THE GROUNDED CIRCUIT CONDUCTOR BUS IS ELECTRICALLY CONNECTED TO THE METALLIC PANEL ENCLOSURE BY THE MAIN BONDING JUMPER.

KNOCKOUTS FOR AIR CONDITIONER FEEDER AND SUBPANEL A FEEDER (TYPICAL)

CABLE TELEVISION INTERFACE

MAIN 200 A

EQUIPMENT GROUNDING CONDUCTOR BUS. THIS BUS IS ELECTRICALLY CONNECTED TO THE METALLIC PANEL ENCLOSURE.

TELEPHONE INTERFACE

GROUNDING NOTES TO BE USED WHEN CONSIDERING THE SERVICE ON THE SAMPLE HOUSE:

THE METALLIC OFFSET NIPPLE BETWEEN THE METER HOUSING AND THE MAIN ELECTRICAL PANEL IS CONNECTED TO THE METER ENCLOSURE THROUGH A CONCENTRIC KNOCKOUT WHERE NOT ALL OF THE CONCENTRIC SECTIONS HAVE BEEN REMOVED.

THE METALLIC OFFSET NIPPLE IS CONNECTED TO THE MAIN PANEL THROUGH A CONCENTRIC KNOCKOUT WHERE ALL OF THE CONCENTRIC SECTIONS HAVE BEEN REMOVED.

THE RNC RISER IS CONNECTED TO THE MAIN PANEL THROUGH A CONCENTRIC KNOCKOUT WHERE NOT ALL OF THE CONCENTRIC SECTIONS HAVE BEEN REMOVED.

Figure 13-16

"WIREMAN'S GUIDE"
INSTALLING BRANCH-CIRCUIT CABLES INTO A PANEL

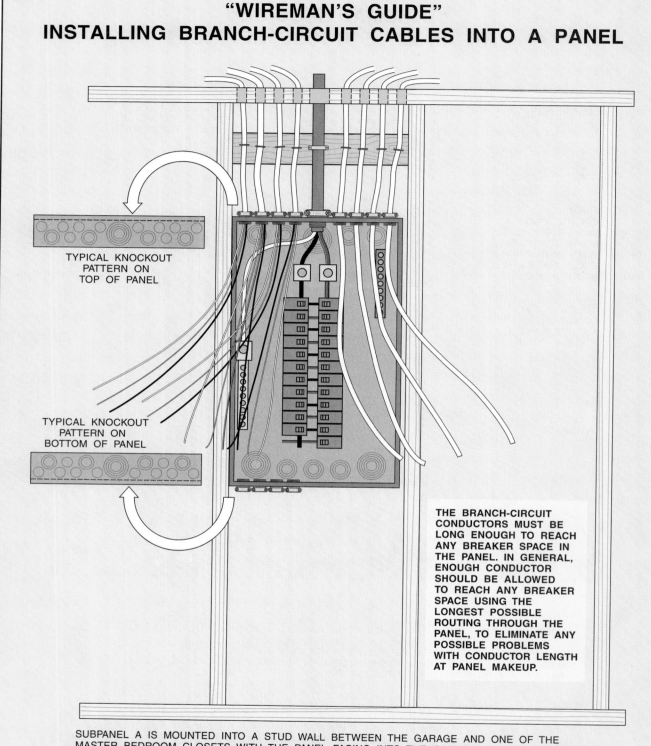

TYPICAL KNOCKOUT
PATTERN ON
TOP OF PANEL

TYPICAL KNOCKOUT
PATTERN ON
BOTTOM OF PANEL

THE BRANCH-CIRCUIT CONDUCTORS MUST BE LONG ENOUGH TO REACH ANY BREAKER SPACE IN THE PANEL. IN GENERAL, ENOUGH CONDUCTOR SHOULD BE ALLOWED TO REACH ANY BREAKER SPACE USING THE LONGEST POSSIBLE ROUTING THROUGH THE PANEL, TO ELIMINATE ANY POSSIBLE PROBLEMS WITH CONDUCTOR LENGTH AT PANEL MAKEUP.

SUBPANEL A IS MOUNTED INTO A STUD WALL BETWEEN THE GARAGE AND ONE OF THE MASTER BEDROOM CLOSETS WITH THE PANEL FACING INTO THE GARAGE. HOLES ARE DRILLED ALONG THE TOP OF THE STUD WALL TO ALLOW THE NMS CABLE ACCESS TO THE PANEL. CABLE CONNECTORS THAT ARE LISTED FOR NM CABLE ARE INSERTED INTO THE PANEL WHERE THE CABLE ENTRY IS TO BE ACCOMPLISHED, AND THE CABLE IS FITTED INTO THE CONNECTOR AND SECURED. ACCORDING TO *312.5(C)*, WHERE A CABLE ENTERS A CABINET, CUTOUT BOX, OR METER SOCKET ENCLOSURE, THE CABLE MUST BE SECURELY FASTENED TO THE ENCLOSURE OR CABINET. MOST CABLE CLAMPS ARE NOT APPROVED FOR MORE THAN ONE CABLE PER CONNECTOR. THE TOP AND BOTTOM SIDES OF THE PANEL HAVE AN ABUNDANCE OF KNOCKOUTS.

Figure 13-17

SUMMARY

Electrical services come in one of two types: overhead and underground. The conductors that provide power for an overhead service are called drop conductors. For underground services they are called lateral conductors. The drop or lateral conductors connect to the dwelling's service conductors at the point of service. The service conductors connect the electric utility's incoming power with the dwelling's service equipment to the main overcurrent-protective and disconnecting means. The main panel distributes the power to the feeders or branch circuits that are a part of the structure. The service disconnect means must be part of or immediately adjacent to the service overcurrent-protective device, and the center of the handle of the disconnect must not be more than 6 ft 7 in. (2 m) above the finished grade. One hundred amperes is the minimum size electrical service that can be installed on a single-family home. The local power provider usually determines the exact location for the electrical service. It also has standards for the proper mounting height for the meter enclosure.

REVIEW

1. According to *Table 300.5* in the Code, the minimum burial depth for underground feeders or service conductors at a dwelling is _____.

   a. 6 in.

   b. 12 in.

   c. 18 in.

   d. 24 in.

2. The maximum number of disconnects for one service is _____.

   a. 1

   b. 3

   c. 6

   d. none of the above—there is no limit as long as they are grouped

3. The grounded circuit conductor is usually _____.

   a. connected to the meter enclosure

   b. isolated from the meter enclosure

   c. connected to the enclosure only when the meter is installed

   d. isolated from the meter enclosure when the meter is installed

4. Service conductors for dwellings are sized according to _____ of the Code.

   a. *Table 310.16*

   b. *Table 310.17*

   c. *Table 310.15(B)(6)*

   d. *Table 310.13*

5. *True or False*: Underground service-entrance conductors sometimes originate at service power pedestals or a ground-mounted transformer.

6. *True or False*: The bonding conductor that connects the meter enclosure to the main panel in the sample house must employ a bonding (or grounding) bushing on both ends of the nipple.

7. *True or False*: The grounded service conductor is allowed to act as a grounding conductor at service-equipment and service raceways.

8. *True or False*: The service overcurrent-protective device and the service disconnecting means are separated from each other by at least 5 ft (1.5 m) at all dwelling services.

9. *True or False*: A mast can be part of an overhead service.

10. *True or False*: The main bonding jumper is located in the meter enclosure.

"WIREMAN'S GUIDE" REVIEW

1. Explain the bonding system on the line side of the service disconnect. Is the grounded service conductor allowed to bond line side equipment and raceways? What is the *NEC®* reference for this rule?

2. Explain why the Code allows service conductors to be installed on the outside surface of a dwelling but not inside the dwelling. Must the service conductors be protected from physical damage?

3. Draw a picture of a meter enclosure. Show all of the components, and label the line and load terminals for all conductor terminations.

4. Explain why most power providers require the lateral conductors that enter a meter enclosure to take the longest route to their terminals.

5. When a power pedestal has an overcurrent-protective device, the conductors from the pedestal to the dwelling become feeders and the pedestal is the service because the service main overcurrent-protective device is at the pedestal. Does the dwelling need to have a main overcurrent-protective device and a main disconnect also? Explain your answer. What section of the Code applies?

CHAPTER 14

Grounding and Grounding Systems

OBJECTIVES

On completion of this chapter, the student will be able to:

☑ State the main reasons for grounding an electrical system.

☑ Explain the differences between the five different grounding systems required at every dwelling service.

☑ State the Code requirements for sizing and installing the different ground system conductors in dwellings.

☑ Identify where grounding bushings or grounding locknuts are required.

☑ Design and install a grounding system for a one-family dwelling.

INTRODUCTION

This chapter is concerned with grounding. Proper grounding is one of the most important tasks of the electrician. Grounding is concerned with almost every aspect of the electrical system, from services to luminaires (fixtures). Much of the study of grounding is concerned with overcurrent, because the ground system is designed to ensure a path of low impedance to ground in the event of a fault. Therefore, a review of Chapter 7 of this book is recommended before continuing with Chapter 14.

14.1: GROUNDING

The act of grounding an electrical system is a two-part process. The first part involves literally connecting the electrical system to the earth. This connection is accomplished by grounding various types of electrodes as specified by the Code. The second part of grounding is concerned with creating a low-impedance path to the earth for the current to follow should a fault occur.

14.2: THE REASONS FOR GROUNDING

KEY TERMS

Low-impedance path to ground A route that is provided for the electrons in the event of a ground fault. The current flow during a fault must be large enough to cause the overcurrent-protective device to open. If the pathway has too large an impedance, the current flow may be limited to the level below that needed to cause the overcurrent-protective device to open.

The main reason for grounding an electrical system is to provide a permanent **low-impedance path to ground** in the event of a fault. The current flow from the fault must be large enough to open the overcurrent-protective device guarding the circuit. If the impedance is too large, current flow will be insufficient to open the overcurrent-protective device. There are other reasons stated by the Code for intentionally grounding electrical systems, and they are listed in *250.4(A)*, but if the other reasons did not exist, there would still be a need for a grounding system to provide a low-impedance fault path.

If the electrical system is not grounded, there is no reference point for the voltage. This means that the electrical system does not have a consistent voltage when referenced to the earth. Figure 14-1 shows a very simple electrical circuit. The circuit is not connected to the ground and therefore the voltage to ground will vary widely from different places in the circuit and at different times. The circuit voltage will be correct, however, because the voltage source is referenced to its own neutral. The voltage to ground reference will change constantly with a change in the number of free electrons that are present at that particular spot on the earth's surface at any one moment in time. This variability in voltage potential makes the low-impedance ground-fault path unreliable to produce a large current flow and must be eliminated.

A grounded electrical system intentionally connects the neutral conductor (the grounded circuit or the grounded service conductor) to the earth to eliminate the varying voltage potential, as shown in Figure 14-2. This connection allows the voltage to be referenced to the ground and thus makes the ground-fault path reliable to produce a large current flow if needed. It also allows for the measurement of the voltage to ground at any point in the system during troubleshooting.

14.3: THE GROUNDING SYSTEMS IN EVERY DWELLING

Five separate and distinct grounding systems are required by the *NEC®* for every dwelling or other structure. These systems all have their own procedures for sizing and installation, and they all have their own functions to perform. However, many of the procedures for installation are the same between systems. These similarities may lead to confusion if the procedures are not organized properly.

1. The Grounding Electrode System: The grounding electrode system is the system that physically connects the dwelling electrical system to the earth. There are two parts to the electrode grounding system: (1) the electrode that is in contact with the earth and (2) the conductor that connects the electrical service equipment to the electrode. When referring to the grounding electrode system, the Code uses the terms *electrode*, *grounding electrode*, or *grounding electrode conductor*.

2. The Main Bonding Jumper: The main bonding jumper connects the grounded circuit bus bar, the service enclosure, and the equipment grounding conductor bus bar to the service enclosure. At the service, all of the equipment grounding conductors and all of the grounded circuit conductors are electrically connected together, and then they are connected to the grounding electrode conductor(s). When referring to this system the Code uses the words *bonding jumper* or *main bonding jumper*. If the Code refers to just *bonding*, it may not be referring to the bonding jumper system. There is another grounding system that is referred to in the Code as simply the *bonding system*; this is discussed as the last item in this listing.

3. The Equipment Grounding System: The equipment grounding system connects all noncurrent-carrying parts of equipment and raceways on the load side of the service that are likely to become energized in the event of a ground fault to the bonding jumper system. This connection is made to provide the low-impedance path to ground for load side equipment and raceways. When the Code is referring to the equipment grounding system, it will use the word "grounding" or "equipment grounding" with an "ing" suffix.

4. The Grounded Circuit Conductor (Neutral) System: The grounded circuit conductor is the conductor that is intentionally connected to ground to produce the necessary voltage reference. The neutral is a current-carrying conductor that is part of the electrical circuit. When referring to the grounded circuit conductor, the Code uses the term *grounded*, with an *ed* suffix.

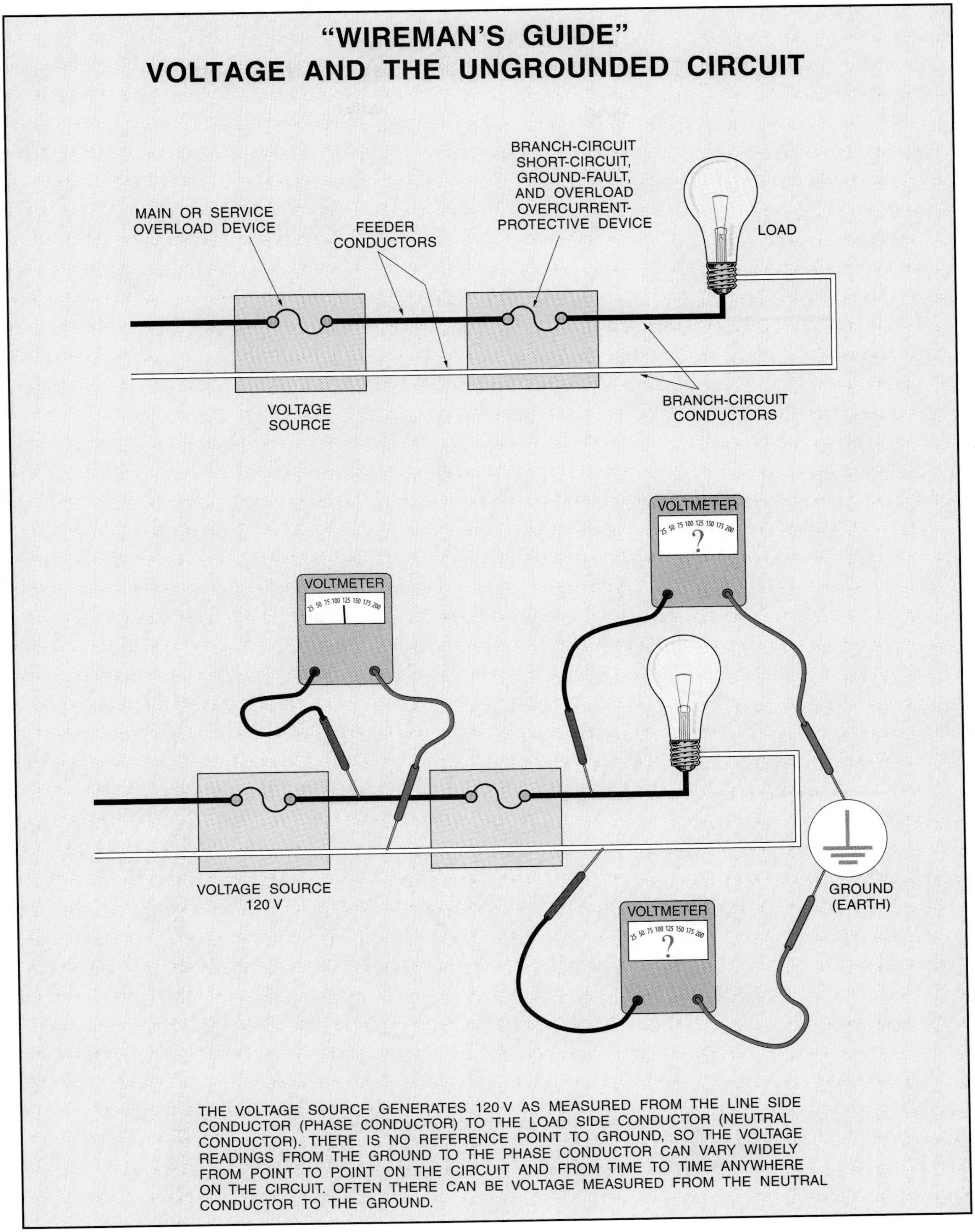

"WIREMAN'S GUIDE"
VOLTAGE AND THE UNGROUNDED CIRCUIT

BRANCH-CIRCUIT
SHORT-CIRCUIT,
GROUND-FAULT,
AND OVERLOAD
OVERCURRENT-
PROTECTIVE DEVICE

MAIN OR SERVICE
OVERLOAD DEVICE

FEEDER
CONDUCTORS

LOAD

VOLTAGE
SOURCE

BRANCH-CIRCUIT
CONDUCTORS

VOLTMETER

25 50 75 100 125 150 175 200

?

VOLTMETER

25 50 75 100 125 150 175 200

VOLTAGE SOURCE
120 V

GROUND
(EARTH)

VOLTMETER

25 50 75 100 125 150 175 200

?

THE VOLTAGE SOURCE GENERATES 120 V AS MEASURED FROM THE LINE SIDE
CONDUCTOR (PHASE CONDUCTOR) TO THE LOAD SIDE CONDUCTOR (NEUTRAL
CONDUCTOR). THERE IS NO REFERENCE POINT TO GROUND, SO THE VOLTAGE
READINGS FROM THE GROUND TO THE PHASE CONDUCTOR CAN VARY WIDELY
FROM POINT TO POINT ON THE CIRCUIT AND FROM TIME TO TIME ANYWHERE
ON THE CIRCUIT. OFTEN THERE CAN BE VOLTAGE MEASURED FROM THE NEUTRAL
CONDUCTOR TO THE GROUND.

Figure 14-1

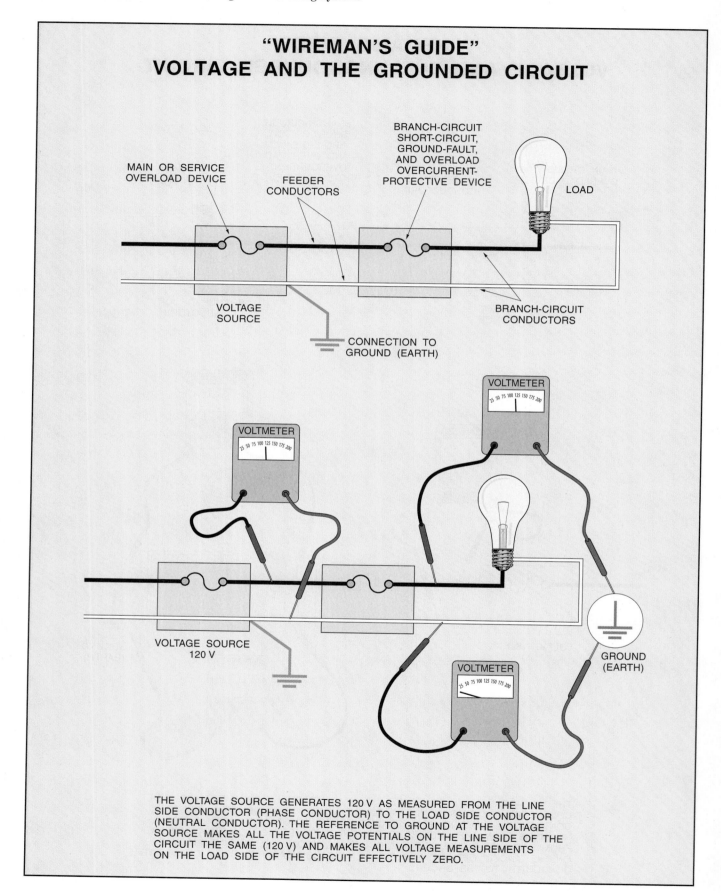

"WIREMAN'S GUIDE"
VOLTAGE AND THE GROUNDED CIRCUIT

MAIN OR SERVICE
OVERLOAD DEVICE

FEEDER
CONDUCTORS

BRANCH-CIRCUIT
SHORT-CIRCUIT,
GROUND-FAULT,
AND OVERLOAD
OVERCURRENT-
PROTECTIVE DEVICE

LOAD

VOLTAGE
SOURCE

CONNECTION TO
GROUND (EARTH)

BRANCH-CIRCUIT
CONDUCTORS

VOLTMETER
25 50 75 100 125 150 175 200

VOLTMETER
25 50 75 100 125 150 175 200

VOLTAGE SOURCE
120 V

VOLTMETER
25 50 75 100 125 150 175 200

GROUND
(EARTH)

THE VOLTAGE SOURCE GENERATES 120 V AS MEASURED FROM THE LINE
SIDE CONDUCTOR (PHASE CONDUCTOR) TO THE LOAD SIDE CONDUCTOR
(NEUTRAL CONDUCTOR). THE REFERENCE TO GROUND AT THE VOLTAGE
SOURCE MAKES ALL THE VOLTAGE POTENTIALS ON THE LINE SIDE OF THE
CIRCUIT THE SAME (120 V) AND MAKES ALL VOLTAGE MEASUREMENTS
ON THE LOAD SIDE OF THE CIRCUIT EFFECTIVELY ZERO.

Figure 14-2

5. The Bonding System: Bonding involves electrically connecting together all other metallic systems installed in the dwelling that are likely to become energized in the event of a fault. For example, the water piping system in the dwelling may become energized if a fault occurs. Bonding also involves the connection of the service-equipment enclosures and raceways for fault current. It is not a part of the electrode system, although they share certain common elements.

Each of these grounding systems, except for the bonding system, is concerned with a specific location in the electrical system, as shown in Figure 14-3. *NEC® 250* is the main source of information about the grounding systems, but other sections of the Code contribute as well.

14.4: THE GROUNDING ELECTRODE SYSTEM

Six items are listed in *250.52(A)* may be used as grounding electrodes. If any of these six items are available at the site, they are to be connected together to form an electrode grounding system. Briefly, these six items are:

1. Metal underground water pipe: The metal water pipe that provides potable water to the dwelling is to be used as a grounding electrode for the electrical system. Figure 14-4 shows the details for using the metal underground water pipe as an electrode ground. If the incoming water service is supplied by non-metallic piping, it cannot be used as a grounding electrode. If the water pipe is used as a grounding electrode, it must be supplemented by at least one other type of electrode. Figure 14-5 (page 340) shows a typical water pipe connection for the electrode conductor.

2. Concrete-encased electrode: Any building with a concrete foundation may have a concrete-encased electrode ground if it was installed. Figure 14-6 (page 341) shows the details of a concrete-encased electrode installation. If this feature is available, it must be used as an electrode, either as the supplemental electrode ground required for water pipes or as the system electrode ground if the water pipe is non-metallic.

3. Ground ring: A ground ring may be available at any dwelling construction site where there are excavations to install foundation walls. Figure 14-7 (page 342) shows the requirements for use of a ground ring as a grounding electrode. If a ground ring has been installed, it can be used as the supplemental electrode for the water pipe, or it may be the system grounding electrode with or without a concrete-encased electrode.

4. Rod or pipe electrode: A rod or pipe that is driven into the ground can be used as a grounding electrode under the conditions shown in Figure 14-8 (page 343). If none of the items listed above is available, or if the water pipe is the only electrode, the rod or pipe electrode can be used as the system electrode ground or as the supplemental electrode to the water pipe. Figure 14-9 (page 344) is a close-up view of a ground rod connection to the electrode conductor.

5. Plate electrode: A metal plate can serve as a grounding electrode if it is installed as shown in Figure 14-10 (page 345). It can supplement the water pipe ground or may be a stand-alone grounding electrode if no other electrode is available.

6. Metal frame of the building or structure: In some localities, dwellings are being constructed using metallic studs and framing members. If the structure has metallic framing members, the framing system must be connected to the service as an electrode. This requirement applies most often to commercial buildings with a metal frame supported by concrete footers.

The grounding electrode conductor connects the electrodes to the dwelling electrical system at the service. The grounding electrode conductor is sized according to *250.66* and is installed as shown in Figure 14-11 (page 346). It is possible to run one grounding electrode conductor to each of the electrodes in series, or to run a separate grounding electrode conductor to each electrode, or to use any combination of these two methods. The grounding electrode conductor will terminate in the service at any location at or ahead of the main overcurrent-protective device and main disconnection switch, but most often it is terminated at the service-equipment grounded service conductor (neutral) bus. The grounding electrode conductor can be solid or stranded or covered, insulated, or bare.

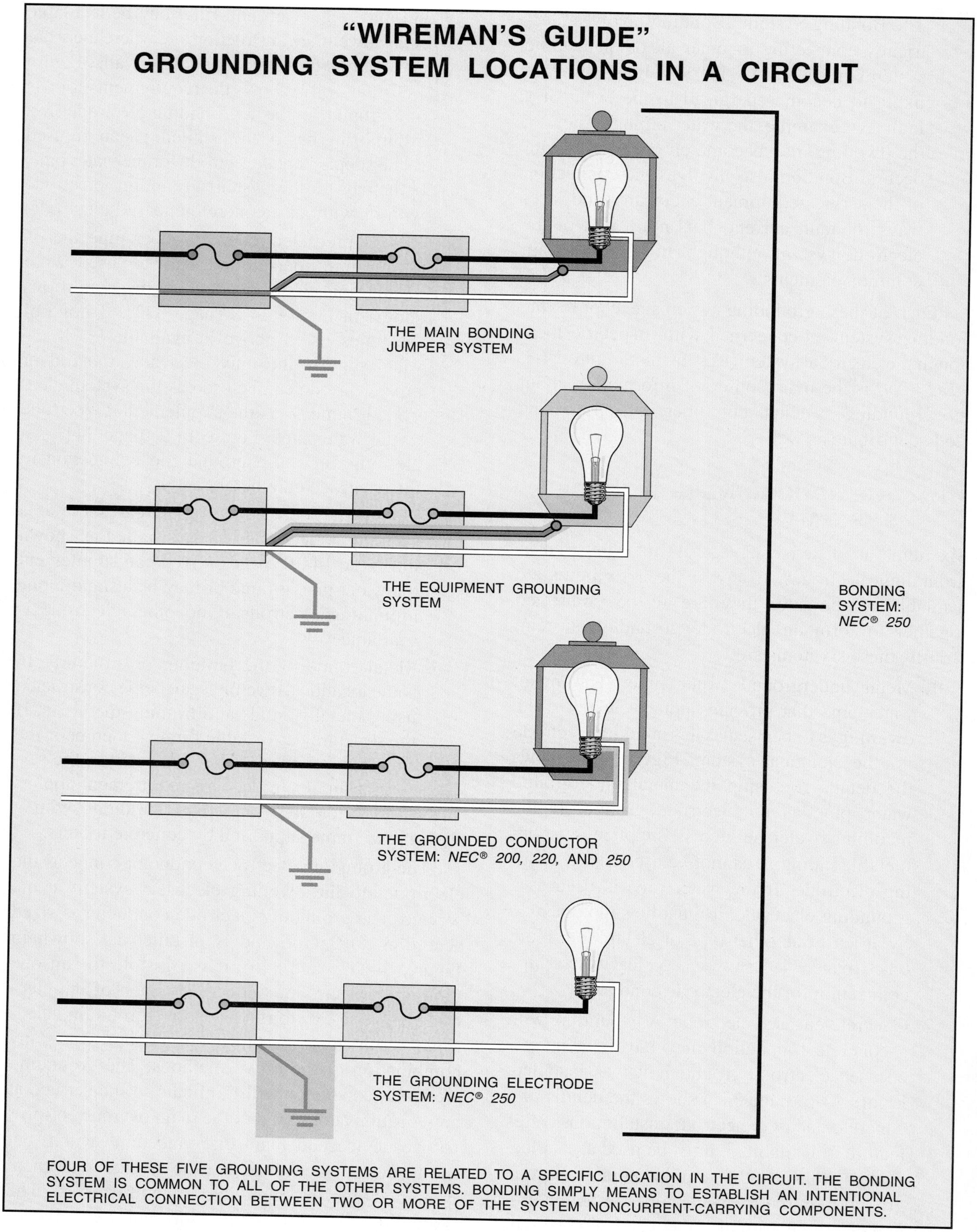

"WIREMAN'S GUIDE"
GROUNDING SYSTEM LOCATIONS IN A CIRCUIT

THE MAIN BONDING
JUMPER SYSTEM

THE EQUIPMENT GROUNDING
SYSTEM

THE GROUNDED CONDUCTOR
SYSTEM: *NEC® 200, 220,* AND *250*

THE GROUNDING ELECTRODE
SYSTEM: *NEC® 250*

BONDING
SYSTEM:
NEC® 250

FOUR OF THESE FIVE GROUNDING SYSTEMS ARE RELATED TO A SPECIFIC LOCATION IN THE CIRCUIT. THE BONDING
SYSTEM IS COMMON TO ALL OF THE OTHER SYSTEMS. BONDING SIMPLY MEANS TO ESTABLISH AN INTENTIONAL
ELECTRICAL CONNECTION BETWEEN TWO OR MORE OF THE SYSTEM NONCURRENT-CARRYING COMPONENTS.

Figure 14-3

"WIREMAN'S GUIDE"
THE INCOMING WATER PIPE AS
A GROUNDING ELECTRODE

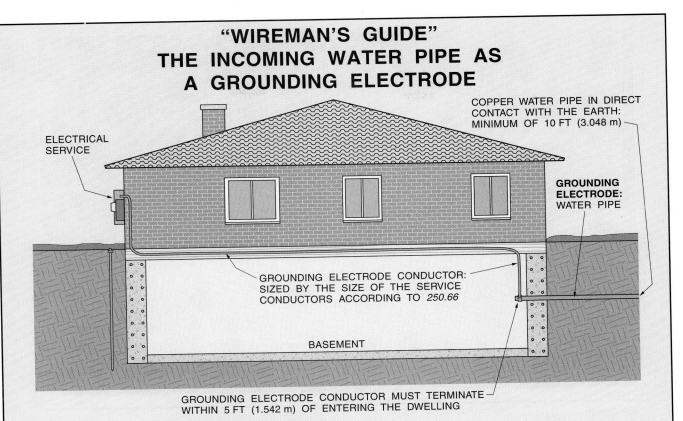

COPPER WATER PIPE IN DIRECT CONTACT WITH THE EARTH: MINIMUM OF 10 FT (3.048 m)

ELECTRICAL SERVICE

GROUNDING ELECTRODE: WATER PIPE

GROUNDING ELECTRODE CONDUCTOR: SIZED BY THE SIZE OF THE SERVICE CONDUCTORS ACCORDING TO 250.66

BASEMENT

GROUNDING ELECTRODE CONDUCTOR MUST TERMINATE WITHIN 5 FT (1.542 m) OF ENTERING THE DWELLING

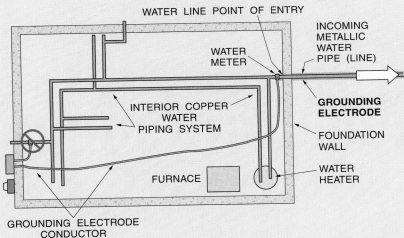

WATER LINE POINT OF ENTRY

INCOMING METALLIC WATER PIPE (LINE)

WATER METER

INTERIOR COPPER WATER PIPING SYSTEM

GROUNDING ELECTRODE

FOUNDATION WALL

FURNACE

WATER HEATER

GROUNDING ELECTRODE CONDUCTOR

THE GROUNDING ELECTRODE IS THE METALLIC WATER PIPE THAT IS IN CONTACT WITH THE EARTH OUTSIDE THE BASEMENT. THE INTERIOR WATER PIPING SYSTEM IS NOT PART OF THE GROUNDING ELECTRODE. EVEN THOUGH THERE MAY BE A WATER PIPE CLOSER TO THE ELECTRICAL SERVICE, IT CANNOT BE USED AS THE GROUNDING ELECTRODE CONDUCTOR FROM THE ELECTRICAL SERVICE TO THE POINT OF ENTRY OF THE WATER LINE.

THE GROUNDING ELECTRODE CONDUCTOR FOR THE COPPER WATER PIPE MUST BE TERMINATED WITHIN 5 FT (1.542 m) OF WHERE THE WATER PIPE ENTERS THE BUILDING. IF THERE ARE ANY INSULATING FITTINGS OR WATER METERS BETWEEN THE TERMINATION OF THE GROUNDING ELECTRODE CONDUCTOR AND THE OUTSIDE EARTH, A BONDING CONNECTION MUST BE INSTALLED AROUND THE FITTING OR METER. THE GROUNDING ELECTRODE CONDUCTOR MUST BE UNSPLICED, OR IF SPLICED USING A NONREVERSIBLE METHOD SUCH AS USE OF HIGH-COMPRESSION CONNECTORS OR EXOTHERMIC WELDING.

THE SIZE OF THE GROUNDING ELECTRODE CONDUCTOR IS DETERMINED FROM THE SIZE OF THE INCOMING SERVICE CONDUCTORS AND 250.66. ALTHOUGH GROUNDING ELECTRODE CONDUCTORS CAN BE ALUMINUM, IN MOST CASES THE AHJ WILL REQUIRE A COPPER CONDUCTOR TO BE USED. THE CLAMP MUST BE LISTED FOR USE AS A WATER PIPE GROUND CLAMP, AND THE CLAMP MUST BE SIZED FOR THE INCOMING WATER PIPE AND THE SIZE OF THE GROUNDING ELECTRODE CONDUCTOR USED.

GROUNDING ELECTRODE CONDUCTOR FROM SERVICE EQUIPMENT

INCOMING COPPER WATER PIPE

BONDING CONNECTION

WATER METER

LISTED PIPE CLAMP TO TERMINATE THE GROUNDING ELECTRODE CONDUCTOR (TYPICAL)

FOUNDATION WALL

Figure 14-4

Figure 14-5 The water pipe grounding electrode conductor connection is made within 5 ft (1.5 m) of the point of entry to the dwelling.

14.5: THE BONDING JUMPER AND THE MAIN BONDING JUMPER

The bonding jumper system electrically connects together all of the grounded circuit conductors, and equipment grounding conductors, to the grounding electrode conductors. The main service bus that terminates the grounded circuit conductors (neutrals) is connected to the service enclosure by the main bonding jumper. The equipment grounding conductor bus is electrically connected to the service enclosure; therefore, the grounded conductors are electrically connected to the grounding conductors through the service enclosure. Because the grounding electrode conductor terminates on the grounded circuit conductor bus, all three types of grounded conductors are thereby connected.

The main bonding jumper can be a wire sized according to *250.28(D),* or it can be a busbar or a green screw included with the service equipment, as shown in Figure 14-12 (page 347).

14.6: THE EQUIPMENT GROUNDING SYSTEM

The equipment grounding conductor system connects all noncurrent-carrying metal components of electrical equipment and raceways with the grounding electrode conductor at the service to provide a low-impedance path to ground. If there is a fault to the metal casing of a saw, for example, the equipment grounding system will provide a path to ground that ensures that enough current will flow to open the overcurrent-protective device.

For installing cable with the sheathing that is not approved as an equipment grounding pathway, as with NM cable, an equipment grounding conductor must be included within the cable sheathing with the phase and neutral conductors to provide the equipment grounding path. The size of the equipment grounding conductor is determined according to *250.122* and is based on the size of the overcurrent-protective device protecting the circuit. See Figure

"WIREMAN'S GUIDE"
CONCRETE-ENCASED ELECTRODE

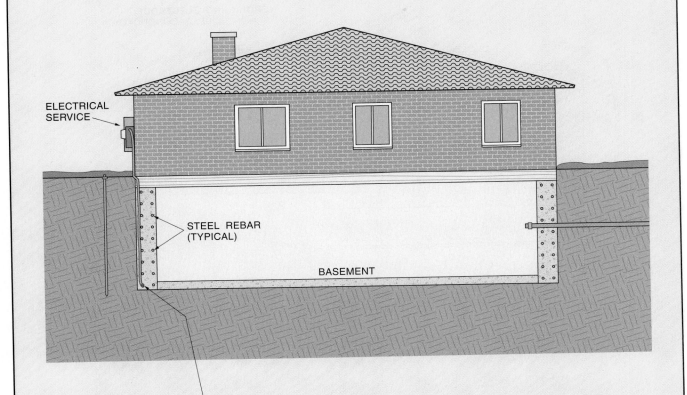

ELECTRICAL SERVICE

STEEL REBAR (TYPICAL)

BASEMENT

THE REBAR IS THE GROUNDING ELECTRODE: GROUNDING ELECTRODE CONDUCTOR BONDED TO REBAR AT THE BOTTOM OF FOUNDATION WALL

THE STEEL REINFORCING BAR IN CONCRETE FOUNDATION IS ALLOWED AS A GROUNDING ELECTRODE AND MUST BE USED IF IT IS AVAILABLE AT THE SITE. A 2 AWG BARE CONDUCTOR IS BONDED TO AT LEAST 20 FT (6 m) OR 1/2-IN. (.153-m)-DIAMETER REBAR AND IS ROUTED TO THE ELECTRIC SERVICE. THE CLAMP MUST BE LISTED FOR THE 2 AWG CONDUCTORS AND APPROVED FOR INSTALLATION IN CONCRETE. AN ALTERNATIVE IS TO INCLUDE AT LEAST 20 FT (6 m) OF 2 AWG COPPER SOLID CONDUCTOR LOCATED AT THE BOTTOM OF THE FOUNDATION WALL AND ROUTED TO END AT THE SERVICE EQUIPMENT. THIS CONDUCTOR MUST BE UNSPLICED, OR IF IT IS SPLICED, THE SPLICE MUST BE ACCOMPLISHED WITH AN IRREVERSIBLE-PRESSURE LUG LISTED FOR THE USE, EXOTHERMIC WELDING, OR OTHER PERMANENT MEANS ACCEPTABLE TO THE AHJ.

Figure 14-6

"WIREMAN'S GUIDE"
GROUND RING

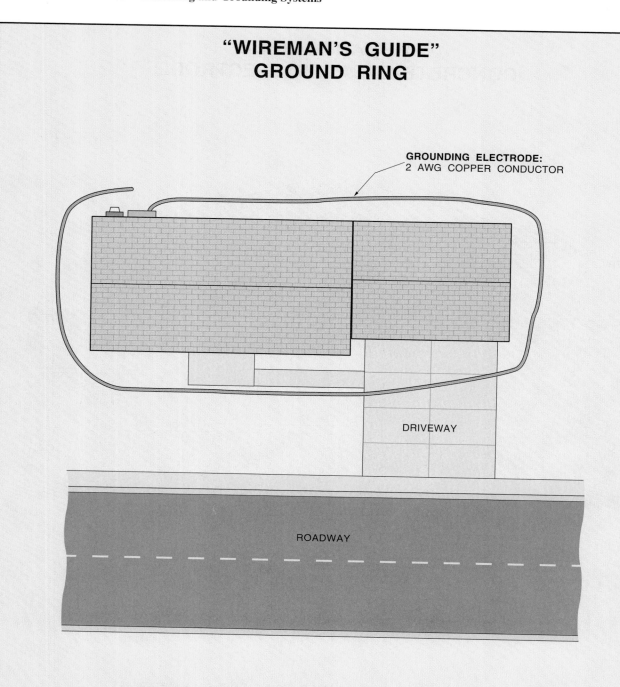

GROUNDING ELECTRODE:
2 AWG COPPER CONDUCTOR

DRIVEWAY

ROADWAY

A GROUND RING IS A GROUNDING ELECTRODE AND A GROUNDING ELECTRODE CONDUCTOR. THE GROUND RING IS COMPOSED OF AT LEAST 20 FT (6.096 m) OF 2 AWG COPPER CONDUCTOR SURROUNDING THE ENTIRE DWELLING. ONE END OF THE CONDUCTOR IS ROUTED TO THE SERVICE EQUIPMENT FOR TERMINATION. THE GROUNDING ELECTRODE MUST BE BURIED AT LEAST 30 IN. (750 mm) BELOW THE SURFACE AND MUST BE UNSPLICED. IF THE GROUNDING ELECTRODE CONDUCTOR IS SPLICED, THE SPLICE MUST BE ACCOMPLISHED USING A NONREVERSIBLE MEANS SUCH AS EXOTHERMIC WELDING OR HIGH-COMPRESSION CONNECTORS. IF A GROUND RING IS TO BE INSTALLED, IT IS BEST TO DO SO WHEN THE FOUNDATION WALLS ARE EXPOSED BEFORE THEY ARE BACKFILLED.

Figure 14-7

"WIREMAN'S GUIDE"
GROUND RODS AND PIPES

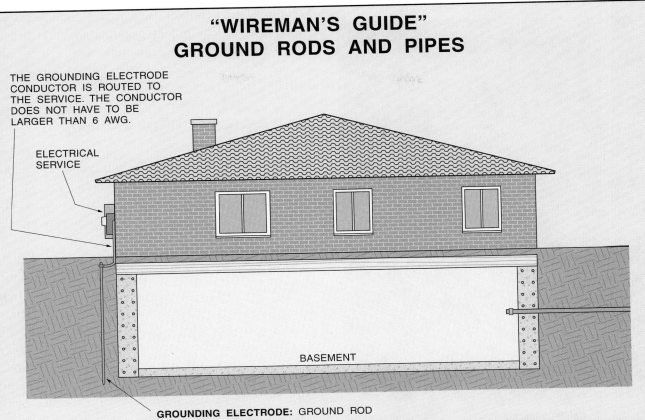

THE GROUNDING ELECTRODE
CONDUCTOR IS ROUTED TO
THE SERVICE. THE CONDUCTOR
DOES NOT HAVE TO BE
LARGER THAN 6 AWG.

ELECTRICAL
SERVICE

BASEMENT

GROUNDING ELECTRODE: GROUND ROD

WHERE NONE OF THE ELECTRODES ARE AVAILABLE, THE CODE REQUIRES THAT ONE OR MORE ELECTRODES BE INSTALLED. THE ROD AND PIPE ELECTRODE CONSISTS OF AT LEAST 8 FT (2.5 m) OF GALVANIZED OR OTHERWISE CORROSION-RESISTANT COATING 3/4 IN. TRADE SIZE (21 mm TRADE SIZE) STEEL OR IRON PIPE. AN ALTERNATIVE IS AN 8 FT (2.5 m) STEEL ROD NO LESS THAN 1/2 IN. (13 mm) IN DIAMETER. THE ROD OR PIPE MUST BE DRIVEN INTO THE GROUND FOR THE ENTIRE LENGTH. A LISTED CLAMP IS THEN USED TO CONNECT THE ROD OR PIPE TO THE GROUNDING ELECTRODE CONDUCTOR, AND THE CONDUCTOR IS ROUTED TO THE SERVICE EQUIPMENT. THE GROUNDING ELECTRODE CONDUCTOR MUST BE UNSPLICED OR, IF SPLICED, CONNECTED USING A NONREVERSIBLE MEANS SUCH AS HIGH-COMPRESSION CONNECTORS OR EXOTHERMIC WELDING. THE GROUNDING ELECTRODE CONDUCTOR FROM THE ROD OR PIPE TO THE SERVICE DOES NOT NEED TO BE LARGER THAN 6 AWG COPPER.

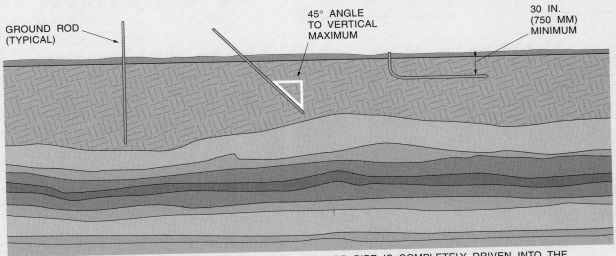

GROUND ROD
(TYPICAL)

45° ANGLE
TO VERTICAL
MAXIMUM

30 IN.
(750 MM)
MINIMUM

IF BEDROCK IS ENCOUNTERED BEFORE THE GROUND ROD OR PIPE IS COMPLETELY DRIVEN INTO THE GROUND, THE ROD OR PIPE CAN BE DRIVEN AT UP TO A 45° ANGLE FROM VERTICAL. IF BEDROCK IS STILL ENCOUNTERED, THE GROUND ROD OR PIPE CAN BE BENT AND PUT INTO A TRENCH AT LEAST 30 IN. (750 mm) DEEP. THE TOP OF THE GROUND ROD MUST BE LEVEL WITH OR BELOW THE GROUND LEVEL, OR THE CLAMP AND CONDUCTOR WILL NEED TO BE PROTECTED AGAINST PHYSICAL DAMAGE.

Figure 14-8

Figure 14-9 A typical ground rod, ground rod clamp, and electrode conductor termination.

14-13 (page 348). The equipment grounding conductor can be covered, insulated, or bare. When the conductor is covered or insulated, it must be either green or green with one or more yellow stripes.

Several wiring methods are also employed as equipment grounding pathways other than conductors in cables. These wiring methods are the metallic conduit and raceway systems listed in *250.118*. The metallic raceways are allowed to serve as the equipment grounding conductor with varying restrictions. With these wiring methods, larger circuits (with a higher ampere rating) require larger wire, which

requires larger conduit. Because the conduit is being used for the equipment grounding conductor, the size of the equipment grounding conductor is automatically increased when there is an increase in circuit size.

14.7: THE GROUNDED CONDUCTOR SYSTEM

Whenever the Code refers to the *grounded* conductor, it is referring to the neutral conductor. As shown in Figure 14-2, the grounding of the load side conductor establishes a common reference point with the ground for voltage regulation.

NEC® 200 details the use of the grounded circuit conductor as well as the use of white as an insulation color. White is reserved for use as a grounded circuit conductor identification, and if a white insulated conductor is used for any other purpose, it must be permanently re-identified at its terminations. *NEC® 220.22* provides information on sizing the grounded circuit conductor for feeders and service conductors. *NEC® 250* includes information on the use of the grounded circuit conductor at services.

The grounded circuit conductor is a current-carrying conductor and is a part of any branch circuit that operates at 120 volts in a residential installation. In a 2-wire circuit, the grounded conductor carries the same current as the ungrounded conductors and therefore must be the same size as the ungrounded conductors. In multiwire branch circuits, the grounded circuit conductor is installed to carry any imbalance between the phase conductors. The maximum possible imbalance between two-phase conductors—for example, phase A and phase B—will be the circuit ampere rating, because all loads on phase A may be off while all loads from phase B may be on at any time. Therefore, the grounded circuit conductor will be the same size as the ungrounded circuit conductors in multiwire branch circuits. If the multiwire branch circuit serves only one 240-volt load, such as for a range, the neutral needs to be sized to carry back the unbalanced current. The unbalanced current will be the load from those components of the load that are 120 volt rated.

On the load side of the service disconnect, the grounded conductors are kept separate from the grounding conductors. The grounded conductors are current-carrying circuit conductors, and to connect

"WIREMAN'S GUIDE"
PLATE ELECTRODE

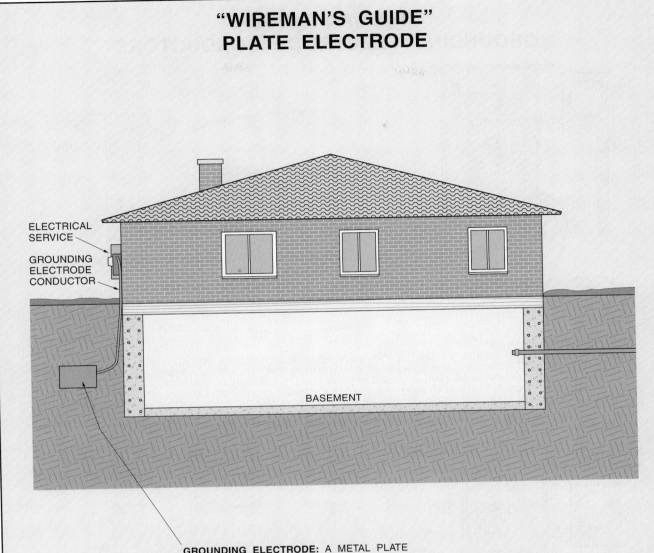

ELECTRICAL
SERVICE

GROUNDING
ELECTRODE
CONDUCTOR

BASEMENT

GROUNDING ELECTRODE: A METAL PLATE
THAT EXPOSES AT LEAST 2 FT² (.186 m²)
OF SURFACE AREA TO THE SOIL

A METALLIC PLATE THAT EXPOSES AT LEAST 2 FT² (.186 m²) OF SURFACE AREA TO
THE EARTH IS ACCEPTABLE AS A GROUNDING ELECTRODE. IF THE PLATE IS STEEL
OR IRON, IT MUST BE AT LEAST ¼ IN. (6.4 m) THICK. IF THE PLATE IS MADE OF
COPPER, IT MUST BE AT LEAST .06 IN. (1.5 mm) THICK. THE PLATE MUST BE BURIED
AT LEAST 30 IN. (750 mm) BELOW THE SURFACE. THE GROUNDING ELECTRODE
CONDUCTOR DOES NOT NEED TO BE LARGER THAN 6 AWG COPPER. THE GROUNDING
ELECTRODE CONDUCTOR MUST BE UNSPLICED, OR IF IT IS SPLICED, THE SPLICE
MUST BE ACCOMPLISHED USING A NONREVERSIBLE MEANS SUCH AS EXOTHERMIC
WELDING OR HIGH-COMPRESSION CONNECTORS.

Figure 14-10

"WIREMAN'S GUIDE"
GROUNDING ELECTRODE CONDUCTORS

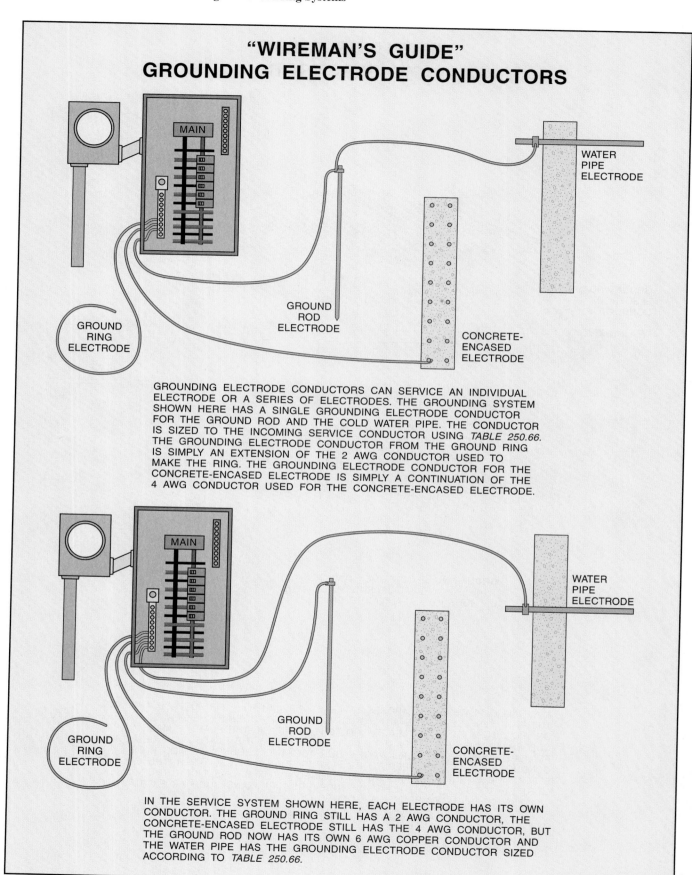

GROUNDING ELECTRODE CONDUCTORS CAN SERVICE AN INDIVIDUAL ELECTRODE OR A SERIES OF ELECTRODES. THE GROUNDING SYSTEM SHOWN HERE HAS A SINGLE GROUNDING ELECTRODE CONDUCTOR FOR THE GROUND ROD AND THE COLD WATER PIPE. THE CONDUCTOR IS SIZED TO THE INCOMING SERVICE CONDUCTOR USING *TABLE 250.66*. THE GROUNDING ELECTRODE CONDUCTOR FROM THE GROUND RING IS SIMPLY AN EXTENSION OF THE 2 AWG CONDUCTOR USED TO MAKE THE RING. THE GROUNDING ELECTRODE CONDUCTOR FOR THE CONCRETE-ENCASED ELECTRODE IS SIMPLY A CONTINUATION OF THE 4 AWG CONDUCTOR USED FOR THE CONCRETE-ENCASED ELECTRODE.

IN THE SERVICE SYSTEM SHOWN HERE, EACH ELECTRODE HAS ITS OWN CONDUCTOR. THE GROUND RING STILL HAS A 2 AWG CONDUCTOR, THE CONCRETE-ENCASED ELECTRODE STILL HAS THE 4 AWG CONDUCTOR, BUT THE GROUND ROD NOW HAS ITS OWN 6 AWG COPPER CONDUCTOR AND THE WATER PIPE HAS THE GROUNDING ELECTRODE CONDUCTOR SIZED ACCORDING TO *TABLE 250.66*.

Figure 14-11

"WIREMAN'S GUIDE"
THE MAIN BONDING JUMPER

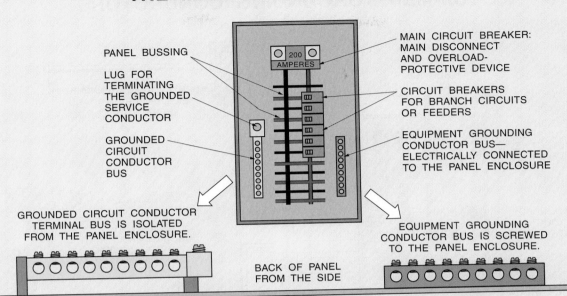

PANEL BUSSING

LUG FOR TERMINATING THE GROUNDED SERVICE CONDUCTOR

GROUNDED CIRCUIT CONDUCTOR BUS

MAIN CIRCUIT BREAKER: MAIN DISCONNECT AND OVERLOAD-PROTECTIVE DEVICE

CIRCUIT BREAKERS FOR BRANCH CIRCUITS OR FEEDERS

EQUIPMENT GROUNDING CONDUCTOR BUS— ELECTRICALLY CONNECTED TO THE PANEL ENCLOSURE

GROUNDED CIRCUIT CONDUCTOR TERMINAL BUS IS ISOLATED FROM THE PANEL ENCLOSURE.

EQUIPMENT GROUNDING CONDUCTOR BUS IS SCREWED TO THE PANEL ENCLOSURE.

BACK OF PANEL FROM THE SIDE

THE GROUNDED CIRCUIT CONDUCTOR BUS IS ISOLATED FROM THE DISTRIBUTION PANEL ENCLOSURE, USUALLY BY NONMETALLIC SUPPORTS. THE EQUIPMENT GROUNDING BUS IS SCREWED (BONDED) TO THE PANEL ENCLOSURE. IN THIS CONFIGURATION, THEY ARE NOT IN ELECTRICAL CONTACT WITH EACH OTHER.

MAIN BONDING JUMPER

SIDE VIEW AND TOP VIEW OF A MAIN BONDING JUMPER AS A BUS

BACK OF PANEL FROM THE SIDE

THE INSTALLATION OF THE MAIN BONDING JUMPER (IN THIS EXAMPLE, A SCREW THAT IS PART OF THE DISTRIBUTION PANEL COMPONENTS) CONNECTS THE GROUNDED CIRCUIT-CONDUCTOR BUS TO THE DISTRIBUTION PANEL ENCLOSURE. THE GROUNDED BUS IS NOW ELECTRICALLY CONNECTED WITH THE GROUNDING BUS. THE MAIN BONDING JUMPER MAY BE A SMALL BUS THAT TERMINATES IN THE GROUNDED CONDUCTOR BUS AND SCREWS (BONDS) TO THE PANEL ENCLOSURE.

MAIN BONDING JUMPER

USE OF A SCREW IS NOT THE ONLY METHOD FOR INSTALLING A MAIN BONDING JUMPER. A CONDUCTOR CAN CONNECT THE GROUNDED CIRCUIT-CONDUCTOR BUS WITH THE EQUIPMENT GROUNDING CONDUCTOR BUS. THE CONDUCTOR MUST BE THE SAME SIZE AS, OR LARGER THAN, THE GROUNDING ELECTRODE CONDUCTOR.

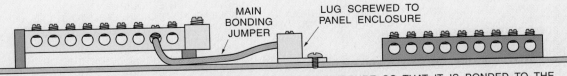

MAIN BONDING JUMPER

LUG SCREWED TO PANEL ENCLOSURE

A LUG CAN BE SECURED TO THE DISTRIBUTION PANEL ENCLOSURE SO THAT IT IS BONDED TO THE ENCLOSURE. A CONDUCTOR, THE SAME SIZE AS OR LARGER THAN THE GROUNDING ELECTRODE CONDUCTOR, RUN FROM THE GROUNDED CIRCUIT CONDUCTOR BUS TO THE LUG CAN BE USED TO BOND THE GROUNDED BUS TO THE GROUNDING BUS.

Figure 14-12

"WIREMAN'S GUIDE"
EQUIPMENT GROUNDING CONDUCTORS

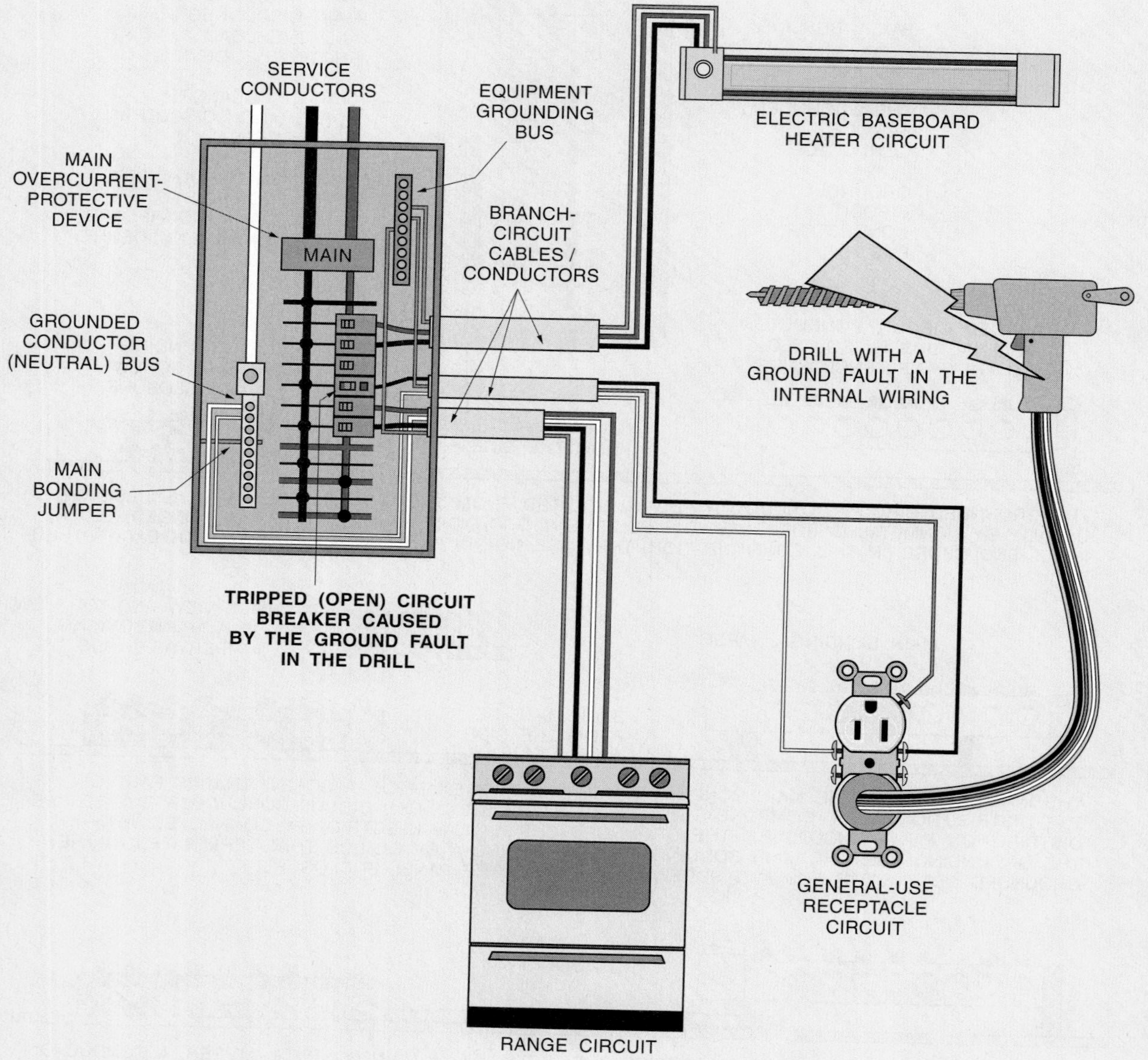

SERVICE CONDUCTORS

EQUIPMENT GROUNDING BUS

MAIN OVERCURRENT-PROTECTIVE DEVICE

MAIN

BRANCH-CIRCUIT CABLES / CONDUCTORS

ELECTRIC BASEBOARD HEATER CIRCUIT

GROUNDED CONDUCTOR (NEUTRAL) BUS

DRILL WITH A GROUND FAULT IN THE INTERNAL WIRING

MAIN BONDING JUMPER

TRIPPED (OPEN) CIRCUIT BREAKER CAUSED BY THE GROUND FAULT IN THE DRILL

GENERAL-USE RECEPTACLE CIRCUIT

RANGE CIRCUIT

THE EQUIPMENT GROUNDING CONDUCTOR IS RUN ALONG WITH THE GROUNDED CIRCUIT CONDUCTOR AND THE UNGROUNDED PHASE CONDUCTOR AND IS CONNECTED TO THE NONCURRENT-CARRYING COMPONENTS OF EQUIPMENT AND RACEWAYS. THE BASEBOARD HEATER FRAME IS CONNECTED TO THE EQUIPMENT GROUNDING CONDUCTOR. THE FRAME OF THE RANGE IS CONNECTED TO THE EQUIPMENT GROUNDING CONDUCTOR. THE GROUNDING PRONG OF THE DUPLEX RECEPTACLE IS ALSO CONNECTED TO THE EQUIPMENT GROUNDING CONDUCTOR. THE OUTER HOUSING AND OTHER NONCURRENT-CARRYING PARTS OF THE DRILL ARE BONDED TO THE EQUIPMENT GROUNDING CONDUCTOR THAT IS PART OF THE SUPPLY CORD-AND-ATTACHMENT PLUG. WHEN THE SUPPLY CORD IS PLUGGED INTO THE RECEPTACLE, THE EQUIPMENT GROUNDING CIRCUIT IS COMPLETE. IF THERE IS A GROUND FAULT WITH THE WIRING INSIDE THE DRILL, THE EQUIPMENT GROUNDING CONDUCTOR WILL HAVE A LOW ENOUGH IMPEDANCE TO GROUND TO CAUSE A LARGE CURRENT FLOW ON THE CIRCUIT. THIS LARGE CURRENT FLOW WILL CAUSE THE OVERCURRENT-PROTECTIVE DEVICE PROTECTING THE CIRCUIT TO OPEN.

Figure 14-13

them to the grounding system would introduce current on the noncurrent-carrying parts of the equipment and raceways. At the service, however, the grounded and the grounding conductors are connected together and therefore are electrically the same. Accordingly, it is permissible to use the grounded conductor system to ground the noncurrent-carrying parts of the service enclosures. The grounded conductor is usually used to bond the meter enclosure to the main panel in most dwellings. The grounded service conductor is bonded to the meter enclosure, and the main panel has the grounded circuit conductor system bonded to the panel enclosure at the neutral bus with the main bonding jumper.

14.8: THE BONDING SYSTEM

Bonding has two major functions in the electrical system. The first is to connect together all of the metallic systems in the dwelling. The second is the connection of all of the equipment and enclosures that make up the service.

14.8.1: Bonding Other Metallic Systems That May Become Energized in the Event of a Fault

The Code requires that all metallic systems in the dwelling—even those that have nothing to do with the electrical system but that could become energized—be bonded to the service or to the equipment grounding conductor from the circuit that is likely to energize the system. These systems include the domestic water system, both hot water and cold water, the mechanical ductwork system, the piping (if metallic) that supplies the gas to the furnace, the water heater, and possibly other appliances and equipment and other types of systems that may be installed in dwellings. *NEC® 250.104* provides details about these system connections.

Bonding the water system is not the same as using the incoming water pipe as an electrode ground. The water piping system must be connected to a service enclosure with a conductor sized to the grounded electrode conductor whether or not it is being used as an electrode. Furthermore, the water heater may not electrically connect the hot water

system to the cold water system. If there is a doubt about the electric continuity between the two systems, a bonding connection between the two water systems should be installed. See Figure 14-14.

The gas piping system inside a dwelling is usually metallic. This system must be bonded to the service enclosures using a conductor sized by the size of the circuit that may potentially energize the circuit. Again, bonding the gas piping system is not the same as using the incoming gas service pipe as an electrode ground.

14.8.2: Bonding the Service Equipment and Enclosures

Bonding together all noncurrent-carrying components of the electrical system ensures that in the event of a fault, the current will have a reliable path to ground. On the load side of the service overcurrent-protective device, this grounding system needs to carry current only long enough to cause the circuit breaker supplying the circuit to open or the fuse to burn out. On the line side of the service overcurrent-protective device, however, there is no overcurrent-protective device to open, so the bonding system must be more substantial to carry the fault currents likely to be imposed. Figure 7-16 presents details of the consequences of faults on service equipment.

Three methods can be employed to bond the service equipment.

1. The conduit and pipe nipples that constitute the service can serve in many cases as the bonding method. Threaded hubs and threaded couplings made up tight can be used as the bonding conductor between service enclosures. Threadless connectors can also be used with RMC, IMC, EMT, and MC. However, standard locknuts and bushings are not approved for the bonding connections and must be of the bonding type or another approved type.

KEY TERMS

Concentric knockout A bonding component in which multiple-size knockouts are arranged in a target-like pattern. In this pattern, the centers of all of the knockouts are in the same location.

"WIREMAN'S GUIDE"
BONDING OTHER METALLIC SYSTEMS

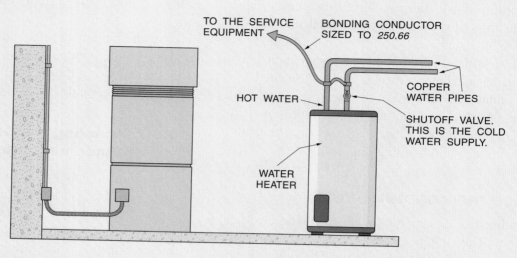

TO THE SERVICE EQUIPMENT

BONDING CONDUCTOR SIZED TO *250.66*

HOT WATER

COPPER WATER PIPES

SHUTOFF VALVE. THIS IS THE COLD WATER SUPPLY.

WATER HEATER

THE DWELLING'S WATER SYSTEM, IF THE PIPING IS METALLIC, MAY BECOME ENERGIZED IN THE EVENT OF A FAULT. THE PIPING SYSTEM MUST BE BONDED TO THE SERVICE EQUIPMENT OR THE ELECTRODE GROUNDING EQUIPMENT. THIS IS NOT THE SAME AS USING THE INCOMING WATER SUPPLY PIPE AS A GROUNDING ELECTRODE. THE SIZE OF THE BONDING CONDUCTOR IS SIZED BY *250.66* FOR THE GROUNDING ELECTRODE CONDUCTOR. THE HOT WATER PIPING SYSTEM MAY NOT BE BONDED TO THE COLD WATER SYSTEM THROUGH THE WATER HEATER. IT IS OFTEN NECESSARY TO BOND THE HOT WATER AND THE COLD WATER SYSTEMS TOGETHER WITH A BONDING CONNECTION.

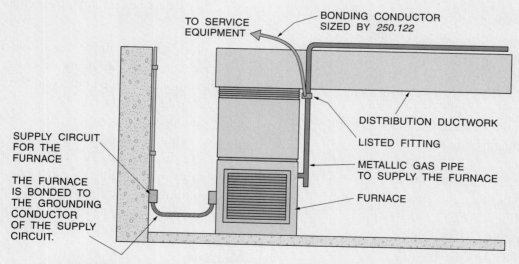

TO SERVICE EQUIPMENT

BONDING CONDUCTOR SIZED BY *250.122*

DISTRIBUTION DUCTWORK

LISTED FITTING

SUPPLY CIRCUIT FOR THE FURNACE

METALLIC GAS PIPE TO SUPPLY THE FURNACE

THE FURNACE IS BONDED TO THE GROUNDING CONDUCTOR OF THE SUPPLY CIRCUIT.

FURNACE

FURNACES ARE USUALLY CONNECTED TO THE EQUIPMENT GROUNDING SYSTEM THROUGH THE SUPPLY CABLE. THE GAS PIPING SYSTEM IN A DWELLING IS ALSO USUALLY METALLIC AND THEREFORE SUBJECT TO BECOMING ENERGIZED. THE GAS LINE THAT SUPPLIES THE FURNACE MUST BE BONDED TO THE SERVICE. THE BONDING CONNECTION IS SIZED ACCORDING TO *TABLE 250.122* FOR THE CIRCUIT THAT IS LIKELY TO ENERGIZE THE GAS PIPE.

Figure 14-14

KEY TERMS, *continued*

Eccentric knockout A bonding component in which multiple-size knockouts are arranged so that all of the various sizes of knockouts share one specific location along the edge of the largest knockout. In this pattern, of all of the knockouts have one edge in the same location.

Made knockout A properly sized knockout that is cut into the enclosure, or a concentric or eccentric knockout from which all of the smaller knockout sections have been removed.

Knockouts present a problem with bonding. Common types of knockouts are the **concentric knockout** and the **eccentric knockout**; their design allows them to serve many different sizes of connector or conduit, as shown in Figure 14-15. If all of the knockout sections are removed, the knockout becomes a **made knockout** (just like a knockout that is field made), and no bonding problem exists. However, if not all of the sections are removed from a concentric or an eccentric knockout, the remaining sections constitute a weak spot in the bonding system, and a bonding connection must be installed around the knockout.

2. A separate bonding jumper can be used to connect the enclosures and equipment. The bonding conductor is sized by the incoming service conductors according to *Table 250.66*. These bonding conductors are most often used for bonding nipples and short conduit runs with bonding locknuts and grounding bushings. Figure 14-16 shows the use of the bonding or grounding bushing.

3. The grounded service conductor can be used to bond the service equipment. The grounded service conductor (neutral) can be used for bonding the service enclosures, risers, switches, and raceways of the service equipment. A split bolt or other approved and listed fitting can be utilized to connect to the grounded conductor without splicing and will allow a tap for connection to the enclosure or the raceway.

14.9: BONDING FOR OTHER SYSTEMS

The electrical system is not the only system in modern dwellings that needs to be grounded. Telephone and television systems also need to be grounded to protect against potentially high voltages between systems in the event of lightning strikes or line surges. It is the responsibility of the electrician to provide access to the grounding electrode system for use by the installers of the other systems.

The Code allows three possible methods to accomplish this bonding:

1. By exposed nonflexible metallic raceway that is part of the service equipment. Many services may not use exposed metallic raceway, such as the sample house service.

2. By exposed grounding electrode conductor. The grounding electrode conductors may be installed inside walls and ceilings and may not be accessible. Even the conductor to the ground rod (if one was installed) may not be accessible because it may be routed inside the wall before it runs outside. Once outside, the conductor may be installed a few in. underground as it runs to the ground rod. However, the 6 AWG grounding electrode conductor that goes to the ground rod is allowed to be fastened directly to the outside wall if it will be free from exposure to physical damage. If a solid 4 AWG conductor is installed, rather than a 6 AWG, it only needs to be protected against severe physical damage. The absence of the risk of severe physical damage allows the installation of the ground rod grounding electrode conductor on the outside wall near the service, and this electrode conductor satisfies the Code requirement and can be employed to ground the other systems.

3. By other approved means for external connection to a grounding conductor or a grounded raceway or equipment. The FPN following *250.94* describes an acceptable method to provide access to the power grounding system. The local AHJ should also be consulted to determine any specific requirements.

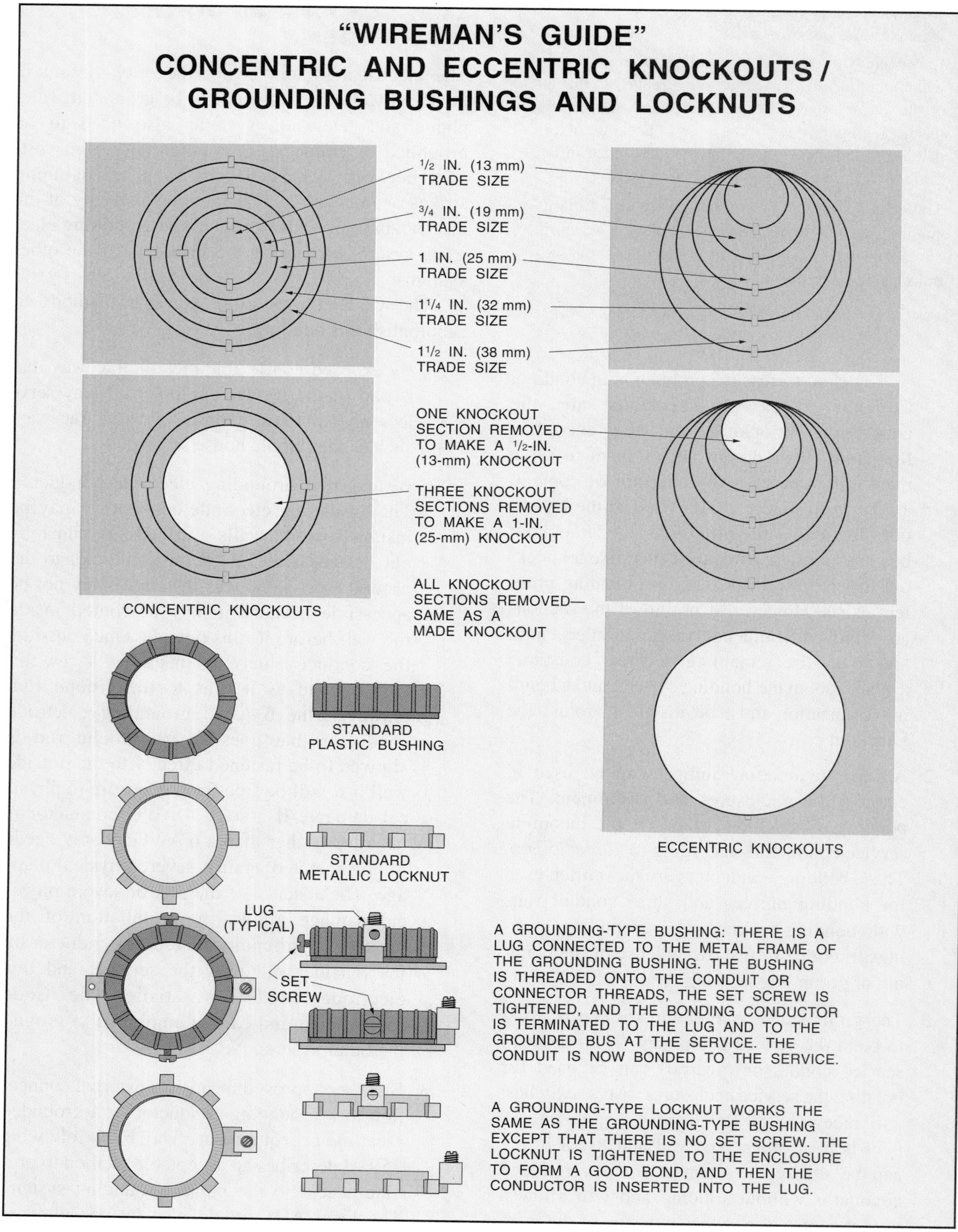

"WIREMAN'S GUIDE"
CONCENTRIC AND ECCENTRIC KNOCKOUTS / GROUNDING BUSHINGS AND LOCKNUTS

¹/₂ IN. (13 mm) TRADE SIZE

³/₄ IN. (19 mm) TRADE SIZE

1 IN. (25 mm) TRADE SIZE

1¹/₄ IN. (32 mm) TRADE SIZE

1¹/₂ IN. (38 mm) TRADE SIZE

ONE KNOCKOUT SECTION REMOVED TO MAKE A ¹/₂-IN. (13-mm) KNOCKOUT

THREE KNOCKOUT SECTIONS REMOVED TO MAKE A 1-IN. (25-mm) KNOCKOUT

ALL KNOCKOUT SECTIONS REMOVED— SAME AS A MADE KNOCKOUT

CONCENTRIC KNOCKOUTS

ECCENTRIC KNOCKOUTS

STANDARD PLASTIC BUSHING

STANDARD METALLIC LOCKNUT

LUG (TYPICAL)

SET SCREW

A GROUNDING-TYPE BUSHING: THERE IS A LUG CONNECTED TO THE METAL FRAME OF THE GROUNDING BUSHING. THE BUSHING IS THREADED ONTO THE CONDUIT OR CONNECTOR THREADS, THE SET SCREW IS TIGHTENED, AND THE BONDING CONDUCTOR IS TERMINATED TO THE LUG AND TO THE GROUNDED BUS AT THE SERVICE. THE CONDUIT IS NOW BONDED TO THE SERVICE.

A GROUNDING-TYPE LOCKNUT WORKS THE SAME AS THE GROUNDING-TYPE BUSHING EXCEPT THAT THERE IS NO SET SCREW. THE LOCKNUT IS TIGHTENED TO THE ENCLOSURE TO FORM A GOOD BOND, AND THEN THE CONDUCTOR IS INSERTED INTO THE LUG.

Figure 14-15

"WIREMAN'S GUIDE"
BONDING AND BONDING BUSHINGS

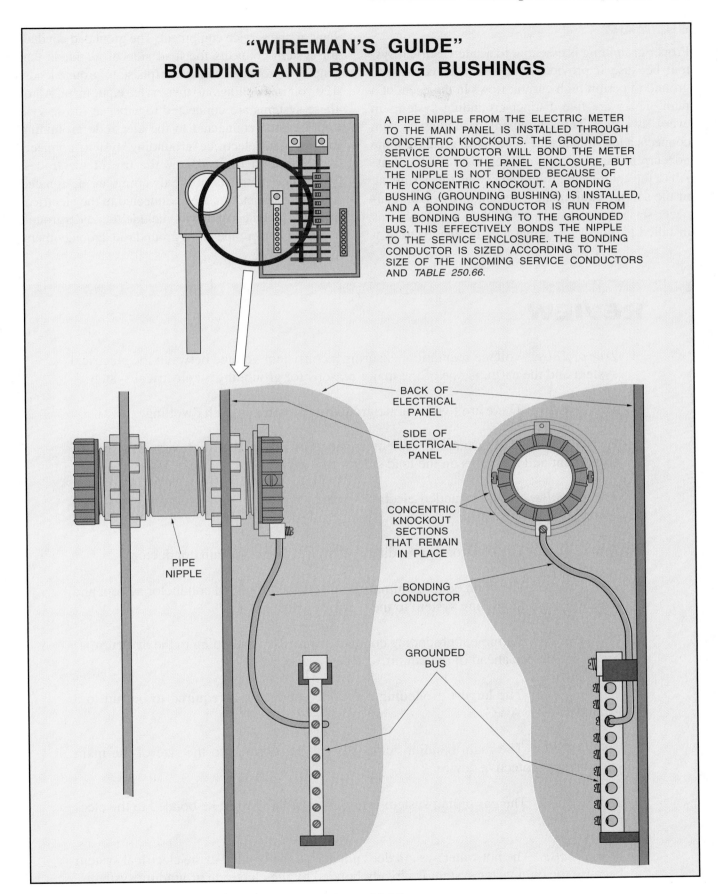

A PIPE NIPPLE FROM THE ELECTRIC METER TO THE MAIN PANEL IS INSTALLED THROUGH CONCENTRIC KNOCKOUTS. THE GROUNDED SERVICE CONDUCTOR WILL BOND THE METER ENCLOSURE TO THE PANEL ENCLOSURE, BUT THE NIPPLE IS NOT BONDED BECAUSE OF THE CONCENTRIC KNOCKOUT. A BONDING BUSHING (GROUNDING BUSHING) IS INSTALLED, AND A BONDING CONDUCTOR IS RUN FROM THE BONDING BUSHING TO THE GROUNDED BUS. THIS EFFECTIVELY BONDS THE NIPPLE TO THE SERVICE ENCLOSURE. THE BONDING CONDUCTOR IS SIZED ACCORDING TO THE SIZE OF THE INCOMING SERVICE CONDUCTORS AND *TABLE 250.66*.

BACK OF
ELECTRICAL
PANEL

SIDE OF
ELECTRICAL
PANEL

CONCENTRIC
KNOCKOUT
SECTIONS
THAT REMAIN
IN PLACE

PIPE
NIPPLE

BONDING
CONDUCTOR

GROUNDED
BUS

Figure 14-16

SUMMARY

Proper grounding is essential to a safe electrical system because it provides a low-impedance path to ground to permit high current flows in the event of a fault. There are five distinct grounding systems in every dwelling. The equipment grounding system connects a low-impedance pathway to ground for the noncurrent-carrying parts of equipment and raceway. All of the equipment grounding conductors terminate in the service equipment. The bonding system connects service equipment, and other metal systems installed in the dwelling that may be energized by a fault, to the service equipment. The grounded conductor system connects the load side of all loads that employ a grounded conductor (phase-to-ground loads: 120 volt in dwellings) to the service equipment. All of these systems are connected together at the service and are then connected to the electrode grounding system. The electrode grounding system connects the service to the earth by the use of electrodes. Various components of the dwelling are designated as electrodes and must be connected to the electrical system. These components include the underground incoming water pipe and a ground rod, among others.

REVIEW

1. *True or False:* Voltage regulation, ensuring a consistent voltage between the electrical system and the earth, is one of the major reasons for grounding an electrical system.

2. *True or False:* There are five separate grounding systems in each dwelling.

3. *True or False:* The grounded system connects to the noncurrent-carrying parts of equipment and raceways on the load side of the service.

4. *True or False:* An ungrounded electrical system produces 120 volts to ground at all times.

5. *True or False:* The electrode grounding system is a branch-circuit conductor.

6. *True or False:* The main bonding jumper connects the grounded conductor system and the equipment grounding system to the service enclosure.

7. *True or False:* Equipment grounding conductors are sized according to the overcurrent-protective device ahead of the circuit.

8. *True or False:* The largest grounding electrode conductor is required to be run to a ground rod is 6 AWG.

9. *True or False:* The main bonding jumper is sized according to the size of the main overcurrent-protective device.

10. *True or False:* The gas piping system inside of a dwelling must be bonded to the electrical system.

11. *True or False:* The hot water system does not need to be bonded to the electrical system because the cold water system is already bonded by the electrode grounding system.

12. *True or False:* The connection to the incoming cold water supply must be accomplished within the first 5 ft (1.5 m) of where the pipe enters the dwelling.

13. *True or False:* Ground rods must be installed vertically under all circumstances.

14. *True or False:* All electrodes must be connected together with one unspliced grounding electrode conductor.

15. *True or False:* Bonding of the service equipment with standard locknuts and bushings provides an adequate path for fault current.

16. All of the following are approved for use as a grounding electrode except
    _____.

    a. incoming cold water pipe
    b. incoming gas pipe
    c. steel rebar inside a foundation of the correct size and length
    d. a ground ring

17. Each of the following is a grounding system found in dwellings except _____.
    a. the grounded conductor system
    b. the electrode grounding system
    c. the equipment grounding system
    d. building steel

18. Bonding service equipment through concentric knockouts is permitted if _____.
    a. the knockout is eccentric instead of concentric
    b. the voltage is 240 volts or less
    c. the service size is not larger than 200 amperes
    d. all of the knockout sections are removed

19. Bonding jumpers at the service must terminate _____.
    a. at the equipment grounding conductor bus
    b. at the grounding electrode conductor bus
    c. at the first accessible point inside the dwelling
    d. at any point ahead of the main disconnect and overcurrent-protective device

20. All of the following voltage readings for a dwelling are correct except _____.
    a. 120 volts from phase A to the equipment grounding conductor
    b. 120 volts from phase A to phase B
    c. 120 volts from phase B to the grounding electrode conductor
    d. 120 volts from phase A to the main bonding jumper

"WIREMAN'S GUIDE" REVIEW

1. Explain why an ungrounded electrical system cannot be relied on to provide a high current flow in the event of a fault.

2. The Code says that the building water piping system cannot be used as a grounding electrode conductor. What does this restriction mean and what procedure is established to ensure that this rule is enforced?

3. Explain why the bonding jumper system at the service equipment is installed under a different set of rules from those for the equipment grounding system on the load side of the service.

4. Explain when it is acceptable to install a ground rod in other than vertical position.

5. Explain the acceptable methods for installation of the grounding electrode conductor. Does the grounding electrode conductor need to be unspliced? If not, what are the acceptable methods for splicing electrode conductors?

CHAPTER 15

Telephone, Television, and Doorbell Systems

OBJECTIVES

On completion of this chapter, the student will be able to:

- ☑ **State the origin of 60-hertz hum and describe ways to avoid problems with hum.**
- ☑ **Distinguish between one-pair (one-line) and two-pair (two-line) telephone jacks, and connect standard telephone outlets.**
- ☑ **Name the different sources of television signals.**
- ☑ **Install television coaxial cable and television jacks and plates.**
- ☑ **Install doorbell circuits, and wire a doorbell transformer, chime, and pushbutton.**

INTRODUCTION

This chapter is concerned with the additional electrical systems that are installed in most houses, other than the service for power and lighting systems. The wiring for telephone and television service is installed at rough-in, in many cases by the electrical contractor. Some low-voltage systems, such as the doorbell circuit, are installed by the electrical contractor as part of the standard wiring in a dwelling; the other systems are often installed by specialty contractors. Many other special systems may be installed in the dwelling, including intruder alarms, intercom systems, sophisticated fire alarm and detection systems, wiring for computer networks, sound systems, and others. The sample house has only telephone, television, and doorbell circuits.

15.1: BROADCASTING 60-HERTZ HUM

KEY TERMS

Induced interference Signal interference that is generated within a conductor by a magnetic field caused by current flow in another conductor that is in close proximity.

60-hertz hum The interference induced by normal electrical power intended for power and lighting loads. Normal household power has a frequency of 60 hertz (60 cycles per second).

Many of the installation rules encountered in installing telephone, television, intercom, sound sys-

tem, and other audio systems are related to **induced interference** from the power and lighting circuits. The alternating-current (ac) power used in dwellings is capable of inducing a signal with a frequency of 60 hertz, or 60 cycles per second (the frequency of the ac power used in North America), into communications circuits. This signal mixes with the audio signal to produce a background hum. This problem used to be much more prevalent with knob-and-tube wiring systems. Modern wiring methods have all but eliminated the **60-hertz hum** if the wiring is installed properly. Figure 15-1 provides more detailed information about broadcast hum.

"WIREMAN'S GUIDE"
60-Hz SIGNAL INDUCTION

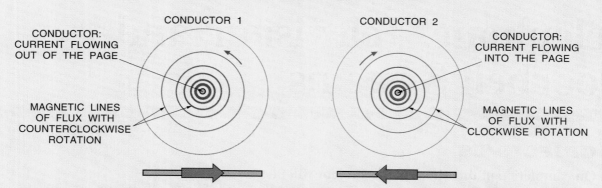

AT A FROZEN INSTANT IN TIME: THE CURRENT FLOW THROUGH THE CONDUCTOR CREATES A MAGNETIC FIELD AROUND THE CONDUCTOR. THE STRENGTH OF THE MAGNETIC FIELD IS DEPENDENT ON THE MAGNITUDE OF THE CURRENT FLOW. NOTICE THAT THE FARTHER FROM THE CONDUCTOR, THE WEAKER THE MAGNETIC FIELD. THE DIRECTION OF ROTATION OF THE MAGNETIC FIELD IS DEPENDENT ON THE DIRECTION OF THE CURRENT FLOW. IN THE CONDUCTOR WITH THE BLUE MAGNETIC FIELD, THE CURRENT FLOW IS INTO THE PAGE, OR FROM LEFT TO RIGHT. FOR THE CONDUCTOR WITH THE RED MAGNETIC FIELD THE CURRENT FLOW IS OUT OF THE PAGE, OR FROM RIGHT TO LEFT. THIS IS 60-Hz AC CURRENT, SO THE MAGNITUDE OF THE MAGNETIC FIELD IS CONSTANTLY CHANGING AND THE DIRECTION OF THE CURRENT FLOW CHANGES EACH $1/120$ OF 1 SEC.

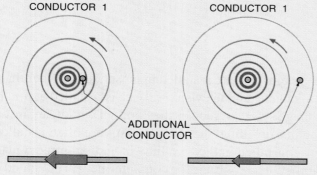

IF AN ADDITIONAL CONDUCTOR IS PLACED CLOSE TO CONDUCTOR 1 (OR CONDUCTOR 2), A CURRENT WILL BE INDUCED TO FLOW IN THE ADDITIONAL CONDUCTOR IN THE OPPOSITE DIRECTION FROM THE CURRENT FLOW IN CONDUCTOR 1. THE MAGNITUDE OF THE INDUCED CURRENT IS DEPENDENT ON THE LEVEL OF CURRENT FLOW IN CONDUCTOR 1 AND THE PROXIMITY OF THE ADDITIONAL CONDUCTOR TO CONDUCTOR 1.

IF THE ADDITIONAL CONDUCTOR HAPPENS TO BE A STEREO, TELEPHONE, OR TELEVISION ANTENNA CABLE, THE INDUCED CURRENT WILL TAKE THE FORM OF 60-Hz HUM.

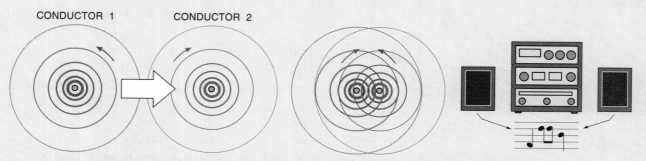

IF A CONDUCTOR 2 IS PLACED IN CLOSE PROXIMITY TO CONDUCTOR 1, THE OPPOSITE MAGNETIC FIELDS WILL CANCEL EACH OTHER OUT, AND VERY LITTLE 60-Hz SIGNAL WILL BE BROADCAST. NOTICE THAT IT IS NOT POSSIBLE TO CANCEL ALL OF THE MAGNETIC FIELD BECAUSE THE CONDUCTORS ARE ALWAYS SEPARATED BY AT LEAST THE THICKNESS OF THE INSULATION.

Figure 15-1

Noise An unwanted signal in a communication system conductor; usually caused by induced interference.

The installing electrician must be sure that the wiring for the power and lighting circuits is installed correctly. If shortcuts are used, problems are inevitable. In older homes in which the knob-and-tube wiring method was used, the phase conductor and the neutral conductor were sometimes separated from each other by as much as 20 to 30 ft (6.0 to 9 m) or more. With this system, the magnetic field caused by the ac flow through the conductors radiates outward (broadcast). The strength of the broadcast signal is directly dependent on the current flow through the circuit. If a telephone wire or a radio happens to be in the immediate area, the alternating magnetic field induces a current to flow in the telephone wire or is received by the radio as a 60-hertz signal. A 60-hertz signal is a deep humming noise. This induced signal, as well as any other unwanted signal induced in the circuit, is sometimes called **noise**.

Modern wiring methods such as those using NM cable include the phase conductors and the neutral conductors in the same cable. The magnetic field created by the phase conductor current flow is canceled out by the opposite magnetic field created by the current flow in the neutral conductor. Therefore, no discernible 60-hertz hum is produced.

15.2: TELEPHONE SYSTEM

The communications revolution has placed increased emphasis on the telephone system as a means of communicating data as well as for voice communications. It is not unusual for residences today to be served by two telephone lines, and some have three lines or more. High-speed computers use the telephone lines for Internet connections that bring the world to the desktop of the average person. The responsibility for installing telephone circuits in dwellings varies widely throughout the country. In many localities, the electrical contractor installs the outlets and cables; in other areas, independent telephone companies or specialty contractors do the installations. *NEC® 800* details the Code requirements for telephone systems.

The requirements for the telephone outlets in the sample house are on Construction Drawing E-4, the main floor telephone and television outlet plan. Six telephone outlets are shown on the plan: one in each bedroom and one each in the garage, the living room, and the kitchen. According to the electrical plan symbols on Construction Drawing E-7, the telephone outlets in the kitchen and the garage are wall mounted. The specifications on Construction Drawing E-1 shows these outlets to be installed at 54 in. (1.4 m) AFF, whereas the others are installed at the same height as for the receptacle outlets at 13 in. (.331 m) AFF.

The outlets are installed with the same types of boxes used for the power and lighting outlets. The telephone module outlet devices are designed to install to a single-gang box flush with the wall finish. Modular outlets are also available that fit the older round boxes installed by the telephone companies before deregulation. These boxes are not complete boxes but are open in the back. In some localities, the installation of a square plaster ring instead of a device box is permitted for mounting telephone jacks. The ring is also open in the back, allowing plenty of room for connections if needed. The local AHJ should be consulted about any requirements for telephone outlet installation.

Network interface The location where the dwelling wiring systems connect to the incoming telephone, cable, or other communication system supply conductor.

The connection to the incoming telephone service is accomplished at some point, either inside or outside the building, as determined by the telephone company. Whether the telephone service is installed overhead or underground does not affect the wiring method for the dwelling. The telephone company usually installs an enclosure, called an interface or a **network interface**, where the incoming service cable is terminated to the dwelling telephone cables. The local telephone company should be consulted if there are questions about the interface or termination location. Homerun cables for the interior wiring will be taken to this interface for connection by the telephone company.

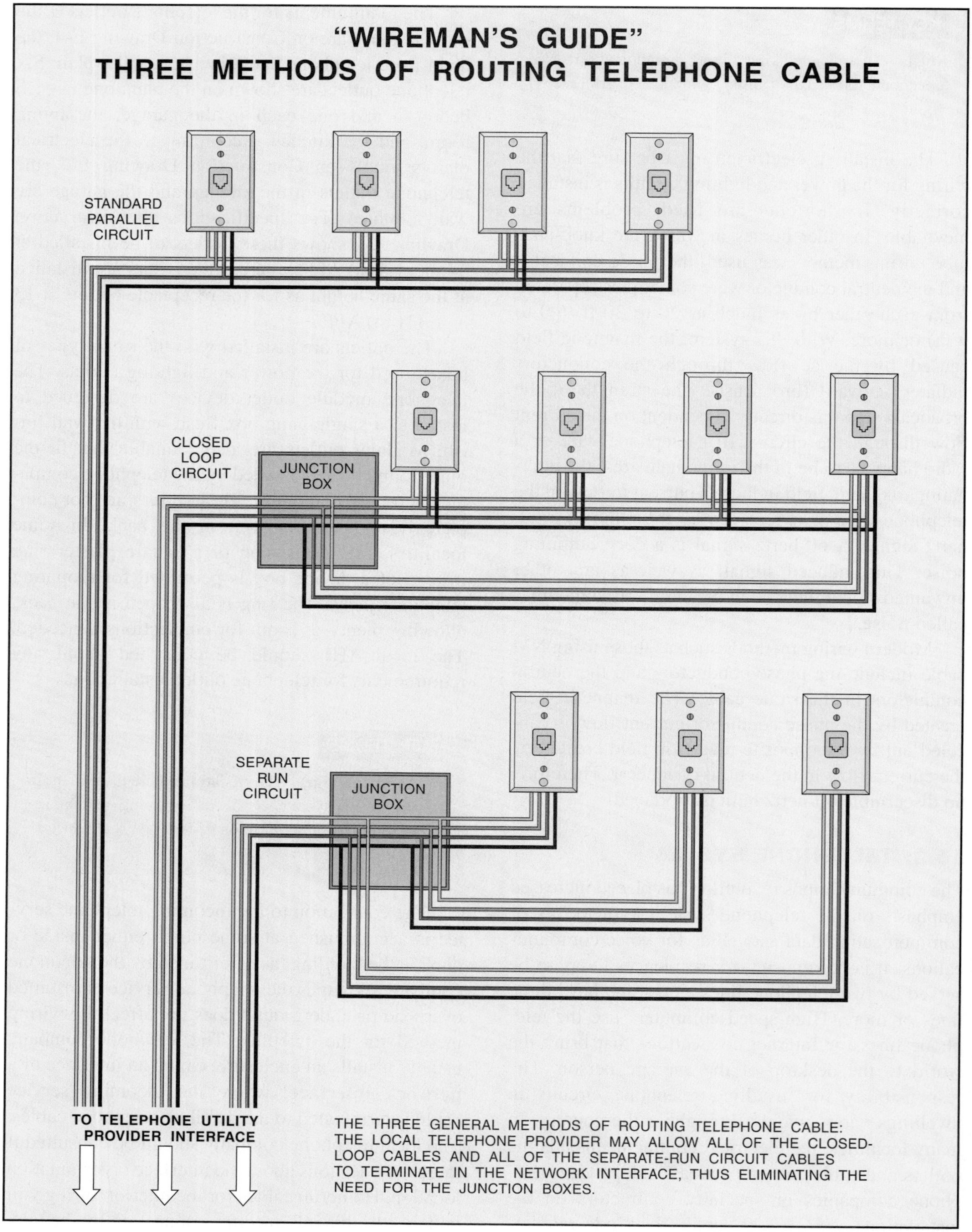

"WIREMAN'S GUIDE"
THREE METHODS OF ROUTING TELEPHONE CABLE

STANDARD
PARALLEL
CIRCUIT

CLOSED
LOOP
CIRCUIT

JUNCTION
BOX

SEPARATE
RUN
CIRCUIT

JUNCTION
BOX

**TO TELEPHONE UTILITY
PROVIDER INTERFACE**

THE THREE GENERAL METHODS OF ROUTING TELEPHONE CABLE:
THE LOCAL TELEPHONE PROVIDER MAY ALLOW ALL OF THE CLOSED-
LOOP CABLES AND ALL OF THE SEPARATE-RUN CIRCUIT CABLES
TO TERMINATE IN THE NETWORK INTERFACE, THUS ELIMINATING THE
NEED FOR THE JUNCTION BOXES.

Figure 15-2

There are three basic methods for routing telephone cables:

1. A standard parallel circuit. The standard installation is one in which all of the outlets are connected to each other from the first to the last, as shown in Figure 15-2. This is usually the least expensive method but is also the least flexible and the most limited for expansion or the addition of a third line.

2. A closed-loop circuit. A closed-loop system is wired the same as for a standard parallel circuit except that the cable is returned to the interface box from the load side of the last outlet. Telephone companies used this method for many years for installing cables inside dwellings. This method allows for the possibility that damage could occur to the cables over the life of the building. If the cable is accidentally cut at some point along the circuit, the system will continue to operate properly.

3. The separate homerun circuit: With the separate homerun circuit, each individual telephone is supplied with its own homerun. The connections for all the homeruns and the incoming service are made in the interface enclosure. This system is the most flexible and allows any telephone in the dwelling to have its own number. It is also usually the most expensive to install.

There are no special rules for installing telephone cable that apply to dwellings listed in the *NEC®*. The rules for drilling and installing telephone cables are the same as for NM cable. In addition to these rules, telephone utilities recommend the following:

- Telephone cables should be kept 12 in. (300 mm) away from power and lighting circuits if the cables are running parallel. Therefore, the same holes should not be used for power and lighting cables and for telephone cables. As shown in Figure 15-1, it is not possible to eliminate all stray magnetic fields from the power and lighting cable, and if the telephone cable is very close to the power cable, a small amount of noise may be induced into the telephone circuit. If the wires need to cross, their intersection should be at right angles to minimize induced noise.

- The cable should be installed using staples specifically designed for cable installation or with insulated staples. A special type of staple gun designed for staples that are rounded on the back end to form the letter "U" is to be used instead of a standard staple gun that uses staples that are flat. The U shape of the staple helps prevent application of too much pressure on the cable, which can cause short circuits or insulation damage. Television cables must not be strapped or taped to pipes or conduits for support.

According to *800.51(D)* and *(E)*, telephone cable for dwellings is to be Type CM or CMX. The standard telephone cable installed in dwellings is a two-pair cable. Each telephone line requires two conductors: the tip conductor and the ring conductor. The conductor pairs of the standard two-pair telephone cable are green (tip) and red (ring) for the first line and yellow (tip) and black (ring) for the second line. Figure 15-3 shows more details about telephone conductor and line pairs. Figure 15-4 shows the typical connections of a two-line modular telephone wall jack. Figure 15-5 (page 364) shows a close-up view of the front and the back of a typical telephone jack box cover.

15.3: TELEVISION SOURCES AND CABLING

KEY TERMS

Television signal   The signal, consisting of instructions for forming sounds and pictures, from a broadcast or cable television system.

There are four television outlets in the sample dwelling, as shown in Construction Drawing E-4. There is one outlet in each of the three bedrooms, plus one outlet in the living room. There is no indication from the plans whether the **television signal** will be provided from a broadcast antenna, a community antenna or cable system (CATV— community antenna television), or a satellite dish. Any of these sources may be available, as shown in Figure 15-6 (page 365). The incoming signal cable, whether from an antenna, a CATV company, or a satellite dish, is routed to the point of entry for connection to the interior television outlet homerun cables.

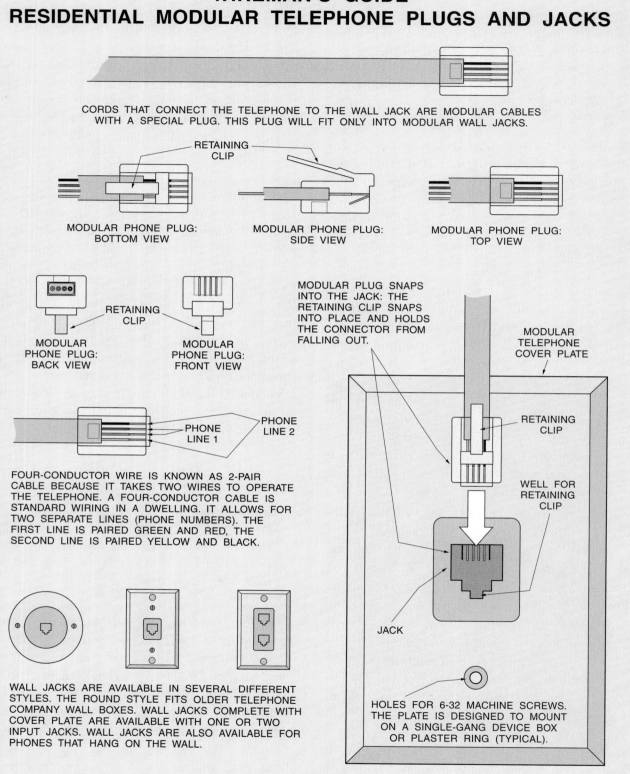

"WIREMAN'S GUIDE"
RESIDENTIAL MODULAR TELEPHONE PLUGS AND JACKS

CORDS THAT CONNECT THE TELEPHONE TO THE WALL JACK ARE MODULAR CABLES WITH A SPECIAL PLUG. THIS PLUG WILL FIT ONLY INTO MODULAR WALL JACKS.

RETAINING CLIP

MODULAR PHONE PLUG: BOTTOM VIEW

MODULAR PHONE PLUG: SIDE VIEW

MODULAR PHONE PLUG: TOP VIEW

RETAINING CLIP

MODULAR PHONE PLUG: BACK VIEW

MODULAR PHONE PLUG: FRONT VIEW

MODULAR PLUG SNAPS INTO THE JACK: THE RETAINING CLIP SNAPS INTO PLACE AND HOLDS THE CONNECTOR FROM FALLING OUT.

MODULAR TELEPHONE COVER PLATE

RETAINING CLIP

WELL FOR RETAINING CLIP

PHONE LINE 2

PHONE LINE 1

JACK

FOUR-CONDUCTOR WIRE IS KNOWN AS 2-PAIR CABLE BECAUSE IT TAKES TWO WIRES TO OPERATE THE TELEPHONE. A FOUR-CONDUCTOR CABLE IS STANDARD WIRING IN A DWELLING. IT ALLOWS FOR TWO SEPARATE LINES (PHONE NUMBERS). THE FIRST LINE IS PAIRED GREEN AND RED, THE SECOND LINE IS PAIRED YELLOW AND BLACK.

WALL JACKS ARE AVAILABLE IN SEVERAL DIFFERENT STYLES. THE ROUND STYLE FITS OLDER TELEPHONE COMPANY WALL BOXES. WALL JACKS COMPLETE WITH COVER PLATE ARE AVAILABLE WITH ONE OR TWO INPUT JACKS. WALL JACKS ARE ALSO AVAILABLE FOR PHONES THAT HANG ON THE WALL.

HOLES FOR 6-32 MACHINE SCREWS. THE PLATE IS DESIGNED TO MOUNT ON A SINGLE-GANG DEVICE BOX OR PLASTER RING (TYPICAL).

OF COURSE, IT IS POSSIBLE TO INSTALL CABLES WITH MORE THAN 2-PAIR CONDUCTORS. OTHER READILY AVAILABLE SIZES OF TELEPHONE CABLE ARE 3-PAIR AND 4-PAIR.

Figure 15-3

"WIREMAN'S GUIDE"
CONNECTING MODULAR TELEPHONE OUTLETS

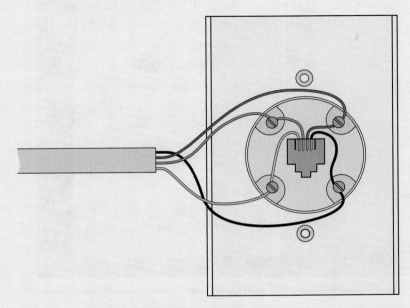

THE CONDUCTORS FROM THE TELEPHONE CABLE CONNECT TO THE TERMINALS ON THE TELEPHONE PLATE, OR MODULAR OUTLET, COLOR TO COLOR (GREEN TO THE GREEN TERMINAL, RED TO THE RED TERMINAL, AND SO ON). MOST TELEPHONE JACKS HAVE SCREW TERMINALS FOR CONNECTION TO THE CONDUCTORS.

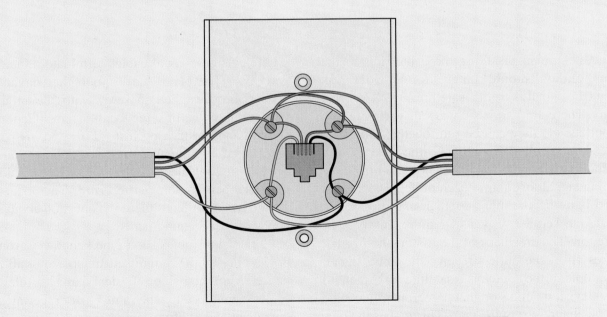

IF THERE IS MORE THAN ONE CABLE INVOLVED WITH THE TERMINATION, THE CONDUCTORS CONNECT IN PARALLEL (COLOR TO COLOR), AS IN THE ONE-CABLE EXAMPLE ABOVE.

Figure 15-4

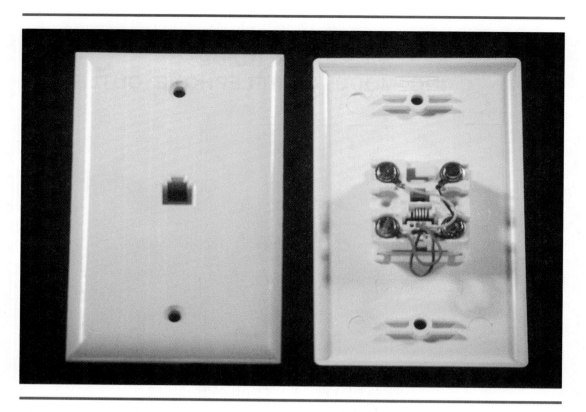

Figure 15-5   A typical modular telephone plate and jack.

The electrician installing the cable runs to the various outlets should take into consideration whether the signal is from a satellite, a CATV, or an antenna. There are two basic ways of wiring television outlets: (1) they can be run together, one after another, in a parallel circuit, or (2) they can be supplied individually with a separate homerun from the incoming signal location to each outlet. The second method is by far the more popular, and most cable and satellite service providers require this type of installation. The drawback of the first method is that it makes all of the outlets receive the same signal. For example, if the source is a satellite dish, then all of the television outlets on that homerun will receive the same channel from the dish. This makes the outlets slaves to the satellite dish settings. If each of the outlets is an individual run from the point of entry, however, each can be connected to whatever source is available in almost any combination. Figure 15-7 presents additional details.

Television coaxial cables terminate in a single-gang box. The plates that support the coaxial connectors are designed to screw into the mounting holes on the box. In some localities, installation of a single-gang plaster ring instead of a box is permitted. Use of a ring makes for easier splicing, or splitting, of the cable because the hollow space inside the wall is open to the plaster ring. The local AHJ should be consulted if there is a question about stringency of local requirements.

Television cables should be kept away from the power and lighting circuits as much as possible and be free from the risk of physical damage. Television cables cannot occupy the same box or raceway as for the power and lighting cable unless there is an effective barrier between the two systems. If the installation requires that the television cables cross power or lighting circuit cables, the crossing should be made at right angles. If the cable is to be run parallel to power and lighting circuits, the systems must be

"WIREMAN'S GUIDE"
SOURCES OF TELEVISION SIGNALS

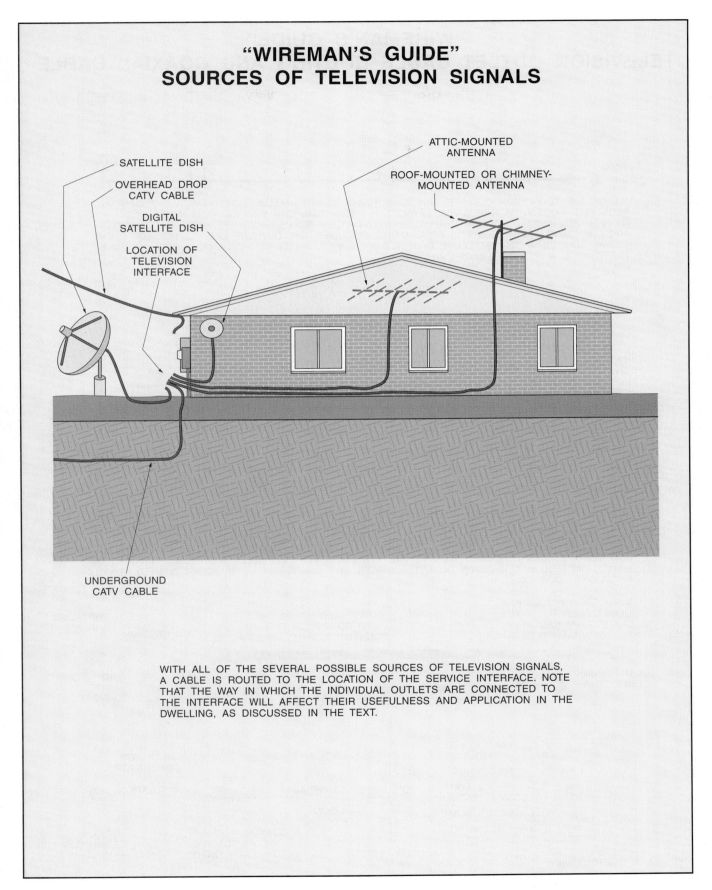

SATELLITE DISH

OVERHEAD DROP
CATV CABLE

DIGITAL
SATELLITE DISH

LOCATION OF
TELEVISION
INTERFACE

ATTIC-MOUNTED
ANTENNA

ROOF-MOUNTED OR CHIMNEY-
MOUNTED ANTENNA

UNDERGROUND
CATV CABLE

WITH ALL OF THE SEVERAL POSSIBLE SOURCES OF TELEVISION SIGNALS,
A CABLE IS ROUTED TO THE LOCATION OF THE SERVICE INTERFACE. NOTE
THAT THE WAY IN WHICH THE INDIVIDUAL OUTLETS ARE CONNECTED TO
THE INTERFACE WILL AFFECT THEIR USEFULNESS AND APPLICATION IN THE
DWELLING, AS DISCUSSED IN THE TEXT.

Figure 15-6

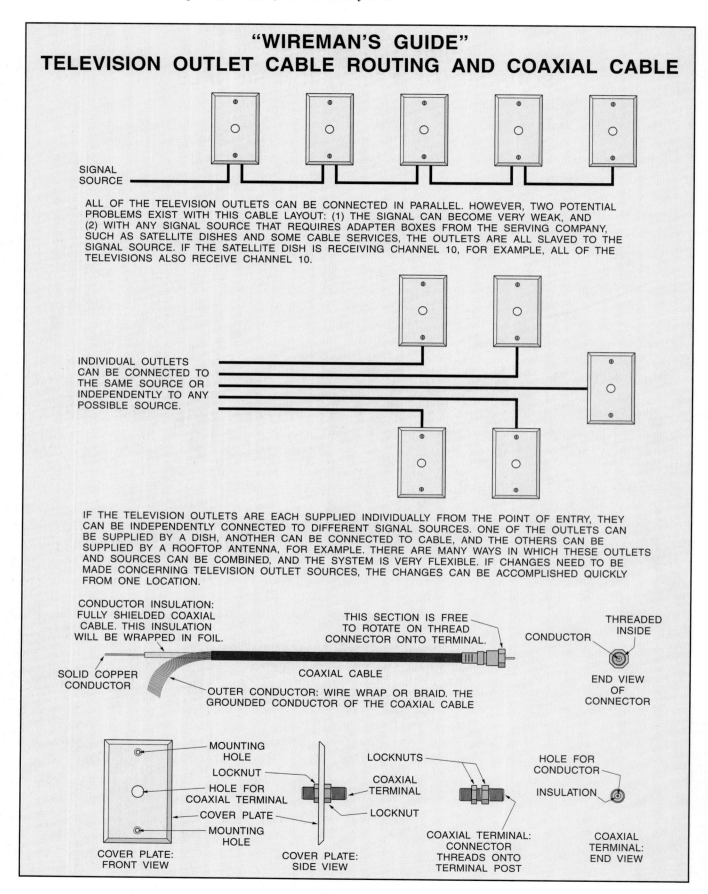

"WIREMAN'S GUIDE"
TELEVISION OUTLET CABLE ROUTING AND COAXIAL CABLE

SIGNAL
SOURCE

ALL OF THE TELEVISION OUTLETS CAN BE CONNECTED IN PARALLEL. HOWEVER, TWO POTENTIAL PROBLEMS EXIST WITH THIS CABLE LAYOUT: (1) THE SIGNAL CAN BECOME VERY WEAK, AND (2) WITH ANY SIGNAL SOURCE THAT REQUIRES ADAPTER BOXES FROM THE SERVING COMPANY, SUCH AS SATELLITE DISHES AND SOME CABLE SERVICES, THE OUTLETS ARE ALL SLAVED TO THE SIGNAL SOURCE. IF THE SATELLITE DISH IS RECEIVING CHANNEL 10, FOR EXAMPLE, ALL OF THE TELEVISIONS ALSO RECEIVE CHANNEL 10.

INDIVIDUAL OUTLETS
CAN BE CONNECTED TO
THE SAME SOURCE OR
INDEPENDENTLY TO ANY
POSSIBLE SOURCE.

IF THE TELEVISION OUTLETS ARE EACH SUPPLIED INDIVIDUALLY FROM THE POINT OF ENTRY, THEY CAN BE INDEPENDENTLY CONNECTED TO DIFFERENT SIGNAL SOURCES. ONE OF THE OUTLETS CAN BE SUPPLIED BY A DISH, ANOTHER CAN BE CONNECTED TO CABLE, AND THE OTHERS CAN BE SUPPLIED BY A ROOFTOP ANTENNA, FOR EXAMPLE. THERE ARE MANY WAYS IN WHICH THESE OUTLETS AND SOURCES CAN BE COMBINED, AND THE SYSTEM IS VERY FLEXIBLE. IF CHANGES NEED TO BE MADE CONCERNING TELEVISION OUTLET SOURCES, THE CHANGES CAN BE ACCOMPLISHED QUICKLY FROM ONE LOCATION.

CONDUCTOR INSULATION:
FULLY SHIELDED COAXIAL
CABLE. THIS INSULATION
WILL BE WRAPPED IN FOIL.

THIS SECTION IS FREE
TO ROTATE ON THREAD
CONNECTOR ONTO TERMINAL.

THREADED
INSIDE

CONDUCTOR

SOLID COPPER
CONDUCTOR

COAXIAL CABLE

OUTER CONDUCTOR: WIRE WRAP OR BRAID. THE
GROUNDED CONDUCTOR OF THE COAXIAL CABLE

END VIEW
OF
CONNECTOR

MOUNTING HOLE
LOCKNUT
HOLE FOR COAXIAL TERMINAL
COVER PLATE
MOUNTING HOLE

LOCKNUTS
COAXIAL TERMINAL
LOCKNUT

HOLE FOR CONDUCTOR
INSULATION

COVER PLATE:
FRONT VIEW

COVER PLATE:
SIDE VIEW

COAXIAL TERMINAL:
CONNECTOR
THREADS ONTO
TERMINAL POST

COAXIAL
TERMINAL:
END VIEW

Figure 15-7

separated by at least 1 ft (300 mm), and more if practicable. Television cables are the most reliable when they are run in their own set of holes in wall studs and floor joists. These rules are a brief summary of some of the separation rules for antennas and CATV systems but includes those rules most relevant for the electrician installing television outlet cables on the construction site. Many of the rules affect the technicians installing the incoming television signal cables and equipment, the CATV cable, or the satellite dish. The *NEC*® has very specific rules concerning separation of television outlet cables from power and lighting circuits, both for the lead conductors from antennas or cable and for cable runs to the individual outlets. Details of these requirements can be found in *810* for radio and television equipment, antennas, and circuits and in *820* for CATV.

Grounding for the television antenna and cable (CATV) systems are virtually identical. The Code says that the grounding connection between the service grounding electrode system and the television system must be made at the nearest accessible location of any acceptable grounding source. The Code requires that provisions be made for bonding other systems to the building's electrode grounding system. The ideal place to connect the television grounding conductor to the building electrode grounding system is at the service equipment. However, the television service location and the power and lighting service locations may not be in close proximity; therefore, the connection to the electrode grounding system must be made at another location. The connection can be made at one of the locations detailed in *810.21* for antenna systems or in *820.40* for CATV systems. These locations are as follows (there is no preference or order to the list; the Code requires connection to the closest one):

1. The building grounding electrode system
2. The incoming metallic water pipe within 5 ft (1.5 m) from the point of entry
3. The accessible means supplied at the service for grounding other systems
4. The metal service riser or raceway
5. The service enclosure for the main disconnection switch
6. The grounding electrode conductor or any metallic conduit protecting the grounding electrode conductor

The size of the television system grounding conductor cannot be smaller than 10 AWG copper for antenna systems. CATV systems require a grounding conductor approximately equal to the current-carrying capacity of the outer conductor of the coaxial cable but not smaller than 14 AWG copper.

15.4: DOORBELL WIRING

KEY TERMS

Pushbutton A momentary-contact button that activates a doorbell chime.

Doorbell chime The chime that makes the sound intended to alert the occupant of a dwelling of the presence of a visitor at the door. The chime is actuated by the pushbutton.

Transformer A device that changes the voltage of an electrical circuit. A transformer can either reduce (step down) or increase (step up) the voltage, but the power output on the secondary side of a transformer will approximately equal the power for the primary side of the transformer.

Low voltage The side of the transformer with the lowest voltage.

According to Construction Drawing E-3, there are doorbell **pushbuttons** at both the front door and the sliding door on the rear patio. The **doorbell chime** is in the hall opposite the stairway. The location of the **transformer** that produces the **low voltage** used by the doorbell circuit is not shown on the plans, so a location needs to be determined by the lead electrician. More detailed information about the doorbell chime, transformers, and circuiting is shown in Figure 15-8.

The doorbell transformer may be installed at any of several locations. The transformer is designed to install on the outside of a metal device or square box, using a ½ in. (13 mm) trade size knockout. The transformer location must be accessible and located to be free of the risk of physical damage. Furthermore, the transformer must be powered by an unswitched outlet. The low-voltage wiring for the doorbell must be kept away from the power and lighting circuits. They cannot occupy the same raceways and cannot terminate in the same outlet box unless a divider is provided to separate the two systems. Bell wire can occupy the same holes as for

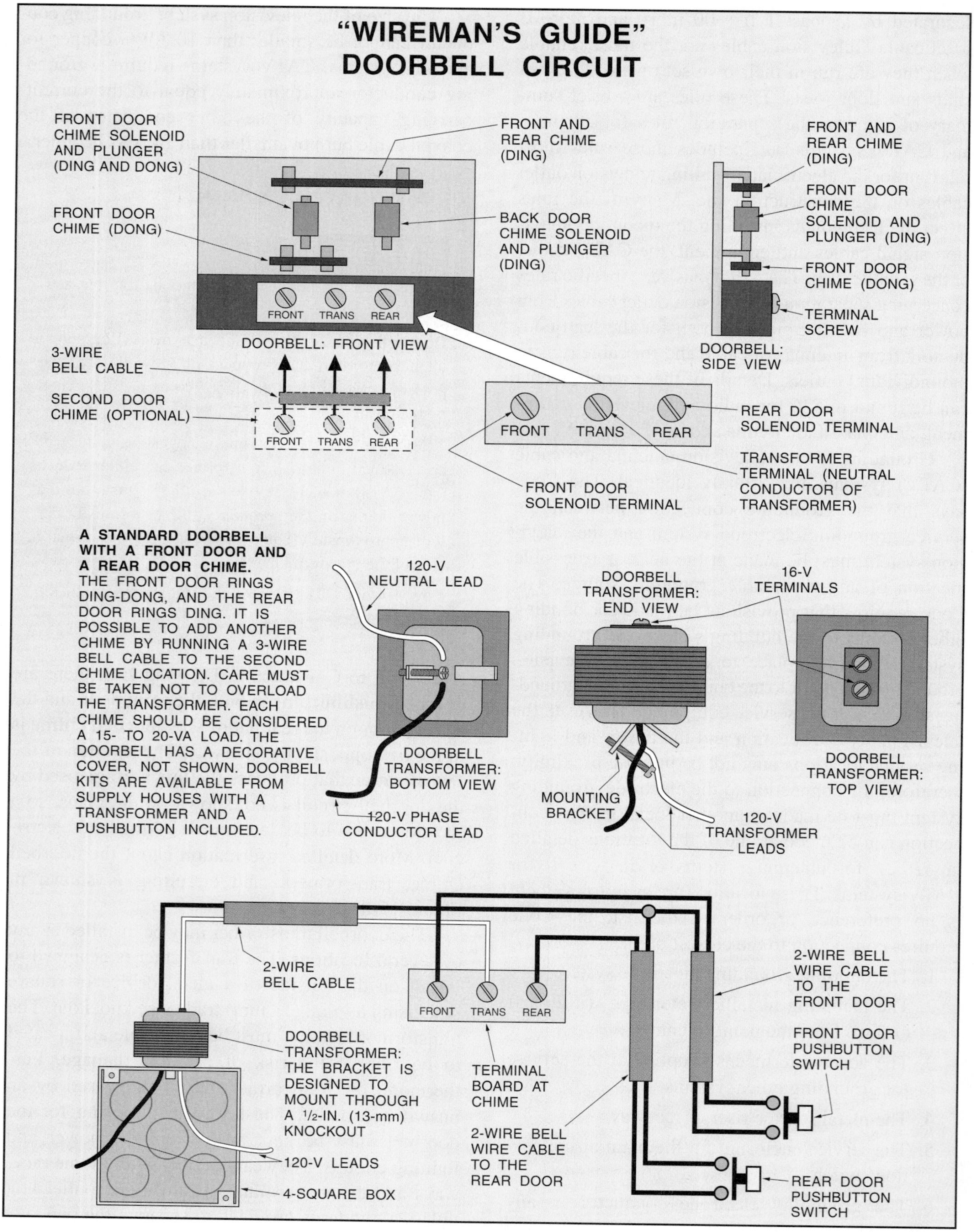

"WIREMAN'S GUIDE" DOORBELL CIRCUIT

FRONT DOOR CHIME SOLENOID AND PLUNGER (DING AND DONG)

FRONT AND REAR CHIME (DING)

FRONT DOOR CHIME (DONG)

BACK DOOR CHIME SOLENOID AND PLUNGER (DING)

FRONT / TRANS / REAR

DOORBELL: FRONT VIEW

FRONT AND REAR CHIME (DING)

FRONT DOOR CHIME SOLENOID AND PLUNGER (DING)

FRONT DOOR CHIME (DONG)

TERMINAL SCREW

DOORBELL: SIDE VIEW

3-WIRE BELL CABLE

SECOND DOOR CHIME (OPTIONAL)

FRONT / TRANS / REAR

FRONT / TRANS / REAR

REAR DOOR SOLENOID TERMINAL

TRANSFORMER TERMINAL (NEUTRAL CONDUCTOR OF TRANSFORMER)

FRONT DOOR SOLENOID TERMINAL

A STANDARD DOORBELL WITH A FRONT DOOR AND A REAR DOOR CHIME. THE FRONT DOOR RINGS DING-DONG, AND THE REAR DOOR RINGS DING. IT IS POSSIBLE TO ADD ANOTHER CHIME BY RUNNING A 3-WIRE BELL CABLE TO THE SECOND CHIME LOCATION. CARE MUST BE TAKEN NOT TO OVERLOAD THE TRANSFORMER. EACH CHIME SHOULD BE CONSIDERED A 15- TO 20-VA LOAD. THE DOORBELL HAS A DECORATIVE COVER, NOT SHOWN. DOORBELL KITS ARE AVAILABLE FROM SUPPLY HOUSES WITH A TRANSFORMER AND A PUSHBUTTON INCLUDED.

120-V NEUTRAL LEAD

DOORBELL TRANSFORMER: BOTTOM VIEW

120-V PHASE CONDUCTOR LEAD

DOORBELL TRANSFORMER: END VIEW

16-V TERMINALS

MOUNTING BRACKET

120-V TRANSFORMER LEADS

DOORBELL TRANSFORMER: TOP VIEW

2-WIRE BELL CABLE

DOORBELL TRANSFORMER: THE BRACKET IS DESIGNED TO MOUNT THROUGH A 1/2-IN. (13-mm) KNOCKOUT

120-V LEADS

4-SQUARE BOX

FRONT / TRANS / REAR

TERMINAL BOARD AT CHIME

2-WIRE BELL WIRE CABLE TO THE REAR DOOR

2-WIRE BELL WIRE CABLE TO THE FRONT DOOR

FRONT DOOR PUSHBUTTON SWITCH

REAR DOOR PUSHBUTTON SWITCH

Figure 15-8

telephone cables and technically can occupy the same holes as for NM cable, but most contractors prefer to keep some separation between the systems.

> **KEY TERMS**
>
> **Inherently limited power source** A power source designed to restrict the power output so that it is maintained at safe levels even in the event of a fault.

The doorbell circuit is an **inherently limited power source** that requires no additional overcurrent protection or disconnection means. The secondary side of the doorbell transformer will produce anywhere from 10 to 24 volts, depending on the manufacturer of the system and the size and type of the chime.

In the sample house, the doorbell transformer is to be installed in a square junction box directly above the switch that controls the luminaires (fix-tures) in the crawl space. This location ensures that the transformer will be free from the threat of physical damage but still be accessible if maintenance is required. The switch for the luminaires (fixtures) is surface mounted on a concrete wall, so it is necessary to install protection for the cables. Usually EMT is used, but other wiring methods are acceptable. This same protection can be expanded for the doorbell transformer. Figure 15-9 shows the type of installation to be installed to power the doorbell transformer. The upper box will provide power to the doorbell transformer. The transformer can mount to one of the side knockouts of the box or it can be installed in a knockout on an otherwise blank cover plate. Caution should be taken in planning to mount a doorbell transformer to the side of a square or device box because some transformers are too large to properly attach to the box. Figure 15-10 shows a typical doorbell transformer installation.

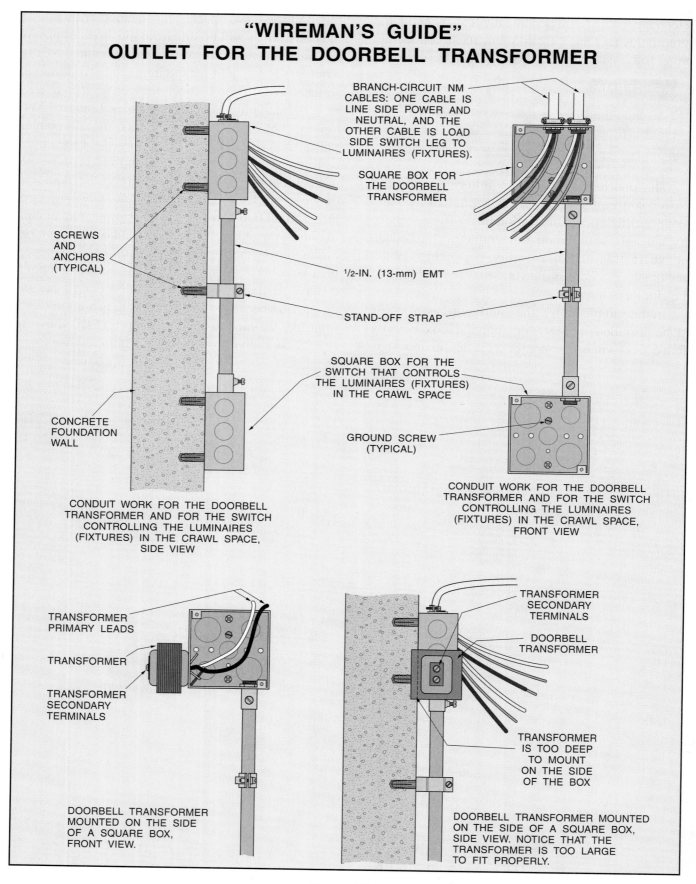

"WIREMAN'S GUIDE"
OUTLET FOR THE DOORBELL TRANSFORMER

BRANCH-CIRCUIT NM CABLES: ONE CABLE IS LINE SIDE POWER AND NEUTRAL, AND THE OTHER CABLE IS LOAD SIDE SWITCH LEG TO LUMINAIRES (FIXTURES).

SQUARE BOX FOR THE DOORBELL TRANSFORMER

SCREWS AND ANCHORS (TYPICAL)

½-IN. (13-mm) EMT

STAND-OFF STRAP

CONCRETE FOUNDATION WALL

SQUARE BOX FOR THE SWITCH THAT CONTROLS THE LUMINAIRES (FIXTURES) IN THE CRAWL SPACE

GROUND SCREW (TYPICAL)

CONDUIT WORK FOR THE DOORBELL TRANSFORMER AND FOR THE SWITCH CONTROLLING THE LUMINAIRES (FIXTURES) IN THE CRAWL SPACE, SIDE VIEW

CONDUIT WORK FOR THE DOORBELL TRANSFORMER AND FOR THE SWITCH CONTROLLING THE LUMINAIRES (FIXTURES) IN THE CRAWL SPACE, FRONT VIEW

TRANSFORMER PRIMARY LEADS

TRANSFORMER

TRANSFORMER SECONDARY TERMINALS

DOORBELL TRANSFORMER MOUNTED ON THE SIDE OF A SQUARE BOX, FRONT VIEW.

TRANSFORMER SECONDARY TERMINALS

DOORBELL TRANSFORMER

TRANSFORMER IS TOO DEEP TO MOUNT ON THE SIDE OF THE BOX

DOORBELL TRANSFORMER MOUNTED ON THE SIDE OF A SQUARE BOX, SIDE VIEW. NOTICE THAT THE TRANSFORMER IS TOO LARGE TO FIT PROPERLY.

Figure 15-9

Figure 15-10   A typical doorbell transformer installation.

SUMMARY

Telephone and television cables should be kept away from power and lighting cables to minimize the possibility of the induction of a 60-hertz hum into the system. The telephone and television outlets can be connected in line, one after the other, or by separate homeruns; the separate homeruns are the preferred method for trouble-free service. The television and telephone systems must be grounded to the dwelling's electrode grounding system, and the Code offers several choices to accomplish the connection. Telephone module jacks are standard with one-pair (one-line) connections but are commonly found with 2-pair (2-line) connections. They are available with three, four, or more lines if necessary. There are several sources for television signals, including satellite dish systems and community antenna systems.

Doorbell systems have three components: (1) transformer, (2) chime, and (3) pushbutton. The transformer should be located where it is free from exposure to physical damage. The doorbell circuit needs a 120-volt supply, and the system may operate on anywhere from 12 to 24 volts. Standard chimes include different chime patterns for two separate pushbuttons for the front and the back door.

REVIEW

1. *True or False:* Doorbell transformers must be connected to the dwelling's electrode grounding system.

2. *True or False:* Telephone and television cables can be installed in the same drilled holes as for power and lighting cables without causing noise induction problems.

3. *True or False:* The doorbell system operates at 120 volts.

4. *True or False:* Most telephone cable contains one pair of wires.

5. *True or False:* The telephone system needs four wires (two pairs) to operate properly.

6. *True or False:* Coaxial cable has three conductors: a phase conductor, a neutral conductor, and a grounding conductor.

7. *True or False:* Television outlets must be installed in series for proper operation.

8. *True or False:* The television system must be bonded to the dwelling's electrode grounding system.

9. *True or False:* It is the responsibility of the electrical contractor to provide access to the electrode grounding system for use with the television and telephone systems service installers.

10. The following locations are acceptable for grounding television and telephone systems except _____.

    a. the incoming cold water pipe

    b. the service enclosure for the main disconnection switch

    c. the equipment grounding conductor terminal on a duplex receptacle

    d. a grounding electrode conductor

11. Doorbell transformers are designed to mount _____.

    a. on a metal device or square box through a ½ in. (16 mm) trade size knockout

    b. on a standard device box with a single-pole switch

    c. in the main electrical panel

    d. on the load end of the equipment grounding conductor system

12. Doorbell cable and power and lighting cable are allowed in the same device box if _____.

    a. the doorbell system voltage is above 16 volts

    b. the doorbell wire is two-pair conductor cable

    c. there is a physical barrier installed in the box to keep the system conductors separated

    d. the power and lighting circuit cable carries less than 15 amperes

13. Noise in a telephone system is caused by _____.

    a. too much current flow in the telephone conductors

    b. broadcast 60-hertz signals from improperly installed power and lighting circuits

    c. voltage surges in the telephone system

    d. leaving the phone off the hook for too long

"WIREMAN'S GUIDE" REVIEW

1. Explain the process by which modern wiring with NM, AC, or MC cable minimizes the induction of noise into audio systems.

2. Explain how the cable routing system can cause television outlets to be enslaved to one source. Why is this not a problem with broadcast television?

3. Draw a simple doorbell circuit for two pushbuttons. How does this wiring system differ from a 3-way switching system?

CHAPTER 16

Trim Switches, Receptacles, and Special Outlets

OBJECTIVES

On completion of this chapter, the student will be able to:

- ☑ State and follow the rules for a safe trim.
- ☑ Explain the reasons for polarity in outlets and how to properly wire receptacle and screw shells.
- ☑ Describe the proper methods for terminating conductors onto devices.
- ☑ Identify device requirements by studying the box makeup.
- ☑ Identify individual branch circuits, and explain the necessity of installing the properly rated receptacle and the correct attachment plug.
- ☑ Explain the proper methods for installing receptacle: split-wired receptacles, GFCI feed-through receptacles, and half-switched receptacles.
- ☑ Identify the proper cover plates for receptacle outlets and switch points.

INTRODUCTION

KEY TERMS

Trim   The set of procedures for installing the switches, receptacles, luminaires (fixtures), and other utilization equipment in preparation for occupancy of the dwelling. Trim also includes checkout and troubleshooting.

This chapter is concerned with **trim**. For the sample house construction, the rough electrical installation is over. It may have been several months since any electricians have been on the site. The walls have been covered, finished, and painted; the kitchen and bathroom cabinets have been installed; the kitchen appliances are on site, and the plumbing fixtures and mechanical systems are either complete or in the process of being completed. The sample house is now ready for electrical trim. Trim involves installing all of the switches, receptacles, luminaires (fixtures), appliances, and other devices to complete the electrical system. The two electrical panels also need to have the circuit breakers installed and the covers replaced. Finally, the entire electrical system needs to be tested to ensure that all is operating as planned. This chapter deals with the trim of the switches and receptacles. Chapter 17 examines the installation of the luminaires (fixtures). Chapter 18 deals with the trim of the appliances and other pieces of electrical equipment in the sample house.

16.1: TRIM SAFETY

Trim involves installing the devices and equipment to finish the dwelling. These terminations are being completed on circuits that have been de-energized and lockout/tag-out procedures employed. However, workers from several other trades are usually working in the dwelling at the same time as the electricians, and they have certain demands for power to run tools and for lighting. It is not unusual for a worker from another trade to attempt to energize circuits in the main panel in an attempt to access power—presenting a potentially deadly situation for anyone working around the circuit. Therefore, when trimming a building, the electrician should always assume that the circuit is energized. Before any work on an outlet, the conductors must be tested with a voltage tester to ensure that the outlet is dead. In addition, the tester should also be tested on a regular basis to ensure that it is working properly.

The best method of minimizing the possibility that the circuit is energized is not to trim the electrical distribution panel until all of the other trim is complete. If the circuits are not connected to circuit breakers, the chance of their being energized is greatly reduced, although not eliminated. The temporary wiring system must be checked to see if unauthorized modifications have been made to the system. The branch-circuit homerun conductors must be checked to ensure that none have been energized or terminated.

Workers in other trades are themselves at risk from potential errors made by electricians. If an outlet is accidentally energized without being trimmed, the energized conductors in the box present a hazard to anyone else working on the site. The electrician should always make sure that other workers are aware that a circuit has been energized, to help prevent injuries from shock or flash burns. All those on the job site should work safely at all times and watch out for their fellow workers.

16.2: COMMON ELEMENTS OF TRIM

The process of installing switches is quite different from powering a furnace. However, some aspects of trim are common to all devices and appliances. For example, the dwelling's electrical plans used at rough-in must be available for the electrician to consult if there is a question about a particular receptacle or switch, to determine the circuit numbers for making up the panels, and for information concerning the locations of boxes in each room.

> ### KEY TERMS
>
> **Buried box** A device or junction box that has been inadvertently covered by sheetrock or other wall covering materials and is not readily available for trim.

Working with other trade workers is also essential at trim. During the wall covering process, the electricians are reliant on the sheetrock installers or other wall finishers to cut the proper-size holes in the wall finish at the box locations. The wall finishers may not cut all of the necessary holes, which can result in a **buried box**. Buried boxes are holes in the electrical system. The electrical problems caused by buried boxes are different with each occurrence, but they constitute an ever-present concern for electricians. If buried boxes are not discovered during the trim process, they will be discovered during checkout and troubleshooting. In addition, it is common practice for wall finishers to use outlet boxes and switch point boxes as edges for cleaning their tools; electrical boxes may be found to be virtually filled with finishing compound and require cleaning out before the device can be installed.

Terminating the conductors is also a common element of trim. Some conductors will terminate under screw terminals, others will terminate in lugs, and others using back-stab terminations. Some loads will connect using solderless pressure connectors or splice caps. Splice caps and pigtailing are detailed in Chapter 9 of this text, and a review of Figures 9-11 and 9-12 should be undertaken before continuing with the trim. Terminating a conductor under a screw is not difficult, but it must be accomplished properly to prevent loose connections. The most important aspect of terminating under a screw is use of the proper size and shape of hook and orientation of the hook to face in the correct direction so that when the screw is tightened, the hook will also be tightened. Figure 16-1 shows proper termination under a screw and some of the common mistakes made with such terminations.

> ### KEY TERMS
>
> **Quick connect** A method of connecting conductors to devices without employing the screw terminals. Many AHJs and contractors do not allow the use of quick connects.

"WIREMAN'S GUIDE"
CONNECTING CONDUCTORS UNDER SCREW TERMINALS

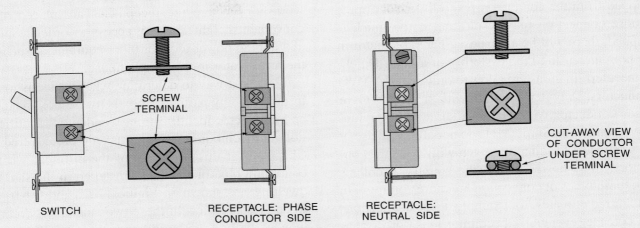

SWITCH

RECEPTACLE: PHASE
CONDUCTOR SIDE

RECEPTACLE:
NEUTRAL SIDE

CUT-AWAY VIEW
OF CONDUCTOR
UNDER SCREW
TERMINAL

SCREW
TERMINAL

THE CODE REQUIRES THAT THE TERMINALS FOR THE GROUNDED CIRCUIT CONDUCTOR (NEUTRAL) BE IDENTIFIED BY A WHITE TERMINAL OR THE WORD "WHITE" OR THE LETTER "W" MARKED NEXT TO THE TERMINAL. STANDARD SWITCHES DO NOT HAVE TERMINALS FOR NEUTRALS.

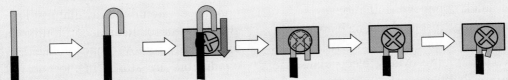

TECHNIQUE: STRIP APPROXIMATELY ³/₄ IN. (20 mm) FROM THE END OF THE CONDUCTOR. FORM A HOOK IN THE CONDUCTOR WITH NEEDLE-NOSE PLIERS OR WIRE STRIPPERS. SLIP THE HOOKED END OF THE CONDUCTOR UNDER THE TERMINAL SCREW, MAKING SURE THAT THE HOOK IS TO THE RIGHT OF THE WIRE. PULL THE HOOK DOWN UNTIL IT SEATS ON THE SHANK OF THE SCREW. SOME ELECTRICIANS WILL AT THIS POINT FURTHER CLOSE THE HOOK SO THE CONDUCTOR WILL NOT SLIP AWAY FROM THE TERMINAL SCREW DURING TIGHTENING. TIGHTEN THE SCREW TO THE MANUFACTURER'S REQUIREMENTS. OFTEN, WHEN THE SCREW IS TIGHTENED, THE HOOK IN THE CONDUCTOR WILL CLOSE FURTHER AS A RESULT OF THE TURNING MOTION OF THE SCREW HEAD.

SOME COMMON MISTAKES MADE IN TERMINATING CONDUCTORS UNDER SCREW TERMINALS:

A: THE HOOK IS NOT CLOSED ENOUGH TO GIVE GOOD CONTACT BETWEEN CONDUCTOR AND TERMINAL.

B: THE HOOK IS TOO SHORT AND WILL NOT PROVIDE THE NECESSARY CONTACT BETWEEN THE TERMINAL SCREW AND THE CONDUCTOR.

C: TOO MUCH INSULATION IS STRIPPED FROM THE END OF THE CONDUCTOR, INCREASING THE CHANCE OF FAULTS IN THE BOX.

D: THE CONDUCTOR HOOK IS DOUBLED OVER ON ITSELF. THE HOOK WILL NOT SEAT ON THE TERMINAL PROPERLY, WHICH CAN PRODUCE LOOSE CONNECTIONS.

E: THE HOOK IS TOO BROAD. THE HOOK CAN SLIP OUT FROM UNDER THE SCREW, AND POOR CONTACT BETWEEN THE TERMINAL AND THE SCREW CAN CAUSE LOOSE CONNECTIONS.

F: THE HOOK IS INSTALLED FACING THE WRONG DIRECTION. AS THE SCREW IS TIGHTENED, THE HOOK WILL LOOSEN, POSSIBLY CAUSING LOOSE OR INADEQUATE CONNECTIONS.

G: NOT ENOUGH INSULATION IS STRIPPED FROM THE CONDUCTOR END. THE SCREW HEAD IS SEATED ON THE CONDUCTOR INSULATION, PRODUCING A LOOSE CONNECTION.

Figure 16-1

Figure 16-2 shows proper termination using the stabbing or **quick connect** method of termination. The most important aspect of terminating with a quick-connect terminal is the length of insulation stripped from the conductor. If too little is stripped, a loose connection may result, and if too much is stripped, there will be an exposed bare conductor in the box, increasing the possibility of a fault. Quick-connect terminations are allowed only with 14 AWG conductors. Any termination with a 12 AWG or larger conductor will require a screw-type or lug-type termination. Some AHJs do not allow the use of quick-connect terminations even with 14 AWG conductors. In some regions of the country, problems with these types of connection are more common. In wet, humid climates, for example, these connections can corrode, eventually becoming loose or disconnecting altogether. In drier areas of the country, problems with use of quick-connect connections are rate. Some contractors have company rules forbidding the use of quick-connect terminations.

All devices must be installed in a neat and workmanlike manner. The switches and the receptacles are to be level and plumb. Many electricians use small torpedo-type bubble levels to ensure proper installations. All sheetrock ears of the device must fit snugly against the outer wall finish, and all cover plates must fit securely against the wall without gaps or openings. If there are problems covering the gaps and openings around cover plates, larger, oversized plates are available from most supply houses.

16.3: CONDUCTOR IDENTIFICATION AND DEVICE SELECTION

The location of the outlet box is the first indicator of whether a particular box is a receptacle outlet, a switch point, or utilization equipment. Because the receptacles are usually located close to the floor, at 13 in. (331 mm) to the bottom of the box according to the specifications for the sample house, any outlet located at that height is expected to be a receptacle. Likewise, any round outlet on the ceiling or wall would be expected to be for a luminaire (fixture) or smoke detector. Boxes located on walls at 44 in. (1.1 m) above the floor and located close to doorways or entries would be expected to be switches; however, these indicators are not correct in all instances. The outlet box on the ceiling may be for

the smoke detector or a luminaire (fixture). There is a receptacle outlet on the ceiling of the garage. Boxes installed at switch height may be receptacles that are above counters or are located in garages or workshops where the receptacles are placed at a convenient height for easy access in order to plug and unplug power tools or other equipment, such as the refrigerator in the sample house.

The best way to determine what device is to be installed in any particular box is by the box makeup and conductor identification. Conductor identification and techniques for box makeup are covered in Chapter 9 of this book. All of the different wiring configurations for the boxes installed in the sample dwelling are shown in Figures 10-5 and 10-16 in Chapter 10. Review of these two chapters is recommended before continuing with the trim. When the conductors are pulled out of the front of the box, as shown in Figure 9-18, only those conductors necessary to connect to the device, luminaire (fixture), or appliance should be available. These conductors should be stored in an accordionlike fashion in order to be easily removed and replaced, while the other conductors are stored in the back of the box and should not need to be accessed at trim. The method of coding the conductors used at rough-in and makeup will provide the information necessary to identify the proper device needed for that particular outlet or switch point.

Figures 16-3 and 16-4 show the wiring as it should appear after being removed from each box for connection to a trim device. The coding on the conductors (conductors with stripped ends, twisted together and marked with black tape) that were left after makeup will provide all of the necessary information for selection of the proper device to install at trim.

16.4: RECEPTACLES

The ampere rating of a receptacle outlet is determined by the ampere rating of the branch circuit. The receptacles and the attachment plugs that plug into the receptacles have different size and blade configurations for different ratings for amperage, voltage, phase, and number of wires. The different configurations do not allow an attachment plug from a higher-rated load, such as a 20-ampere cord-and-plug connected heater, to be plugged into a receptacle not rated for that load, such as a 15-ampere-rated

"WIREMAN'S GUIDE"
USING QUICK CONNECTS ON DEVICES

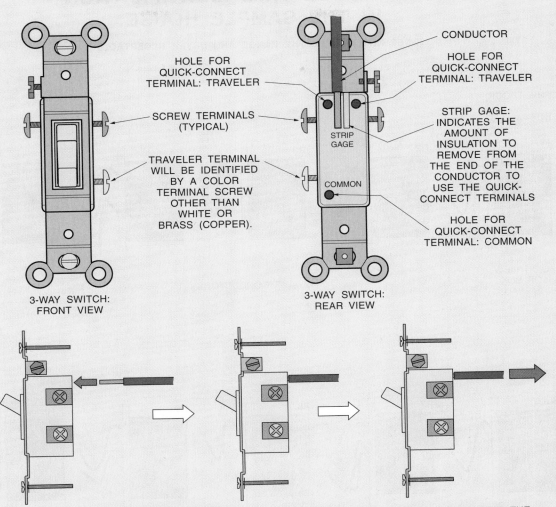

HOLE FOR QUICK-CONNECT TERMINAL: TRAVELER

SCREW TERMINALS (TYPICAL)

TRAVELER TERMINAL WILL BE IDENTIFIED BY A COLOR TERMINAL SCREW OTHER THAN WHITE OR BRASS (COPPER).

3-WAY SWITCH: FRONT VIEW

CONDUCTOR

HOLE FOR QUICK-CONNECT TERMINAL: TRAVELER

STRIP GAGE: INDICATES THE AMOUNT OF INSULATION TO REMOVE FROM THE END OF THE CONDUCTOR TO USE THE QUICK-CONNECT TERMINALS

STRIP GAGE

COMMON

HOLE FOR QUICK-CONNECT TERMINAL: COMMON

3-WAY SWITCH: REAR VIEW

TECHNIQUE: STRIP THE CORRECT LENGTH OF INSULATION FROM THE CONDUCTOR. STAB THE BARE CONDUCTOR INTO THE PROPER HOLE ON THE BACK OF THE DEVICE. GIVE A SHARP TUG ON THE CONDUCTOR TO ENSURE THAT THE QUICK-CONNECT TERMINALS ARE ENGAGED.

ENSURE THAT THE PROPER LENGTH OF INSULATION IS STRIPPED FROM THE CONDUCTOR. IF TOO LITTLE IS STRIPPED, THE QUICK CONNECTS WILL NOT MAKE GOOD CONTACT, RESULTING IN A LOOSE CONNECTION. IF TOO MUCH INSULATION IS STRIPPED, THERE WILL BE A LENGTH OF UNINSULATED CONDUCTOR IN THE BOX, INCREASING THE POSSIBILITY OF A FAULT.

NOT ALL DEVICES HAVE PROVISIONS FOR QUICK-CONNECTING CONDUCTORS.

THE AHJ MAY NOT ALLOW THE USE OF QUICK-CONNECT TERMINALS.

QUICK CONNECTS CAN BE USED ONLY WITH 14 AWG SOLID COPPER CONDUCTORS.

DO NOT QUICK CONNECT ALUMINUM CONDUCTORS.

DO NOT QUICK CONNECT STRANDED CONDUCTORS.

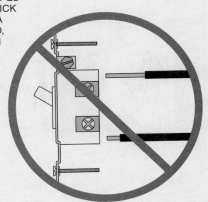

Figure 16-2

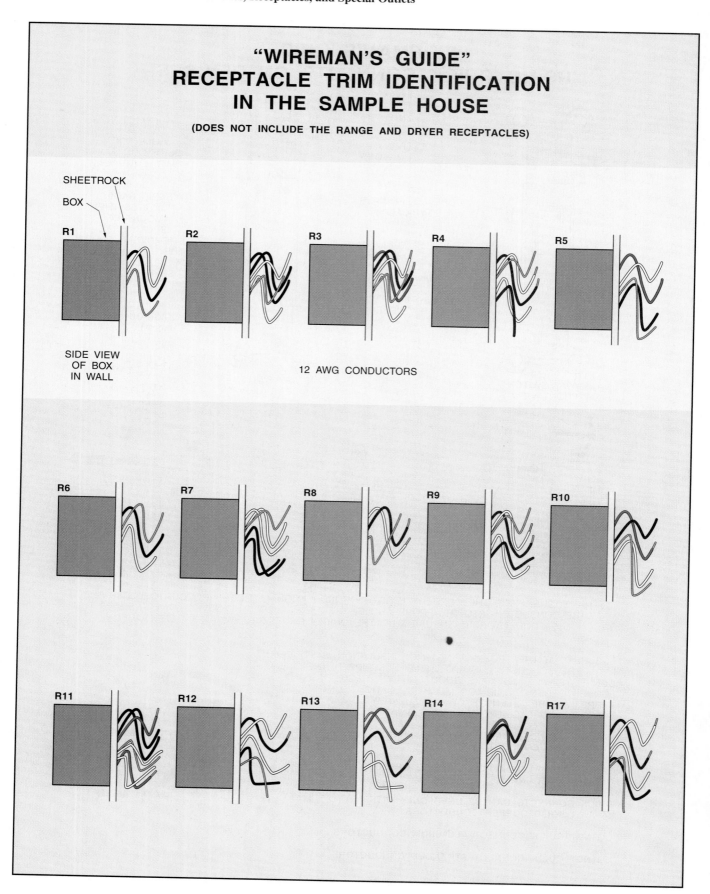

Figure 16-3

"WIREMAN'S GUIDE"
SWITCH TRIM IDENTIFICATION

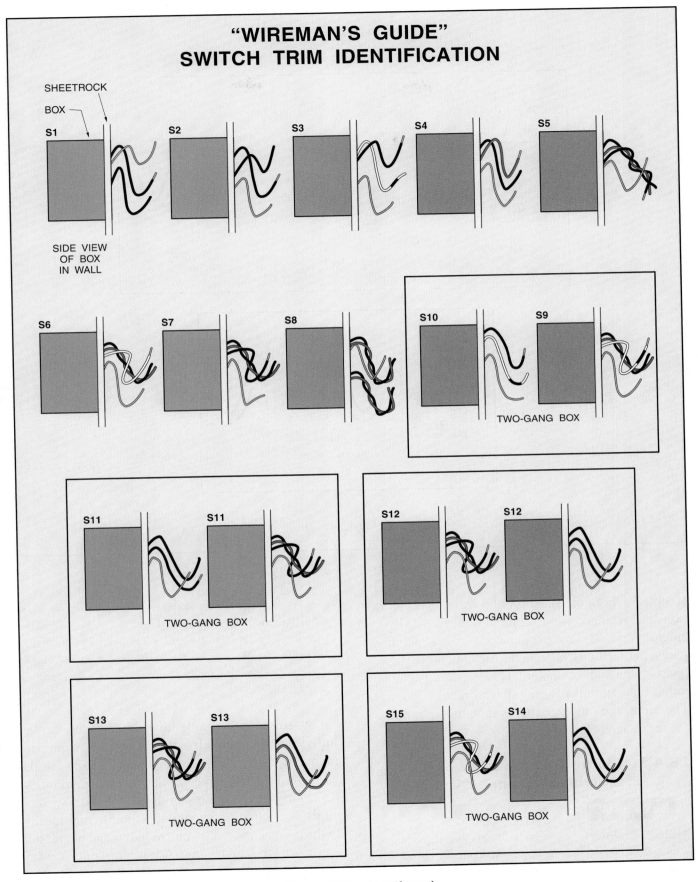

Figure 16-4 *(continues)*

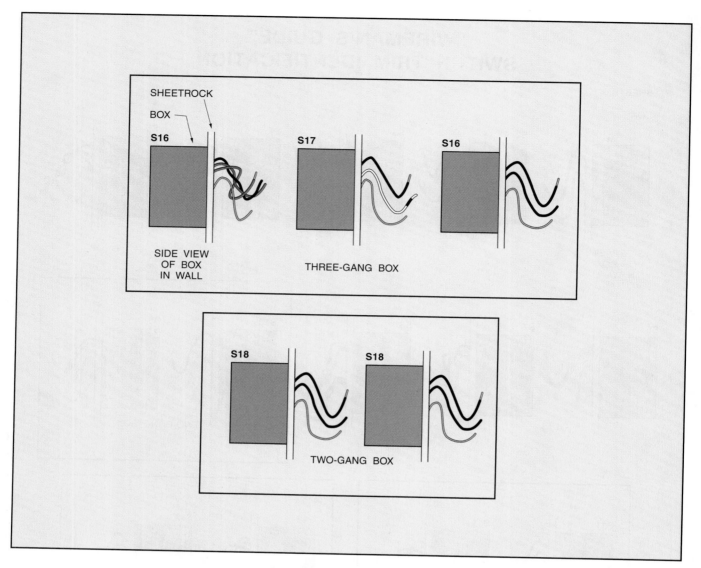

Figure 16-4   *(continued)*

receptacle. Figure 16-5 shows how these configurations are arranged for 15- and 20-ampere, 120-volt, single-phase, 3-wire receptacles and attachment plugs. In dwellings, all general-use outlets are single-phase outlets, all of the outlets are 120 volts, and all require an equipment grounding conductor contact. Therefore, the only variable is the ampere rating of the receptacle.

16.4.1: Single Receptacles on Individual Branch Circuits

> **KEY TERMS**
>
> **Single receptacle**   A yoke with only one receptacle installed. Most yokes have two receptacles installed and are called duplex receptacles.

> **Individual branch circuit**   (See *NEC® Article 100 [Branch Circuit—Individual]*): A branch circuit that supplies one piece of utilization equipment. An individual branch circuit usually requires the use of a single receptacle.

A **single receptacle** that is installed on an **individual branch circuit** must have an ampere rating of not less than the ampere rating of the branch circuit. The receptacle outlet for circuit 24 from subpanel A of the sample house is just such an outlet. The circuit powering the sump pump located in the basement is a 20-ampere individual circuit that serves only the sump pump. According to the mechanical and appliance schedule, Construction Drawing M-1, the sump pump has a ¾ horsepower motor.

"WIREMAN'S GUIDE"
NONINTERCHANGEABILITY OF RECEPTACLES AND ATTACHMENT PLUGS

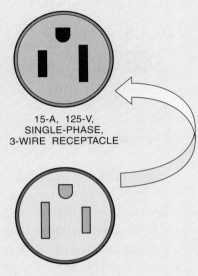

15-A, 125-V,
SINGLE-PHASE,
3-WIRE RECEPTACLE

15-A, 125-V,
SINGLE-PHASE,
3-WIRE ATTACHMENT PLUG

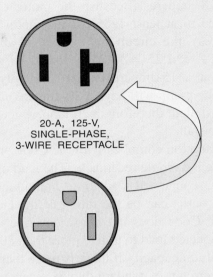

20-A, 125-V,
SINGLE-PHASE,
3-WIRE RECEPTACLE

20-A, 125-V,
SINGLE-PHASE,
3-WIRE ATTACHMENT PLUG

A 20-A, 125-V,
SINGLE-PHASE, 3-WIRE ATTACHMENT
PLUG WILL NOT FIT INTO A
15-A RECEPTACLE.

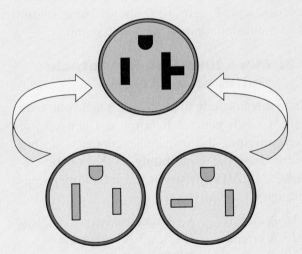

EITHER A 15-A OR 20-A,
125-V, SINGLE-PHASE,
3-WIRE ATTACHMENT PLUG WILL FIT
INTO A 20-A RECEPTACLE.

THE INTERCHANGEABILITY OF ATTACHMENT PLUGS OF ONE VOLTAGE, AMPERAGE, PHASE, OR WIRING CONFIGURATION IS CONTROLLED BY THE ARRANGEMENT OF THE PLUG CONTACT BLADES, THE SHAPE OF THOSE BLADES, AND THE SIZE OF THE ATTACHMENT PLUG AND RECEPTACLE. ALL TYPES AND SIZES OF RECEPTACLES AND ATTACHMENT PLUGS ARE AVAILABLE IN A TWIST-LOCK TYPE THAT TWISTS INTO PLACE AND DOES NOT EASILY DISCONNECT. THERE ARE NO TWIST-LOCK-TYPE RECEPTACLES OR ATTACHMENT PLUGS IN THE SAMPLE HOUSE.

Figure 16-5

Because the sump pump is listed as a motor load rather than an appliance load, the ampere rating of the pump must be obtained from *Table 430.148* based on the nameplate horsepower rating of the motor. A ¾ horsepower motor at 120 volts has an ampere rating of 13.8 amperes. Because the motor circuit must be sized to at least 125% of the motor FLA (13.8 amperes), the circuit must be a 20-ampere circuit (13.8 A × 1.25 = 17.25 A load). To guard against use of a 15-ampere circuit to supply this equipment, the pump will have a 20-ampere attachment plug installed; therefore, a 20-ampere receptacle must be installed.

This installation complies with the Code only if the receptacle is a single receptacle. The load of one cord-and-plug connected load to an individual circuit and one outlet can be the rating of the circuit. However, the Code does not allow any one cord-and-plug connected load to utilize more than 80 percent of the circuit capacity if there is more than one outlet on the circuit. A standard duplex receptacle is considered as two outlets. Because the ampere rating of the sump pump is over 80 percent of the circuit rating (20 A × .80 = 16 A), the outlet must be a single receptacle. Figure 16-6 shows some of the details of the installation for the sump pump.

16.4.2: More than One Receptacle Outlet on a Circuit

When there is more than one receptacle outlet on a circuit, as with general lighting and general-use receptacle circuits, the installed receptacles do not necessarily have the same rating as that of the circuit. *Table 210.21(B)(3)* shows the following receptacle ratings for circuits:

15-A circuit:	15-A receptacles
20-A circuit:	15- or 20-A receptacles
30-A circuit:	30-A receptacles
40-A circuit:	40- or 50-A receptacles
50-A circuit:	50-A receptacles

With these receptacle ratings, two conditions must be satisfied:

1. No one cord-and-plug connected load can use more than 80 percent of the circuit rating. This means that any one 15-ampere receptacle cannot serve more than a single 12-ampere load, and a 20-ampere receptacle cannot serve more than a single 16-ampere load.

2. If any fixed-in-place utilization equipment is supplied by the circuit, the total connected load cannot exceed 50 percent of the circuit ampere rating. For the purposes of this section, luminaires (fixtures) are not considered as fastened-in-place utilization equipment.

16.4.3: Polarity

KEY TERMS

Polarity The relative location of the phase and grounded conductors on an outlet device. Polarity must be properly maintained for safe operation of utilization equipment.

The blades of an attachment plug and the slots of a receptacle are designated for specific conductors. There are a specific slot and blade for the grounded circuit (neutral) conductor, specific slots and blades for the ungrounded conductor or conductors, and a specific slot and blade for the equipment grounding conductor. This system of using various configurations of slots and blades is called **polarity**. To maintain polarity, care must be taken to ensure that the proper conductors are connected to the correct slots and blades. Polarity must be maintained; otherwise, the electrical system can become unnecessarily dangerous.

NEC® 200.10 details the requirements for polarized connections for receptacles, attachment plugs and connectors, screw shells and screw shells with leads, and appliances. In each case, the Code requires that the grounded circuit connection be positively identified so that there is no confusion about which terminal or lead is the grounded conductor terminal. The grounded terminals must be identified by the use of a white metal or by a white outer finish, by the word "white" or the letter "W" located adjacent to the terminal, or by a white or gray conductor lead. The grounded circuit conductor is sometimes referred to as the identified conductor. *NEC® 200.11* forbids the connection of any device or equipment in such a manner as to reverse the intended polarity. Figures 16-7 and 16-8 provide more information about polarity.

16.4.4: Standard Duplex Receptacle

Many boxes in the sample house contain a standard duplex receptacle. An example of a standard duplex installation is shown Figure 10-8 as receptacle 7.11. The rough conductor identification for box 7.11

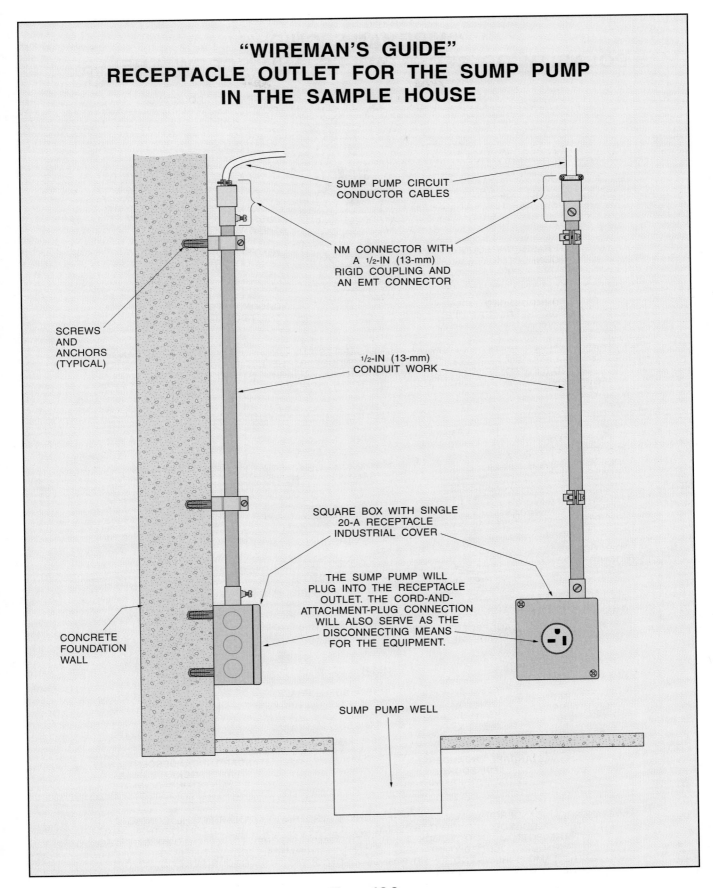

"WIREMAN'S GUIDE"
RECEPTACLE OUTLET FOR THE SUMP PUMP IN THE SAMPLE HOUSE

SUMP PUMP CIRCUIT CONDUCTOR CABLES

NM CONNECTOR WITH A 1/2-IN (13-mm) RIGID COUPLING AND AN EMT CONNECTOR

SCREWS AND ANCHORS (TYPICAL)

1/2-IN (13-mm) CONDUIT WORK

SQUARE BOX WITH SINGLE 20-A RECEPTACLE INDUSTRIAL COVER

THE SUMP PUMP WILL PLUG INTO THE RECEPTACLE OUTLET. THE CORD-AND-ATTACHMENT-PLUG CONNECTION WILL ALSO SERVE AS THE DISCONNECTING MEANS FOR THE EQUIPMENT.

CONCRETE FOUNDATION WALL

SUMP PUMP WELL

Figure 16-6

"WIREMAN'S GUIDE"
POLARITY OF RECEPTACLES AND SCREW SHELLS

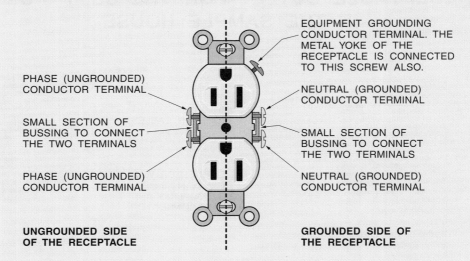

EQUIPMENT GROUNDING CONDUCTOR TERMINAL. THE METAL YOKE OF THE RECEPTACLE IS CONNECTED TO THIS SCREW ALSO.

PHASE (UNGROUNDED) CONDUCTOR TERMINAL

NEUTRAL (GROUNDED) CONDUCTOR TERMINAL

SMALL SECTION OF BUSSING TO CONNECT THE TWO TERMINALS

SMALL SECTION OF BUSSING TO CONNECT THE TWO TERMINALS

PHASE (UNGROUNDED) CONDUCTOR TERMINAL

NEUTRAL (GROUNDED) CONDUCTOR TERMINAL

UNGROUNDED SIDE OF THE RECEPTACLE

GROUNDED SIDE OF THE RECEPTACLE

RECEPTACLES HAVE A POLARITY: ONE SIDE CARRIES THE UNGROUNDED CONDUCTOR AND THE OTHER SIDE CARRIES THE GROUNDED CONDUCTOR. THE ESTABLISHED POLARITY MUST BE MAINTAINED FOR EVERY RECEPTACLE AND EVERY ATTACHMENT PLUG. THE UTILIZATION EQUIPMENT IS MANUFACTURED ACCORDING TO THIS POLARITY.

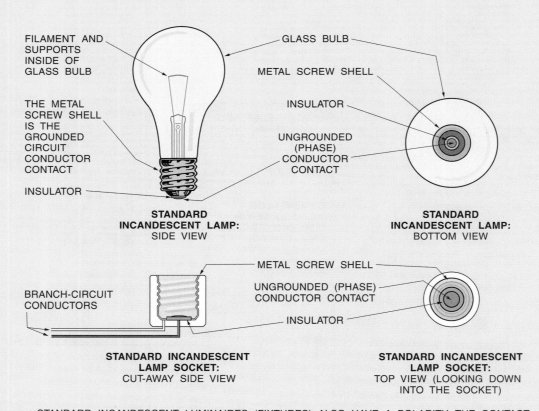

FILAMENT AND SUPPORTS INSIDE OF GLASS BULB

GLASS BULB

METAL SCREW SHELL

INSULATOR

THE METAL SCREW SHELL IS THE GROUNDED CIRCUIT CONDUCTOR CONTACT

UNGROUNDED (PHASE) CONDUCTOR CONTACT

INSULATOR

STANDARD INCANDESCENT LAMP: SIDE VIEW

STANDARD INCANDESCENT LAMP: BOTTOM VIEW

METAL SCREW SHELL

UNGROUNDED (PHASE) CONDUCTOR CONTACT

BRANCH-CIRCUIT CONDUCTORS

INSULATOR

STANDARD INCANDESCENT LAMP SOCKET: CUT-AWAY SIDE VIEW

STANDARD INCANDESCENT LAMP SOCKET: TOP VIEW (LOOKING DOWN INTO THE SOCKET)

STANDARD INCANDESCENT LUMINAIRES (FIXTURES) ALSO HAVE A POLARITY. THE CONTACT FOR THE UNGROUNDED CONDUCTOR IS THE SMALL CONTACT AT THE BOTTOM OF THE LAMP AND AT THE BOTTOM OF THE SOCKET. WHEN THE LAMP IS SCREWED ALL THE WAY INTO THE SOCKET, THE UNGROUNDED CONTACTS ARE CLOSED. THE METAL SCREW SHELL—THE PART THAT THE LAMP SCREWS INTO—IS THE GROUNDED CONTACT. THE GROUNDED CONDUCTOR IS CONNECTED TO THE SCREW SHELL.

Figure 16-7

"WIREMAN'S GUIDE"
POLARITY AT RECEPTACLES AND ATTACHMENT PLUGS

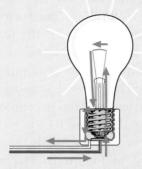

WITH THE CORRECT POLARITY, THE CURRENT WILL RUN THROUGH THE WIRE, THROUGH THE UNGROUNDED (PHASE) CONDUCTOR CONTACTS, UP THE SUPPORT, THROUGH THE FILAMENT, AND DOWN THE OTHER SUPPORT TO THE SCREW SHELL. THE SCREW SHELL IS THE GROUNDED (NEUTRAL) CONTACT, AND THE CURRENT IS CARRIED BACK TO THE GROUNDED BUS IN THE DISTRIBUTION PANEL. THE LAMP IS ILLUMINATED.

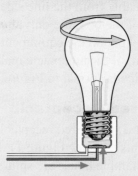

WHEN THE LAMP IS REMOVED FROM THE SOCKET, THE CIRCUIT IS DISCONNECTED AT THE UNGROUNDED CONDUCTOR CONTACT. COMING INTO CONTACT WITH THE SCREW SHELL DURING LAMP REMOVAL PRESENTS NO HAZARD OF ELECTROCUTION.

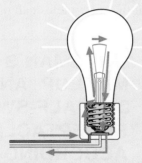

IF THE POLARITY IS REVERSED, IF THE PHASE CONDUCTOR IS CONNECTED TO THE SCREW SHELL AND THE NEUTRAL CONDUCTOR IS CONNECTED TO THE CONTACT AT THE BOTTOM OF THE SOCKET, THE CURRENT WILL FLOW THROUGH THE CONDUCTOR TO THE SCREW SHELL. THE CURRENT THEN FLOWS THROUGH THE SUPPORT, THROUGH THE FILAMENT, AND DOWN THE OTHER SUPPORT TO THE CONTACTS AT THE BOTTOM OF THE SOCKET. THIS CONTACT IS CONNECTED TO THE GROUNDED BUS IN THE DISTRIBUTION PANEL. THE LAMP IS ILLUMINATED.

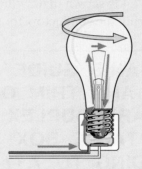

WHEN THE LAMP IS UNSCREWED, THE CIRCUIT WILL OPEN AT THE CONTACTS LOCATED AT THE BOTTOM OF THE SOCKET, AND THE LAMP WILL BE EXTINGUISHED. HOWEVER, THE SCREW SHELL CONTINUES TO BE ENERGIZED, AND ANYONE CONTACTING THE SCREW SHELL, THUS PROVIDING A PATH TO GROUND, WILL BE SUBJECT TO ELECTROCUTION.

THIS TYPE OF EVENT CAN ALSO OCCUR IN ELECTRONIC EQUIPMENT, SUCH AS STEREOS OR TELEVISIONS, THAT USES THE CHASSIS OF THE EQUIPMENT AS THE GROUNDED SIDE OF THE CIRCUIT. THE SAME TYPE OF REQUIREMENT IS MADE FOR SCREW SHELL BASE FUSES WITH THE LOAD SIDE OF THE FUSE WIRED TO THE SCREW SHELL.

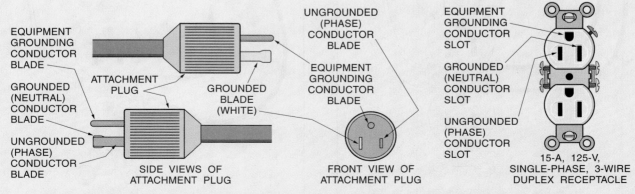

CORD CAPS AND RECEPTACLES MAINTAIN THE POLARITY BY THE SIZE AND CONFIGURATION OF THE BLADES.

Figure 16-8

is shown in drawing R7 in Figure 10-4. At trim, the conductors for receptacle 7.11 will appear as shown in drawing R12 in Figure 16-3. Notice that for a standard duplex with one cable from the line side and one cable to the load side of the box, the only makeup required is to connect and pigtail the equipment grounding conductors. A detail of the receptacle makeup and connection to the device is shown in Figure 16-9.

16.4.5: Split-Wired Receptacle

The functioning of a split-wired receptacle is covered in Chapter 11 of this text (see Figure 11-15). Several receptacles in the sample house are split-wired:

- Two duplex receptacle outlets in the master bedroom are half-switched. These receptacles are marked 7.4 and 7.6 in Figure 10-8. The conductor identification system used during rough-in for receptacle 7.4 is shown in drawing R12 in Figure 10-4. At trim, the conductors for recepta-

cle 7.4 appear as shown in drawing R12 in Figure 16-3. A detailed drawing of the makeup and the connection to the receptacle is shown in Figure 16-10. Notice that the bussing between the two phase conductor terminals has been removed to make the receptacles independent of each other. Notice also that the neutral side bussing has not been removed.

The second split-wired receptacle in the master bedroom is the box marked 7.6 in Figure 10-5. Drawing R14 in Figure 16-3 shows rough identification for the conductors, and drawing R14 in Figure 16-3 shows the conductors as they appear when pulled out of the box. A detailed drawing of the makeup and connections in receptacle box R14 is shown in Figure 16-11. Notice that the bussing between the two phase conductor terminals has been removed so that the receptacles can operate independently.

"WIREMAN'S GUIDE" MAKEUP AND TRIM OF STANDARD DUPLEX RECEPTACLE BOX ACCORDING TO *7.11*

Figure 16-9

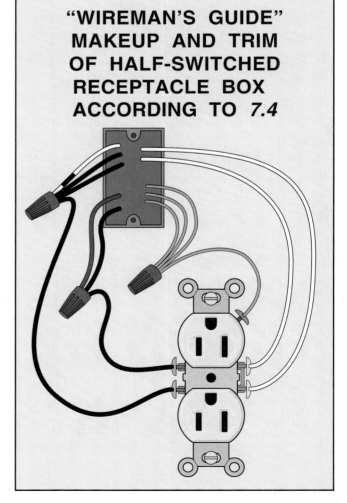

"WIREMAN'S GUIDE" MAKEUP AND TRIM OF HALF-SWITCHED RECEPTACLE BOX ACCORDING TO *7.4*

Figure 16-10

• There is a split-wired receptacle in the living room. It is identified as receptacle 3.12 in Figure 10-10. The receptacle is half-switched from the three-way switching system with switch points at the main entry and at the entry to the hallway. The rough conductor identification is shown as drawing R14 in Figure 10-7 and the wiring after makeup is shown as drawing R14 in Figure 16-3. There is a detailed drawing of the makeup and connection of receptacle 3.12 in Figure 16-11. Notice that this is the same makeup and trim as for receptacle 7.6 in the master bedroom.

• The remaining split-wired receptacle is in the homerun box for the dishwasher and disposal circuits. It is shown as the receptacle box in Figure 11-14. This duplex receptacle will be split between two separate circuits. Notice that the bussing between the phase conductor terminals has been removed to allow the receptacles to operate independently of each other. These circuits will share the grounded circuit conductor, however, so the bussing on the grounded conductor side remains intact. Because this split duplex receptacle is supplied from two circuits, the circuit breakers that these two homeruns connect to in the distribution panel must be connected together so that if one trips or is turned off, the other must open automatically also. For this requirement, the homeruns must occupy spaces next to each other, and the circuit breaker must have a common internal trip mechanism or a tie handle. There is a 12-2 cable from this box to the two-gang switch to the right of the kitchen sink. This switch is to control the disposal receptacle. A detailed drawing of the makeup and connection is shown in Figure 16-12.

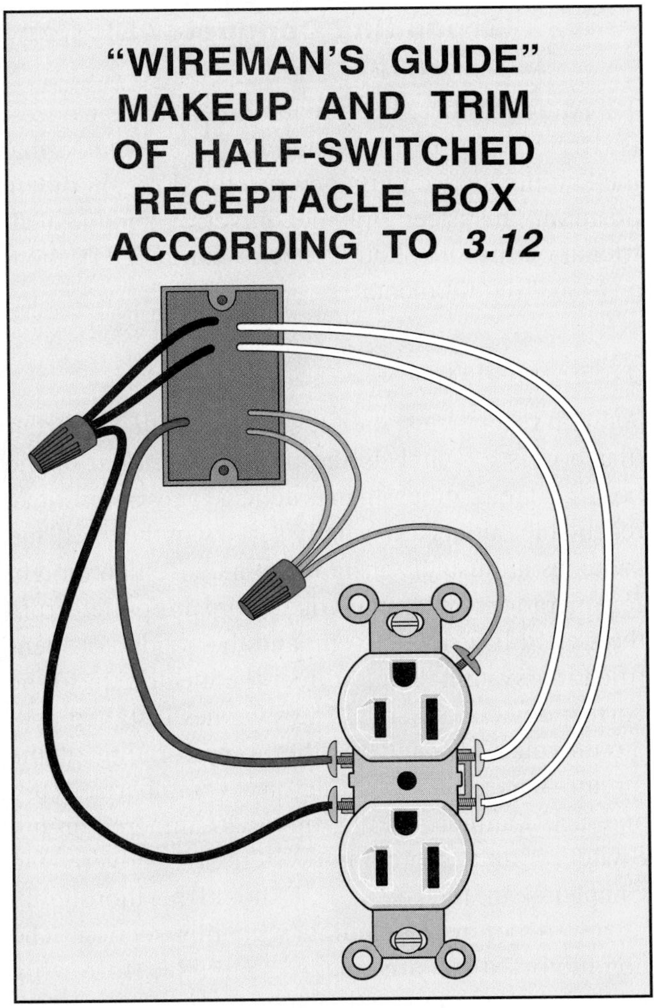

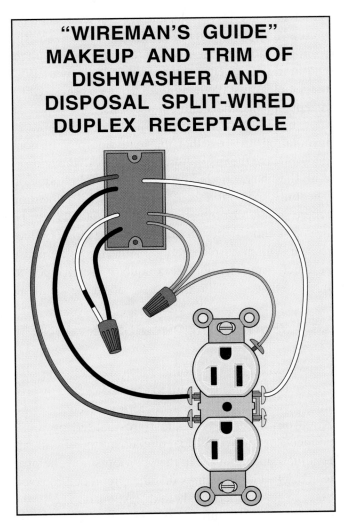

**"WIREMAN'S GUIDE"
MAKEUP AND TRIM
OF HALF-SWITCHED
RECEPTACLE BOX
ACCORDING TO *3.12***

Figure 16-11

**"WIREMAN'S GUIDE"
MAKEUP AND TRIM OF
DISHWASHER AND
DISPOSAL SPLIT-WIRED
DUPLEX RECEPTACLE**

Figure 16-12

The receptacle box behind the refrigerator is supplied by two circuits, but only one of the circuits is actually connected to the receptacle. Therefore, it is not a split-wired receptacle.

16.4.6: Ground-Fault Circuit-Interrupter (GFCI) Circuits

GFCI circuits are covered in Chapter 3 of this text and specifically in Figure 3-38, and a review of that discussion is recommended before completing study of this chapter.

In the sample house, 18 receptacle outlets on all or part of six different circuits are GFCI protected. The three bathroom receptacle outlets are protected from a feed-through receptacle connection in the first outlet in the master bedroom. The small-appliance receptacle outlets are also covered by feed-through receptacles at the homerun boxes for each small-appliance circuit. Notice that the refrigerator outlet is not GFCI protected. The other receptacle outlets required to have GFCI protection—at the front outdoors, at the back outdoors, in the garage, and in the basement—all will have their own GFCI receptacles installed. Figure 16-13 shows GFCI receptacle terminals and connections.

The identification for the GFCI receptacle installation requires that the load side pair of conductors have the insulation stripped from the ends of both the phase conductor and the neutral conductor. The only exception is if the box is in a location where a GFCI-protective device is likely to be required and the box contains no load side conductors, in which case the ends of the line side phase conductor and the neutral conductor are stripped to indicate that a GFCI receptacle needs to be installed.

16.4.7: Arc-Fault Circuit-Interrupter (AFCI)

All outlets in bedrooms, not just receptacle outlets, in dwellings are required to be protected by arc-fault circuit-interrupter protection. Because the AFCI device protects against arcs caused by loose connections or by physical damage to the circuit, all terminations at outlets in the bedrooms must be protected. The Code requires that the circuit be protected before the first terminal. If the AFCI device is installed at the first outlet in the bedroom, the line side termination on that outlet will not be AFCI protected.

In the sample house, two bedroom circuits require AFCI protection. Circuit 5 from subpanel A supplies bedroom outlets and lighting exclusively, except that the homerun box is the outlet in the hallway. An AFCI feed-through receptacle is installed in the hallway to protect the bedroom outlets connected to the load side of the device. The second circuit, circuit 7, requires an AFCI circuit breaker. The homerun box is in the master bedroom and the line side terminals of the receptacle need to be protected.

The identification of an AFCI receptacle is the same as for the GFCI receptacle. The load side conductors are stripped at the ends of both the phase conductor and the neutral conductor. If there are no load side conductors, the line side conductors are stripped to indicate that an AFCI receptacle must be installed.

16.4.8: Orientation of the Equipment Grounding Conductor Slot on a Receptacle

Although there is no requirement in the Code, it is common practice to install the receptacle so that the equipment grounding conductor slot is in the up position. This is certainly not universal, and the electrician on the site should check with the AHJ about any local requirements.

16.5: SWITCHES

As with receptacles, the first indication that a particular box is a switch is the physical location of the box. In areas other than kitchens and bathrooms, receptacle outlets are usually set low in the wall, and switches are usually installed at a convenient height for operation. When the conductors are pulled from the switch box, the makeup and the conductor identification system should direct the electrician to the correct device to install. The switches installed on a circuit must be rated for the ampere rating of the circuit—a 15-ampere rating on a 15-ampere circuit, and a 20-ampere rating on a 20-ampere circuit. Switches and switching system are covered in Chapter 4 of this book and should be thoroughly reviewed before continuing with this section. The conductors that are made available, following makeup and for connection at trim, for all the switch boxes in the sample house are shown in Figure 16-4.

"WIREMAN'S GUIDE"
GROUND-FAULT CIRCUIT-INTERRUPTER (GFCI) RECEPTACLE CONNECTIONS

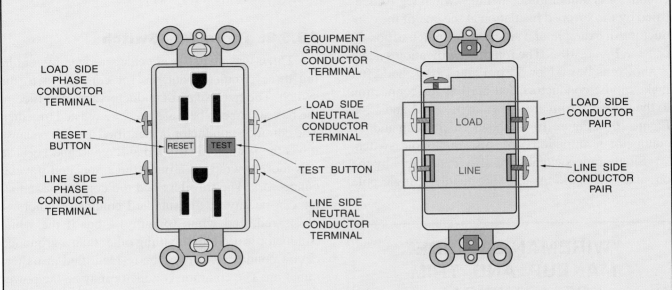

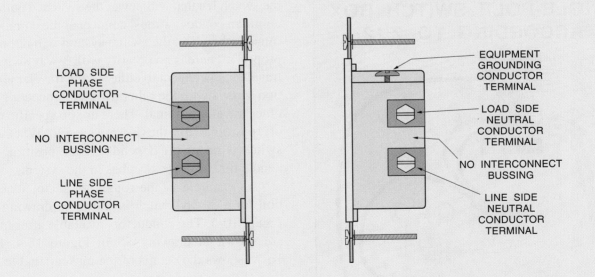

LINE SIDE OF THE GFCI PROVIDES POWER TO THE RECEPTACLE. THE RECEPTACLE IS AUTOMATICALLY PROTECTED AGAINST GROUND FAULTS WHEN IT IS PROPERLY CONNECTED TO THE CIRCUIT. WHEN LOAD SIDE CONDUCTORS ARE CONNECTED TO THE TERMINALS MARKED LOAD, EVERY DEVICE, APPLIANCE, OR OTHER EQUIPMENT ON THE LOAD SIDE OF THE GFCI WILL ALSO HAVE GFCI PROTECTION. THE PROTECTION OF LOAD SIDE COMPONENTS IS SOMETIMES REFERRED TO AS A FEED-THROUGH. A GFCI RECEPTACLE TAKES UP A LOT OF ROOM IN THE BOX.

Figure 16-13

16.5.1: Single-Pole Switches

There are several single-pole switches in the sample house. The single-pole switch is identified by the two ungrounded conductors to connect to the switch, the power supply for the switch, and the switch leg to power the switched outlet. The power conductor is unidentified, and the switch leg is identified by the stripped insulation at the end of the conductor. An example of a single-pole switch is box 51 or 52 in Figure 10-5. The rough identification system is shown as box 51 or 52 in Figure 10-5, and after makeup, the conductors that are left for connection to the switch at trim are shown in boxes 51 and 52 in Figure 16-4. Figure 16-14 shows the makeup and the conductor terminations for this single-pole switch.

Single-pole switches are also identifiable by the on and off markings on the handle. Single-pole switches must be installed so that the switch handle will be in the down position when the switch is open (off). Single-pole and double-pole switches have specific on and off positions. Three-way and 4-way switches do not have any specific up or down position for on or off but depend on the position of the other switches in the system.

16.5.2: Three-Way Switch

Three-way switches are characterized by requiring three circuit conductors for connection to the switch. The two traveler conductors are identified in the box at makeup because they are twisted together. The common conductor is identified as such because it has the insulation stripped at the end and because it is loosely wrapped around the associated traveler conductors. The travelers and the common conductor can be any of the standard conductor colors—black, red, and white (or gray)—but if the white conductor is used for anything other than a grounded circuit conductor, it must be re-identified at its termination. The common conductor may be the power supply to the 3-way switching system or it may be the switch leg that will power the switched outlet. The common conductor must connect at the common terminal; otherwise, the switch system will not operate properly. There is no polarity with 3-way switches; a traveler can connect to either of the traveler termination screws with the other traveler connecting to the other traveler terminal. There are no specific on and off positions for the 3-way switch. Whether the switch is on or off depends on the position of the handle for the other switches in the system.

An example of the rough conductor identification for a 3-way switch is shown in drawing S7 in Figure 10-5. The conductors available after makeup are shown in drawing S5 in Figure 16-4. Figure 16-15 shows the details of the makeup and termination of these conductors.

16.5.3: Four-Way Switch

Four-way switches are identified by the two sets of travelers. These travelers are in pairs—an incoming pair and an outgoing pair. Each pair is twisted together to maintain the association of one traveler with the other. There is also another set of conductors in the box associated with the 4-way switching

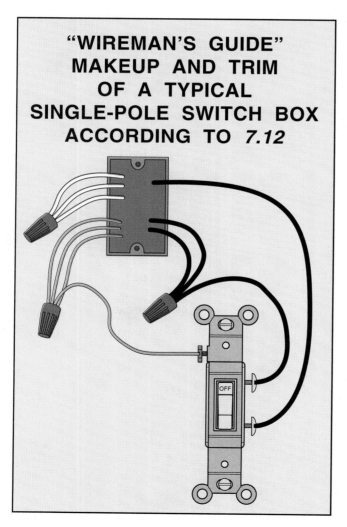

"WIREMAN'S GUIDE" MAKEUP AND TRIM OF A TYPICAL SINGLE-POLE SWITCH BOX ACCORDING TO 7.12

OFF

Figure 16-14

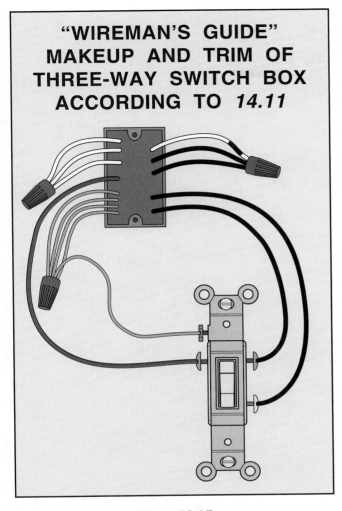

Figure 16-15

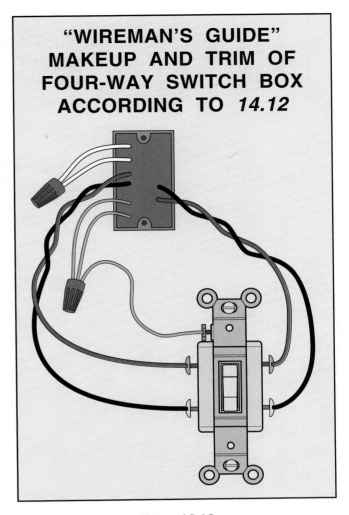

Figure 16-16

system, which are connected together but only feed through the box and are not connected to the switch. The traveler pairs can be of any color, but if white is used for a traveler, it must be re-identified to indicate its purpose. Care must be taken to determine how the traveler pairs connect to the 4-way switch. With some brands of 4-way switches, the travelers are paired by the top and the bottom of the switch, whereas with others, they are paired by the right side and the left side of the switch. A wiring diagram accompanies all 4-way switches to aid in this process.

There is one 4-way switch in the sample house, box 14.12 in Figure 10-12. The rough identification for the conductors in box 14.12 is shown in drawing S8 in Figure 10-8, and the conductors that connect to the switch at trim are shown in drawing S8 in Figure 16-3. Figure 16-16 shows the makeup of the box and the termination of each conductor to the 4-way switch.

16.6: COVER PLATES

Many varieties of cover plates are available. Cover plates are manufactured in bare metal, chrome, steel, stainless steel, brushed aluminum, and other metallic finishes. Painted metal cover plates are available in several standard colors—ivory, white, and brown—and can be easily spray-painted to match any wall color. Cover plates are also manufactured of nonmetallic materials such as plastic or vinyl. Figure 16-17 shows some of the variations in device cover plates.

The cover plate should fit snugly against the wall with no gaps that would allow accidental contact with the device or the wiring inside the box. The device also has to properly fit the cover. If the device is installed too far back into the box, the cover will not seat properly. The sheetrock ears of the device should seat snugly against the finished wall to allow

"WIREMAN'S GUIDE" COVER PLATES

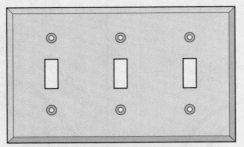

THREE-GANG SWITCH: IVORY, METAL, OR NONMETALLIC. COVER PLATES ARE AVAILABLE WITH VIRTUALLY ANY NUMBER OF GANGS.

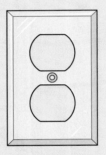

COVER PLATES ARE AVAILABLE IN CHROME, BRUSHED ALUMINUM, AND BRASS.

ALL SIZES AND TYPES OF COVER PLATES ARE AVAILABLE IN PAINTED WHITE, IVORY, GRAY, BROWN, AND OTHER COLORS.

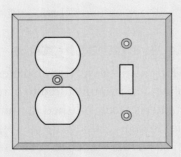

COVER PLATES ARE AVAILABLE IN ALL OF THE VARIOUS COMBINATIONS OF SWITCHES, RECEPTACLES, AND GFCI OR AFCI RECEPTACLES.

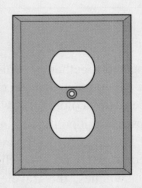

COVER PLATES ARE AVAILABLE IN LARGE SIZES IF NEEDED. IF ONE COVER IN A ROOM IS OVERSIZED, THE REST OF THE COVERS MAY NEED TO BE OVERSIZED TO PRESENT A UNIFORM APPEARANCE.

NORMAL RISE

DEEP RISE

STANDARD AND OVERSIZED PLATES ARE AVAILABLE WITH A DEEP RISE OR A NORMAL RISE.

Figure 16-17

the cover plate to fit properly. If the device yoke cannot seat against the finished wall because the hole was cut too big during installation of the wallboard or sheetrock, provisions must be made to stabilize the device and hold it secure in the proper location. The cover plate should never be used to stabilize the device if the ears are not firmly seated against the wall.

SUMMARY

The identification of the type of device to be installed in which box is determined from the location of the box and the coding on the wires left in the box at makeup (during rough-in) for connection to the device, appliance, or luminaire (fixture). Single receptacles on individual branch circuits must be rated for the ampere rating of the branch circuit. On circuits where more than one receptacle is installed, the *NEC*® allows for a 15-ampere receptacle to be installed on a 20-ampere circuit. The ampere rating of a switch must be at least as high as the circuit ampacity. Devices must be installed so that the sheetrock ears seat snugly against the finished wall. Cover plates must cover the hole in the finished wall completely so there are no gaps to allow accidental access to the device.

REVIEW

1. A half-switched duplex receptacle must have the bussing removed from _____ in order to operate properly.
   a. the phase conductor side
   b. the grounded conductor side
   c. both the phase conductor and the grounded conductor sides
   d. none of the above—no buss on the switch needs to be removed for a half-switched receptacle

2. In connecting conductors to screw terminals on devices, the open side of the hook should be to the _____ .
   a. top
   b. left
   c. right
   d. bottom

3. A white terminal on a device identifies it as _____ .
   a. a phase conductor terminal
   b. a grounded conductor terminal
   c. an equipment grounding conductor terminal
   d. a traveler conductor terminal

4. The following statements are true about quick-connect, or back-stabbed, connections except _____ .
   a. Only 14 AWG copper conductors are allowed to be connected in this manner.
   b. Only 12 AWG copper conductors are allowed to be connected in this manner.

c. Only 14 AWG aluminum conductors are allowed to be connected in this manner.

d. Only 12 AWG aluminum conductors are allowed to be connected in this manner.

5. Concerning 15-ampere and 20-ampere-rated cord-and-attachment plugs and receptacles, only the following statement is true: _____.

a. A 20-ampere rated attachment plug will fit a 15-ampere-rated receptacle.

b. A 15-ampere rated attachment plug will fit only a 15-ampere-rated receptacle.

c. A 15-ampere rated attachment plug will fit only a 20-ampere-rated receptacle.

d. A 20-ampere rated attachment plug will fit only a 20-ampere-rated receptacle.

6. Single receptacles on individual branch circuits must be _____.

a. rated to the ampere draw of the load

b. sized to the ampere rating of the branch circuit

c. must be sized to the ampere rating of the branch-circuit conductors

d. must be sized to the ampere rating of the attachment plug

7. *True or False:* The screw shell of a screw shell-type luminaire (fixture) must be connected to the grounded circuit conductor.

8. *True or False:* The center slot on the face of a 15-ampere-rated receptacle is for the connection of the grounded circuit conductor.

9. *True or False:* For a split-wired duplex receptacle supplied by a three-wire cable (plus equipment grounding conductor), the phase conductor side bus must be removed to allow the receptacles to operate independently.

10. *True or False:* The common screw on a 3-way switch is colored white and is marked with either the word "white" or the letter "W."

11. The common terminal on a 4-way switch is _____.

a. colored white

b. colored brass

c. colored green

d. not colored—there is no common on a 4-way switch

12. *True or False:* All circuits should be turned off at the distribution panel during trim.

13. *True or False:* When restoring power to a temporary circuit, the electrician should notify all other workers on the site before actually energizing the circuit.

14. *True or False:* All branch circuits can be assumed to be de-energized during trim because permanent power has not been connected.

15. *True or False:* Live circuits during trim are a danger to only the electricians.

"WIREMAN'S GUIDE" REVIEW

1. Explain why maintaining polarity is an important part of trimming an electrical system.

2. Explain how to maintain polarity with Edison-based fuseholders.

3. Explain the method used by the electrical industry to ensure that the proper receptacle is used with any given cord-and-attachment-plug connection.

4. Study the rough conductor coding shown in Figures 10-4 and 10-5. Compare them with the conductors shown for connection to the device shown in Figures 16-3 and 16-4. Make a drawing of the conductors' makeup as they appear in, for example, Figure 16-9 or 16-15.

CHAPTER 17

Luminaire (Fixture) Trim

OBJECTIVES

On completion of this chapter, the student will be able to:

☑ Explain the concerns related to the production of heat as well as light by the luminaires (fixtures).

☑ Identify the three major ways in which artificial light is produced: incandescent, fluorescent, and high-intensity discharge.

☑ Name the general sizes and shapes of popular incandescent lamps and the various sizes of lamp bases.

☑ Name the general sizes and terminals for popular types of fluorescent lamps.

☑ State the uses of compact fluorescent lamps.

☑ Describe the different installation environmental limitations, such as wet locations, enforced for installing luminaires (fixtures).

☑ Name the types of residential recessed luminaires (fixtures), and select the proper lamps and trims for each.

INTRODUCTION

This chapter is concerned with the trim of the luminaires (fixtures) in the sample house. The first practical application of electricity was to produce light. Today, the lighting industry has many lighting products that will satisfy almost any lighting need. Lighting for dwellings is usually the easiest type of lighting installation because the luminaires (fixtures) used in dwellings are mass produced, distributed, and stocked in many regions to supply the millions of homes built in this country. The ready availability of these luminaires (fixtures) also allows the installing electrician to maintain familiarity with various installation procedures and manufacturers' requirements. Most of the rooms in dwellings have 8-ft (2.5-m) ceilings that can be easily reached using a 6-ft ladder. In addition, most rooms are relatively small, so the energy demands of, and the physical size of, the luminaires (fixtures) are relatively small; therefore, the luminaires (fixtures) are relatively easier to install. In addition, most of the lighting installed is for general illumination instead of special task lighting, so special circuits or mounting methods usually are not needed.

17.1: THE NATURE OF LIGHTING

The amount of light from a point source that falls on any surface is directly proportional to the absolute intensity of the light and inversely proportional to the square of the distance from that light source. Artificial lighting sources direct light into beams, or flood light to a given area, but the intensity still diminishes over the distance traveled. See Figure 17-1. The major problem associated with lighting using electricity is the amount of heat that is also created along with the light. This heat can be very intense and can cause fires by damaging the insulation

"WIREMAN'S GUIDE"
FOOT CANDLES AND LUMENS

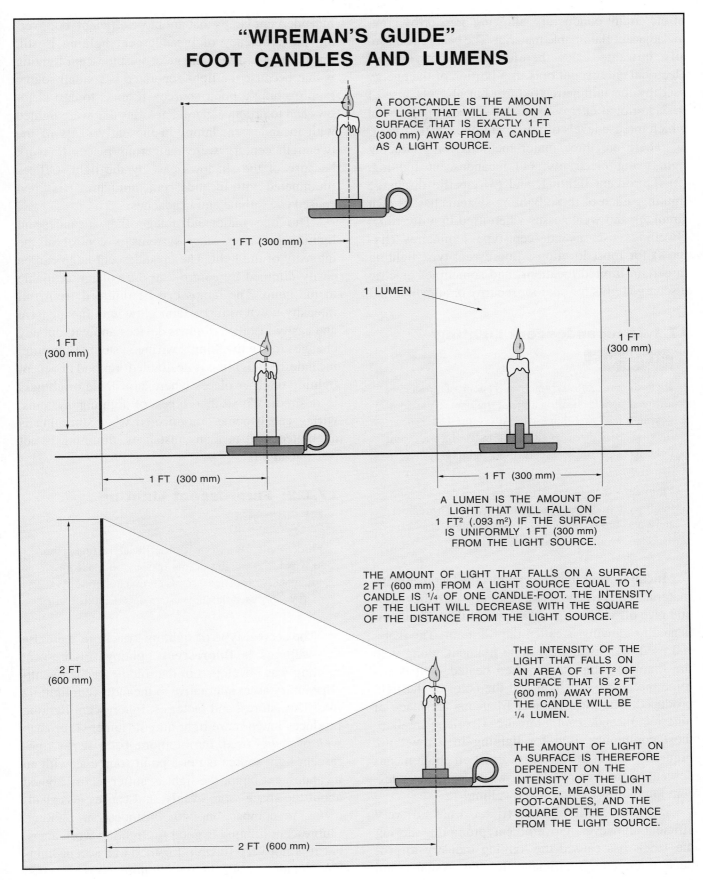

A FOOT-CANDLE IS THE AMOUNT OF LIGHT THAT WILL FALL ON A SURFACE THAT IS EXACTLY 1 FT (300 mm) AWAY FROM A CANDLE AS A LIGHT SOURCE.

1 FT (300 mm)

1 FT (300 mm)

1 LUMEN

1 FT (300 mm)

1 FT (300 mm)

1 FT (300 mm)

A LUMEN IS THE AMOUNT OF LIGHT THAT WILL FALL ON 1 FT² (.093 m²) IF THE SURFACE IS UNIFORMLY 1 FT (300 mm) FROM THE LIGHT SOURCE.

THE AMOUNT OF LIGHT THAT FALLS ON A SURFACE 2 FT (600 mm) FROM A LIGHT SOURCE EQUAL TO 1 CANDLE IS 1/4 OF ONE CANDLE-FOOT. THE INTENSITY OF THE LIGHT WILL DECREASE WITH THE SQUARE OF THE DISTANCE FROM THE LIGHT SOURCE.

THE INTENSITY OF THE LIGHT THAT FALLS ON AN AREA OF 1 FT² OF SURFACE THAT IS 2 FT (600 mm) AWAY FROM THE CANDLE WILL BE 1/4 LUMEN.

THE AMOUNT OF LIGHT ON A SURFACE IS THEREFORE DEPENDENT ON THE INTENSITY OF THE LIGHT SOURCE, MEASURED IN FOOT-CANDLES, AND THE SQUARE OF THE DISTANCE FROM THE LIGHT SOURCE.

2 FT (600 mm)

2 FT (600 mm)

Figure 17-1

of the circuit conductors, allowing arcs, or by igniting adjacent flammable material. The heat is also usually unwanted and therefore represents wasted electrical energy, not only as a portion of the energy used by the luminaire (fixture) but also because of the additional energy expended on cooling areas in which intense levels of artificial light are used.

There are three major methods for producing light using electricity: (1) incandescent lighting, (2) fluorescent lighting, and (3) electric discharge lighting. Each of these lighting systems has its own strengths and weaknesses when used in a dwelling. Dwellings use incandescent-type luminaires (fixtures) for most locations, fluorescent-type lighting in certain limited locations, and sometimes electric discharge lights for outdoor security or area lighting.

17.1.1: Incandescent Lighting

KEY TERMS

Incandescent Referring to a system of producing artificial light by heating a metal filament to the point at which it glows, thus giving off light. This type of lighting also produces a lot of heat. Incandescent lighting is the most widely used lighting system in dwellings.

Filament The metal strip in an incandescent lamp (bulb) system that is heated until it glows. The temperature and the composition of the filament determine the intensity and the color of the light.

Incandescent lighting exploits the fact that many materials can be heated to a point at which they glow and give off visible light. The part of the lamp that is heated to glowing is called the **filament**. The material used to construct the lamp filament, along with the filament temperature when heated, will determine the amount of light and the color of the light produced. The most popular filaments are made of tungsten. Incandescent luminaires (fixtures) are those most commonly used for lighting in a dwelling. Figure 17-2 presents more information about incandescent lamps. Figure 17-3 shows a typical incandescent luminaire (fixture) in a dwelling.

Incandescent luminaires (fixture) are not very efficient, however. Government prompting during the 1990s has caused the lighting industry to produce more energy-efficient lighting systems and incandescent lamps that are more efficient, but overall, the efficiency of incandescent lighting is still poor. Another disadvantage of incandescent lighting is that because the light generated is a small source (approaching a point source), it tends to cast shadows and to produce glare. Factories that were lighted with incandescent lamps in the early years of the twentieth century were dangerous places to work because of the shadows cast by the light. Offices illuminated with incandescent luminaires (fixtures) caused eye fatigue and headaches.

The one major advantage that incandescent lighting has over other systems is control of the intensity of the light. The incandescent lamp can be easily dimmed by controlling the voltage available to the lamp. The lamps can be dimmed from full intensity down to just a faint glow with inexpensive and easily installed dimmer devices and without any changes to the dwelling's wiring system. Therefore, incandescent lighting is desirable for mood or accent lighting or other places where adjustable brightness is desired. With other types of lighting systems, wiring can become complicated when dimming is attempted, and in many instances, dimming is not possible at all.

17.1.2: Fluorescent Lighting

KEY TERMS

Fluorescent Referring to an artificial lighting system in which the coating on the inside of a glass tube is stimulated to produce light. Fluorescent lighting is the most widely used lighting system outside of dwellings.

The second type of lighting system in wide use in dwellings is **fluorescent** lighting. Fluorescent lighting was developed in the middle of the twentieth century as an alternative to incandescent lighting in offices, stores, and factories. Fluorescent lighting produces much more light output (lumens) for each watt of power used. Furthermore, because the fluorescent light source is not a point source as with an incandescent lamp, the light is softer, is distributed about the space more evenly, and causes less shadowing in comparison with incandescent lighting. Fluorescent lighting is good for lighting work areas, but it is not very flexible for mood or accent lighting. Not all types of fluorescent luminaires (fixtures)

"WIREMAN'S GUIDE" INCANDESCENT LAMPS

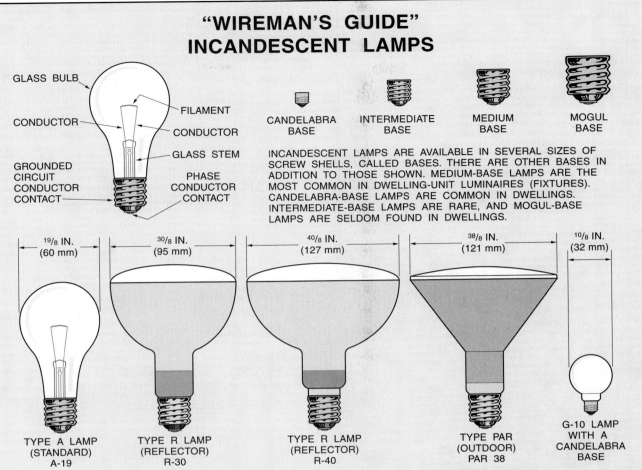

GLASS BULB

FILAMENT

CONDUCTOR

CONDUCTOR

GLASS STEM

GROUNDED CIRCUIT CONDUCTOR CONTACT

PHASE CONDUCTOR CONTACT

CANDELABRA BASE

INTERMEDIATE BASE

MEDIUM BASE

MOGUL BASE

INCANDESCENT LAMPS ARE AVAILABLE IN SEVERAL SIZES OF SCREW SHELLS, CALLED BASES. THERE ARE OTHER BASES IN ADDITION TO THOSE SHOWN. MEDIUM-BASE LAMPS ARE THE MOST COMMON IN DWELLING-UNIT LUMINAIRES (FIXTURES). CANDELABRA-BASE LAMPS ARE COMMON IN DWELLINGS. INTERMEDIATE-BASE LAMPS ARE RARE, AND MOGUL-BASE LAMPS ARE SELDOM FOUND IN DWELLINGS.

$19/8$ IN. (60 mm)

$30/8$ IN. (95 mm)

$40/8$ IN. (127 mm)

$38/8$ IN. (121 mm)

$10/8$ IN. (32 mm)

TYPE A LAMP (STANDARD) A-19

TYPE R LAMP (REFLECTOR) R-30

TYPE R LAMP (REFLECTOR) R-40

TYPE PAR (OUTDOOR) PAR 38

G-10 LAMP WITH A CANDELABRA BASE

INCANDESCENT LAMPS ARE AVAILABLE IN SEVERAL DIFFERENT BULB SHAPES. THE SHAPES SHOWN HERE ARE THE ONES MOST OFTEN FOUND IN DWELLINGS. LAMPS ARE SIZED BY THE WATTAGE RATING AND BY THE SIZE AND SHAPE OF THE GLASS BULB SURROUNDING THE FILAMENT. EACH OF THESE LAMP TYPES IS AVAILABLE IN SEVERAL WATT (W) RATINGS, SUCH AS 60 W, 75 W, OR 150 W, ALTHOUGH NOT ALL LAMP TYPES ARE AVAILABLE IN ALL WATT RATINGS. THE SIZE OF THE LAMP IS DETERMINED BY THE DIAMETER OF THE GLASS BULB MEASURED IN $1/8$ IN. (3.175 mm) INCREMENTS. FOR EXAMPLE, AN A-19 LAMP HAS A BULB THAT IS $19/8$ IN. ($2^3/8$ IN.) (60 mm) IN DIAMETER. AN R-40 LAMP HAS A DIAMETER OF $40/8$ IN. (5 IN.) (127 mm). THE COMPLETE DESCRIPTION OF THE LAMP INCLUDES THE WATT RATING, THE BULB SHAPE, AND THE BULB SIZE—FOR EXAMPLE, 75 W R-40 OR 40 W G-10. INCANDESCENT LAMPS PRODUCE ABOUT 20 LUMENS PER WATT OF POWER USED.

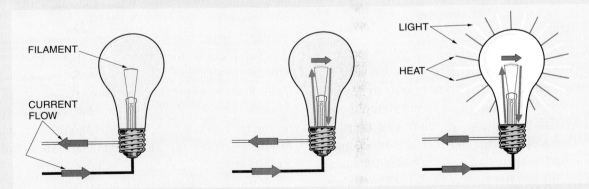

FILAMENT

CURRENT FLOW

LIGHT

HEAT

INCANDESCENT LAMPS PRODUCE LIGHT BECAUSE THE FILAMENT IS HEATED BY THE CURRENT UNTIL IT IS SO HOT THAT IT GLOWS. BY TAILORING THE MATERIAL THE FILAMENT IS MADE FROM, THE COLOR OF THE LIGHT CAN BE CONTROLLED TO SOME EXTENT. UNFORTUNATELY, THIS METHOD OF PRODUCING LIGHT ALSO PRODUCES A LOT OF HEAT. THIS WASTES ELECTRICITY AND REDUCES THE LIFE OF THE LAMP.

Figure 17-2

Figure 17-3 A typical chain-hung incandescent luminaire (fixture).

can be dimmed, and special wiring is required for the types that can. They also do not perform well if installed in a cold location. Fluorescent lighting can reduce the costs of lighting dramatically both in lower monthly utility bills and in reduced maintenance costs and is therefore very popular for commercial applications. Because fluorescent luminaires (fixtures) produce less heat, the lamps last 2 to 5 times longer than do incandescent lamps.

Fluorescent lighting employs glass tubes that are coated with fluorescent chemicals on the inside surface. See Figure 17-4. These fluorescent chemicals emit visible light when excited by ultraviolet radia-

tion. The glass tube contains argon gas and a little mercury, and when an electrical arc is struck down the length of the tube, the current excites the mercury inside the tube, causing it to emit ultraviolet radiation. This ultraviolet radiation strikes the fluorescent chemicals lining the tube, causing them to emit visible light. The tube becomes illuminated over its entire length except for a small portion at either end of the tube. By changing the chemicals that line the glass tube, different colors of light can be obtained to some degree. Figure 17-5 shows a typical fluorescent luminaire (fixture) installed in the kitchen of a dwelling. Notice that the luminaire (fixture) is enclosed in a decorative box so that the glass tubes are hidden from direct view.

> **KEY TERMS**
>
> **Ballast** The component in a fluorescent lighting system that controls the flow of current to the lamps.
>
> **Semiconductor** An electronic component that can be a conductor or an insulator, depending on its electrical state. The state can be changed by the introduction of a control current.

The operation of the fluorescent lighting system is more complex than that of incandescent lighting because of the difficulty in controlling the current flow through the glass tube. The resistance of the atmosphere inside the tube is relatively high under conditions of cold; therefore, large voltage is required to overcome the resistance so that the current will begin to flow through the tube. When the lamp strikes (begins to light), the resistance drops substantially, and the current flow increases as a result. If this current flow increase were not limited by some process, the current flow would reach unacceptable levels, and the system would be too dangerous to use. Fluorescent luminaires (fixtures) employ a **ballast** to control the current flow. The ballast also produces the voltage boost to get the process started. Figure 17-6 presents additional details about ballasts and fluorescent luminaires (fixtures).

Ballasts were originally made from copper coils and iron-core autotransformers that limited the current by working at saturation. With this system, a lot of electrical energy was converted to heat within the ballast, creating operating inefficiencies; however, the lighting system was still more efficient than incandescent lighting. The advances in **semiconductor**

"WIREMAN'S GUIDE"
FLUORESCENT LAMPS (TUBES)

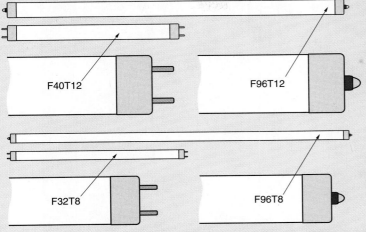

F40T12

F96T12

F32T8

F96T8

FLUORESCENT TUBES COME IN SEVERAL STANDARD SIZES, LENGTHS, AND SHAPES. THE MOST COMMON LENGTHS OF STRAIGHT TUBES ARE 18 IN. (457 mm), 24 IN. (610 mm), 48 IN. (1.22 m), AND 96 IN. (2.44 m). THEY ARE ALSO AVAILABLE IN A U-SHAPED TUBE. 96-IN. (2.44-m) TUBES HAVE A SINGLE-PIN CONNECTION TO THE FIXTURE. OTHER SIZE TUBES HAVE A 2-PIN CONNECTION METHOD.

FLUORESCENT TUBES ARE ALSO AVAILABLE IN DIFFERENT DIAMETERS. LIKE INCANDESCENT LAMPS, FLUORESCENT LAMPS ARE MEASURED IN 1/8 IN. (3-mm) UNITS AND CARRY THE PREFIX "T" IN THE PART NUMBER. THE LAMPS USED IN DWELLINGS WILL BE EITHER T-12 [12/8 IN. (38 mm)] OR T-8 [8/8 IN. (25 mm)].

T-12 LAMPS ARE TO BE USED WITH CORE AND COIL WIRE BALLASTS ONLY.
T-8 LAMPS ARE TO BE USED WITH ELECTRONIC BALLASTS ONLY.
USING THE WRONG LAMP OR THE WRONG BALLAST WILL SHORTEN THE LIFE OF BOTH COMPONENTS.

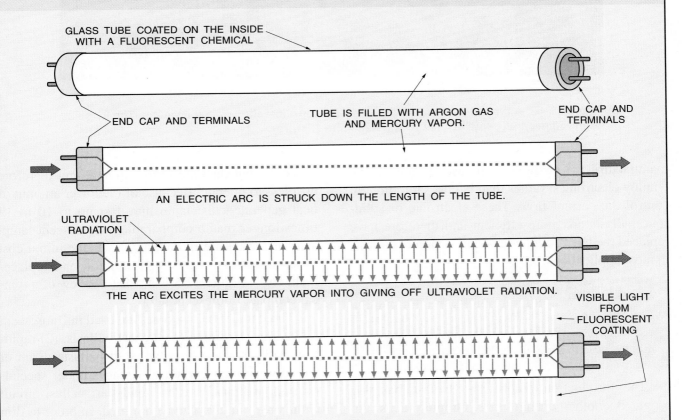

GLASS TUBE COATED ON THE INSIDE WITH A FLUORESCENT CHEMICAL

END CAP AND TERMINALS

TUBE IS FILLED WITH ARGON GAS AND MERCURY VAPOR.

END CAP AND TERMINALS

AN ELECTRIC ARC IS STRUCK DOWN THE LENGTH OF THE TUBE.

ULTRAVIOLET RADIATION

THE ARC EXCITES THE MERCURY VAPOR INTO GIVING OFF ULTRAVIOLET RADIATION.

VISIBLE LIGHT FROM FLUORESCENT COATING

FLUORESCENT LIGHT IS PRODUCED BY AN ELECTRIC ARC THAT PRODUCES ULTRAVIOLET RADIATION. THE ULTRAVIOLET RADIATION CAUSES THE FLUORESCENT MATERIAL TO PRODUCE VISIBLE LIGHT. BY CONTROLLING THE MATERIAL COMPOSITION OF THE COATING, THE COLOR OF THE LIGHT CAN BE CONTROLLED TO SOME DEGREE. THIS METHOD OF LIGHTING PRODUCES MORE LIGHT AND LESS HEAT FOR EACH WATT OF ENERGY USED. WHEREAS AN INCANDESCENT LAMP WILL PRODUCE ABOUT 14 LUMENS PER WATT OF POWER USED, A FLUORESCENT LAMP PROVIDES ABOUT 80 TO 100 LUMENS PER WATT OF POWER USED.

Figure 17-4

Figure 17-5 A typical fluorescent luminaire (fixture) installation in a dwelling. The lens has been removed to allow access to the luminaire (fixture).

manufacturing have allowed the use of ballasts that employ electronics, rather than ballast saturation, to control the current flow. These electronic or solid-state ballasts are more efficient and therefore have replaced the use of copper coil and iron-core ballasts in most applications.

> **KEY TERMS**
>
> **Compact fluorescent** Referring to a type of fluorescent lamp designed to fit into luminaires (fixtures) manufactured to accommodate incandescent lamps.

The introduction of solid-state ballasts has also provided another change in lighting systems. As shown in Figure 17-7, **compact fluorescent** lamps are replacing incandescent lamps in many luminaires (fixtures). Compact fluorescent lamps are available in several wattage ratings; provide a cooler, more efficient lighting system; and can be installed directly into many existing luminaires (fixtures) with screw shells. The compact fluorescent lamps cost more to purchase, but they are more efficient and therefore cost less to operate. Because of the lower amount of heat generated, the lamp may last up to 10 to 12 times longer than a comparable incandescent lamp. This increased life more than offsets the initial cost in savings from energy consumption and replacement costs. Figure 17-8 is a close-up view of a typical compact fluorescent lamp.

Three major ballast types are used in fluorescent luminaires (fixtures): (1) preheat ballasts, (2) rapid-start ballasts, and (3) instant-start ballasts. Each of these circuits has certain advantages under specialized applications, but the rapid-start ballast circuit is the only one that can be dimmed. In order to dim fluorescent lighting, a special ballast is required. In addition, the wiring from the dimmer to the luminaire (fixture) requires three wires, as opposed to the two wires required by incandescent luminaires (fixtures) and nondimming fluorescent luminaires (fixtures). Changes to the electrical system may be required if replacement of incandescent luminaires (fixtures)

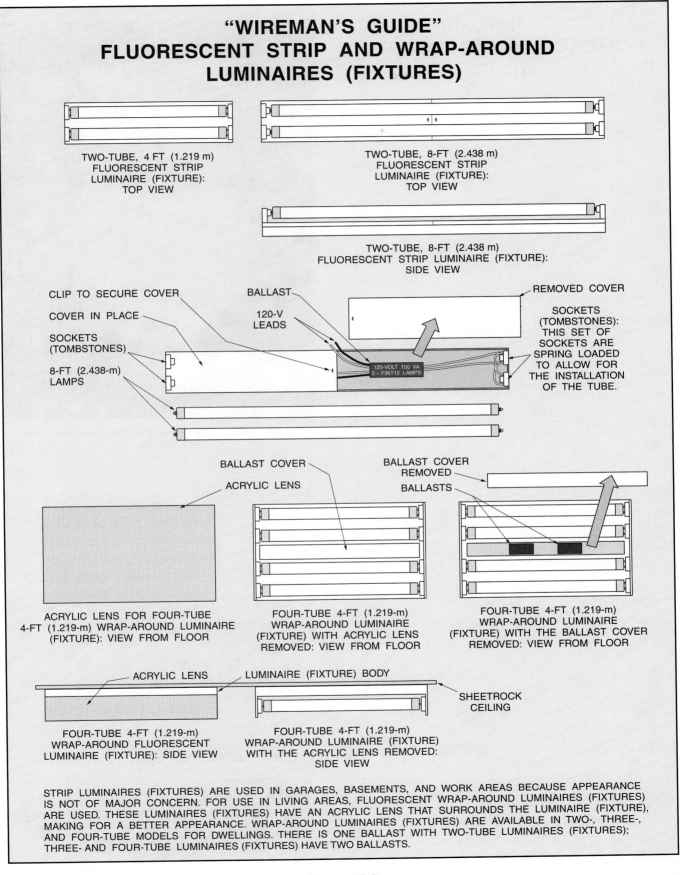

Figure 17-6

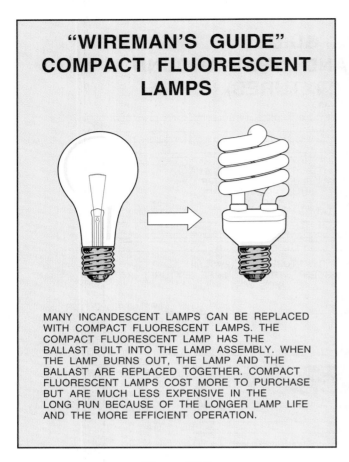

MANY INCANDESCENT LAMPS CAN BE REPLACED WITH COMPACT FLUORESCENT LAMPS. THE COMPACT FLUORESCENT LAMP HAS THE BALLAST BUILT INTO THE LAMP ASSEMBLY. WHEN THE LAMP BURNS OUT, THE LAMP AND THE BALLAST ARE REPLACED TOGETHER. COMPACT FLUORESCENT LAMPS COST MORE TO PURCHASE BUT ARE MUCH LESS EXPENSIVE IN THE LONG RUN BECAUSE OF THE LONGER LAMP LIFE AND THE MORE EFFICIENT OPERATION.

Figure 17-7

Figure 17-8 A compact fluorescent lamp.

with fluorescent luminaires (fixtures) is undertaken and dimming is a requirement for the new lighting system. Even with the special equipment and wiring, dimming over the entire range of lighting intensity cannot be achieved. The difficulty in dimming fluorescent luminaires (fixtures) is not a major shortcoming in the use of fluorescent lighting for offices, factories, or stores, but in other locations where mood or accent lighting is employed, their use is limited.

The ballast and the lamps of a fluorescent luminaire (fixture) must be matched exactly; otherwise, the luminaire (fixture) will not operate properly. Each ballast is marked with the number and size of the lamps that it is designed to supply. Improper matching of ballasts and lamps can cause reduced life for the ballast and the lamp, poor light quality, high operating temperatures and poor efficiencies, or even total failure of the luminaire (fixture). All luminaires (fixtures) with ballasts that are installed in dwellings employ Class P ballasts. Class P ballasts have a switch internal to the ballast that opens

the circuit to the luminaire (fixture) if the ballast temperature becomes too high. These types of ballasts have improved the fire safety of the fluorescent luminaires (fixtures), which was less than optimal when these luminaires (fixtures) were first introduced.

Another disadvantage of fluorescent lighting is the possibility of ballast failure as well as lamp failure. Whereas an incandescent lamp may fail, necessitating maintenance or replacement, the fluorescent luminaire (fixture) may have either a ballast failure or a lamp failure, or both. The maintenance electrician must check out both the lamps and the ballast to ensure that the problem has been corrected.

17.1.3: Electric Discharge Lighting

KEY TERMS

Electric discharge   Referring to an artificial lighting system in which light is given off by electrons as they become bound to the atoms within the lamp. This is a very efficient system for lighting large areas.

The third method of producing artificial light from electricity is called **electric discharge** lighting, or sometimes high-intensity discharge (HID) lighting. This type of lighting is used for highways, parking lots, and other locations where a large area needs to be illuminated. This type of lighting has limited applications in dwellings and therefore is not considered in detail here. Figure 17-9 presents more information about HID lighting. Figure 17-10 shows a typical HID luminaire (fixture) used for lighting roadways.

17.2: MANUFACTURER'S INSTALLATION INSTRUCTIONS

Each luminaire (fixture) comes complete with a set of installation instructions. The luminaire (fixture) is a listed piece of electrical equipment that was tested and approved for a given set of conditions by a testing agency, and the installation of any luminaire (fixture) in locations for which the luminaire (fixture) has not been tested is a violation of the *NEC®*. A long list of specific locations and conditions must be satisfied before a luminaire (fixture) can be installed. Here is a partial listing of some of those locations and conditions:

- Suitable for wet locations
- Suitable for damp locations
- Suitable for use in suspended ceilings
- Suitable for poured concrete
- Wall mount only
- Ceiling mount only

Other markings include the maximum watt rating and shape of the lamps and the temperature rating of the supply conductors. If the luminaire (fixture) is a recessed can, the marking will include Type IC if the can is rated for installation in direct contact with insulation. Otherwise, a recessed luminaire (fixture) must be kept clear of insulation so that it can cool.

It is very important that the luminaire (fixture) be installed according to the manufacturer's instructions and the listing. The instructions may include directions about box supports, clearances from flammable materials, or listings of approved accessories. The definitions of damp, wet, and other conditions or installation locations are detailed in *100 Definitions* in the Code.

17.3: LUMINAIRES (FIXTURES) IN THE SAMPLE HOUSE

The luminaire (fixture) schedule for the sample house is found in Construction Drawing E-5. All of the luminaires (fixtures) are surface mounted, except for Type C and Type D. Type C and Type D luminaires (fixtures) are recessed cans; in the sample house, there is one on the back porch and another in the master bedroom closet. The recessed can housing assemblies were installed and connected to the branch-circuit wiring at rough-in.

In the sample house, all luminaires (fixtures) use incandescent lamps except for Type N luminaires (fixtures). The luminaires (fixtures) mounted on the garage ceiling are two-tube, surface-mounted, Type N fluorescent luminaires (fixtures). The only way to determine if a luminaire (fixture) uses incandescent lamps from the luminaire (fixture) schedule is to look at the last column to the right in Construction Drawing E-5 titled Lamps. The luminaires (fixtures) with lamps that stipulate a designated-shape letter code and a power rating (wattage) are incandescent lamps. For example, luminaire (fixture) Type B shows that the luminaire (fixture) requires three 60-watt A-19 lamps. An A-19 lamp is the standard shape for an incandescent lamp (see Figure 17-2), and 60 watt is a standard power rating for a lamp.

The final makeup for all of the luminaires (fixtures) and smoke detectors is shown in Figure 17-11 (page 408), except for the recessed luminaires (fixtures) that were connected at rough-in. The rough installation of the conductors and boxes is covered in Chapter 10 of this text. Figure 10-6 shows the conductor identification for each box before makeup.

17.3.1: Surface-Mounted Incandescent Luminaires (Fixtures)

The majority of the luminaires (fixtures) in the sample house are surface-mounted incandescent. There are wall-mounted luminaires (fixtures) in each of the bathrooms above the counters and a wall-mounted luminaire (fixture) in the stairway to the basement. The luminaires (fixtures) on the outside of the sample dwelling are wall mounted, except for the recessed can above the patio. Two of these outdoor luminaires (fixtures) are installed on brick, and two are installed on shiplap siding. Some of the

"WIREMAN'S GUIDE"
HIGH-INTENSITY DISCHARGE (HID) LIGHTING

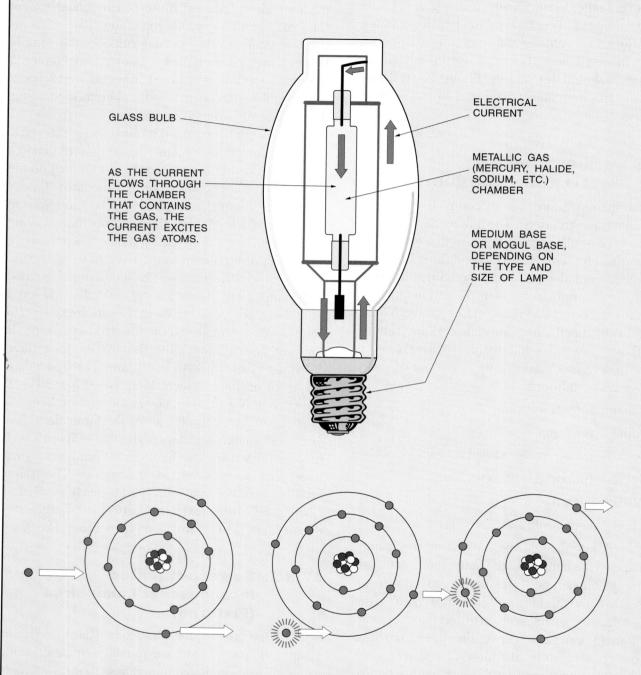

GLASS BULB

ELECTRICAL CURRENT

AS THE CURRENT FLOWS THROUGH THE CHAMBER THAT CONTAINS THE GAS, THE CURRENT EXCITES THE GAS ATOMS.

METALLIC GAS (MERCURY, HALIDE, SODIUM, ETC.) CHAMBER

MEDIUM BASE OR MOGUL BASE, DEPENDING ON THE TYPE AND SIZE OF LAMP

AS THE CURRENT FLOWS THROUGH THE CHAMBER WITH THE METALLIC GAS, IT EXCITES THE GAS. AN ELECTRON FROM THE CURRENT FLOW DISPLACES AN ELECTRON FROM THE GAS ATOM. THE ELECTRON GIVES OFF A SMALL AMOUNT OF LIGHT WHEN IT SETTLES INTO ITS ORBIT, AND THE DISPLACED ELECTRON THEN TRAVELS TO ANOTHER ATOM. THE FREE ELECTRON DISPLACES ONE ELECTRON FROM A GAS ATOM, AND AS THE ELECTRON FALLS INTO THE ORBIT, IT WILL ALSO GIVE OFF A SMALL AMOUNT OF ENERGY IN THE FORM OF LIGHT. THIS PROCESS TAKES PLACE MANY BILLIONS OF TIMES EACH SECOND, THEREBY PRODUCING VISIBLE LIGHT.

Figure 17-9

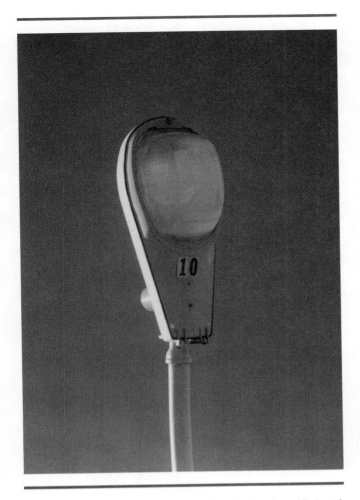

Figure 17-10 An example of a luminaire (fixture) with an HID lamp and ballast.

problems to consider in installing outdoor luminaires (fixtures) are shown in Figure 17-12. It is important to note that the luminaire (fixture) must be approved for installation in a wet location but that it does not have to be watertight. Watertight means that water cannot enter the luminaire (fixture). Installation outdoors usually requires the luminaire (fixture) to be only weatherproof, which means that the entry of water will not cause damage to the luminaire (fixture) or interfere with its successful operation.

The remainder of the luminaires (fixtures) in the sample house are indoors and ceiling mounted. In most instances, the luminaire (fixture) is mounted to the outlet box. Because the luminaire (fixture) is removable (mounted with screws), it is not considered a permanent part of the structure and is allowed to cover the outlet box. Most interior walls and ceil-

ings provide a smooth finished wall surface for the luminaire (fixture) to seat against when installed.

It is sometimes difficult to determine the polarity of the wiring in some less expensive luminaires (fixtures). The two circuit fixture wires may both be black instead of one black and one white. Every effort must be undertaken to identify which conductor is for the grounded circuit conductor. It may be identified by a groove or series of grooves or by ridges in the surface of the insulation, or it may be identified by other type of marking. As a final resort, it may be necessary to test the polarity with an ohmmeter to determine the grounded circuit conductor lead for the luminaire (fixture).

17.3.2: Recessed Incandescent Luminaires (Fixtures)

There are two recessed can luminaires (fixtures) in the sample house. One recessed can is located in the ceiling of the rear patio cover, and the other is located in one of the master bedroom closets. The recessed cans themselves—metal housings installed in the ceiling and connected to the switch leg cable at rough-in—are identical. They have different trims, however: a waterproof trim for outdoors location and an open black baffle trim for the closet location. See Figures 17-13 and 17-14 (page 410 and page 411). The waterproof trim is a closed trim that has a glass or plastic lens enclosing the lamp and eliminating access to the lamp without removing or lowering the entire trim. The fact that the lamp is enclosed means that the heat generated by the lamp will be trapped within the luminaire (fixture). The temperature inside the housing may become too great for the wiring; therefore, the lamp size and type are restricted by the manufacturer. According to the luminaire (fixture) schedule on Construction Drawing E-5, the outdoor luminaire (fixture) uses a 60-watt A-19 lamp. The baffle trim in the bedroom closet is an open trim; therefore, air can circulate to keep the housing cooler. The recessed luminaire (fixture) in the closet uses a 75-watt, R-30 lamp.

According to the specifications in Construction Drawing E-1, the general contractor or the builder is to supply all luminaires (fixtures) and lamps, except for recessed can luminaires (fixtures), which the electrical contractor is to supply. Because the can housings were installed and connected at rough-in, all that remains at trim is to install the trims and the lamps.

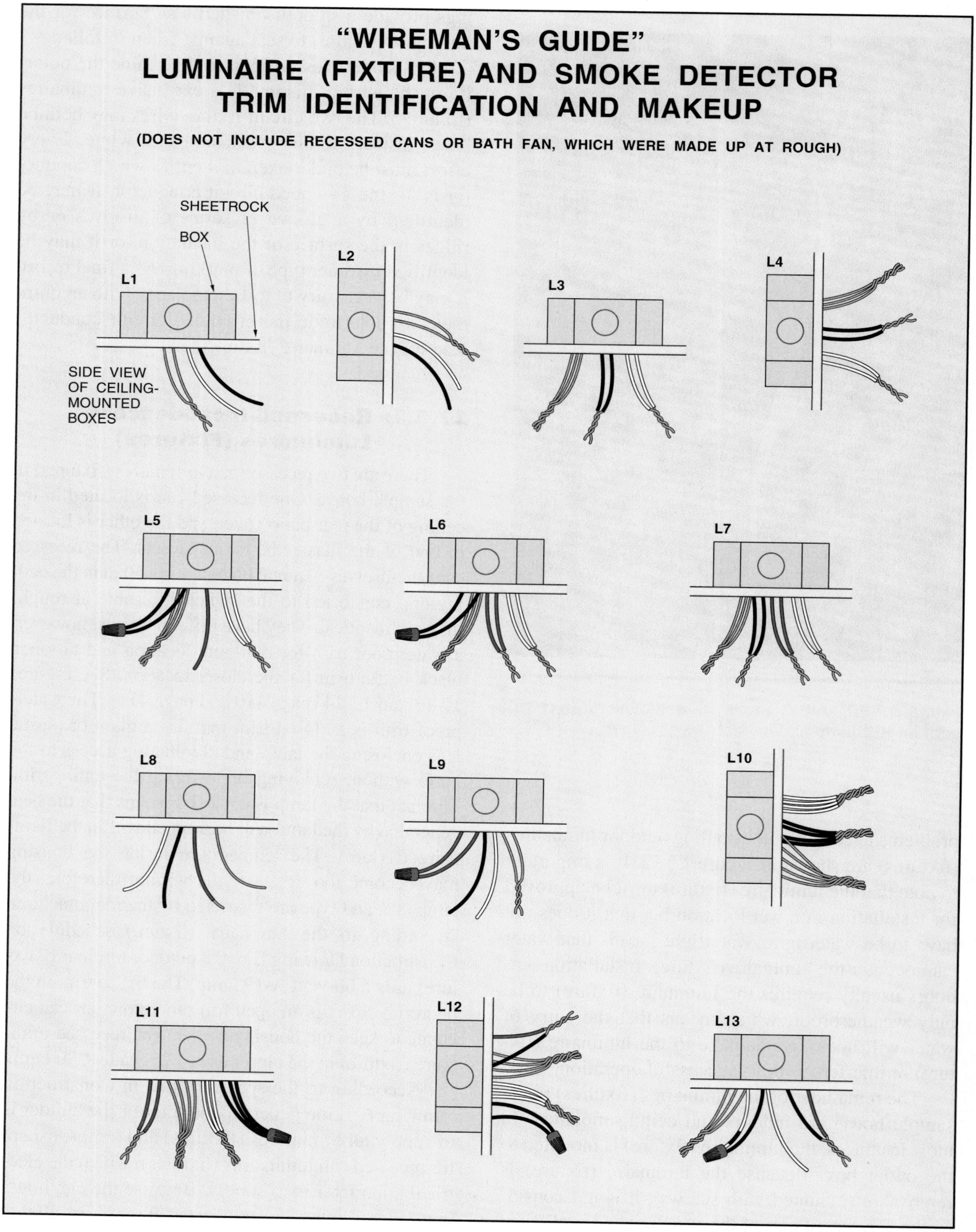

"WIREMAN'S GUIDE"
LUMINAIRE (FIXTURE) AND SMOKE DETECTOR
TRIM IDENTIFICATION AND MAKEUP

(DOES NOT INCLUDE RECESSED CANS OR BATH FAN, WHICH WERE MADE UP AT ROUGH)

SHEETROCK

BOX

L1

SIDE VIEW
OF CEILING-
MOUNTED
BOXES

L2

L3

L4

L5

L6

L7

L8

L9

L10

L11

L12

L13

Figure 17-11

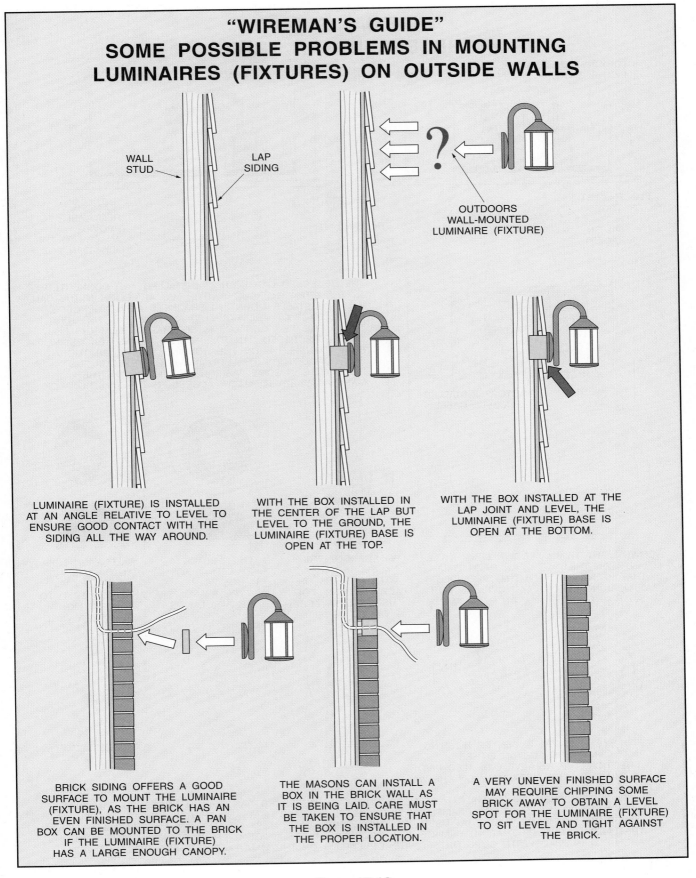

"WIREMAN'S GUIDE"
SOME POSSIBLE PROBLEMS IN MOUNTING LUMINAIRES (FIXTURES) ON OUTSIDE WALLS

WALL STUD

LAP SIDING

?

OUTDOORS WALL-MOUNTED LUMINAIRE (FIXTURE)

LUMINAIRE (FIXTURE) IS INSTALLED AT AN ANGLE RELATIVE TO LEVEL TO ENSURE GOOD CONTACT WITH THE SIDING ALL THE WAY AROUND.

WITH THE BOX INSTALLED IN THE CENTER OF THE LAP BUT LEVEL TO THE GROUND, THE LUMINAIRE (FIXTURE) BASE IS OPEN AT THE TOP.

WITH THE BOX INSTALLED AT THE LAP JOINT AND LEVEL, THE LUMINAIRE (FIXTURE) BASE IS OPEN AT THE BOTTOM.

BRICK SIDING OFFERS A GOOD SURFACE TO MOUNT THE LUMINAIRE (FIXTURE), AS THE BRICK HAS AN EVEN FINISHED SURFACE. A PAN BOX CAN BE MOUNTED TO THE BRICK IF THE LUMINAIRE (FIXTURE) HAS A LARGE ENOUGH CANOPY.

THE MASONS CAN INSTALL A BOX IN THE BRICK WALL AS IT IS BEING LAID. CARE MUST BE TAKEN TO ENSURE THAT THE BOX IS INSTALLED IN THE PROPER LOCATION.

A VERY UNEVEN FINISHED SURFACE MAY REQUIRE CHIPPING SOME BRICK AWAY TO OBTAIN A LEVEL SPOT FOR THE LUMINAIRE (FIXTURE) TO SIT LEVEL AND TIGHT AGAINST THE BRICK.

Figure 17-12

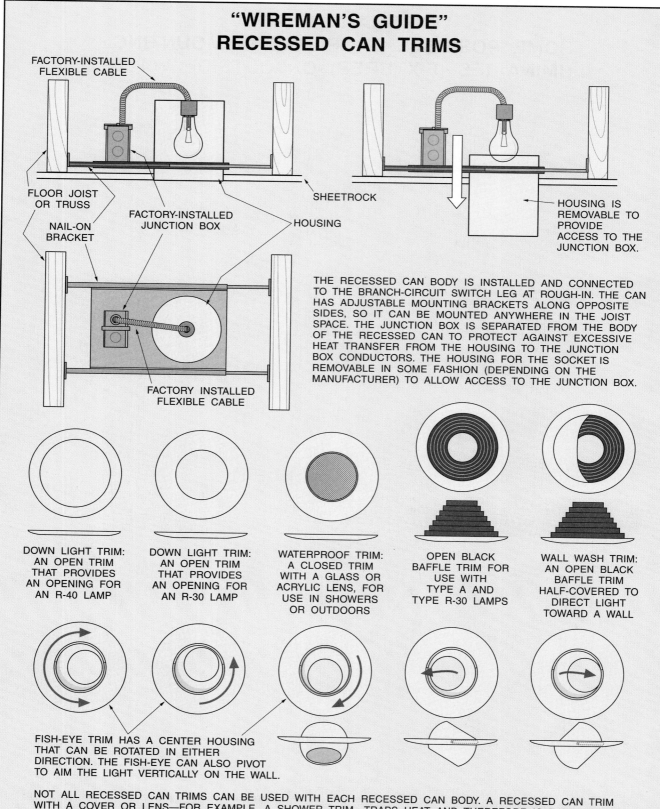

"WIREMAN'S GUIDE" RECESSED CAN TRIMS

FACTORY-INSTALLED FLEXIBLE CABLE

FLOOR JOIST OR TRUSS

NAIL-ON BRACKET

FACTORY-INSTALLED JUNCTION BOX

SHEETROCK

HOUSING

HOUSING IS REMOVABLE TO PROVIDE ACCESS TO THE JUNCTION BOX.

FACTORY INSTALLED FLEXIBLE CABLE

THE RECESSED CAN BODY IS INSTALLED AND CONNECTED TO THE BRANCH-CIRCUIT SWITCH LEG AT ROUGH-IN. THE CAN HAS ADJUSTABLE MOUNTING BRACKETS ALONG OPPOSITE SIDES, SO IT CAN BE MOUNTED ANYWHERE IN THE JOIST SPACE. THE JUNCTION BOX IS SEPARATED FROM THE BODY OF THE RECESSED CAN TO PROTECT AGAINST EXCESSIVE HEAT TRANSFER FROM THE HOUSING TO THE JUNCTION BOX CONDUCTORS. THE HOUSING FOR THE SOCKET IS REMOVABLE IN SOME FASHION (DEPENDING ON THE MANUFACTURER) TO ALLOW ACCESS TO THE JUNCTION BOX.

DOWN LIGHT TRIM: AN OPEN TRIM THAT PROVIDES AN OPENING FOR AN R-40 LAMP

DOWN LIGHT TRIM: AN OPEN TRIM THAT PROVIDES AN OPENING FOR AN R-30 LAMP

WATERPROOF TRIM: A CLOSED TRIM WITH A GLASS OR ACRYLIC LENS, FOR USE IN SHOWERS OR OUTDOORS

OPEN BLACK BAFFLE TRIM FOR USE WITH TYPE A AND TYPE R-30 LAMPS

WALL WASH TRIM: AN OPEN BLACK BAFFLE TRIM HALF-COVERED TO DIRECT LIGHT TOWARD A WALL

FISH-EYE TRIM HAS A CENTER HOUSING THAT CAN BE ROTATED IN EITHER DIRECTION. THE FISH-EYE CAN ALSO PIVOT TO AIM THE LIGHT VERTICALLY ON THE WALL.

NOT ALL RECESSED CAN TRIMS CAN BE USED WITH EACH RECESSED CAN BODY. A RECESSED CAN TRIM WITH A COVER OR LENS—FOR EXAMPLE, A SHOWER TRIM—TRAPS HEAT AND THEREFORE IS RATED FOR A LOWER WATTAGE LAMP THAN THAT USED FOR OPEN TRIMS. CARE MUST BE TAKEN IN MATCHING A TRIM TO THE LAMP TO ENSURE THAT NO DAMAGE WILL OCCUR AS A RESULT OF OVER-TEMPERATURE. THE TRIMS ARE HELD IN PLACE WITH CLIPS OR SPRINGS, DEPENDING ON THE MANUFACTURER AND STYLE OF TRIM.

Figure 17-13

Figure 17-14 Two examples of recessed can trims. The trim to the right is a black baffle trim; to the left is a wall wash trim.

17.3.3: Fluorescent Luminaires (Fixtures)

The fluorescent luminaires (fixtures) on the garage ceiling are two-tube, 4-ft (1.2-m) strip lights. They are connected to the ceiling outlet box via a knockout in the rear wall of the luminaire (fixture). The luminaire (fixture) is held to the ceiling with expansion fasteners or screws into a roof truss. Figure 17-4 presents additional information about the fluorescent luminaires (fixtures) like the ones to be installed in the garage.

SUMMARY

The intensity of the light that falls on any given surface is measured in lumens and is dependent on the absolute intensity of the light source and the distance the surface is away from the source. There are three principal types of artificial lighting: incandescent lighting, in which light is produced when a fila-

ment is heated until it glows; fluorescent lighting, in which light is produced when excited atoms strike a fluorescent coating inside a glass tube; and high-intensity discharge lighting, in which light is produced by exciting atoms held under pressure inside a small glass enclosure.

Luminaires (fixtures) must be installed according to the manufacturer's instructions, and only in those locations for which they are listed. There are many different listings for luminaire (fixture) locations, including wet locations and wall mount only. Incandescent luminaires (fixtures) are approved for a maximum wattage and shape of lamp; adherence to these requirements is essential to prevent overheating of the luminaire (fixture). Special attention to installation is required with recessed luminaires (fixtures) because of the possibility of trapping heat inside a luminaire (fixture) with a closed lens. The ballast of the fluorescent and HID luminaires (fixtures) must be exactly matched to the lamps, or unsatisfactory lighting will result.

REVIEW

1. Each of the following is one of the three major methods of producing artificial light for dwellings except _____ .
   a. incandescent
   b. radiant
   c. fluorescent
   d. high-intensity discharge

2. In luminaires (fixtures) that employ a ballast to control current flow, the ballast must be matched with all of the following except _____ .
   a. the size (watt rating) of the lamp(s)
   b. the number of lamps powered
   c. the voltage rating of the lamp
   d. the P rating of the ballast

3. Luminaires (fixtures) that are approved for wet locations can be installed in all of the following locations except _____ .
   a. underwater
   b. in wet locations
   c. in damp locations
   d. in dry locations

4. Compact fluorescent lamps are designed to replace _____ lamps.
   a. bulky fluorescent tube
   b. HID
   c. incandescent
   d. dimmable fluorescent tube

5. Incandescent lamps are rated by all of the following except _____ .
   a. the size of the base
   b. the wattage
   c. the lamp shape and size
   d. the luminaire (fixture) into which it is to be installed

6. *True or False:* Incandescent lamps burn cooler than most fluorescent lamps.

7. *True or False:* High-intensity discharge lighting is used extensively within dwellings.

8. *True or False:* Type P fluorescent ballasts have a thermal protector that opens the circuit to the luminaire (fixture) if the temperature becomes too great.

9. *True or False:* Incandescent lighting is popular in dwellings because it can be easily dimmed.

10. *True or False:* Some types of fluorescent luminaires (fixtures) can be dimmed, but special ballasts and wiring are required.

"WIREMAN'S GUIDE" REVIEW

1. Discuss the different lengths of fluorescent lamps and the size of the luminaires (fixtures). Do all lamps have the same configuration of terminals?

2. Why must the ballast and the lamps be matched in a fluorescent luminaire (fixture)? Remember that the ballast is designed to control the current flow to the lamps.

3. What can be done to increase the life of an incandescent lamp? Do you think that there is a direct relationship between the brightness of an incandescent lamp and the heat that the lamp produces?

4. Discuss where it could be practical to install HID lighting for a dwelling.

5. Study the box rough conductor coding for makeup for luminaires (fixtures) in Figure 10-6. Compare those drawings with the drawing of the conductors that are intended to be connected to the luminaire (fixture) at trim in Figure 17-11. Be ready to discuss the details of each box makeup. Why do some of the conductors shown in Figure 17-11 have splice caps and others do not?

CHAPTER 18

Appliance Connections

OBJECTIVES

On completion of this chapter, the student will be able to:

☑ Explain the importance of connecting appliances according to the manufacturer's installation instructions.

☑ State the connection requirements for the appliances in the sample house, such as the dishwasher, the garbage disposal, the sump pump, the microwave, the furnace, the baseboard heater, and the air conditioner.

☑ Explain the procedures for protecting the supply cables from physical damage in wiring utilization equipment.

☑ Identify the requirements of the *NEC®* for disconnecting means for appliances.

INTRODUCTION

The final items needing connection in the sample house are the appliances. Most of these appliances are in the kitchen, and most of them are cord-and-attachment-plug connected. Some of the connections have been introduced elsewhere in this book. For example, many aspects of the location of the range receptacle outlet are detailed in Figures 11-10 and 11-12. The placement of the receptacle outlet for the sump pump is covered in Figure 16-6. Six appliances—the range, garbage disposal, dishwasher, and microwave oven in the kitchen; the sump pump in the basement; and the dryer in the main-level laundry area—are installed using rules for cord-and-attachment-plug connection. The furnace in the basement, the air conditioner outdoors, and the electric baseboard heater in the garage are permanently connected. A thorough review of Chapter 11 of this book is recommended before continuing with this chapter.

With this chapter, the electrical installation for the sample house is completed. Chapter 19 examines the checkout and troubleshooting of the dwelling's electrical system for proper operation.

18.1: MANUFACTURER'S INSTALLATION INSTRUCTIONS

KEY TERMS

Appliance (See *NEC® Article 100*): A type of utilization equipment that is usually manufactured for residential use and is mass produced rather than being individually designed and constructed.

As with luminaires (fixtures), all **appliances** must be installed in strict accordance with the manufacturer's installation instructions that accompany the appliance or equipment. The appliances are listed assemblies and as such are tested and listed as safe only under the conditions stipulated in the installation instructions. Any other use may lead the AHJ to reject the installation.

Obviously, the installing electricians must plan for the proper connection of the appliances at rough-in. The location of the outlet and the type of connection—hard wire with permanent wiring method versus cord-and-attachment plug—must be anticipated at rough-in for a successful installation at trim. The location of any needed disconnecting means or additional overcurrent-protective devices must also be anticipated at rough-in.

18.2: CORD-AND-PLUG CONNECTED APPLIANCES

For each of the appliances that are cord-and-attachment-plug connected, the attachment plug serves as the required appliance disconnecting means. For some of the appliances, such as the dishwasher and the sump pump, the cord attachment also serves as a means to damp any vibration from the appliance that would cause some permanent wiring methods to suffer damage over time.

18.2.1: Dishwasher

> **KEY TERMS**
>
> **Pigtail** (referring to connecting utilization equipment): A cord-and-attachment-plug assembly that is used to connect utilization equipment to receptacle outlets.

The dishwasher in the sample house is located to the left of the kitchen sink. It is powered by the split-wired receptacle outlet installed in the cabinet beneath the sink. The dishwasher usually has a junction box or wiring compartment for the connection of the supply conductors that is accessed from the front of the appliance. To access the junction box usually requires the removal of the bottom plate or grill and the removal of the junction box cover, as shown in Figure 18-1. The junction box has an equipment grounding terminal and usually a black and a white lead for connection to the supply conductors. The supply cable is an attachment plug and cord, sometimes called a **pigtail**, that must be listed for use with a dishwasher and sized according to the length requirements stipulated in *422.16(B)(2)*.

A hole must be made into the back sidewall of the kitchen cabinet that supports the sink. The pigtail is routed through the hole to the duplex receptacle that is provided in the cabinet to provide power to the dishwasher and disposal. The receptacle is a split-wired duplex receptacle; one receptacle is supplied by circuit 8, from subpanel A in the sample house electrical panel, to power the dishwasher, and the other receptacle is supplied by circuit 10 to power the garbage disposal (see Figure 16-12). The

garbage disposal receptacle is switched to provide on-off control for the appliance. The power to the dishwasher is not switched because the dishwasher has a built-in on-off control.

Because of the location of the junction box for the dishwasher supply conductors, it makes little difference whether the dishwasher is wired before or after it has been installed in place under the countertop. However, the routing of the pigtail into the sink cabinet is easier if it can be installed while the dishwasher is being installed.

18.2.2: Garbage Disposal

The garbage disposal in the sample house is installed under the kitchen sink. Supply conductors are terminated in the split-wired duplex receptacle located in the sink support cabinet. The receptacle outlet for the disposal is switched by a single-pole switch just to the right of the sink above the countertop. The disposal attachment cord and plug must be plugged into the switched receptacle.

Because the wiring compartment is usually located on the bottom of the disposal and because the mounting location under the sink severely limits access to the disposal, it is much easier to connect the pigtail to the disposal before the appliance is mounted into place. The lead electrician should coordinate the installation of the disposal with the plumbing contractor or the general contractor or the builder. Figure 18-2 shows the wiring of a kitchen garbage disposal with a pigtail.

18.2.3: Microwave Oven

The microwave oven in the sample house is mounted directly above the range. In addition to functioning as a microwave oven, it also functions as a nonducted (filtered) exhaust fan for the range. The rough installation is covered in Chapter 11 of this text and is detailed in Figure 11-13. The microwave oven is specifically designed for installation above the range and has the supply cord-and-attachment plug exiting the appliance on the top surface, as shown in Figure 18-3 (page 418). Mounting hardware and instructions are included with the appliance packaging from the factory.

The receptacle outlet for the microwave oven is located in the wall-mounted kitchen cabinet directly above the range. The supply cable has been stubbed

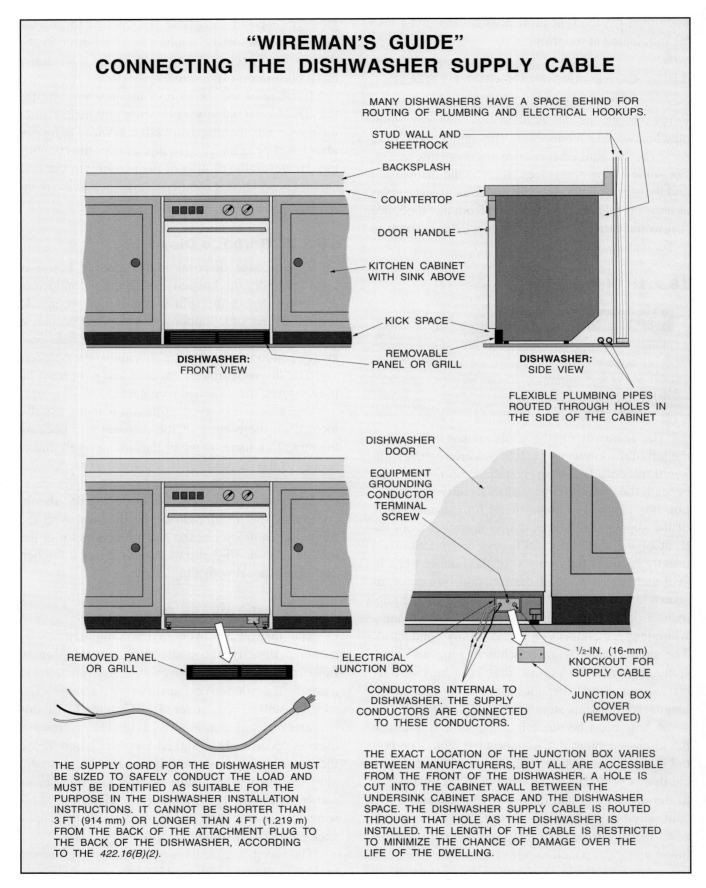

"WIREMAN'S GUIDE"
CONNECTING THE DISHWASHER SUPPLY CABLE

MANY DISHWASHERS HAVE A SPACE BEHIND FOR ROUTING OF PLUMBING AND ELECTRICAL HOOKUPS.

STUD WALL AND SHEETROCK

BACKSPLASH

COUNTERTOP

DOOR HANDLE

KITCHEN CABINET WITH SINK ABOVE

KICK SPACE

REMOVABLE PANEL OR GRILL

DISHWASHER: FRONT VIEW

DISHWASHER: SIDE VIEW

FLEXIBLE PLUMBING PIPES ROUTED THROUGH HOLES IN THE SIDE OF THE CABINET

DISHWASHER DOOR

EQUIPMENT GROUNDING CONDUCTOR TERMINAL SCREW

REMOVED PANEL OR GRILL

ELECTRICAL JUNCTION BOX

CONDUCTORS INTERNAL TO DISHWASHER. THE SUPPLY CONDUCTORS ARE CONNECTED TO THESE CONDUCTORS.

1/2-IN. (16-mm) KNOCKOUT FOR SUPPLY CABLE

JUNCTION BOX COVER (REMOVED)

THE SUPPLY CORD FOR THE DISHWASHER MUST BE SIZED TO SAFELY CONDUCT THE LOAD AND MUST BE IDENTIFIED AS SUITABLE FOR THE PURPOSE IN THE DISHWASHER INSTALLATION INSTRUCTIONS. IT CANNOT BE SHORTER THAN 3 FT (914 mm) OR LONGER THAN 4 FT (1.219 m) FROM THE BACK OF THE ATTACHMENT PLUG TO THE BACK OF THE DISHWASHER, ACCORDING TO THE *422.16(B)(2)*.

THE EXACT LOCATION OF THE JUNCTION BOX VARIES BETWEEN MANUFACTURERS, BUT ALL ARE ACCESSIBLE FROM THE FRONT OF THE DISHWASHER. A HOLE IS CUT INTO THE CABINET WALL BETWEEN THE UNDERSINK CABINET SPACE AND THE DISHWASHER SPACE. THE DISHWASHER SUPPLY CABLE IS ROUTED THROUGH THAT HOLE AS THE DISHWASHER IS INSTALLED. THE LENGTH OF THE CABLE IS RESTRICTED TO MINIMIZE THE CHANCE OF DAMAGE OVER THE LIFE OF THE DWELLING.

Figure 18-1

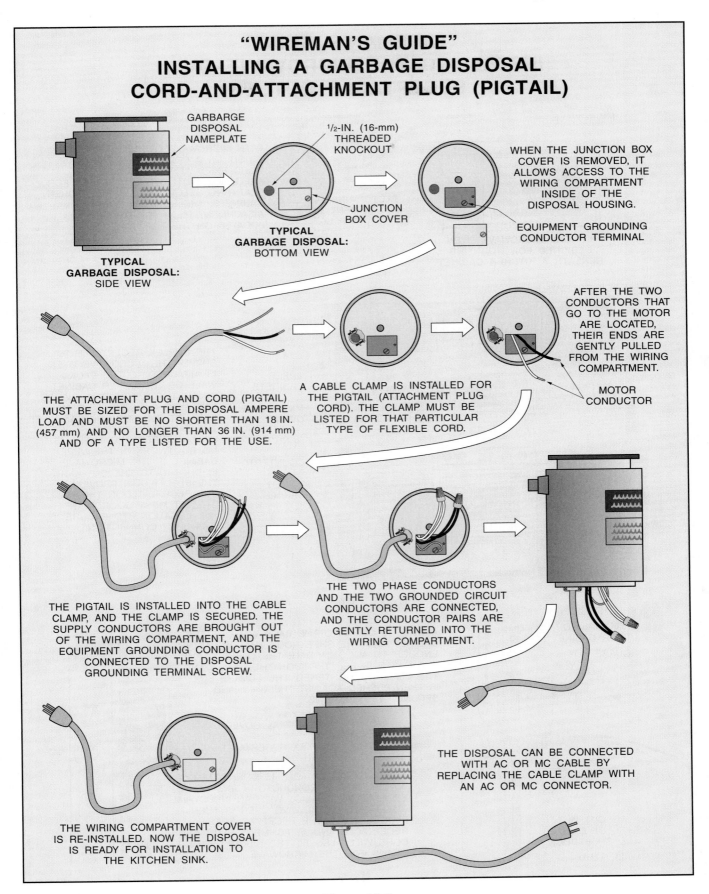

"WIREMAN'S GUIDE"
INSTALLING A GARBAGE DISPOSAL
CORD-AND-ATTACHMENT PLUG (PIGTAIL)

GARBARGE DISPOSAL NAMEPLATE

½-IN. (16-mm) THREADED KNOCKOUT

JUNCTION BOX COVER

TYPICAL GARBAGE DISPOSAL: BOTTOM VIEW

TYPICAL GARBAGE DISPOSAL: SIDE VIEW

WHEN THE JUNCTION BOX COVER IS REMOVED, IT ALLOWS ACCESS TO THE WIRING COMPARTMENT INSIDE OF THE DISPOSAL HOUSING.

EQUIPMENT GROUNDING CONDUCTOR TERMINAL

AFTER THE TWO CONDUCTORS THAT GO TO THE MOTOR ARE LOCATED, THEIR ENDS ARE GENTLY PULLED FROM THE WIRING COMPARTMENT.

MOTOR CONDUCTOR

THE ATTACHMENT PLUG AND CORD (PIGTAIL) MUST BE SIZED FOR THE DISPOSAL AMPERE LOAD AND MUST BE NO SHORTER THAN 18 IN. (457 mm) AND NO LONGER THAN 36 IN. (914 mm) AND OF A TYPE LISTED FOR THE USE.

A CABLE CLAMP IS INSTALLED FOR THE PIGTAIL (ATTACHMENT PLUG CORD). THE CLAMP MUST BE LISTED FOR THAT PARTICULAR TYPE OF FLEXIBLE CORD.

THE PIGTAIL IS INSTALLED INTO THE CABLE CLAMP, AND THE CLAMP IS SECURED. THE SUPPLY CONDUCTORS ARE BROUGHT OUT OF THE WIRING COMPARTMENT, AND THE EQUIPMENT GROUNDING CONDUCTOR IS CONNECTED TO THE DISPOSAL GROUNDING TERMINAL SCREW.

THE TWO PHASE CONDUCTORS AND THE TWO GROUNDED CIRCUIT CONDUCTORS ARE CONNECTED, AND THE CONDUCTOR PAIRS ARE GENTLY RETURNED INTO THE WIRING COMPARTMENT.

THE WIRING COMPARTMENT COVER IS RE-INSTALLED. NOW THE DISPOSAL IS READY FOR INSTALLATION TO THE KITCHEN SINK.

THE DISPOSAL CAN BE CONNECTED WITH AC OR MC CABLE BY REPLACING THE CABLE CLAMP WITH AN AC OR MC CONNECTOR.

Figure 18-2

"WIREMAN'S GUIDE"
INSTALLING THE MICROWAVE OVEN
RECEPTACLE OUTLET

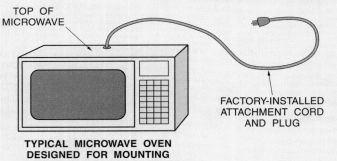

TOP OF MICROWAVE

FACTORY-INSTALLED ATTACHMENT CORD AND PLUG

TYPICAL MICROWAVE OVEN DESIGNED FOR MOUNTING BENEATH A KITCHEN CABINET

WITH MICROWAVE OVENS INTENDED TO SIT ON A COUNTERTOP, THE SUPPLY CORD EXITS AT THE REAR OF THE APPLIANCE. WITH OVENS INTENDED FOR MOUNTING BELOW THE UPPER KITCHEN CABINETS, THE SUPPLY CORD-AND-ATTACHMENT PLUG ARE USUALLY LOCATED ON THE TOP.

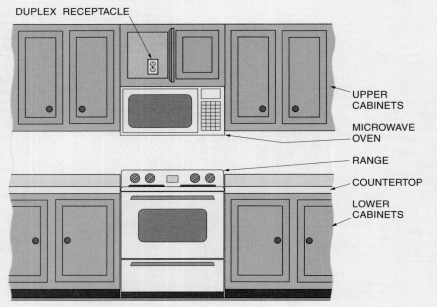

DUPLEX RECEPTACLE

UPPER CABINETS

MICROWAVE OVEN

RANGE

COUNTERTOP

LOWER CABINETS

THE DUPLEX RECEPTACLE TO SUPPLY THE MICROWAVE OVEN IS ATTACHED TO THE BACK WALL OF THE UPPER CABINET ABOVE THE RANGE. THE OVEN'S FAN ALSO FUNCTIONS AS AN EXHAUST FAN FOR THE RANGE. THE FAN IS UNVENTED (IT IS JUST FILTERED), SO THERE IS NO DUCTWORK IN THE UPPER CABINET. IF THE MICROWAVE IS DUCTED TO THE OUTDOORS, THE ELECTRICIAN PROVIDING THE RECEPTACLE IN THE CABINET MUST COORDINATE THE LOCATIONS TO ENSURE A PROPER CONNECTION.

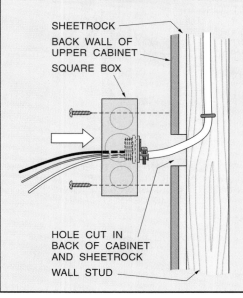

SHEETROCK

BACK WALL OF UPPER CABINET

SQUARE BOX

HOLE CUT IN BACK OF CABINET AND SHEETROCK

WALL STUD

A HOLE IS CUT INTO THE BACK WALL OF THE CABINET AND THE SHEETROCK THAT IS LARGE ENOUGH TO ACCOMMODATE THE CABLE CLAMP. THE CABLE IS FED THROUGH THE CLAMP AND THE CLAMP IS TIGHTENED. THE SQUARE BOX IS THEN PUSHED AGAINST THE WALL AND SECURED WITH SCREWS.

INDUSTRIAL COVER

DUPLEX RECEPTACLE

A DUPLEX RECEPTACLE IS CONNECTED TO THE SUPPLY CONDUCTORS AND IS SECURED TO THE BOX WITH AN INDUSTRIAL COVER. THE MICROWAVE OVEN IS THEN PLUGGED INTO THE RECEPTACLE. THE RECEPTACLE AND ATTACHMENT PLUG AND CORD WILL ALSO BE THE DISCONNECTING MEANS FOR THE MICROWAVE.

Figure 18-3

out of the wall at an elevation that ensures that the wire will be behind the cabinet. The cabinet installer must make a hole into the back of the cabinet and route the cable through that hole when the cabinet is screwed to the wall. A square box is installed in the cabinet, covering where the wire is stubbed out; this is the box for the receptacle outlet. A hole is drilled in the bottom of the cabinet in the location where the supply cord-and-attachment plug from the microwave oven will enter the cabinet. This arrangement will also allow the receptacle and the attachment cord and plug to function as the disconnecting means for the appliance.

18.2.4: Household Electric Range

The range is located along the east wall of the kitchen, as shown in Figure 11-9. The power supply cable is routed as shown in Figure 11-10 from subpanel A. The box is a deep square box with a two-gang plaster ring. The range is plugged into the receptacle using the access provided by the removal of the storage drawer beneath the range.

The range receptacle must be a 4-wire receptacle. The grounded circuit conductor system must be kept separate from the equipment grounding conductor system everywhere on the load side of the service equipment. If the grounded circuit conductor is bonded to the frame of the range from the factory, the bond must be removed and the two systems isolated from each other. A 4-wire pigtail is also necessary. The slot and blade configuration of the 4-wire receptacle and attachment cord eliminates the possibility of plugging a 3-wire attachment plug into a 4-wire receptacle.

The range circuit encompasses circuits 9 and 11 from subpanel A. There is a 40-ampere circuit breaker protecting the range circuit and 8-3 (plus equipment grounding conductor) AWG copper Type NM cable (a 4-wire cable), which, according to the ampacity tables in *310.16,* can carry 40 amperes. However, the receptacle and attachment plug and cord set are 50-ampere rated, as allowed by *210.21(B)(3).*

18.2.5: Electric Clothes Dryer

The dryer outlet in the sample house is located in the laundry area just outside the master bedroom

entry, as shown in Figure 11-8. The conductors were routed from subpanel A, circuits 13 and 15, down the back wall of the laundry area to a two-gang plastic box mounted 46 in. (1.17 m) above the floor.

The dryer receptacle must be a 4-wire receptacle. The frame of the dryer must be electrically separate from the grounded circuit conductor. If the dryer frame is bonded to the grounded circuit conductor terminal, it must be disconnected and isolated. The frame of the dryer will be grounded by the equipment grounding conductor system. As with the range, the pigtail for the dryer must also be a 4-wire pigtail. The attachment plug configuration and the receptacle configuration will not be compatible if a 3-wire pigtail is installed.

The dryer circuit is installed using 10-3 (plus equipment grounding conductor) copper Type NM cable. This cable is rated for 30 amperes as required by the Code. The receptacle and the cord-and-attachment-plug set are also rated at 30 amperes.

18.2.6: Sump Pump

The sump pump is located in the basement of the sample house along the north wall. The pump is supplied by a 20-ampere circuit from subpanel A, circuit 24. As detailed in Chapter 16, the cable for the sump pump must be brought down on the surface of a concrete basement wall. Figure 16-6 shows the details of the conduit work that is required to protect the cable against physical damage.

Because of the load of the sump pump (see Chapter 11), the receptacle for the sump pump is a single, 20-ampere, 125-volt-rated receptacle. The receptacle is mounted to an industrial cover and then installed into the outlet box on the wall. The receptacle does not have to have GFCI protection according to *210.8(A)(5), Exception No. 2.*

18.3: PERMANENTLY CONNECTED APPLIANCES

For the furnace, the air conditioner, and the electric baseboard heater, permanent wiring methods for the supply conductors are used. The wiring methods employed with the air conditioner and the furnace need to be flexible because of the potential for vibration during operation. The vibrations will cause the fittings to come loose if a rigid wiring method is employed.

18.3.1: Furnace

The furnace is supplied from circuit 17 from subpanel A, as shown in Figure 11-16. The furnace is located close to the south wall of the basement; therefore, the supply conductors are run down that wall to a square junction box. The cable must be protected against physical damage, as shown in Figure 11-17. The furnace also requires a disconnecting means. If the blower motor is thermally protected against overload overcurrent, the furnace can be connected through a single-pole switch, which can function as the required disconnecting means. Many furnace manufacturers, however, require that a fuse be installed on the circuit to protect the furnace equipment. If a fuse is required, it can be easily installed using a device known as an SSU, which combines a single-pole switch and an Edison-based fuseholder in the same device. For more about furnace hookups, see Figure 18-4. Figure 18-5 shows a typical furnace hookup in a dwelling. Notice the use of the FMC to protect the conductors.

After being set in place, the furnace can be connected to the electrical system using a permanent wiring method. See Figure 18-6. There is usually a wiring compartment within the furnace that allows access to the furnace wiring system by removal of a small cover. This feature is not universal, however, and it may be necessary to screw a junction box to the side of the furnace. The furnace internal wiring must be connected to the equipment grounding conductor system by the flexible conduit or cable. It is recommended that a separate equipment grounding conductor be used as the grounding method even if the metal sheath of the flexible wiring method is approved as an equipment grounding conductor, such as with AC cable. It is essential that the grounding method installed be reliable because of the extensive ductwork system that may become energized in the event of a fault.

In many regions of the country, the furnace must be installed at rough to provide heating for the house during the remainder of the construction process. The compound used with sheetrock finishing, or the plaster used for the wall finish, must be protected from freezing during the drying process. During the winter months, the general contractor may choose to run the furnace rather than providing temporary portable heaters. Special arrangements may need to be made with the local AHJ to get the necessary releases before the gas and electricity services can be connected. For this purpose, the power for the furnace is obtained from the temporary construction power panel; usually a permanent connection to the temporary panel, rather than a cord-and-attachment plug, is used to deter tampering.

18.3.2: Air Conditioner

The air conditioner is supplied from the main electrical panel, circuits 1 and 3. A 10-2 AWG supply cable (plus an equipment grounding conductor) is routed from the back of the main panel east in the outside wall and stubbed out at the planned location of the air conditioner, as shown in Figures 11-23 and 11-24.

The air conditioner is required to have a disconnect switch. In the sample house, the electrical panel where the circuit originates is within sight from the air-conditioning equipment. According to *440.14*, the disconnecting switch must be within sight from the air-conditioning equipment, so the circuit breaker in the main electrical panel can serve as the required disconnecting means. This holds true if the air conditioner does not require fuse protection. As shown in Figure 11-24, the transition from cable wiring method to conduit wiring methods can be accomplished with a simple junction box. However, as with furnaces, many air conditioner manufacturers require a fuse in the electrical supply circuit. In such cases, a fused disconnect switch, sized to a minimum 115 percent of the nameplate ampere rating of the air-conditioning equipment, or the branch-circuit selection current from the nameplate, whichever is the larger, is installed. The nameplate branch-circuit selection current (minimum circuit amperes) is 28.8 for the sample house, according to Construction Drawing M-1. The nameplate calls for a 35-ampere set of fuses. Therefore, a 60-ampere (standard size) waterproof disconnect switch, with 35-ampere fuses, will be installed where the cable is stubbed through the wall, as shown in Figure 18-7 (page 424). Figure 18-8 (page 425) shows a typical air conditioner hookup for a dwelling unit. Notice the disconnect on the outside wall of the dwelling.

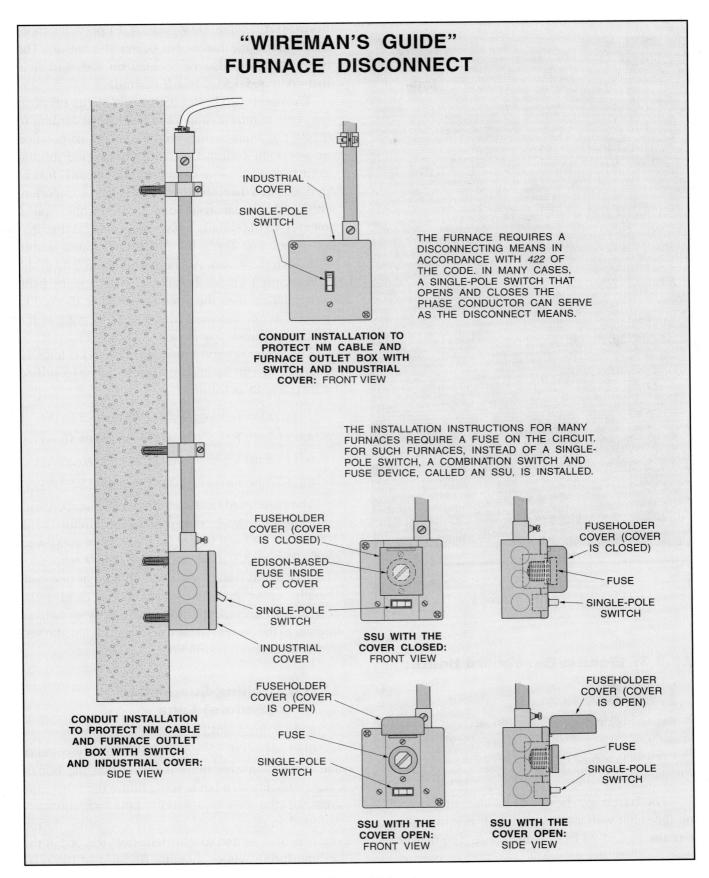

"WIREMAN'S GUIDE"
FURNACE DISCONNECT

INDUSTRIAL COVER

SINGLE-POLE SWITCH

THE FURNACE REQUIRES A DISCONNECTING MEANS IN ACCORDANCE WITH *422* OF THE CODE. IN MANY CASES, A SINGLE-POLE SWITCH THAT OPENS AND CLOSES THE PHASE CONDUCTOR CAN SERVE AS THE DISCONNECT MEANS.

CONDUIT INSTALLATION TO PROTECT NM CABLE AND FURNACE OUTLET BOX WITH SWITCH AND INDUSTRIAL COVER: FRONT VIEW

THE INSTALLATION INSTRUCTIONS FOR MANY FURNACES REQUIRE A FUSE ON THE CIRCUIT. FOR SUCH FURNACES, INSTEAD OF A SINGLE-POLE SWITCH, A COMBINATION SWITCH AND FUSE DEVICE, CALLED AN SSU, IS INSTALLED.

FUSEHOLDER COVER (COVER IS CLOSED)

EDISON-BASED FUSE INSIDE OF COVER

SINGLE-POLE SWITCH

INDUSTRIAL COVER

FUSEHOLDER COVER (COVER IS CLOSED)

FUSE

SINGLE-POLE SWITCH

SSU WITH THE COVER CLOSED: FRONT VIEW

CONDUIT INSTALLATION TO PROTECT NM CABLE AND FURNACE OUTLET BOX WITH SWITCH AND INDUSTRIAL COVER: SIDE VIEW

FUSEHOLDER COVER (COVER IS OPEN)

FUSE

SINGLE-POLE SWITCH

FUSEHOLDER COVER (COVER IS OPEN)

FUSE

SINGLE-POLE SWITCH

SSU WITH THE COVER OPEN: FRONT VIEW

SSU WITH THE COVER OPEN: SIDE VIEW

Figure 18-4

Figure 18-5   A typical furnace hookup. Note the use of FMC and the furnace disconnect switch.

18.3.3: Electric Baseboard Heater

KEY TERMS

In-line thermostat A type of thermostat that is installed directly into a piece of space-heating equipment rather than on the wall.

The baseboard heater is located in the garage on the same wall as subpanel A. It is powered from circuits 20 and 22 from subpanel A. The thermostat that controls the heater is installed in one of the ends of the heater and is referred to as an **in-line**

thermostat. Figure 18-9 (page 426) provides more details about the baseboard heater thermostat. The thermostat may also be located on the wall in a flush-mounted device box if desired.

The baseboard heater thermostat has an off position. This feature is important because according to *422.34(C)*, a unit switch with a marked off position can serve, in conjunction with the service disconnecting means, as a disconnecting means for an appliance in a dwelling unit. This is not so important with the particular installation in the sample house, however. In the sample house, the electrical distribution panel that feeds the heater is located within sight from the heater. According to *422.31(B)*, if the branch-circuit circuit breaker is within sight from the appliance, then the circuit breaker is allowed to function as the disconnecting means as well as the overcurrent-protective device.

The heater is approximately 8 ft (2.5 m) long. In general, 240-volt baseboard heaters can be identified by their length as follows:

2 ft (600 mm) long:	500 W rated (2.08 A)
4 ft (1.2 m) long:	1000 W rated (4.16 A)
6 ft (1.8 m) long:	1500 W rated (6.25 A)
8 ft (2.5 m) long:	2000 W rated (8.33 A)

The baseboard heater circuit is a 15-ampere circuit. The only load connected to the circuit is the 2000-watt, 240-volt heater. At 240 volts a 2000-watt load will draw 8.3 amperes (2000 W/240 V = 8.33 A). The baseboard heater must be supplied at one end. Usually, either end of the heater can receive the power, with only slight modifications to the wiring internal to the heater. These modifications are allowed by the installation instructions.

18.3.4: Ceiling-Suspended (Paddle) Fans

Although no ceiling-suspended paddle fans are installed in the sample house, they are very popular, and there is always the possibility that the homeowner may install a fan at some future time. In many areas of the country, paddle fans are routinely installed in every room to provide cooling during summer months and to distribute heat trapped at the ceiling during winter months. Because of the relative high weight and vibration from the turning

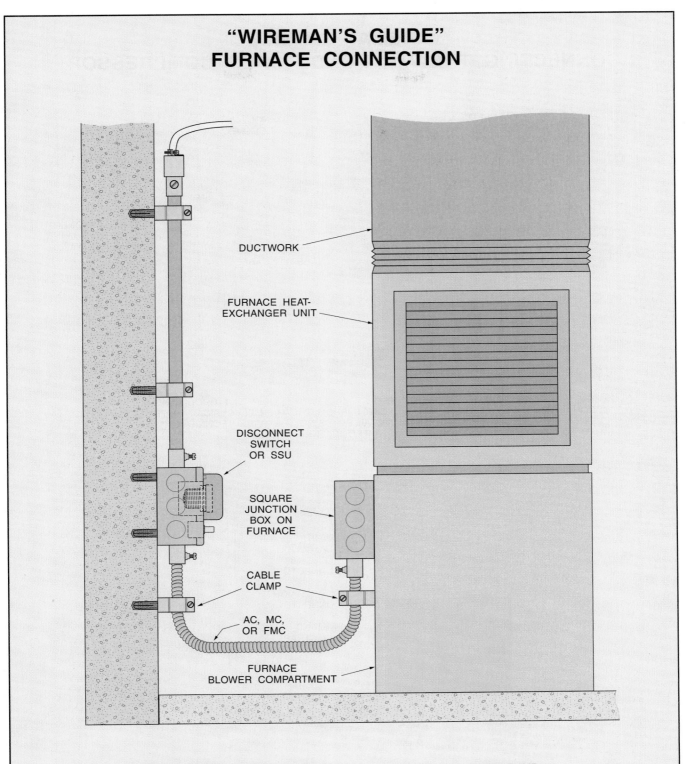

"WIREMAN'S GUIDE"
FURNACE CONNECTION

DUCTWORK

FURNACE HEAT-
EXCHANGER UNIT

DISCONNECT
SWITCH
OR SSU

SQUARE
JUNCTION
BOX ON
FURNACE

CABLE
CLAMP

AC, MC,
OR FMC

FURNACE
BLOWER COMPARTMENT

THE FURNACE IS CONNECTED WITH A FLEXIBLE WIRING METHOD TO DAMP
THE VIBRATION FROM THE BLOWER MOTOR. IT IS SOMETIMES NECESSARY TO
INSTALL A JUNCTION BOX ONTO THE FURNACE TO PROVIDE MAKEUP SPACE
FOR THE CONDUCTORS. USUALLY THE CONNECTING CABLE CAN TERMINATE
ON THE FURNACE, AND THE CONDUCTORS ARE MADE UP INSIDE THE FURNACE
WIRING COMPARTMENT.

Figure 18-6

"WIREMAN'S GUIDE"
CONNECTING THE AIR CONDITIONER COMPRESSOR

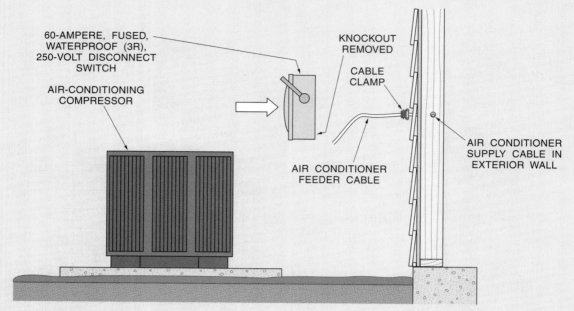

THE AIR CONDITIONER IS SUPPLIED FROM A 10-2 (WITH EQUIPMENT GROUNDING CONDUCTOR) AWG COPPER TYPE NM CABLE FROM THE MAIN ELECTRICAL DISTRIBUTION PANEL. THE DISCONNECT SWITCH IS INSTALLED USING THE SAME TECHNIQUE AS FOR THE MICROWAVE OVEN, ILLUSTRATED IN FIGURE 18-3.

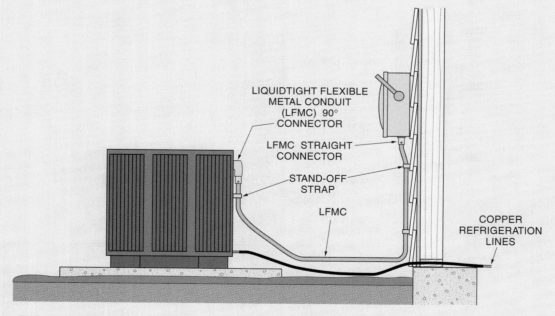

THE AIR CONDITIONER IS SUPPLIED USING 10 AWG COPPER CONDUCTORS FROM THE DISCONNECT SWITCH TO THE COMPRESSOR UNIT. A METHOD USING LFMC OR OTHER WIRING MAY BE EMPLOYED. GOOD WIRING PRACTICE SUGGESTS THAT THE WIRING METHOD SHOULD INVOLVE USE OF SOME FLEXIBLE CONDUIT AND THAT THE CONDUIT SHOULD CONTAIN AN EQUIPMENT GROUNDING CONDUCTOR EVEN IF THE CONDUIT IS APPROVED AS AN EQUIPMENT GROUNDING CONDUCTOR. THE COMPRESSOR UNIT IS CONNECTED TO THE FURNACE HEAT-EXCHANGER UNIT, AND THEREBY THE ENTIRE DUCTWORK SYSTEM, BY COPPER REFRIGERATION LINES.

Figure 18-7

Figure 18-8 A typical residential air conditioner installation.

motor and blades, care must be taken in installing ceiling-suspended fans to ensure that they do not fall from the ceiling during operation. Figure 18-10 shows some of the requirements for the installation of ceiling-supported paddle fans. Ceiling fans can employ a special mounting box that straddles a joist or rafter, thereby providing a firm support as well as room within the box to make up the conductors. Figure 18-11 (page 428) shows an example of this type of box.

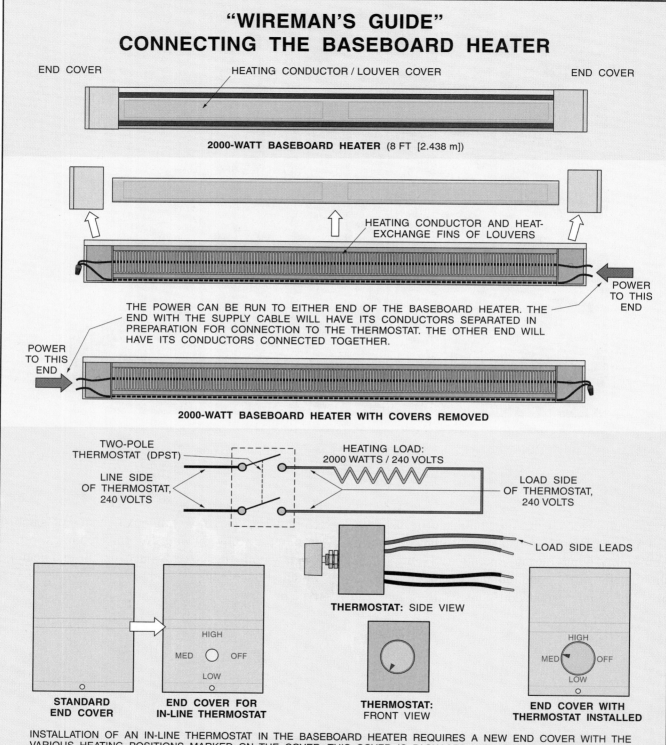

"WIREMAN'S GUIDE"
CONNECTING THE BASEBOARD HEATER

END COVER   HEATING CONDUCTOR / LOUVER COVER   END COVER

2000-WATT BASEBOARD HEATER (8 FT [2.438 m])

HEATING CONDUCTOR AND HEAT-
EXCHANGE FINS OF LOUVERS

POWER
TO THIS
END

THE POWER CAN BE RUN TO EITHER END OF THE BASEBOARD HEATER. THE END WITH THE SUPPLY CABLE WILL HAVE ITS CONDUCTORS SEPARATED IN PREPARATION FOR CONNECTION TO THE THERMOSTAT. THE OTHER END WILL HAVE ITS CONDUCTORS CONNECTED TOGETHER.

POWER
TO THIS
END

2000-WATT BASEBOARD HEATER WITH COVERS REMOVED

TWO-POLE
THERMOSTAT (DPST)

HEATING LOAD:
2000 WATTS / 240 VOLTS

LINE SIDE
OF THERMOSTAT,
240 VOLTS

LOAD SIDE
OF THERMOSTAT,
240 VOLTS

LOAD SIDE LEADS

THERMOSTAT: SIDE VIEW

STANDARD
END COVER

END COVER FOR
IN-LINE THERMOSTAT

HIGH

MED   OFF

LOW

THERMOSTAT:
FRONT VIEW

HIGH

MED   OFF

LOW

END COVER WITH
THERMOSTAT INSTALLED

INSTALLATION OF AN IN-LINE THERMOSTAT IN THE BASEBOARD HEATER REQUIRES A NEW END COVER WITH THE VARIOUS HEATING POSITIONS MARKED ON THE COVER. THIS COVER IS PACKAGED WITH THE THERMOSTAT. THE LINE SIDE HEATER LEADS ARE CONNECTED TO THE THERMOSTAT LINE SIDE LEADS OR TERMINALS, AND THE LOAD SIDE TERMINALS OR LEADS ARE CONNECTED TO THE LOAD SIDE HEATER LEADS. THIS THERMOSTAT HAS THE TWO-LINE SIDE LEADS AS THE BLACK CONDUCTORS AND TWO-LOAD SIDE LEADS AS RED CONDUCTORS. ALL MANUFACTURERS HAVE DIFFERENT SYSTEMS OF COVERS AND THERMOSTAT MOUNTING. THE MANUFACTURER'S INSTALLATION INSTRUCTIONS MUST BE CHECKED BEFORE INSTALLATION OF THE HEATER. NOT ALL THERMOSTATS HAVE LEADS COLOR-CODED IN THIS MANNER. THE THERMOSTAT INSTALLATION INSTRUCTIONS MUST ALSO BE CHECKED BEFORE IT IS CONNECTED TO THE CIRCUIT. THERMOSTATS CAN ALSO BE MOUNTED ON THE WALL IN FLUSH-MOUNTED DEVICE BOXES.

Figure 18-9

"WIREMAN'S GUIDE"
CEILING FAN INSTALLATION

CEILING-MOUNTED PADDLE FANS HAVE TWO MAJOR DRAWBACKS: (1) THEY ARE RELATIVELY HEAVY AND (2) BECAUSE THEY MOVE, THEY VIBRATE. *NEC® 314.27(D)* REQUIRES THAT ANY BOX USED TO MOUNT A CEILING PADDLE FAN BE LISTED FOR THE USE AND FOR THE WEIGHT OF THE FAN. ONLY BOXES THAT ARE TESTED AND LISTED FOR SUPPORT OF PADDLE FANS CAN BE USED TO INSTALL A PADDLE FAN. *NEC® 422.18* FURTHER STATES THAT IF THE PADDLE FAN WEIGHS 35 LB (16 kG) OR LESS, THE FAN CAN BE SUPPORTED BY THE BOX, BUT IF THE FAN WEIGHS MORE THAN THAT, IT MUST BE SUPPORTED INDEPENDENTLY OF THE BOX.

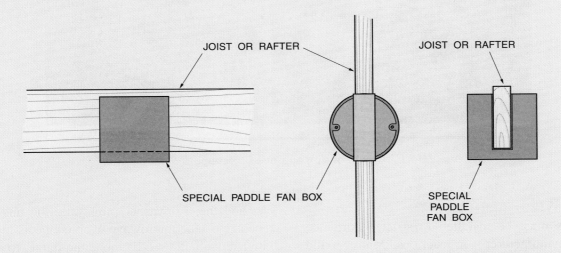

ONE TYPE OF PADDLE FAN BOX STRADDLES THE JOIST, RAFTER, OR TRUSS. IT ALLOWS ACCESS TO THE BOTTOM OF THE JOIST, RAFTER, OR TRUSS SUPPORTING THE PADDLE FAN. IT ALSO DISTRIBUTES THE WEIGHT OF THE FAN MORE EVENLY THROUGHOUT THE BOX, ELIMINATING ANY SAGGING THAT A SIDE NAIL-ON BOX CAN ALLOW.

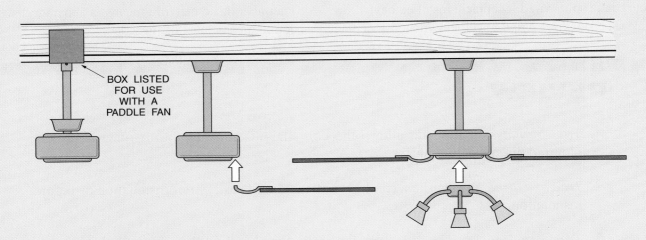

THE PADDLE FAN MOUNTING HARDWARE IS INSTALLED ACCORDING TO THE MANUFACTURER'S INSTRUCTIONS. IT IS EXTREMELY IMPORTANT THAT THE FAN SUPPORTS BE PROPERLY INSTALLED; OTHERWISE, THE FAN COULD FALL FROM THE CEILING, CAUSING SEVERE INJURY. THE FAN IS THEN ASSEMBLED, AND ANY ACCESSORIES, SUCH AS A LUMINAIRE (FIXTURE), ARE INSTALLED BEFORE THE FAN IS ENERGIZED. ALL MANUFACTURER'S INSTRUCTIONS MUST BE FOLLOWED INSTALLING A PADDLE FAN.

Figure 18-10

Figure 18-11 A type of box approved to support a paddle fan.

SUMMARY

Appliances must be installed in strict accordance with the manufacturer's instructions. Many of the appliances are allowed by the Code to be cord-and-attachment-plug connected, as well as receptacle connected. The *NEC*® lists some length restrictions on the pigtails for disposals and dishwashers. The two appliances connect with a cord-and-attachment plug to the split-wired receptacle installed below the sink. The disposal plugs into the switched receptacle, and the dishwasher plugs into the unswitched receptacle. The microwave oven connects inside the kitchen cabinet directly above the range, to a receptacle outlet that was installed inside the cabinet (see Chapter 18). The furnace, the air conditioner, and the baseboard heater are hard wired (a permanent wiring method) rather than cord-and-attachment-plug connected. Because there is no attachment plug to act as a disconnecting means, a separate disconnecting means must be provided.

REVIEW

1. *True or False:* The Code allows the cord-and-attachment plug to function as the required disconnecting means.

2. *True or False:* Dishwashers can be wired almost as easily after installation as before installation.

3. *True or False:* The pigtail for the disposals should be installed before the disposal is installed.

4. *True or False:* In the sample house, the disconnecting means for the microwave oven is the branch-circuit overcurrent-protective device.

5. *True or False:* Range receptacles are accessible by removal of a storage drawer.

6. *True or False:* A disposal pigtail must be at least 5 ft (1.5 m) long.

7. *True or False:* The dryer in the sample house is connected using FMC.

8. *True or False:* The receptacle outlet functions as overload overcurrent protection for the sump pump in the sample house.

9. *True or False:* The actual method used to connect an appliance does not have to be determined until trim because any rough installation will meet the needs of trim.

10. *True or False:* If the installation instructions for a furnace show a fused disconnect switch, it is possible to substitute the branch-circuit overcurrent-protective device for the fuse.

11. *True or False:* The electric baseboard heater is required to be installed on a 20-ampere circuit.

12. *True or False:* The pigtail for the range can have either a 3-wire or a 4-wire attachment plug.

13. *True or False:* The attachment plug for the dryer, if installed, must be a 3-wire device.

14. *True or False:* The electric baseboard heater is required to have a separate overload overcurrent-protective device to protect the heater against overloads.

"WIREMAN'S GUIDE" REVIEW

1. Explain why the disposal should be wired with a pigtail before it is installed below the kitchen sink. Why is this not always true for a dishwasher?

2. Explain why fuse protection may be required even though the furnace blower motor is thermally protected. What load should the fuse be sized to if it is required to be installed?

3. Explain why the air conditioner for the sample house does not have to have a separate disconnect switch.

4. Describe how to rewire an electric baseboard heater if it is necessary because of the end of the heater into which the power supply terminates. Are the modifications usually part of the manufacturer's installation instructions?

5. Explain why a 3-wire cable (plus equipment grounding conductor) is required for the electric range and the dryer, whereas only a 2-wire cable (plus equipment grounding conductor) is needed for the air conditioner and the electric baseboard heater.

6. If the sump pump arrives at the job site wired as a 240-volt load (and it cannot be rewired on the site), explain what you can do to supply 240 volts to the pump and still meet all Code requirements.

CHAPTER 19

Checkout and Troubleshooting

OBJECTIVES

On completion of this chapter, the student will be able to:

- ☑ Connect the dwelling's permanent wiring system to the temporary power panel.
- ☑ Explain the dangers to all workers on the job site of turning on the circuits for checkout.
- ☑ Describe how a voltage tester operates.
- ☑ Identify the procedures to follow for checkout of receptacles and for checkout of luminaires (fixtures) and their associated switching systems.
- ☑ Identify the procedures to follow for checkout of utilization equipment or appliances.
- ☑ Perform all of the required tests to ensure proper checkout.
- ☑ Describe and apply the troubleshooting technique known as "halving the circuit."

INTRODUCTION

This chapter is concerned with checkout and troubleshooting. Theoretically, these processes should not be needed if everything was completed properly during rough-in and trim-out and if no damage was done to the electrical system by other trades. Unfortunately, in the real world, mistakes happen. Circuit damage is sometimes suffered at the hands of the other trades. Sometimes the equipment installed is faulty. Sometimes make-up mistakes exist, and sometimes travelers are crossed with commons. These things happen from time to time. The intent of this book has been to allow the field electrician to minimize the number of mistakes, and to limit the number of outside events that affect the installation. "Experience is the best teacher" is as true today as it has always been. However, it is also the most expensive way of learning.

KEY TERMS

Checkout The process of ensuring that all electrical systems within a dwelling are functioning properly following completion of trim.

Troubleshooting procedure The process for efficiently discovering and repairing circuits and utilization equipment that are not operating properly.

Two separate processes are involved in the final step in providing a functional and safe electrical system. The first is **checkout**. Checkout determines if there is a problem with the just-completed electrical system. The second, **troubleshooting**, is the process of acquiring the information necessary to effect repairs.

19.1: CONNECTING THE TEMPORARY SERVICE TO THE DWELLING SERVICE

In order to complete the installation, it is necessary to energize the just-completed electrical system. The local AHJ will want to check the various outlets to ensure that they have been wired according to the *NEC®*, local ordinances, and good wiring practices. Many inspectors test every receptacle, every luminaire (fixture), and every piece of equipment for grounding and for proper operation before a final inspection is completed.

During checkout and troubleshooting, the service connection to the dwelling's electrical system is still temporary. The local power-providing utility will connect the permanent power after all inspections have been completed and approved by the AHJ. The power to energize the house comes from the temporary construction panel. However, the temporary connection still has to be a safe connection. Connecting the temporary power to the building's permanent electrical system potentially energizes all of the service equipment, including the meter housing. This can create dangers for the utility line workers when they arrive to connect the permanent power. It also presents a hazard to anyone working in the immediate area if the opening for the meter is not effectively closed and protected. Faulty circuits can draw tremendous amounts of current in the event of a fault, causing flash burns to anyone working nearby.

19.2: WHERE TO TERMINATE THE TEMPORARY FEED TO THE DWELLING SERVICE

The easiest and safest location to connect the temporary feeders for the sample house is into a set of circuit breakers in the main panel. The circuit that is used to supply the house is connected to an overcurrent-protective device at the temporary electrical panel. This makes the conductors that are temporarily supplying the dwelling feeder conductors. The circuit breakers installed in the main electrical panel should be specially installed to provide the temporary power rather than backfeeding an exiting set of breakers, if possible. See Figure 19-1. This system allows for the most flexibility and safety because the entire feed to the house could be totally discon-

nected at the dwelling service or at the temporary construction panel. Furthermore, it allows for all of the dwelling's branch circuits to be energized at one time or to be energized individually. It also allows for the isolation of the service-entrance conductors and the meter enclosure by simply having the dwelling's main circuit breaker in the open (off) position. Additionally, it keeps any unnecessary wiring out of the meter enclosure, as required by most power-providing utilities.

19.3: CHECKOUT

In order to determine if the installation is complete, each outlet must be thoroughly tested. This process is called checkout. Plug testers are available to help with this process for receptacles, but the best tester is a hand-held voltage tester like the one pictured in Figure 19-2. This type of tester is versatile; it can test receptacles, junctions, equipment terminals, sockets, and other types of electrical connections. It should be a required tool for every field electrician.

The set of electrical plans used at rough-in and trim-out of the electrical system can be very helpful for checkout. The plans should show which outlets are installed on what circuits, as installed at rough-in. These are called "as-built" drawings and are similar to the wiring plans for the sample house in Figures 1-8 and 1-9.

Only one branch circuit should be energized at any one time during checkout. By comparing the outlets on that circuit as shown on the as-built plans with the outlets on the energized circuit, the electrician is able to determine that all outlets on that circuit are working properly. If an outlet that is supposed to be energized according to the as-built drawings is in fact not energized, troubleshooting procedures can be started. Every outlet on that circuit should be tested to determine that voltage (120 volts) is present, that the outlet has proper polarity, and that the equipment grounding system is connected.

19.3.1: Checkout of Receptacles

Receptacle outlets should be tested by using a voltage tester and the following procedure:

1. Test between the phase conductor and the grounded circuit conductor. There should be 120 volts present.

"WIREMAN'S GUIDE"
CONNECTING THE SAMPLE HOUSE SERVICE
TO TEMPORARY POWER

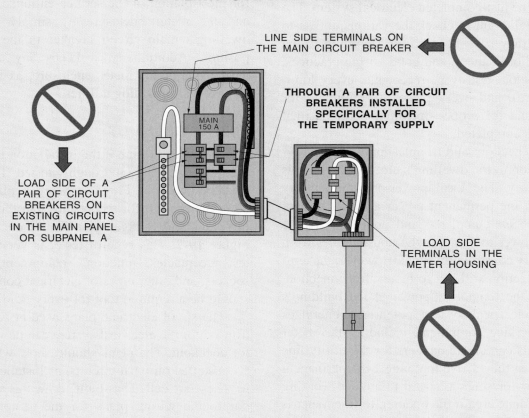

LINE SIDE TERMINALS ON
THE MAIN CIRCUIT BREAKER

THROUGH A PAIR OF CIRCUIT
BREAKERS INSTALLED
SPECIFICALLY FOR
THE TEMPORARY SUPPLY

MAIN
150 A

LOAD SIDE OF A
PAIR OF CIRCUIT
BREAKERS ON
EXISTING CIRCUITS
IN THE MAIN PANEL
OR SUBPANEL A

LOAD SIDE
TERMINALS IN THE
METER HOUSING

LOAD SIDE TERMINALS IN THE METER HOUSING: A CONNECTION AT THE LOAD SIDE TERMINALS OF THE METER HOUSING WILL ALLOW FOR THE ENERGIZING OF ALL OF THE CIRCUITS IN THE SAMPLE HOUSE. HOWEVER, IT ALSO MAKES THE METER ENCLOSURE A HAZARD. IT IS ALSO PROBABLY NOT ALLOWED BY THE LOCAL POWER UTILITY. THIS TYPE OF CONNECTION SHOULD BE USED AS A LAST RESORT ONLY AND SHOULD BE DISCONNECTED AS SOON AS THE CHECKOUT IS COMPLETE.

LINE SIDE TERMINALS OF THE MAIN CIRCUIT BREAKER: A CONNECTION AT THE LINE SIDE TERMINALS OF THE DWELLING MAIN CIRCUIT BREAKER IS JUST AS BAD AS A CONNECTION IN THE METER HOUSING. IN FACT, IT IS THE SAME PLACE ELECTRICALLY. THE SERVICE-ENTRANCE CONDUCTORS WILL CARRY THE CURRENT TO THE LOAD SIDE TERMINALS IN THE METER HOUSING. THIS CONNECTION IS TO BE MADE ONLY AS A LAST RESORT AND MUST BE DISCONNECTED ON THE COMPLETION OF CHECKOUT.

LOAD SIDE OF A PAIR OF EXISTING CIRCUIT BREAKERS: THERE IS A TECHNIQUE CALLED BACKFEEDING. THE TEMPORARY FEEDERS CAN BE TERMINATED THROUGH THE LOAD SIDE OF THE CIRCUIT BREAKER. THIS ALLOWS FOR EASY DISCONNECTING OF THE POWER IN THE EVENT OF A PROBLEM. HOWEVER, UNLESS THE CONDUCTORS THAT ARE INTENDED FOR THAT CIRCUIT BREAKER ARE DISCONNECTED, THE LOAD CONNECTED TO THE CIRCUIT BREAKERS WILL BE ENERGIZED ALL THE TIME. IF THE CONDUCTORS ARE REMOVED, THE EQUIPMENT WILL NOT BE OPERABLE FOR CHECKOUT.

LOAD SIDE OF A PAIR OF CIRCUIT BREAKERS INSTALLED IN THE PANEL SPECIFICALLY FOR BACKFEEDING THE BUSSING: IF A PAIR OF CIRCUIT BREAKERS ARE INSTALLED INTO THE MAIN PANEL AND THE TEMPORARY FEED IS CONNECTED TO THESE BREAKERS, THE POWER WILL FEED THE BUSSING OF THE MAIN PANEL THROUGH THE BREAKER. IF THE MAIN CIRCUIT BREAKER, AND ALL OF THE BRANCH-CIRCUIT BREAKERS, ARE OFF (OPEN), THE POWER WILL BE CONTAINED TO JUST THE PANEL BUSSING. EACH CIRCUIT CAN THEN BE CLOSED (ON) ONE AT A TIME FOR CHECKOUT. BECAUSE CHECKOUT DOES NOT TAKE A LOT OF POWER, THE BACK-FED CIRCUIT BREAKERS COULD BE 20-AMPERE OR 30-AMPERE BREAKERS, THEREBY PROVIDING A LEVEL OF PROTECTION ABOVE THAT OF THE MAIN BREAKER.

Figure 19-1

"WIREMAN'S GUIDE"
A TYPICAL VOLTAGE TESTER

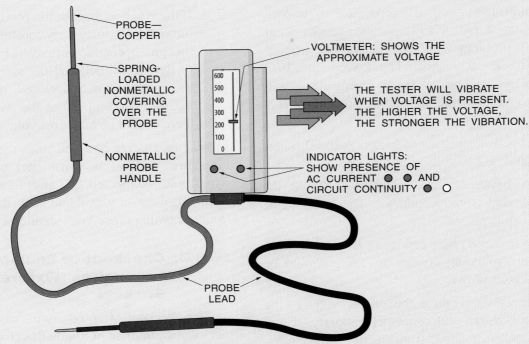

PROBE—
COPPER

SPRING-
LOADED
NONMETALLIC
COVERING
OVER THE
PROBE

NONMETALLIC
PROBE
HANDLE

VOLTMETER: SHOWS THE
APPROXIMATE VOLTAGE

THE TESTER WILL VIBRATE
WHEN VOLTAGE IS PRESENT.
THE HIGHER THE VOLTAGE,
THE STRONGER THE VIBRATION.

INDICATOR LIGHTS:
SHOW PRESENCE OF
AC CURRENT ● ● AND
CIRCUIT CONTINUITY ● ○

PROBE
LEAD

TYPICAL VOLTAGE TESTER
SEVERAL BRANDS OF VOLTAGE TESTERS ARE
AVAILABLE. NOT ALL VOLTAGE TESTERS HAVE THE
SAME FEATURES. THE MORE EXPENSIVE MODELS HAVE
FEATURES LACKING ON THE LOWER PRICED MODELS.
IT IS RECOMMENDED TO OBTAIN A VOLTAGE TESTER
WITH AT LEAST THE FUNCTIONS SHOWN HERE.

THE VOLTAGE TESTER MUST ALWAYS BE
TESTED ON A KNOWN VOLTAGE SOURCE
BEFORE IT IS USED FOR TESTING ON THE
CIRCUIT. THE VOLTAGE TESTER MAY HAVE
BEEN DAMAGED SINCE ITS LAST USE.

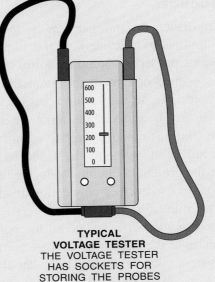

**TYPICAL
VOLTAGE TESTER**
THE VOLTAGE TESTER
HAS SOCKETS FOR
STORING THE PROBES
WHEN THE TESTER
IS NOT IN USE.

Figure 19-2

2. Test between the phase conductor and the equipment grounding conductor. There should be 120 volts present.

If both of these tests produce a 120-volt reading on the voltage tester (by meter reading, by vibration, and by the indicator lights), the receptacle is properly powered and wired. If any other test results occur, there is a problem with the system, and troubleshooting of the circuit should begin. See Figure 19-3.

If the receptacle is a 240-volt receptacle, as with the dryer and the range, the testing procedures is as follows:

1. Test between phase conductors A and B. There should be 240 volts present.

2. Test between phase conductor A and the grounded circuit conductor. There should be 120 volts present.

3. Test between phase conductor B and the equipment grounding conductor. There should be 120 volts present.

If all three tests produce the proper voltage, the receptacle is wired and connected properly. If any other outcome is obtained from any of the three tests, there is a problem with the system, and troubleshooting procedures should begin.

19.3.2: Checkout of Luminaires (Fixtures) or Devices with Sockets

Most electricians do not perform a complete checkout of luminaires (fixtures). If the luminaire (fixture) operates properly when the control switching system is opened or closed, the luminaire (fixture) is wired properly. Many times it is assumed that it is wired properly because of the requirement for the luminaire (fixture) manufacturer to identify the grounded circuit conductor and the Code requirement to have the grounded circuit conductor identified in the box. Color-to-color connection at trim is usually very reliable.

However, if any doubt exists about the polarity of the connection or the continuity of the equipment grounding system, the following tests should be undertaken:

1. Test between the screw shell of the lampholder and a grounded noncurrent-carrying part of the luminaire (fixture). There should be no (zero) voltage present. There should be an indication of continuity on the voltage tester indicator lights. If a voltage of 120 volts is present, there is a polarity, or possibly another problem with the luminaire (fixture) wiring or connection.

2. If there is doubt about the power supply to the luminaire (fixture), very carefully test between the phase conductor terminal at the bottom of the screw shell and the screw shell. When the control switch is closed, there should be 120 volts present. When the switch is opened, there should be no (zero) voltage present.

If these results are obtained from the two tests, the luminaire (fixture) is considered to be properly wired and connected. If any other results are obtained, there is a problem with the system, and troubleshooting procedures should begin.

19.3.3: Checkout of Equipment and Luminaires (Fixtures) without Sockets

For checkout of an otherwise working piece of electrical equipment or luminaires (fixtures) without sockets (such as fluorescent luminaires [fixtures]), testing is limited to ensuring that the equipment grounding system is properly connected. The phase conductor terminal or splice must be accessed for this test.

- Test between the phase conductor (or closed switch leg connection) and any noncurrent-carrying metal part of the luminaire (fixture) or equipment. The proper voltage (120 volts) should be present.

If the proper voltage is present, the luminaire (fixture) or equipment is connected and wired properly. Any other voltage reading indicates a problem with the system, and implementing troubleshooting procedures is necessary.

19.3.4: Checkout of Switching Systems

In dwellings, switching systems usually operate luminaires (fixtures), which makes them easy to check out. Single-pole switching systems are easily checked simply by opening and closing the switch. If the luminaire (fixture), or the switched receptacle, goes on and off as it is supposed to, the system is wired and connected properly.

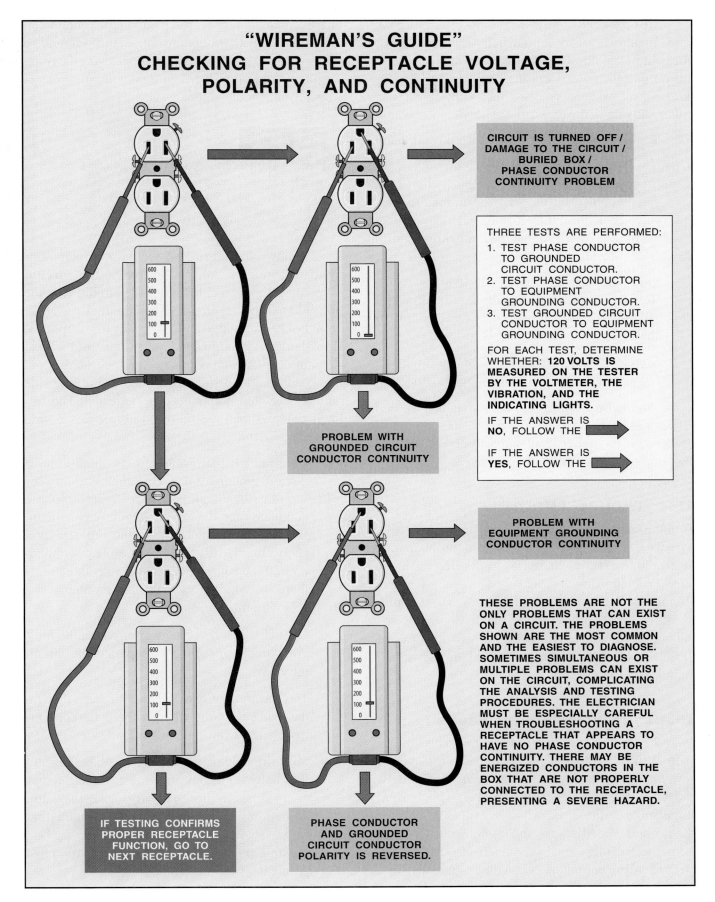

"WIREMAN'S GUIDE"
CHECKING FOR RECEPTACLE VOLTAGE, POLARITY, AND CONTINUITY

CIRCUIT IS TURNED OFF /
DAMAGE TO THE CIRCUIT /
BURIED BOX /
PHASE CONDUCTOR
CONTINUITY PROBLEM

THREE TESTS ARE PERFORMED:

1. TEST PHASE CONDUCTOR
   TO GROUNDED
   CIRCUIT CONDUCTOR.
2. TEST PHASE CONDUCTOR
   TO EQUIPMENT
   GROUNDING CONDUCTOR.
3. TEST GROUNDED CIRCUIT
   CONDUCTOR TO EQUIPMENT
   GROUNDING CONDUCTOR.

FOR EACH TEST, DETERMINE
WHETHER: **120 VOLTS IS
MEASURED ON THE TESTER
BY THE VOLTMETER, THE
VIBRATION, AND THE
INDICATING LIGHTS.**

IF THE ANSWER IS
NO, FOLLOW THE

IF THE ANSWER IS
YES, FOLLOW THE

PROBLEM WITH
GROUNDED CIRCUIT
CONDUCTOR CONTINUITY

PROBLEM WITH
EQUIPMENT GROUNDING
CONDUCTOR CONTINUITY

THESE PROBLEMS ARE NOT THE
ONLY PROBLEMS THAT CAN EXIST
ON A CIRCUIT. THE PROBLEMS
SHOWN ARE THE MOST COMMON
AND THE EASIEST TO DIAGNOSE.
SOMETIMES SIMULTANEOUS OR
MULTIPLE PROBLEMS CAN EXIST
ON THE CIRCUIT, COMPLICATING
THE ANALYSIS AND TESTING
PROCEDURES. THE ELECTRICIAN
MUST BE ESPECIALLY CAREFUL
WHEN TROUBLESHOOTING A
RECEPTACLE THAT APPEARS TO
HAVE NO PHASE CONDUCTOR
CONTINUITY. THERE MAY BE
ENERGIZED CONDUCTORS IN THE
BOX THAT ARE NOT PROPERLY
CONNECTED TO THE RECEPTACLE,
PRESENTING A SEVERE HAZARD.

IF TESTING CONFIRMS
PROPER RECEPTACLE
FUNCTION, GO TO
NEXT RECEPTACLE.

PHASE CONDUCTOR
AND GROUNDED
CIRCUIT CONDUCTOR
POLARITY IS REVERSED.

Figure 19-3

Three-way and 4-way switching systems are not so elementary. These multiswitch systems must be checked out in all of their possible combinations to ensure that they are operating properly. For a standard 3-way system, such as one with switches at either end of a hallway, use the following procedure:

1. Position either switch so that the luminaire (fixture) is energized.

2. At switch A (which may be either switch), test the switch to ensure that the luminaire (fixture) goes off and on as the switch is toggled. Leave the switch so that the luminaire (fixture) is energized.

3. Go to the other end of the hallway to switch B. Toggle the switch to ensure that the luminaire (fixture) operates properly. Leave switch B so that the luminaire (fixture) is off.

4. Toggle switch A to ensure that the luminaire (fixture) is operating properly. See Figure 19-4.

If the luminaire (fixture) is being switched on and off as it should, the system is wired and connected properly. If any other results are observed, troubleshooting procedures should be initiated.

19.4: TROUBLESHOOTING

Troubleshooting is the term used to describe the process of identifying the electrical problem and determining exactly where it is in the electrical system. Problems in electrical systems can be very complex and can have multiple causes and symptoms. In some ways, good troubleshooting is as much art as science. However, the vast majority of the electrical problems encountered in a dwelling are relatively simple and easy to detect.

19.4.1: Troubleshooting Faults

When there is a fault on a circuit, it is obvious as soon as the circuit breaker is switched on. The huge current flow through the system will immediately, and violently, trip the circuit breaker. This current surge subjects the electrical system to stresses that should be avoided whenever possible, so the cable from the faulting circuit should be located in the electrical distribution panel and all conductors—the

phase conductor, the grounded circuit conductor, and the equipment grounding conductor—should be removed from the circuit breaker and the bussing. Testing is accomplished with a continuity tester (there is one on the voltage tester).

KEY TERMS

Halving the circuit A specific technique used for troubleshooting circuit problems that involves repeatedly splitting the troubled circuit into halves.

As shown in Figure 19-5 the basic principle for troubleshooting a fault is to **halve the circuit**. Disconnect the load side half of the circuit from the line side half of the circuit at some outlet or switch point box. Test the line and load sides for continuity. If the fault is on the line side, then none of the load side outlets or switch points need to be tested. If the fault is on the load side, then none of the line side outlets or switch points need to be tested. The following steps should be followed in troubleshooting a circuit for faults:

1. Determine if the problem is a ground fault or a short circuit by checking the continuity of the three conductors. Test between the phase conductor and the grounded circuit conductor. If there is continuity, then the problem is a short circuit. Test between the phase conductor and the equipment grounding conductor. If there is continuity, then the problem is a ground fault.

2. Locate a device box at approximately midpoint in the circuit, and separate the circuit into halves.

3. Test the continuity on the line side and then on the load side. Note whether the fault (continuity) is on the line side or the load side.

4. Split the half of the circuit that has the fault into halves and repeat the test. Repeat for as many times as necessary to locate the fault. Sometimes the box may be located on the first try, and sometimes it will be at the farthest possible point, but over time this is the most efficient way to locate the fault.

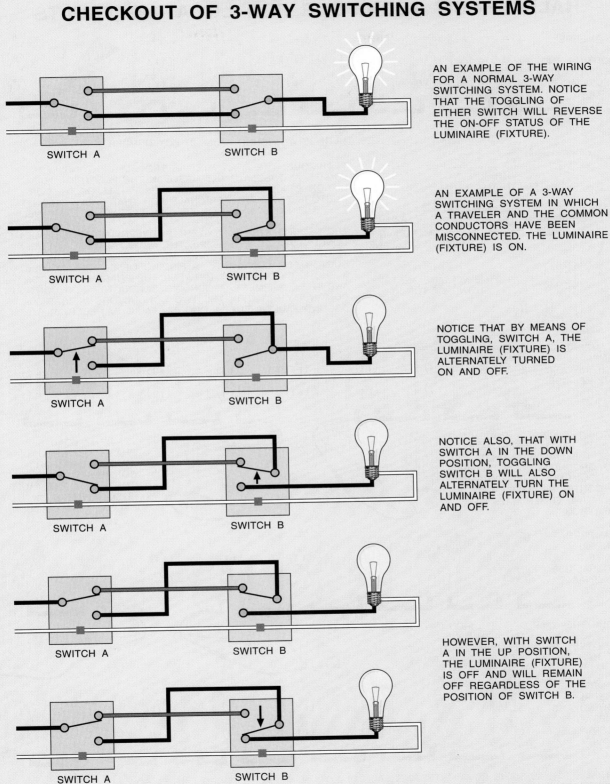

"WIREMAN'S GUIDE"
CHECKOUT OF 3-WAY SWITCHING SYSTEMS

AN EXAMPLE OF THE WIRING FOR A NORMAL 3-WAY SWITCHING SYSTEM. NOTICE THAT THE TOGGLING OF EITHER SWITCH WILL REVERSE THE ON-OFF STATUS OF THE LUMINAIRE (FIXTURE).

AN EXAMPLE OF A 3-WAY SWITCHING SYSTEM IN WHICH A TRAVELER AND THE COMMON CONDUCTORS HAVE BEEN MISCONNECTED. THE LUMINAIRE (FIXTURE) IS ON.

NOTICE THAT BY MEANS OF TOGGLING, SWITCH A, THE LUMINAIRE (FIXTURE) IS ALTERNATELY TURNED ON AND OFF.

NOTICE ALSO, THAT WITH SWITCH A IN THE DOWN POSITION, TOGGLING SWITCH B WILL ALSO ALTERNATELY TURN THE LUMINAIRE (FIXTURE) ON AND OFF.

HOWEVER, WITH SWITCH A IN THE UP POSITION, THE LUMINAIRE (FIXTURE) IS OFF AND WILL REMAIN OFF REGARDLESS OF THE POSITION OF SWITCH B.

IN ORDER TO ENSURE THAT THE 3-WAY AND 4-WAY SWITCHING SYSTEMS ARE WORKING PROPERLY, IT IS NECESSARY TO TEST THE SYSTEM IN ALL POSSIBLE SWITCH POSITION CONFIGURATIONS.

Figure 19-4

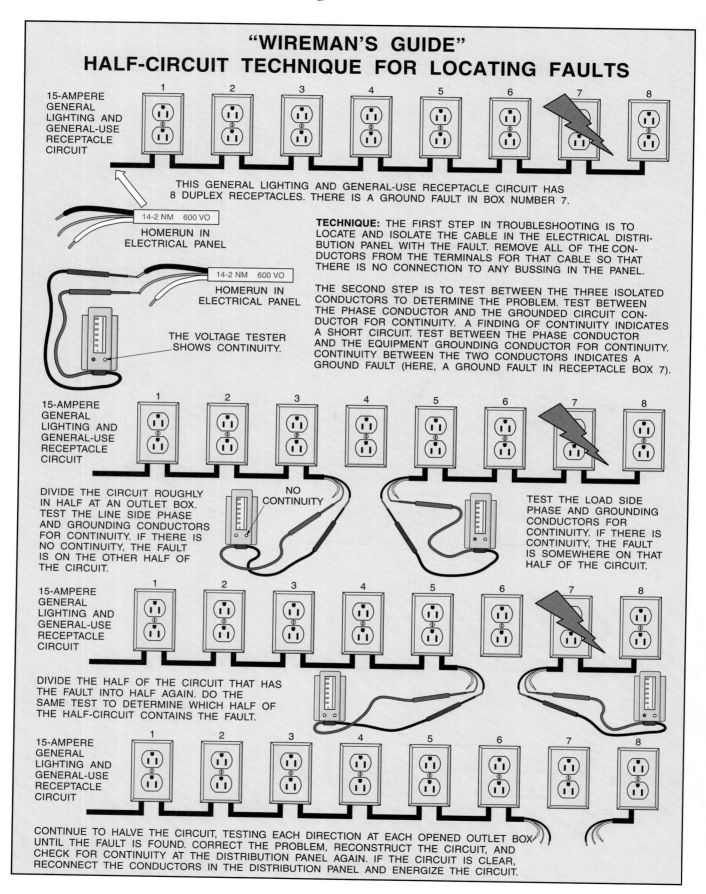

"WIREMAN'S GUIDE"
HALF-CIRCUIT TECHNIQUE FOR LOCATING FAULTS

15-AMPERE GENERAL LIGHTING AND GENERAL-USE RECEPTACLE CIRCUIT

THIS GENERAL LIGHTING AND GENERAL-USE RECEPTACLE CIRCUIT HAS 8 DUPLEX RECEPTACLES. THERE IS A GROUND FAULT IN BOX NUMBER 7.

14-2 NM   600 VO

HOMERUN IN ELECTRICAL PANEL

14-2 NM   600 VO

HOMERUN IN ELECTRICAL PANEL

THE VOLTAGE TESTER SHOWS CONTINUITY.

TECHNIQUE: THE FIRST STEP IN TROUBLESHOOTING IS TO LOCATE AND ISOLATE THE CABLE IN THE ELECTRICAL DISTRIBUTION PANEL WITH THE FAULT. REMOVE ALL OF THE CONDUCTORS FROM THE TERMINALS FOR THAT CABLE SO THAT THERE IS NO CONNECTION TO ANY BUSSING IN THE PANEL.

THE SECOND STEP IS TO TEST BETWEEN THE THREE ISOLATED CONDUCTORS TO DETERMINE THE PROBLEM. TEST BETWEEN THE PHASE CONDUCTOR AND THE GROUNDED CIRCUIT CONDUCTOR FOR CONTINUITY. A FINDING OF CONTINUITY INDICATES A SHORT CIRCUIT. TEST BETWEEN THE PHASE CONDUCTOR AND THE EQUIPMENT GROUNDING CONDUCTOR FOR CONTINUITY. CONTINUITY BETWEEN THE TWO CONDUCTORS INDICATES A GROUND FAULT (HERE, A GROUND FAULT IN RECEPTACLE BOX 7).

15-AMPERE GENERAL LIGHTING AND GENERAL-USE RECEPTACLE CIRCUIT

DIVIDE THE CIRCUIT ROUGHLY IN HALF AT AN OUTLET BOX. TEST THE LINE SIDE PHASE AND GROUNDING CONDUCTORS FOR CONTINUITY. IF THERE IS NO CONTINUITY, THE FAULT IS ON THE OTHER HALF OF THE CIRCUIT.

NO CONTINUITY

TEST THE LOAD SIDE PHASE AND GROUNDING CONDUCTORS FOR CONTINUITY. IF THERE IS CONTINUITY, THE FAULT IS SOMEWHERE ON THAT HALF OF THE CIRCUIT.

15-AMPERE GENERAL LIGHTING AND GENERAL-USE RECEPTACLE CIRCUIT

DIVIDE THE HALF OF THE CIRCUIT THAT HAS THE FAULT INTO HALF AGAIN. DO THE SAME TEST TO DETERMINE WHICH HALF OF THE HALF-CIRCUIT CONTAINS THE FAULT.

15-AMPERE GENERAL LIGHTING AND GENERAL-USE RECEPTACLE CIRCUIT

CONTINUE TO HALVE THE CIRCUIT, TESTING EACH DIRECTION AT EACH OPENED OUTLET BOX UNTIL THE FAULT IS FOUND. CORRECT THE PROBLEM, RECONSTRUCT THE CIRCUIT, AND CHECK FOR CONTINUITY AT THE DISTRIBUTION PANEL AGAIN. IF THE CIRCUIT IS CLEAR, RECONNECT THE CONDUCTORS IN THE DISTRIBUTION PANEL AND ENERGIZE THE CIRCUIT.

Figure 19-5

KEY TERMS

Feed-through The process by which a load that is installed on a troubled circuit allows the passage of a signal that is part of a troubleshooting test. Feed-through can give false indications under many testing procedures.

One circuit condition may cause false readings on the continuity tests. This condition is called **feed-through**, or sometimes, "read-through." If there is a load connected to the circuit, the current from the continuity tester can flow through the load and indicate a fault (phase conductor to grounded circuit conductor) reading, as shown in Figure 19-6. Many general-use receptacle and general lighting branch circuits include both receptacle outlets and lighting outlets on the same circuit. Before testing, all switching systems must be in the off position. In addition, the electrician should ensure that there are no portable luminaires (fixtures) connected to the circuit and should check for equipment that other trade workers may have plugged into the circuit.

Faults are not always located in a device or switch point box. Damage to cables inside the walls is also a possibility. Such damage will show up dur-ing the testing procedure. Eventually, the checkout procedure will demonstrate a device box with continuity on the load side and the next device box with continuity on the line side. In this case, the damaged cable must be abandoned and a new cable installed.

19.4.2: Troubleshooting Opens

Opens can occur on a circuit for several reasons. Damage inflicted on the cable can be so severe that the cable is cut in half. All of the conductors in the cable may be cut, or only the phase conductor or the grounded circuit conductor may be damaged; in either case, there is an open in the wiring system. Opens can also occur as the result of poor makeup or poor trim, in which all of the necessary conductors are not connected to the devices. Opens can also be caused by buried boxes. If a buried box contains conductors requiring connection to the device for continuity, such as many receptacle outlets, an open in the circuit will result. However, if there is continuity to ground on the load side of the open circuit, it indicates that the equipment grounding conductors are continuous and that the problem is probably a buried box. Opens may also occur when a necessary cable between two boxes is inadvertently omitted during rough-in.

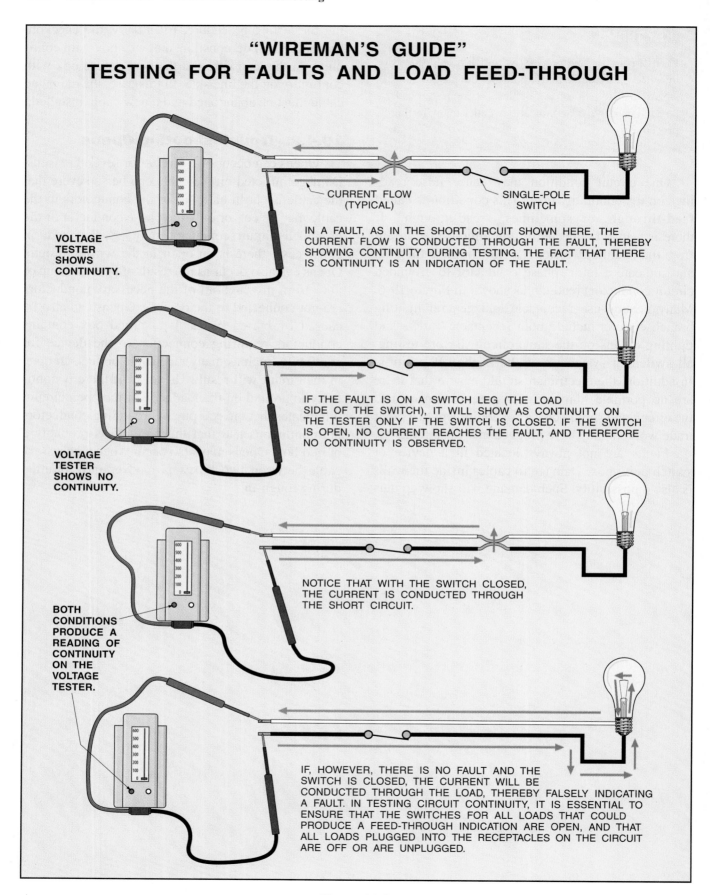

"WIREMAN'S GUIDE"
TESTING FOR FAULTS AND LOAD FEED-THROUGH

CURRENT FLOW
(TYPICAL)

SINGLE-POLE
SWITCH

VOLTAGE
TESTER
SHOWS
CONTINUITY.

IN A FAULT, AS IN THE SHORT CIRCUIT SHOWN HERE, THE
CURRENT FLOW IS CONDUCTED THROUGH THE FAULT, THEREBY
SHOWING CONTINUITY DURING TESTING. THE FACT THAT THERE
IS CONTINUITY IS AN INDICATION OF THE FAULT.

VOLTAGE
TESTER
SHOWS NO
CONTINUITY.

IF THE FAULT IS ON A SWITCH LEG (THE LOAD
SIDE OF THE SWITCH), IT WILL SHOW AS CONTINUITY ON
THE TESTER ONLY IF THE SWITCH IS CLOSED. IF THE SWITCH
IS OPEN, NO CURRENT REACHES THE FAULT, AND THEREFORE
NO CONTINUITY IS OBSERVED.

BOTH
CONDITIONS
PRODUCE A
READING OF
CONTINUITY
ON THE
VOLTAGE
TESTER.

NOTICE THAT WITH THE SWITCH CLOSED,
THE CURRENT IS CONDUCTED THROUGH
THE SHORT CIRCUIT.

IF, HOWEVER, THERE IS NO FAULT AND THE
SWITCH IS CLOSED, THE CURRENT WILL BE
CONDUCTED THROUGH THE LOAD, THEREBY FALSELY INDICATING
A FAULT. IN TESTING CIRCUIT CONTINUITY, IT IS ESSENTIAL TO
ENSURE THAT THE SWITCHES FOR ALL LOADS THAT COULD
PRODUCE A FEED-THROUGH INDICATION ARE OPEN, AND THAT
ALL LOADS PLUGGED INTO THE RECEPTACLES ON THE CIRCUIT
ARE OFF OR ARE UNPLUGGED.

Figure 19-6

SUMMARY

The connection of the dwelling's permanent electrical system must be connected to the temporary construction panel to provide power for checkout of the system. The supply to the dwelling should be a feeder from the temporary power panel to the main panel in the dwelling. All receptacles and switching systems should be completely checked out for faults and opens. A good voltage tester with continuity test capabilities is a very useful tool for checking out all aspects of a wiring system. If any problems are encountered, troubleshooting procedures need to be started. The basic troubleshooting technique is called halving the circuit and involves breaking the faulty circuit into half until the cause of the problem is discovered. Feed-through is a condition encountered when there are loads connected to the circuit, such as luminaires (fixtures). Although no fault is actually present, the condition will cause the continuity tester to register what appears to be a fault. Instead, the tester is reading continuity through the load.

REVIEW

1. The supply conductors from the temporary construction power to the dwelling permanent wiring system should be _____ .

   a. feeders

   b. branch circuits

   c. service conductors

   d. switch legs

2. In testing a receptacle, the test indicates 120 volts from the phase conductor to the grounded circuit conductor and continuity between the phase conductor and the equipment grounding conductor. The likely cause is _____ .

   a. an open equipment grounding system

   b. reversed polarity at the receptacle

   c. an open grounded circuit conductor

   d. a short circuit at the panel

3. In testing a receptacle outlet, the measurement is 120 volts between the phase conductor and the grounded circuit conductor and there is no reading at all between the phase conductor and the equipment grounding conductor. The likely cause is _____ .

   a. an open equipment grounding conductor

   b. reverse polarity at the receptacle

   c. an open grounded circuit conductor

   d. a short circuit at the panel

4. Using the halve-the-circuit system, the maximum number of outlets that must be checked in order to find the trouble (assuming that the trouble is caused by bad makeup in a box) in a circuit with 16 outlets is _____ .

a. 4

b. 6

c. 8

d. 10

5. *True or False:* All circuits should be on at the same time for checkout of a dwelling's electrical system.

6. *True or False:* In troubleshooting faults, the circuit should be left connected to the circuit breaker and 120-volt power used to test for the fault.

7. *True or False:* In troubleshooting luminaires (fixtures), all possible combinations of on and off must be tested to ensure a properly working system.

8. *True or False:* A voltage tester with a continuity tester can be used to test for voltage and continuity but it cannot check for opens or faults.

9. *True or False:* If the neutral and the switch leg to an incandescent luminaire (fixture) are not yet connected to a switch, but the luminaire (fixture) and the lamp have been installed, and a continuity check between the neutral and the switch leg produces an indication of continuity, there must be a short circuit somewhere on the circuit.

10. *True or False:* A voltage tester can be used to check for polarity in a screw shell socket.

11. *True or False:* If testing a duplex receptacle shows a reading for the top half of the receptacle of 120 volts between the phase conductor and the equipment grounding conductor but no reading between the grounded circuit conductor and the phase conductor, the problem is probably an open phase conductor.

12. *True or False:* A plug tester is more accurate and more versatile than a voltage tester.

"WIREMAN'S GUIDE" REVIEW

1. Explain how read-through (feed-through) can produce false readings on a continuity tester. How can the problem be resolved?

2. Explain the procedure for checking a live receptacle outlet with a voltage tester. How many tests does it take to ensure that the receptacle is installed and wired properly?

3. Make a drawing of a 4-way switching system in which the third conductor pair (the pair of conductors other than the travelers) has been confused with a traveler pair at the 4-way switch. Assume that the third conductor pair is the switch leg from one of the 3-way switches. What sort of results will be noted during checkout of the system?

KEY TERMS

60-hertz hum   The interference induced by normal electrical power intended for power and lighting loads. Normal household power has a frequency of 60 hertz (60-cycles per second.

American Wire Gauge   One of the two wire sizing systems employed by the *NEC*® to define the size of a conductor. The AWG system assigns a gauge number, such as 12 or 00, to standard sizes of conductors. The largest conductor defined under the AWG system is 0000 (4/0).

Ampacity   (See *NEC*® *Article 100*): The amount of current that a conductor is allowed to carry continuously without heating to a point at which damage occurs to the conductor or the conductor's insulation.

Appliance   (See *NEC*® *Article 100*): A type of utilization equipment that is usually manufactured for residential use and is mass-produced rather than being individually designed and constructed.

Approved   (See *NEC*® *Article 100*): Referring to equipment, devices, raceways, and other electrical materials that are acceptable to the AHJ for installation and use.

Arc-fault circuit-interrupter (AFCI)   A device capable of detecting arc faults that will shut off the flow of current to the circuit when an arc fault is detected.

Attachment plug-and-cord   (See *NEC*® *Article 100*): An assembly of a male cord cap (plug) and the *nec*essary cable intended to con*nec*t utilization equipment to a receptacle outlet.

Authority having jurisdiction   (See *NEC*® *Article 100*): The local governmental authority that is charged with the regulation of construction projects. The AHJ may be a city, county, or sometimes a state organization, but it is always answerable to a legi

Automatic   (See *NEC*® *Article 100*): Capable of operating without direct human action through a controller. Automatically operated equipment is inherently dangerous to maintain because of the possibility of its suddenly starting at any time.

Available fault current   The maximum amount of current available if a fault occurs. The available current is limited by the amount of current the power provider's generator can produce, the impedance in the delivery system circuiting, and the let-through

Ballast   The component in a fluorescent lighting system that controls the flow of current to the lamps.

Bathroom   (See *NEC*® *Article 100*): A nonhabitable room that contains a washbasin and at least one other bathroom fixture such as a toilet, shower, or tub.

Box makeup   The act of preparing the conductors contained in a device box for the installation of a device. Makeup is accomplished with the intent of making subsequent installation of the device as easy as possible.

Boxes   Housings in the electrical circuit that contain splices and terminations. Boxes can be metallic or nonmetallic and may house devices or simply contain conductors, but they provide a barrier between the electrical system of the structure and the living or working space of that structure.

Branch circuit   (See *NEC*® *Article 100*): A circuit that has overcurrent protection, other than supplemental overcurrent protection, on the line side only.

Building permit   A document issued by the AHJ, usually for a fee, that allows the construction of, or addition to, a building or other structure. In most areas of the country, it is not legal to begin construction before a permit has been obtained.

Buried box   A device or junction box that has been inadvertently covered by sheetrock or other wall covering materials and is not readily available for trim.

Cable A group of conductors that are associated with each other by being twisted together or covered with an outer jacketing that provides electrical energy to the load. Cables may have a metallic or a nonmetallic outer covering, or sheathing, and come in many different sizes, types, and styles for use according to the intended installation environment and the nature of the load served.

Calculated load The load of utilization equipment or outlets as determined by procedures allowed or required to be employed by the *NEC®*. For example, the general lighting load is calculated from the total floor area of a building, or a motor load is determined to be 125% of the motor's FLA rating.

Change order Authorization to proceed with a change—an addition or a deletion—to scheduled work. In many cases a change order takes the place of a work order.

Checkout The process of ensuring that all electrical systems within a dwelling are functioning properly following completion of trim.

Circuit breaker (See *NEC® Article 100*): A device that is sensitive to both the heating caused by current flow and the magnetic field created by current flow. The device will open the circuit if the current flow becomes too great. Circuit breakers are usually housed in circuit-breaker panels, although enclosures for individual circuit breakers are available.

Circuit overutilization A condition that causes overcurrent on the circuit caused by too many loads being connected in parallel, and operating at the same time, for one circuit to supply. The circuit is operating properly, but each load connected in parallel reduces the overall impedance of the circuit.

Circular mil The unit of measure used for defining wire size by the *NEC®*. A circular mil is equal to the diameter of a conductor in thousandths of inches, squared. One circular mil equals approximately .0007 square inch.

Common terminal One of the types of terminals on a 3-way switch. The common terminal is connected, depending on the position of the toggle handle, to one of the two traveler conductors at any point in time.

Compact fluorescent Referring to a type of fluorescent lamp designed to fit into luminaires (fixtures) manufactured to accommodate incandescent lamps.

Concentric knockout A bonding component in which multiple-size knockouts are arranged in a targetlike pattern. In this pattern, the centers of all of the knockouts are in the same location.

Conductor (See *NEC® Article 100*): A metallic bar or wire designed and intended to carry electrical current. It can be insulated, covered, or bare. Other components of an electrical circuit, such as metallic raceways, may also carry current in the event of a fault, but such components are not considered conductors because they are not intended to carry current during normal operations.

Conduit body (See *NEC® Article 100*): A type of raceway fitting that allows for a rapid change in direction of wiring, or for the tapping of a raceway, without the use of a box. Conduit bodies have removable covers that allow access to the interior of the fitting to facilitate the installation of the conductors.

Connected load The load, measured in amperes or watts (volt-amperes), that is actually required by a load to operate properly. Connect load is the nameplate load rating of an appliance or other type of electrical utilization equipment.

Construction drawings A series of drawings, sometimes called blueprints, that show the intended design of a building or other structure. There are usually different construction drawings for each of the different trades involved in the construction process.

Construction power Temporary electrical power that is provided for use by the trade workers and others working on the construction site. The power is usually provided and maintained by the electrical contractor.

Construction process Collectively, the procedures followed by the various trades for the successful completion of the construction project. The construction process is usually coordinated by a general contractor or a construction management firm.

Control (See *NEC® Article 100*): Turning an electrical circuit on or off, or regulation of the speed or level of operation of utilization equipment. A controller is the device that performs the control function.

Counter current/counter electromagnetic force (CEMF) The current generated within a turning motor that is in opposition to the current flow drawn from the source.

Current-limiting fuses A class of fuses that limit the maximum amount of current that can flow through a circuit regardless of the available fault current. These types of fuses are designed to operate very quickly in the event of a fault.

Cut notches Sections along the edge of a joist or stud for the installation of cables or raceways. Notches must be cut with a saw, and usually the opening must be covered with a plate to protect the cable or raceway. Notches must meet the requirements of the *NEC®* and the UBC.

Demand factor (See *NEC® Article 100*): Any of the multipliers used in calculating demand loads.

Demand load The load remaining after the factors allowed by the *NEC®* for a number of like or similar loads have been employed. For example, a demand load reduction is allowed for the total load for the ranges of an apartment building on the theory that not all ranges will be in use at the same time.

Device (See *NEC® Article 100*): Electrical equipment that does not use electrical energy. Devices such as switches are intended to control electrical circuits, but, as in the case of receptacles to provide access to the electrical system, do not use any electrical energy themselves.

Device boxes Electrical boxes intended to house and make available devices such as switches and receptacles.

Disconnect switch An enclosure that houses a switch mechanism and sometimes also fuses. The switch is designed to disconnect a portion of the circuit or a load from the electrical supply and when fuses are included, the switch can provide overcurrent protection.

Doorbell chime The chime that makes the sound intended to alert the occupant of a dwelling of the presence of a visitor at the door. The chime is actuated by the pushbutton.

Double-pole switch A switch that is capable of controlling the current from two separate circuits at the same time. The switch has terminals for two separate phase conductors and two separate switch leg conductors and resembles two single-pole switches connected together.

Drilled holes Holes that are drilled in framing members for the installation of electrical conductors or cables. Holes that are drilled in studs or joists must meet the requirements of the *NEC®* and the UBC.

Drip loop A loop formed in the wire connecting the drop conductors to the service conductors to prevent water from entering the weatherhead.

Eccentric knockout A bonding component in which multiple-size knockouts are arranged so that all of the various sizes of knockouts share one specific location along the edge of the largest knockout. In this pattern, of all of the knockouts have one edge in the same location.

Electric discharge Referring to an artificial lighting system in which light is given off by electrons as they become bound to the atoms within the lamp. This is a very efficient system for lighting large areas.

Electrode The conductor or other material that physically connects the electrical system to the earth, or ground. Several types of electrodes are recognized by the *NEC®*; included are underground metallic water piping systems, rod and pipe electrodes, ground rings, and concrete-encased electrodes.

Environmental airspaces   Airspaces within a structure that are intended as part of the structure's heating and cooling systems. These areas are used to circulate the air through the furnace or air conditioner and then return the air to the rooms of the building. Wiring within these areas is restricted because of the possibility of the rapid spreading of fire through the air-handling system should a fire occur.

Fault   The term used by the *NEC®* to define a circuit that is not operating properly, so that current flow is using an unintended pathway.

Feed-through   The process by which a load that is installed on a troubled circuit allows the passage of a signal that is part of a troubleshooting test. Feed-through can give false indications under many testing procedures.

Feeder   (See *NEC® Article 100*): A circuit that has overcurrent protection on both the line side and the load side.

Filament   The metal strip in an incandescent lamp (bulb) system that is heated until it glows. The temperature and the composition of the filament determine the intensity and the color of the light.

Fine print note (FPN)   A type of entry in the *NEC®* that provides explanatory information but is not formally enforceable as part of the *NEC®*.

Flag note   An icon used on construction drawings that alerts the construction team to the existence of additional information or requirements listed elsewhere on the drawings.

Fluorescent   Referring to an artificial lighting system in which the coating on the inside of a glass tube is stimulated to produce light. Fluorescent lighting is the most widely used lighting system outside of dwellings.

Four-way switch   A type of switch used in conjunction with two 3-way switches to allow for control of a switched outlet or outlets from more than two switching locations. Control can be obtained by adding as many 4-way switching locations as desired, but the system must include a 3-way switch at either end.

Fuse   A device that is sensitive to heating caused by circuit current flow. If the current becomes too high, the heating will cause the fuse link to open, thus eliminating the current flow through the circuit.

General lighting outlets   Outlets intended for general use as lighting outlets for fixed-in-place luminaires (fixtures). Specialized task lighting is not included in this category.

General use snap switch   (See *NEC® Article 100* [Switch — General Use Snap]): The term used by the *NEC®* to describe a switch used for single-pole, double-pole (2-pole), 3-way, or 4-way outlet control in a dwelling or other structure. General use snap switches come in various grades of quality, such as standard grade, specification grade, and hospital grade.

General-use receptacle outlets   Receptacle outlets installed for general use as outlets for portable lighting (table lamps) as well as general-use appliances, such as televisions or stereo equipment. Kitchen equipment and other specialized utilization equipment are not included in this category.

Ground fault   A flaw in an electrical circuit in which some current is escaping and is flowing to ground using a pathway other than the one intended. Ground faults pose an electrocution hazard.

Ground-fault circuit-interrupter (GFCI)   A device that will detect a ground fault and then open the circuit in response to that fault. GFCIs are available as receptacle devices or circuit breaker devices.

Habitable   Used for normal living functions such as eating, sleeping, and general living, but excluding unfinished portions of a dwelling, bathrooms, hallways, and closets.

Halving the circuit   A specific technique used for troubleshooting circuit problems that involves repeatedly splitting the troubled circuit into halves.

Impedance The total opposition to current flow in a circuit, consisting of resistance, inductive reactance, and capacitive reactance.

In-line thermostat A type of thermostat that is installed directly into a piece of space-heating equipment rather than on the wall.

Incandescent Referring to a system of producing artificial light by heating a metal filament to the point at which it glows, thus giving off light. This type of lighting also produces a lot of heat. Incandescent lighting is the most widely used lighting

Individual branch circuit (See *NEC® Article 100* [Branch Circuit—Individual]): A branch circuit that supplies one piece of utilization equipment. An individual branch circuit usually requires the use of a single receptacle.

Induced interference Signal interference that is generated within a conductor by a magnetic field caused by current flow in another conductor that is in close proximity.

Inherently limited power source A power source designed to restrict the power output so that it is maintained at safe levels even in the event of a fault.

Inspection The process whereby the AHJ enforces established installation and construction standards. The AHJ periodically reviews the progress of construction and either approves or rejects the quality of the construction and ensures compliance with minimum standards. These reviews are accomplished by physically inspecting the building, and the person who performs the inspections is usually referred to as an *inspector.*

Installation instructions The directions provided by the manufacturer concerning the procedures to be followed in preparing electrical equipment, devices, or other materials for use. These instructions must be closely followed in installing electrical materials; otherwise, unsatisfactory operation may result in fire or other safety hazards.

Insulated Describing a conductor covered with nonconductive material that is thick enough to be recognized by the *NEC®* as insulation. Insulation keeps the electrons confined to the conductor, preventing ground faults.

Insulation type There are many different types of insulations designed for specific locations and load types. Insulation type is usually designated by a letter code—for example, XHHW.

International System of Units (SI) The measurement system, sometimes called the metric system in the United States, defined by the use of meters and kilograms, that uses base 10 unit divisions (SI units). The SI system is widely employed in almost all areas of the world except in the United States. The *NEC®* uses the SI system as the primary measurement system, with the English system, defined by the use of feet and pounds, used as a secondary system.

Interrupt rating (See *NEC® Article 100*): The rating of an overcurrent-protective device that defines the maximum amount of current the device can pass without failing. The overcurrent-protective device must have an interrupt rating that is at least as high as the available fault current.

Island countertop space Countertop space located above an island counter, which has no countertop wall space.

Isolation Removal from the electrical circuit. An isolation switch disconnects utilization equipment from a circuit to ensure safety in maintenance or to interrupt the flow of power to the equipment.

Joist A framing member that makes up part of a flooring system.

Kitchen countertop The top surface of a kitchen counter. Access to the electrical system must be available for electrical appliances that are used on kitchen countertops.

Labeled (See *NEC® Article 100*): Referring to materials and equipment that have been found to meet certain requirements for safety and function by a testing agency recognized by the AHJ. An identification label, marking, or decal from the testing agency

Lighting outlet boxes Boxes that are designed to supply outlets for luminaires (fixtures). These boxes are usually round in shape, can be metallic or nonmetallic, and have 8-32 threaded holes for the connection of the luminaire (fixture) support hardware.

Listed (See *NEC® Article 100*): Referring to materials and equipment that have been found to meet certain requirements for safety and function by a testing agency recognized by the AHJ. The materials are placed on a list to identify them as having been tested.

Locked-rotor current The amount of current that a motor draws when it is not turning (the rotor is locked). This current flow is limited by the resistance of the motor windings only and is approximately equal to the amount of current the motor requires at start (starting overcurrent).

Loose connection A flaw in a connection between two conductors, or between a conductor and a terminal, that allows for series arc faults. Tight connections prevent series arcs and the associated problem of generation of high levels of heat.

Low voltage The side of the transformer with the lowest voltage.

Low-impedance path to ground A route that is provided for the electrons in the event of a ground fault. The current flow during a fault must be large enough to cause the overcurrent-protective device to open. If the pathway has too large an impedance, the current flow may be limited to the level below that needed to cause the overcurrent-protective device to open.

Luminaire (fixture) (See *NEC® Article 100*): An electrical appliance intended to provide lighting to a designated space. Formerly known as "lighting fixture."

Made knockout A properly sized knockout that is cut into the enclosure, or a concentric or eccentric knockout from which all of the smaller knockout sections have been removed.

Mandatory Rules Describing the rules and procedures listed in the *NEC®* with the terms "shall" and "shall not." These rules must be followed to comply with the requirements of the *NEC®*.

Maximum operating temperature Each insulation type has a maximum operating temperature rating that should not be exceeded. If the maximum operating temperature of the conductor insulation is exceeded for a long period, the insulation can be damaged, usually by drying out and cracking, allowing electrons to escape the circuit.

Metric designator A dimension corresponding to trade size, as employed for the equipment and raceways, measured using the SI system.

Motor starting overcurrent The temporary overcurrent in a circuit associated with starting a motor from rest. The high current flow results because the motor produces no counter EMF while at rest, and, on starting, the motor's counter EMF is insufficient to limit the circuit current flow to normal levels. Once the motor reaches full speed, the overcurrent condition disappears.

Multiwire branch circuit (See *NEC® Article 100*): (Branch circuit—multiwire) More than one branch circuit sharing a grounded circuit conductor. This is usually referred to as a "3-wire homerun" in residential work or a "full boat" in commercial and industrial work. Extreme care must be employed in installing multiwire branch circuits to avoid overloading the neutral conductor.

National Electrical Code® (***NEC®***) A set of minimum rules for the design and installation of electrical systems and devices published by the National Fire Protection Association as document NFPA 70. It is the most widely applied standard for electrical installations in the United States, however, it is not recognized in all areas of the country.

National Fire Protection Association (NFPA) The organization devoted to fire safety and prevention that publishes, along with many other documents, the *National Electrical Code®*.

Network interface The location where the dwelling wiring systems connect to the incoming telephone, cable, or other communication system supply conductor.

Noise   An unwanted signal in a communication system conductor; usually caused by induced interference.

Nonmetallic-sheathed cable   Electrical cable with 90°C-rated conductors installed in an outer sheath of nonmetallic material. Use of this type of cable is the most popular and widely used wiring method for dwellings.

Notch plate   A metal plate that is installed over a notch or over a hole that is closer than 1¼ in. (32 mm) from the edge of a framing member in order to protect the cables or raceway.

Occupational Safety and Health Administration (OSHA)   A branch of the federal government that is charged with improving workplace safety and encouraging the establishment of safe workplace practices.

Opening   A term used to describe a receptacle outlet or a switch point.

Operating overcurrents   Overcurrents that can be present on a circuit as the result of the nature of the connected loads, such as motor loads. As the motor is placed under load, the amount of counter EMF is reduced because of the slower speed of the motor. Current flow to the motor from the source increases. This increased current flow, or operating overcurrent, is a temporary condition; when the motor speed returns to normal, the circuit current flow will also return to normal.

Other Loads—All Occupancies   The loads for all utilization equipment other than general lighting and general-use receptacle outlet loads.

Outlet   (See *NEC® Article 100*): A place in the electrical system where access is provided for the use of utilization equipment, such as appliances and luminaires (fixtures). Places of access to the electrical system intended for control, such as switch points, are not considered outlets.

Overcurrent   (See *NEC® Article 100*): Too much current in a circuit. Overcurrent has four possible causes: short circuits, ground faults, overloads, and circuit overutilization.

Overcurrent protective device: A device, consisting of fuses or circuit breakers, that will eliminate the flow of power if the current becomes too large for too long a period of time.

Overload   (See *NEC® Article 100*): A condition in which too much current is flowing through a circuit although the circuit is operating properly (as it should). This condition results during the normal operation of certain types of loads—motor loads, for example—that utilize reactance as the major component of their impedance.

Parallel   A pattern of connection of the components of an electrical system in which the components are connected next to each other and obtain power from the same source. The term can refer to conductors, devices, loads, or other electrical components, or to current flow, as in the case of parallel arc faults.

Peninsula countertop space   Countertop space located above a peninsula, which has no countertop wall space except where the peninsula portion of the countertop is adjacent to a countertop that is not part of the peninsula space and that has countertop wall space.

Permissive Rules   Referring to the rules and procedures listed in the *NEC®* that are allowed but that are not strictly required. These rules are identified in the *NEC®* with the terms "shall be permitted" and "shall not be required."

Personal Protective Equipment (PPE)   Safety equipment, such as hardhats, gloves, safety glasses, and work boots, that is provided by individual construction workers for their own use.

Phase conductor   One of the terms used to describe the nongrounded current-carrying conductor of an electrical circuit. The phase conductor is sometimes also called the hot conductor.

Pigtail   (as referring to splices) An extra conductor that is added to a splice for connection to a device.

Pigtail   (referring to connecting utilization equipment): A cord-and-attachment-plug assembly that is used to connect utilization equipment to receptacle outlets.

Plaster ring   An accessory that is used with square boxes to allow them to be used in flush installations.

Polarity   The relative location of the phase and grounded conductors on an outlet device. Polarity must be properly maintained for safe operation of utilization equipment.

Pushbutton   A momentary-contact button that activates a doorbell chime.

Quick connect   A method of connecting conductors to devices without employing the screw terminals. Many AHJs and contractors do not allow the use of quick connects.

Raceway   (See *NEC® Article 100*): A pathway specifically designed and installed to house conductors for protection against physical damage. Raceways may be metallic or nonmetallic and are usually in the shape of a pipe, or conduit, although some raceways

Rafter   A framing member that makes up part of a roof support system.

Receptacle   (See *NEC® Article 100*): A device that allows repeated and easy access to an outlet for cord-and-plug connected utilization equipment. Receptacles have many slot configurations for differences in ampere rating, voltage rating, and phase configuration.

Receptacle outlet   (See *NEC® Article 100*): An outlet in the electrical system that has a receptacle installed to provide access to the system for cord-and-plug connected loads.

Schedule   A layout, usually in the form of a table, that is part of the construction drawings and provides detailed information concerning materials, devices, or equipment to be installed as part of the construction process. For example, detailed information about the various luminaires (fixtures) to be installed is often conveyed using a schedule.

Scope of work   A construction document that describes the work that is to be accomplished and the companies or trades that are to complete the work. This document provides a framework for the other construction documents.

Semiconductor   An electronic component that can be a conductor or an insulator, depending on its electrical state. The state can be changed by the introduction of a control current.

Series   A pattern of connection of the components of an electrical system in which the components are connected before or after the other components but only one common path for current flow is provided. The term can refer to conductors, devices, loads, or other electrical components, or to current flow, as in the case of series arc faults.

Service   (See *NEC® Article 100*): The raceways, conductors, and equipment that receive the electrical power from the local power-providing utility. It is important to note that electrical services are, for all intents and purposes of the *NEC®*, unprotected against overcurrent due to faults.

Service conductors   (See *NEC® Article 100*): Those conductors from the service point (point of service) to the main, or first, overcurrent-protective device. For all intents and purposes, these conductors are not protected against faults.

Service drop   (See *NEC® Article 100*): Overhead wiring system from the power utility's equipment to the point of service. The service-drop conductors connect to the service conductors (service-entrance conductors).

Service lateral   (See *NEC® Article 100*): Underground wiring system from the power utility's equipment to the point of service. Lateral conductors often terminate in the meter enclosure.

Service point   (See *NEC® Article 100*): The location at which the power utility's conductors (drop or lateral conductors) connect to the dwelling wiring. Overhead drop conductors usually terminate at the service conductors from the riser or mast. Lateral conductors usually terminate in the meter enclosure.

Shop drawings/cut sheets   Drawings, usually provided by the manufacturer of equipment or materials to be installed in the building or structure, that show details of fabrication, such as dimensions, color, and materials used.

Short circuit A circuit that is not operating properly because the current is bypassing the load(s) and returning to ground using the circuit conductors. Short circuits do not pose an electrocution hazard.

Short conduit body A type of conduit body designed for use with smaller sizes of raceways but with limited bending space of the conductors. The length of short conduit bodies is considerably less than the length of a regular conduit body of the same raceway trade size.

Single receptacle A yoke with only one receptacle installed. Most yokes have two receptacles installed and are called duplex receptacles.

Single-pole switch A switch containing terminals for only one phase conductor and one switch leg conductor.

Specifications A construction document that usually controls the quality of the construction installations. It may list the brand names of the materials to be used during construction, certain procedures to be employed, and directions about communications between the various construction trades and the management team.

Splice The act of connecting two individual conductors together to form one continuous conductor, or the location of that connection. Splices usually occur in boxes and are accomplished using proper methods and materials.

Square boxes Electrical boxes that are square in shape and can be metallic or nonmetallic. Square boxes can be used for flush and surface installations by employing plaster rings or industrial covers and are the most common type of box used in commercial electrical work.

Stud A framing member that makes up a part of a wall. Framed walls also usually have a top plate and a bottom plate.

Submittals Cut sheets or shop drawings submitted to an architect or engineer for approval prior to the equipment's installation in the building or structure. Submittals are usually required by the building specifications.

Supplemental overcurrent-protective device A device providing overcurrent protection that is installed to protect a circuit load or component but is not intended to protect the circuit against overcurrent. Many pieces of electronic equipment are protected by fuses internal to the equipment. These fuses are considered supplemental overcurrent-protective devices because they protect the equipment, or part of the equipment, but not the circuit.

Switch leg conductor The conductor that carries current from the single-pole or double-pole switch or from the last 3-way switch on a system to the switched outlet, such as for a luminaire (fixture).

Switch point A point of access to the electrical system intended for the installation of a control device, such as a single-pole switch. Switch points are not considered outlets because they do not provide access for utilization equipment.

Symbols Icons used on construction drawings to represent various design features such as switches, receptacles, and luminaires (fixtures).

Television signal The signal, consisting of instructions for forming sounds and pictures, from a broadcast or cable television system.

Temporary lighting Lighting that is installed in some buildings under construction to provide access and egress lighting for the workers but usually does not include specialized task lighting that may be required by some trades.

Three-way switch A type of switch that is controlled from two distinct locations. Each 3-way switch must have terminals for two travelers and one common conductor, in addition to any terminals for equipment grounding conductors. The term 3-way may also refer to a switching system.

Toggle handle The switch handle of a standard single-pole, 3-way, 4-way, or double-pole (2-pole) switch that is the means for actuating the switch from an on to an off position.

Trade size A system employed in the *NEC®* to define certain standard sizes of electrical equipment and raceways. For example, ½-in. (16-mm) internal diameter trade size conduit can actually measure between .526 in. (13.36 mm) and .660 in. (16.76 mm).

Transformer A device that changes the voltage of an electrical circuit. A transformer can either reduce (step down) or increase (step up) the voltage, but the power output on the secondary side of a transformer will approximately equal the power for the primary side of the transformer

Traveler One of the types of terminals in a 3-way or 4-way switching system. Only one traveler from each traveler pair will carry current at any given point in time, and the current is transferred from one traveler to the other by moving the position of the toggle handle of any switch in the system.

Trim The set of procedures for installing the switches, receptacles, luminaires (fixtures), and other utilization equipment in preparation for occupancy of the dwelling. Trim also includes checkout and troubleshooting.

Troubleshooting procedure The process for efficiently discovering and repairing circuits and utilization equipment that are not operating properly.

Uniform Building Code (UBC) A set of minimum rules for the installation of building systems and the construction of new buildings. It is the most widely applied standard for construction in the United States, however, it is not recognized in all areas of the country.

Utilization equipment (See *NEC® Article 100*): Equipment that requires electricity to function. Utilization equipment includes appliances and luminaires (fixtures) but does not include raceways, boxes, or devices.

Wall countertop space Wall space that is located above countertops. Countertop wall space is measured horizontally along the wall where it meets the countertop.

Wall space Space measured horizontally (linear space) along a wall in a dwelling or other structure. Doorways, fireplace openings, and sliding panels of glass doors are not considered wall space, although fixed panels of glass doors are considered wall space.

Waterproof boxes Electrical boxes designed to be used in wet and damp areas, such as outdoors, where moisture can enter the raceway system or the boxes themselves, thereby causing faulting problems.

Wire nut/splice cap A fitting allowing for a solderless method of splicing conductors.

Wiring method A specific set of materials and installation procedures used to deliver electrical energy from one place to another. Conductors, boxes, raceways or cables, assemblies, straps, clamps, and many other fittings may be used.

Work order An authorization to proceed with scheduled work. Work orders usually involve work that has been priced and contracted or that has been agreed to be completed on a "time and materials" basis.

Yoke Also called a strap. The metal or nonmetallic assembly that contains one or more devices, such as switches or receptacles. The device is attached to the yoke, or strap, which is screwed to the device box. The yoke or strap holds the device in place.

INDEX

Note: Page numbers in **bold type** reference non-text material, such as tables and figures.